CONSTANTS

Mathematical Constants

$\pi = 3.14159$
$e = 2.72$
$\ln 2 = 0.693$
$\log_{10} e = 0.434$
$\sqrt{2} = 1.41$
$\sqrt{3} = 1.73$
1 degree = 0.0175 radian
1 radian = 57.3°

Physical Constants

electron rest mass (m_e) = 9.108×10^{-31} kg
neutron rest mass (m_n) = 1.675×10^{-27} kg = $1839\, m_e$
protron rest mass (m_p) = 1.672×10^{-27} kg = $1836\, m_e$
speed of light (c) = 2.998×10^{8} m/s
Boltzmann's constant (k) = 1.380×10^{-23} J/molecule-°K
electronic charge (e) = 1.602×10^{-19} C
Planck's constant (h) = 6.625×10^{-34} J-s
speed of sound in dry air (at STP) = 331 m/s
density of air (at STP) = 1.29 kg/m³

BASIC DIGITAL ELECTRONICS WITH MSI APPLICATIONS

BASIC DIGITAL ELECTRONICS WITH MSI APPLICATIONS

JOHN A. DEMPSEY College of San Mateo, De Anza College

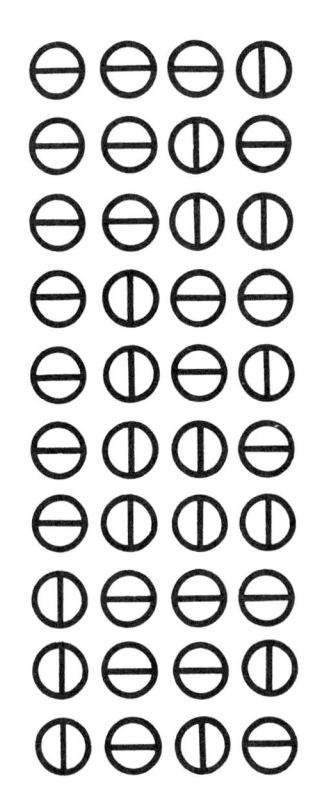

ADDISON-WESLEY PUBLISHING COMPANY
Reading, Massachusetts · Menlo Park, California
London · Amsterdam · Don Mills, Ontario · Sydney

Third printing, April 1979

Copyright © 1977 by Addison-Wesley Publishing Company, Inc. Philippines copyright 1977 by Addison-Wesley Publishing Company, Inc.

All rights reserved. No part of this publication may be reproduced, stored in a retrieval system, or transmitted, in any form or by any means, electronic, mechanical, photocopying, recording, or otherwise, without the prior written permission of the publisher. Printed in the United States of America. Published simultaneously in Canada. Library of Congress Catalog Card No. 75-9009.

ISBN 0-201-01478-5
BCDEFGHIJ-MA-79

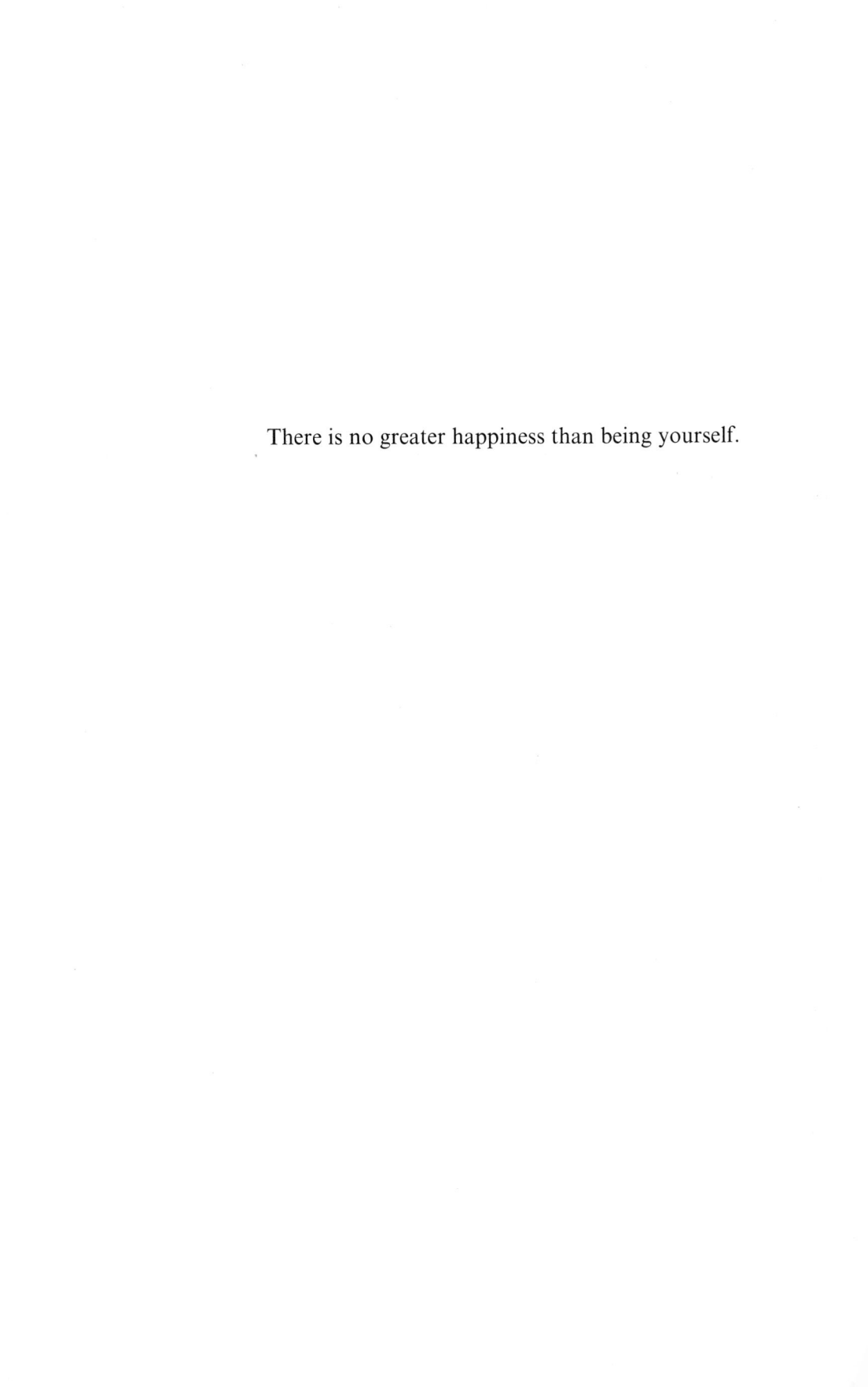

There is no greater happiness than being yourself.

PREFACE

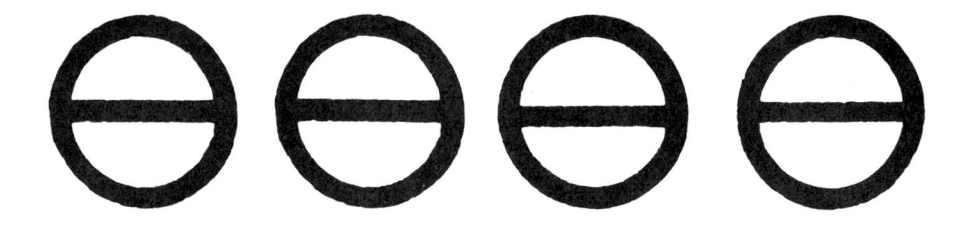

This is a textbook on the basic principles of digital integrated circuit devices and their applications in digital systems. It is written for the electronic technologist who will help develop, install, and maintain digital systems. This text should also be of interest to those wishing to learn computer electronics, and to applied scientists and engineers who want a practical understanding of digital electronics. The practicing engineer and technician who want to update their knowledge of digital electronics will also find the book useful.

The level of this book is suited for the first course in digital electronics at the community college, technical institute, four-year technology school, or four-year college.

Some significant features of this book are: emphasis on integrated circuit (IC) devices of medium complexity (MSI) and large complexity (LSI), extensive use of manufacturer's specification sheets, systems orientation, discussion of digital systems used in industry (including schematics and troubleshooting techniques), the use of the TTL, CMOSL, and ECL families, and laboratory orientation, with availability of a companion laboratory manual.

The corequisite for understanding this book is a course in ac circuits (basic circuit principles and laboratory applications). Thus the text material can be taught to second-semester freshmen in most technical programs. A course on semiconductors is not a prerequisite; however, Chapter 3, on logic gate circuits, should be covered more slowly if the students have not had such a course.

The increased complexity of functions performed by a single digital integrated circuit has made the systems viewpoint a necessity. Therefore, this textbook emphasizes systems on a chip, i.e, MSI and LSI devices. A fundamental course in digital electronics forms a foundation for all future digital courses, just as an ac-dc course is fundamental to most analog courses. One cannot hope to cover basic principles *and* specialized applications adequately in a one-semester course. For this reason, rather than orient this book specifically toward digital computers, communications, or other specialized areas, only the basic subject matter is taught here. Subsequent specialized digital courses can then cover their topics in depth.

Not all basic digital electronic topics can be covered in a one-semester course. Some important topics, such as interfacing, layout, and troubleshooting instrumentation, are left for laboratory experiments.

Many examples are provided which show practical applications of digital devices in areas such as the computer, instrumentation, industrial controls, medical, communications, and consumer electronics. Manufacturer's specification sheets are discussed in detail. The examples, as well as the problems at the end of each chapter, concentrate on the principles and their practical applications. The problems range from easy to challenging and the difficulty of the examples and problems gradually increases through the book. The description of a microcomputer and other digital systems are given in the appendixes. Many of the problems and examples are keyed to these systems. The instructor and student can find much additional information and a source of additional problems in these systems. Problems and examples emphasize reading digital schematics and troubleshooting digital systems.

For ease in teaching, chapter sections are kept as short as possible, with only a few key ideas in each section. Problems at the end of each chapter are keyed to the chapter sections. Some topics, such as flip-flops and counters, are introduced in early chapters and then treated in detail in later chapters. This allows early use of the basic digital building blocks and a means for reemphasizing complex topics.

Starred sections (*) in various chapters can be omitted with very little, if any, loss in comprehension in future sections. The starred sections can be returned to later in the course if time permits. Considerable time can also be saved by omitting the more difficult examples. The first eight chapters can be covered in a one-semester, three-hour-per-week course; they can be covered in one quarter if the starred sections are omitted.

For those who wish a very short introduction to digital electronics, with many examples, Chapters 1, 2, 4, and 5 can be covered. Those who wish to emphasize digital circuits can stress Chapter 3. Chapter 3 is very detailed and should be taken slowly (at least 6 hours), or it can be omitted and returned to later. It is put after Chapter 2 for those who wish to establish an early basis for laboratory learning that goes beyond simply verifying the logic operation of IC devices and for those who wish to emphasize the circuit aspects of digital electronics.

Laboratory experience with digital ICs is an important ingredient for learning this subject. The text is therefore organized with the laboratory in mind, rather than being put in the best academic order, so students will be able to perform meaningful experiments beginning with Chapter 1. A companion laboratory manual, *Experimentation with Digital Electronics,* is available. It contains several experiments for each chapter and provides practical applications as opposed to merely demonstrating the functional behavior of ICs.

Each chapter contains many references which are generally accessible and easy to read. The references provide a background for keeping up to date when working in industry. The appendixes contain IC device characteristic curves, schematics, reference data, and aids to reading schematics and troubleshooting. Answers to selected (mostly odd-numbered) problems are also included.

I would like to express my appreciation to the administrations at the College of San Mateo and De Anza College for the opportunity to develop this book.

Thanks to Paul Emerick of De Anza College for reading the entire manuscript and for his many excellent suggestions. My students have contributed many ideas to this book. I was fortunate to have a cheerful, expert manuscript typist, Dorothy Gundy.

This book is surely not without errors. Suggestions, comments, and corrections are solicited and should be sent in care of the publisher.

Woodside, California   J. D.
January 1977

CONTENTS

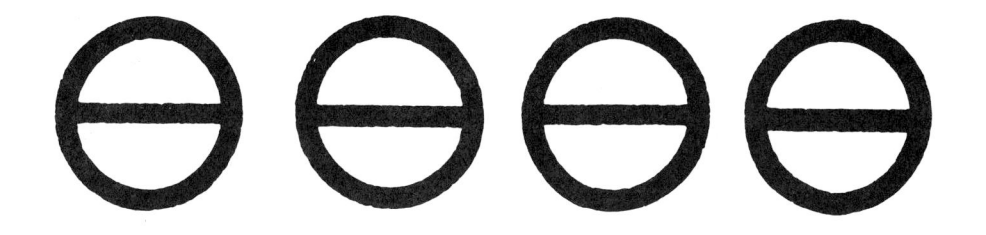

Chapter 1 Introduction to Digital Principles

1.1	Introduction	1
1.2	Principles of digital electronics	4
1.3	Principles of logic gates and flip-flops	6
	1.3.1 The logic gate	6
	1.3.2 The flip-flop	8
1.4	Comparison of digital and analog circuits	10
1.5	Digital integrated circuits	12
1.6	Conversion between decimal and binary numbers	16
	1.6.1 Conversion of binary numbers to decimal	17
	1.6.2 Conversion of decimal numbers to binary	18
1.7	Binary addition	19
	Summary	20

Chapter 2 Logic Gates

2.1	Introduction	23
2.2	The OR gate	23
2.3	The AND gate	26
2.4	The NOT gate	29
2.5	The NAND gate	32
2.6	Specification sheets	34
2.7	The NOR gate	37
	Summary	40

Chapter 3 Digital Circuits

3.1	Introduction	44
3.2	Semiconductor diodes	44
3.3	Bipolar transistors	46
3.4	The emitter follower configuration	49
3.5	Switching times for bipolar transistors	49
3.6	The TTL family	50
	3.6.1 Introduction	50
	3.6.2 The standard 54/74 series NAND gate	52
	*3.6.3 Transfer curves	55
	3.6.4 Characteristics of TTL Devices	56

xii   CONTENTS

        3.6.5 Loading effects 61
      *3.6.6 Open collector devices 65
      *3.6.7 Three-state devices 66
      *3.6.8 The Schottky series 68
      *3.6.9 The low-power Schottky series 70
*3.7 MOS transistors 71
*3.8 The CMOSL family 74
      3.8.1 Basic circuits 74
      3.8.2 Characteristics of CMOSL devices 76
*3.9 The ECL Family 81
      3.9.1 The differential amplifer 82
      3.9.2 The 10K series OR/NOR gate 83
      3.9.3 Characteristics of ECL devices 84
Summary 89

Chapter 4   Logic Gate Applications

4.1 Introduction 95
4.2 Applications using AND, OR, and NOT gates 95
4.3 Cascaded and AOI gates 98
      4.3.1 Cascaded gates 98
      4.3.2 AOI gates 100
4.4 EXCLUSIVE-OR and EXCLUSIVE-NOR gates 102
4.5 From truth table to logic diagram 104
4.6 Logic rules 107
4.7 NAND/NOR implementation 111
4.8 Miscellaneous applications 113
      4.8.1 Frequency dividers 113
      4.8.2 Pulse-shaping circuits 115
    *4.8.3 Applications using RC elements 117
    *4.8.4 The strobed RS latch 118
Summary 119

Chapter 5   MSI Devices and Applications

5.1 Introduction 124
*5.2 The digital comparator 125
5.3 Waveform representation of information 129
*5.4 The parity checker/generator 130
5.5 The digital multiplexer/data selector 133
      5.5.1 Cascading multiplexers 137
      5.5.2 Pulse generation 137
      5.5.3 Function generation 138
5.6 Introduction to binary codes 139
5.7 The decoder 141
5.8 The demultiplexer 144
5.9 Encoders 148
Summary 149

Chapter 6   Small-Scale Solid State Display Systems

6.1	Introduction	155
6.2	Discrete LEDs	156
	6.2.1 Basic principles	156
	6.2.2 Applications of LEDs	158
6.3	Single-digit seven-segment LED displays	163
6.4	Seven-segment display decoder/drivers	167
6.5	Multidigit seven-segment displays	168
*6.6	Multiplexed multidigit seven-segment displays	172
	6.6.1 Basic principles	172
	6.6.2 Applications	174
*6.7	LED dot matrix displays	176
	6.7.1 Single-character displays	176
	6.7.2 Multicharacter displays	179
	Summary	180

Chapter 7   Schmitt Triggers, Monostables, and Clocks

7.1	Introduction	184
*7.2	The Schmitt trigger	184
7.3	Pulse characteristics	188
7.4	The monostable	190
	7.4.1 The nonretriggerable monostable	193
	7.4.2 The retriggerable monostable	193
7.5	The astable multivibrator	198
7.6	Clocks	198
	7.6.1 Introduction	198
	7.6.2 Logic gate clocks	199
	7.6.3 IC clocks	201
	*7.6.4 Crystal controlled clocks	204
	Summary	205

Chapter 8   Flip-Flops

8.1	Introduction	210
8.2	The *RS* latch	210
8.3	The strobed latch	215
	8.3.1 The strobed *RS* latch	215
	8.3.2 The strobed *D* latch	216
8.4	The *D* flip-flop	220
8.5	The *JK* flip-flop	224
*8.6	Methods of clocking flip-flops	228
	8.6.1 The simplified master/slave flip-flop	229
	8.6.2 The simplified edge triggered flip-flop	233
	8.6.3 The capacitively coupled flip-flop	236
8.7	Races and clock skew	237
	Summary	239

Chapter 9 Counters and Registers

- 9.1 Introduction . . . 246
- 9.2 Asynchronous binary counters . . . 247
  - 9.2.1 Basic principles . . . 247
  - 9.2.2 Characteristics of asynchronous counters . . . 248
- 9.3 Synchronous binary counters . . . 251
- 9.4 Synchronous and asynchronous up/down-counters . . . 253
  - 9.4.1 Basic principles . . . 253
  - 9.4.2 Applications . . . 255
- 9.5 Synchronous and asynchronous modulo counters . . . 258
  - 9.5.1 Basic principles . . . 258
  - *9.5.2 Developing modulo counters . . . 259
- 9.6 Electrical characteristics of IC counters . . . 267
- 9.7 Cascading counters . . . 270
- *9.8 Rate multipliers . . . 272
- 9.9 Registers . . . 272
  - 9.9.1 Shift register principles . . . 273
  - 9.9.2 Shift register applications . . . 275
  - *9.9.3 Cascaded shift registers . . . 276
  - *9.9.4 MOS shift registers . . . 277
  - *9.9.5 Shift register memories . . . 277
- *9.10 Shift register counters . . . 279
- Summary . . . 281

Chapter 10 Semiconductor Memories

- 10.1 Introduction . . . 289
- 10.2 The read-only memory (ROM) . . . 290
  - 10.2.1 Basic principles of ROMs . . . 290
  - 10.2.2 Electrical characteristics of ROMs . . . 293
  - 10.2.3 Memory expansion . . . 295
  - 10.2.4 Random logic generation . . . 296
  - *10.2.5 Microprogramming . . . 299
  - *10.2.6 Table look-up . . . 300
  - *10.2.7 Code conversion and generation . . . 301
  - *10.2.8 Character generation . . . 302
- *10.3 Programmable logic arrays (PLAs) . . . 304
- 10.4 The read/write memory . . . 308
  - 10.4.1 Bipolar memory cells . . . 309
  - 10.4.2 MOS memory cells . . . 311
  - 10.4.3 Bipolar memories . . . 311
  - 10.4.4 Static MOS memories . . . 316
  - 10.4.5 Dynamic MOS memories . . . 318
  - *10.4.6 Memory expansion . . . 323
- *10.5 Miscellaneous memories . . . 327
- Summary . . . 329

Appendixes

A	Summary of Boolean Algebra Definitions, Laws, and Rules	335
B	The Powers of 2	337
C	Schematic Diagrams and Troubleshooting Aids	338
D	Digital IC Device Characteristics	343
E	Optoelectronics	352
F	Synthesized Signal Generator—Description and Logic Diagrams . . .	354
G	Microcomputer—Description and Logic Diagrams	367

Answers to Selected Problems   389

Index   405

1: INTRODUCTION TO DIGITAL PRINCIPLES

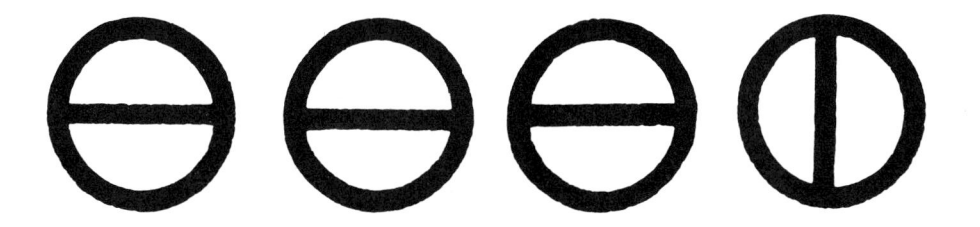

1.1 INTRODUCTION

How many examples of digital circuits have you seen today? You may have seen a pocket calculator (Fig. 1.1), a digital wristwatch (Fig. 1.2), a TV with a digital channel selector or digital channel readout, a digital voltmeter (Fig. 1.3), a digital communication system, a digital computer (Fig. 1.4), or any of a host of other examples. In fact, digital circuits are used extensively in nearly all fields where electronic circuits are needed—for example, the computer, instrumentation, medical, consumer, communications, and controls fields. And their use is rapidly increasing. The reasons for this trend will be discussed in this chapter.

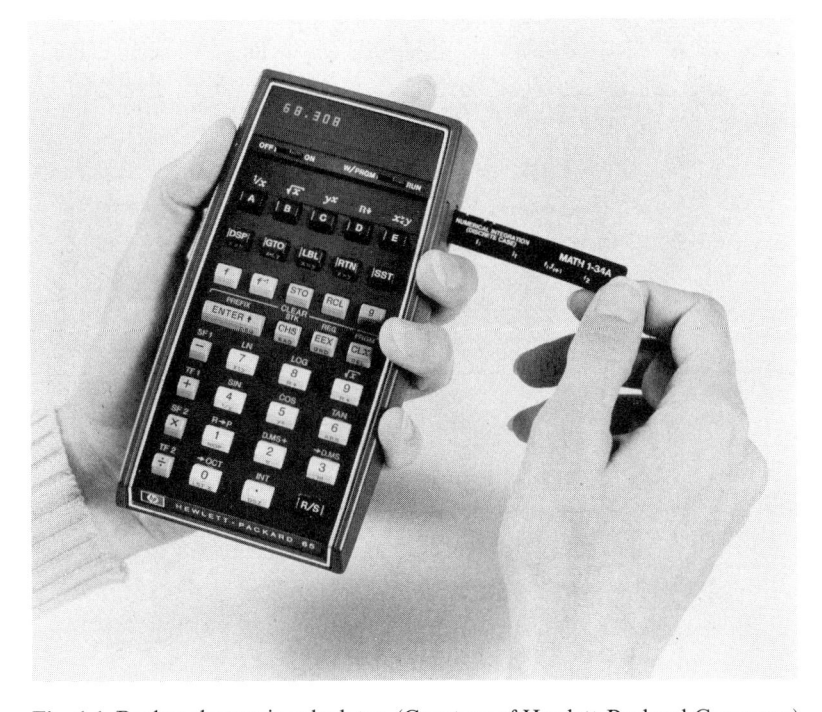

Fig. 1.1 Pocket electronic calculator. (Courtesy of Hewlett-Packard Company.)

1

Fig. 1.2 Electronic wristwatch. (Courtesy of Hughes Aircraft Company.)

Fig. 1.3 Digital multimeter. (Courtesy of Hewlett-Packard Company.)

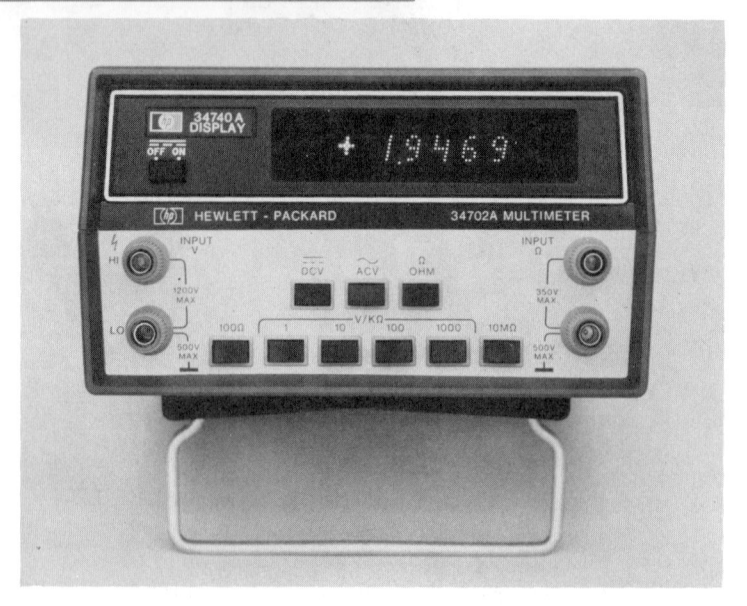

Fig. 1.4 A modern digital computer and peripheral equipment. (Courtesy of Hewlett-Packard Company.)

The technician, technologist, or engineer who works in any field which uses electronic circuits must now have a thorough background in digital electronics. An understanding of digital electronics is just as fundamental and important as an understanding of ac and dc circuits. The purpose of this book is to provide a solid background in digital electronics and some of its applications.

In this chapter we discuss the principles of digital electronics, introduce two basic digital devices (logic gates and flip-flops), and compare analog and digital circuits. Then we give a brief introduction to digital integrated circuits. Finally, in the last two sections, we discuss the conversion between binary and decimal numbers and binary addition.

4 INTRODUCTION TO DIGITAL PRINCIPLES

1.2 PRINCIPLES OF DIGITAL ELECTRONICS [1]

An **electronic system** is a collection of electronic circuits connected together to perform some specified function. Some examples of electronic systems are digital computers, oscilloscopes, electronic watches, and electronic calculators. An **analog electronic system** deals with continuously varying quantities such as temperature or pressure. If continuous (analog) quantities are converted to voltages, called **analog voltages**, these voltages will also vary continuously. Analog devices, such as amplifiers and filters, can then be used to modify analog voltages. Figure 1.5 shows a simple analog system that consists of a temperature sensor, a transducer, which converts the continuous temperature variations into analog voltages, an amplifier, which amplifies these voltages, and a recorder to record the temperature.

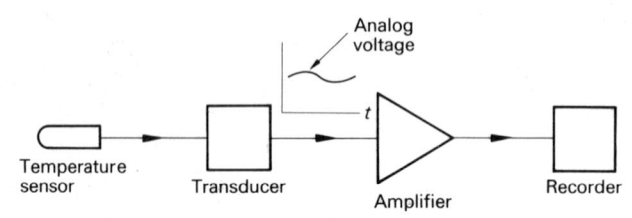

Fig. 1.5 Analog electronic temperature-monitoring system.

A **digital electronic system** deals with discrete quantities such as the decimal digits (0, 1, 2, ..., 9) and alphabetic characters ($A, B, \ldots, Z$). These discrete quantities can be represented by voltages called **digital voltages** or **voltage levels**. Digital voltages can vary, but only in discrete steps. Figure 1.6 shows a simple digital system that counts the number of pills put into a pill bottle. Each time a pill falls through the light beam the counter advances by one count. The counter has ten counts 0, 1, ..., 9, represented by the ten digital voltages 0 V, 1 V, ..., 9 V, respectively. Only the discrete voltages 0 V, 1 V, etc., occur. A voltage of $1\frac{1}{2}$ V never occurs because there are never $1\frac{1}{2}$ pills in the bottle.

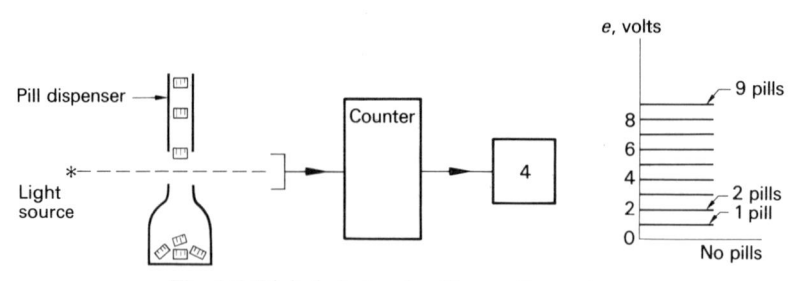

Fig. 1.6 Digital electronic pill-counting system.

Many electronic systems use both digital and analog circuits within the same system. Also, many electronic systems can be constructed from either digital or analog circuits. This book deals primarily with digital circuits and digital systems.

The minimum number of voltage levels a digital circuit can have and still be useful is two. An electronic system that has only two voltage levels is called a **binary** or **two-state system**. The advantages of binary systems over digital systems with more than two states are: (1) many electronic components such as lights, relays, diodes, and switches have only two states; (2) two states are more easily distinguished; and (3) arithmetic, logic, counting, and storing operations are simpler to implement.

Some familiar devices that have only two possible conditions, or states, are: a light switch (it is either open or closed), a light bulb (it is either on or off), a door bell (it either rings or is silent), and an ideal diode (it is either conducting or nonconducting).

The two states of a binary system are called the **HIGH** (H) state and the **LOW** (L) state. We use the symbols V_H to represent the HIGH voltage state and V_L to represent the LOW voltage state. For example, in a binary electronic system the LOW state might be represented by 0 V, that is, $V_L = 0$ V, and the HIGH state by +5 V, that is, $V_H = +5$ V, as illustrated in Fig. 1.7(a).

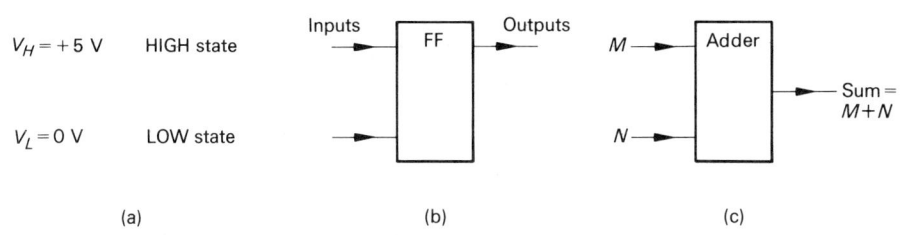

Fig. 1.7 Representations of (a) HIGH and LOW voltage levels, (b) a memory, (c) an adder.

A **binary circuit**, also called a **device**, can be represented by a symbol such as the one shown in Fig. 1.7(b). The shape of the symbol and a name (FF) placed within the symbol tell us this is a flip-flop, i.e., a memory device. Symbols are useful because the human mind comprehends ideas best by means of graphic images. Thus we try to understand a device by forming a mental picture of the mechanism involved. For example, an atom is "pictured" as a circle (the nucleus) surrounded by dots (orbiting electrons).

The memory circuit, represented by the symbol shown in Fig. 1.7(b), has two input lines and one output line to represent circuit inputs and the output. Input lines are usually drawn on the left of the symbol, and output lines on the right. Sometimes arrows are drawn to clearly distinguish input lines from output lines. For binary circuits, input and output lines have only one of two voltages on them, V_H or V_L.

With the trend toward more complex digital circuits, symbols are especially important for representing devices in a simple way. For example, the circuit for a decimal adder is very complicated, but the functional operation of the adder can be represented by a simple, easily understood symbol as shown in Fig. 1.7(c). The adder has two inputs, and one output for the sum.

One great advantage of digital electronics is the ease with which complex systems can be designed and built by interconnecting simple digital devices. The connections for input and output lines obey certain rules which will be developed in detail throughout this text. The rule for the input and output lines of a battery, for example, is that the positive and negative terminals are normally not directly connected to each other.

The emphasis in this text is on the functions performed by digital devices and on how to interconnect these devices to form useful digital systems. Thus we may treat the binary *symbols* as system components, without dwelling on the internal electronic components of the devices. We shall discuss circuit details only to the extent needed for understanding the capabilities, limitations, and applications of each device.

Next let's look at two binary devices used in most digital systems, the logic gate and the flip-flop.

1.3 PRINCIPLES OF LOGIC GATES AND FLIP-FLOPS

1.3.1 The Logic Gate

One of the important operations performed in a digital system is making decisions—for example, "Should this warning light be turned on or off?" "Should this relay be opened or closed?" This decision-making process is called **logic**. **Binary logic** applies to decisions in which there can be only two possible outcomes, such as a light that is either on or off, or a statement that is either true or false. An **AND decision** is a decision whose outcome is true only if all the variables which go into making that decision are true. For example, consider the logic decision, "Will the car start?" The decision is binary in nature—the car will either start or not start. Two conditions, or variables, that might be considered are also binary in nature: "Is the gear shift in neutral (yes or no)?" *and* "Is the key in the start position (yes or no)?" The process for making this decision can be represented graphically as shown in Fig. 1.8(a). The AND symbol represents the AND decision. If both the gear shift is in neutral *and* the key in the start position then the car will start. All possible combinations of inputs can be conveniently shown in a table, called a **truth table**, as shown in Fig. 1.8(b). The words *yes* and *no* used in this truth table represent the two states of the input and output variables. Any two words, such as the HIGH and LOW terms already introduced, could be used equally well in the truth table.

1.3 PRINCIPLES OF LOGIC GATES AND FLIP-FLOPS

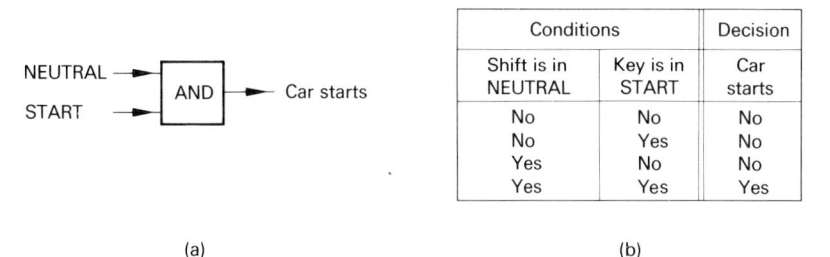

Fig. 1.8 The AND decision: (a) symbol; (b) truth table.

Conditions		Decision
Shift is in NEUTRAL	Key is in START	Car starts
No	No	No
No	Yes	No
Yes	No	No
Yes	Yes	Yes

A **logic gate**, or simply a **gate**, is an electronic circuit that makes decisions, i.e., performs logic, on its binary inputs. Some examples of logic gates, which are described in Chapter 2, are AND, OR, and NOT gates. Each gate has its own schematic symbol. The standard symbol for an AND gate is shown in Fig. 1.9(a). The inputs are labeled A and B, and the output is labeled F. For instance, in our car-starting example A could represent "the shift is in neutral." These shorthand labels are called binary variables. **Binary (logic) variables** are variables that have only one of two states at any one time. The two states are called HIGH (H) and LOW (L). Binary variables can be used to represent the input conditions or the resultant output decisions. The truth table for an AND gate using the alphabetic letters A, B, and F is shown in Fig. 1.9(b). This generalized truth table represents not only the car-starting decision, but any AND decision involving two input variables. Note that the output of an AND gate is HIGH only when both inputs are HIGH.

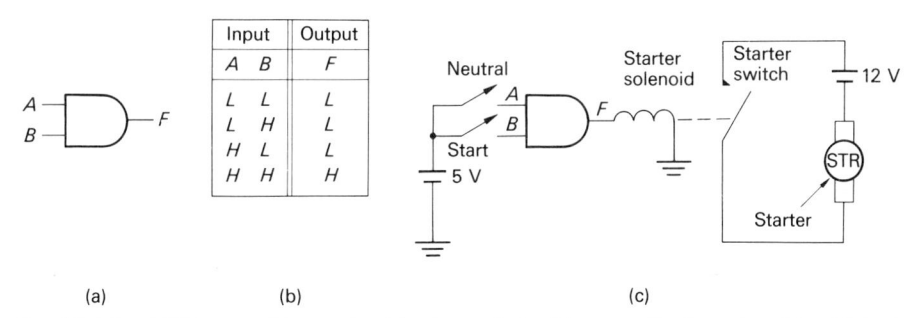

Input		Output
A	B	F
L	L	L
L	H	L
H	L	L
H	H	H

Fig. 1.9 The AND gate: (a) standard logic symbol; (b) generalized truth table; (c) logic diagram for Example 1.1.

Consider the following application of an AND gate.

Example 1.1 Draw the logic diagram for a control circuit that will prevent an automobile from starting unless the gear shift is in neutral and the ignition key is turned to the start position.

Solution. An AND gate can be used to make the decision, "Is the shift in neutral *and* the key in the start position?" as shown in Fig. 1.9(c). Mechanical switches open or close, and thus apply 0 V (LOW) or +5 V (HIGH), to the AND gate inputs A and B. The output F of the AND gate is connected to a starter solenoid. When the AND gate output is HIGH (+5 V), the solenoid causes the starter switch to close. If either or both inputs are LOW (i.e., the shift is not in neutral or the key is not in start) the output of the AND gate is LOW, the starter switch is open, and the car cannot be started. If both inputs are HIGH, the output of the AND gate will be HIGH, which energizes the solenoid, causing the starter switch to close. ∎

1.3.2 The Flip-Flop

The second major device used in most digital systems is the flip-flop. A **flip-flop** is a binary device that remembers its inputs even after the inputs are removed. For example, a fire alarm may be designed so that it continues to ring a bell even after the condition which caused it to ring (the fire) is removed. The alarm bell has two states, ON or OFF. A logic gate, by comparison, has no memory; that is, once its inputs are removed it will not retain its output state.

Figure 1.10(a) shows the logic symbol for one type of flip-flop called an **RS latch**. The inputs are labeled with the binary variables S (set) and R (reset); the output is labeled Q. The *RS* latch has the following properties, as summarized in the truth table of Fig. 1.10(b). When both input conditions (states) are LOW the output is HIGH. The device is then in what is called the **indeterminate** state, because the output may become either HIGH or LOW if both inputs return to HIGH simultaneously; this state is therefore generally avoided. When the R input is LOW and the S input is HIGH the output becomes HIGH. When the output of any device is HIGH, and the device is not in an indeterminate state, the device is said to be **set**. Thus when R is LOW and S is HIGH, the latch is set. When R is HIGH and S is LOW, the output becomes LOW. When the output of any device is LOW the device is said to be **reset**. Finally, when S is HIGH and R is HIGH, the output does not change, i.e., if it was LOW it stays LOW and if it was HIGH it stays HIGH.

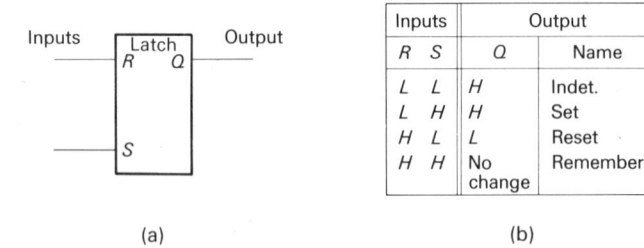

Inputs		Output	
R	S	Q	Name
L	L	H	Indet.
L	H	H	Set
H	L	L	Reset
H	H	No change	Remember

(a)   (b)

Fig. 1.10 The *RS* latch: (a) logic symbol; (b) truth table.

1.3 PRINCIPLES OF LOGIC GATES AND FLIP-FLOPS

Let's consider a practical application of an *RS* latch.

Example 1.2 An automobile starter motor has a thermal overload switch that opens when the temperature of the starter exceeds the safe limit. Design a control circuit that will cause a red warning light on the driver's panel to turn on and to stay on even after the starter temperature has dropped to a safe level. (The red light can be turned off only by pushing a reset button.)

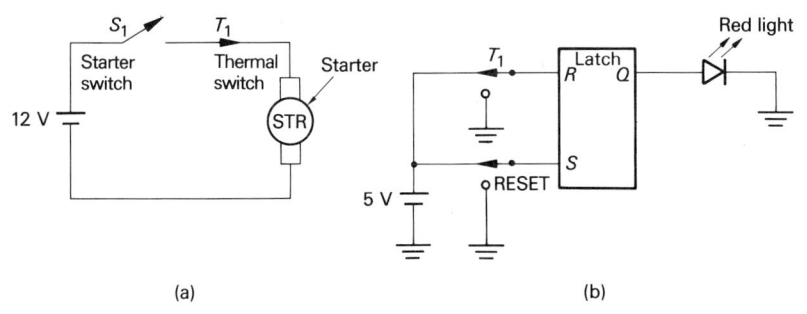

Fig. 1.11 Example 1.2: (a) starter circuit; (b) logic diagram for the warning system.

Solution. Figure 1.11(a) shows the starter motor circuit with thermal overload switch T_1 and starter switch S_1. A latch can be used in the warning circuit to remember that T_1 has opened, possibly indicating a serious problem, even after the motor has cooled and T_1 closes. Figure 1.11(b) shows the logic diagram for the warning system. When the output Q of the latch is LOW the red light is off; when Q becomes HIGH the light turns on. Initially suppose T_1 is closed so R is HIGH, and the reset switch is temporarily put in the reset position, so S is LOW. Referring to Fig. 1.10(b), we see that when $R = H$ and $S = L$, the latch resets, i.e., Q is LOW; initially, therefore, the light is off. Then, while R is HIGH, if a thermal overload occurs T_1 opens so R goes LOW and the latch sets ($S = H, R = L$). This causes Q to go HIGH, turning on the warning light. When the motor cools and T_1 closes, R returns HIGH, but the inputs are now $S = H$ and $R = H$, hence the output does not change; that is, Q remains HIGH. Thus the latch has remembered the thermal overload problem. The warning light can be turned off by momentarily depressing the reset switch. ∎

Simple devices like latches and logic gates can be combined in many ways to form complex digital systems. For example, they can perform jobs such as solving mathematical problems, controlling industrial machines, and measuring physical quantities. The control of a motor having two thermal switches, an AND gate, and a latch is developed in the problems at the end of this chapter.

1.4 COMPARISON OF DIGITAL AND ANALOG CIRCUITS

Digital circuits have several significant advantages over analog circuits which are responsible for the rapidly increasing use of digital circuits. The purpose of this section is to compare the advantages and disadvantages of digital and analog circuits.

Consider first a simple analog circuit that uses a voltage divider to reduce the analog voltage output of a temperature transducer by a factor of 2. Figure 1.12(a) shows the voltage divider, whose input at a particular instant of time is 10 V, and a voltmeter, which displays the output as 5 V.

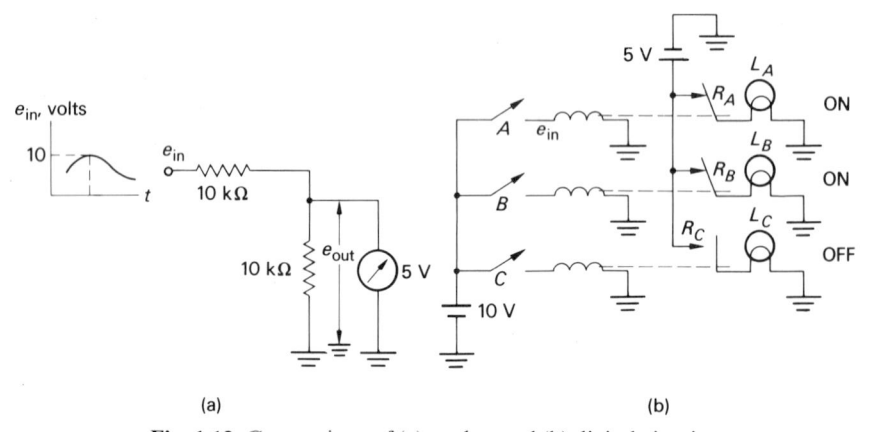

Fig. 1.12 Comparison of (a) analog and (b) digital circuits.

Now consider the measurement of a quantity that is digital in nature. Figure 1.12(b) shows a digital circuit that will determine whether parts on an assembly line have passed points A, B, and C. When a part passes point A it closes switch A, thereby activating the relay and closing switch R_A. The 5 V battery then supplies current to lamp A, turning it on. The relay is activated only when its input voltage is 2 V or more; that is, any voltage between 0 V and 2 V will not close the relay switch. The relay is of the latching type, which means that switch R_A stays closed even after the input A is removed. Thus the relay remembers that a part has passed point A. The relay can also be reset (opened) using a separate input not shown. When lamps A and B are on, the assembly line operator knows parts have passed points A and B on the line.

Now let's compare these two circuits in a general way, realizing there are exceptions to these comparisons.

Noise Unwanted voltages introduced into circuits from sources such as lightning, arcing motors, or power lines are called **noise voltages**. A noise voltage of $+1$ V in the analog circuit of Fig. 1.12 results in an output voltage of $e_{out} = \frac{1}{2}e_{in} = \frac{1}{2}(10 \text{ V} + 1 \text{ V}) = 5.5 \text{ V}$, or an error in the output of $(5.5 - 5)/5 = 0.1 = 10\%$.

In the digital circuit, if switch A is open (no part has passed point A) then a 1-V noise voltage will produce an e_{in} of 1 V, which is not sufficient to activate the relay. The lamp therefore stays off and *no* error results. If switch A is closed, applying 10 V to relay A, then 1 V of noise still has no effect on relay A. Thus digital circuits are less prone to errors due to noise voltages than analog circuits are.

Cost The components used in an analog system, such as precision resistors, amplifiers, and filters, are generally quite expensive. Digital system components, such as solid state switches, logic gates, and memories, are relatively simple devices that are mass produced in integrated circuit (IC) form and are therefore inexpensive.

Readouts Digital readouts, such as lamps, are easier to read and can be read with less chance of error than analog readouts. For example, in Fig. 1.12(a), different observers might read the voltmeter as 5.95 V, 5.05 V, or 5.1 V. Most observers would read the lamps in Fig. 1.12(b) correctly as being either on or off. Furthermore, the voltmeter is a more complicated, costly readout device than a lamp. (Note that the digital system requires three readouts, compared to only one in the analog system.)

Duplication of Circuits To build several analog circuits like the one in Fig. 1.12(a) that will produce identical results requires the use of precise components. Digital systems such as the one shown in Fig. 1.12(b) require only devices that switch for voltages greater than a certain voltage and lamps that are either on or off. Thus digital circuits can usually be duplicated with greater ease and less cost than analog circuits.

Even a complex digital system is made up of many relatively simple IC components that are easily assembled and duplicated. A complex analog system, on the other hand, includes many complex devices, its components are harder to assemble, and the circuit is correspondingly more difficult to duplicate.

Storage and Manipulation of Information By use of latching relays, the digital circuit in Fig. 1.12 can easily remember (store) for a long period of time the fact that a part has passed a certain point on the assembly line. It is relatively difficult for an analog circuit to store information for long periods of time. Manipulation of information, for example by addition, filtering, and delaying, is much easier to do in digital systems than in analog systems. Sequential operations, such as those performed by a numerically controlled milling machine (drill, then ream and then tap), are readily performed by digital systems.

Reliability and Maintenance Most digital systems are made primarily from integrated circuits, which as a rule provide fault-free operation (reliability) for long periods of time. Analog systems are prone to such problems as leaky capacitors, burned out filaments, and overheating, and thus are less reliable than digital systems.

Digital systems are usually easier and cheaper to maintain because there are fewer devices to calibrate, clean, and tune. They are in general more immune to shock and dirt than analog systems. Errors in data in a digital system are usually easier to detect and correct than in analog systems.

Troubleshooting Digital systems with a few inputs and outputs are easy to troubleshoot because all possible combinations of inputs can be tried, and each input or output has only two states. Individual ICs can be easily tested by comparing them with known good ICs. Large digital systems are most easily tested with automatic test equipment. Analog systems usually have many voltage levels and more test points to check than digital systems.

Variables Most naturally occurring variables, such as pressure and temperature, are analog in nature. When digital systems are used to measure or process analog variables, then analog-to-digital converters (ADCs) are required. Digital systems can, however, readily manipulate the time variable. For example, radar signals picked up by an aircraft antenna over a period of several seconds can be compressed into a few milliseconds and repeatedly processed or displayed every few milliseconds.

Human Factors Decimal numbers are much easier for humans to read, remember, and manipulate than binary numbers. Hence binary-to-decimal converters are frequently required for reading information out of digital systems. Analog devices such as voltmeters can be read out directly in decimal. Binary arithmetic is cumbersome.

1.5 DIGITAL INTEGRATED CIRCUITS

Digital circuits come in two forms, discrete circuits and integrated circuits. **Discrete circuits** are made from separate components such as diodes, transistors, capacitors, and resistors. A discrete diode and transistor are shown in Fig. 1.13(a, b).

The second form of digital circuits is the **integrated circuit** (IC), which has all the diodes, transistors, and resistors formed on one semiconducting base called a **chip**. Although some digital applications use discrete circuits, the majority of applications use ICs. A very readable discussion of IC fabrication technology is found in the references [2].*

A **monolithic IC** is an electronic circuit fabricated on one chip that cannot be divided without permanently destroying the designed function of the circuit. Connecting pins are attached to terminal pads on the IC chip; the chip is then enclosed in a protective case called a **package**. In one type of package, called a **dual in-line package** (DIP), the connecting pins are arranged in two (dual) lines as shown in Fig. 1.13(c). Digital ICs commonly have a total of 14, 16, 18, 24, 36, or 40 pins. The DIP shown in Fig. 1.13(c) contains four logic gates in one package. The pin-numbering system shown is the standard one for a 16-pin DIP. This DIP IC would take up about $\frac{1}{2}$ cm^3 of space, or only about 10% of the volume occupied by an equivalent digital circuit made from discrete components.

---

* The bracket indicates references found at the end of the chapter.

1.5 DIGITAL INTEGRATED CIRCUITS 13

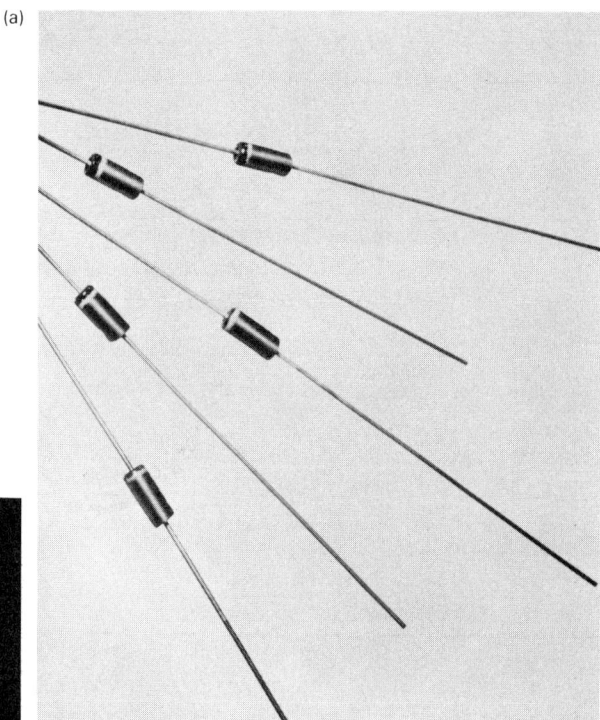

Fig. 1.13 Digital circuits: (a) diodes; (b) transistor; (c) DIP integrated circuit and DIP pin numbering.

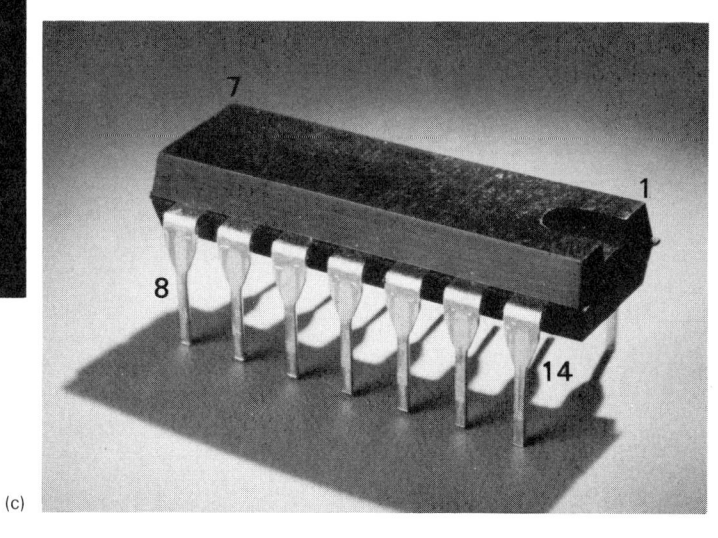

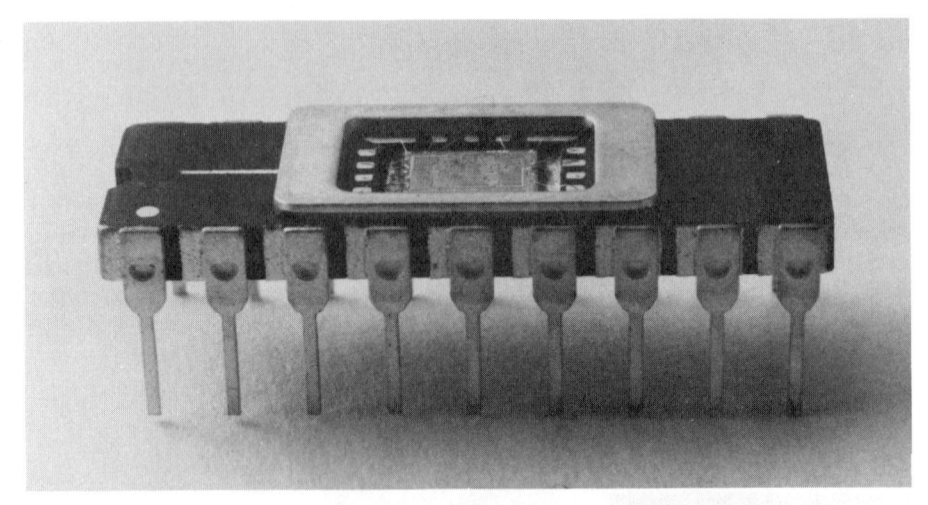

Fig. 1.14 18-pin DIP with chip exposed. (Courtesy of Intel Corporation.)

Figure 1.14 shows a photograph of an 18-pin DIP with the chip exposed. Notice the wires which connect the chip outputs to the terminal pads. These pads are in turn connected to external IC pins.

Digital ICs are broadly classified into three groups according to their complexity. A monolithic IC containing 1 to 12 logic gates is classified as a **small-scale integration** (SSI) device. An IC containing 13 to 99 logic gates is in the **medium-scale integration** (MSI) class, and an IC containing more than 99 gates is in the **large-scale integration** (LSI) class.

Figure 1.15 is a photomicrograph of the LSI chip shown in Fig. 1.14. This chip is the main portion of a microcomputer; it is called a **microprocessor**. Most of the devices shown on the photomicrograph will be studied in this text as separate SSI and MSI devices. The great complexity of ICs such as this is the main reason why the emphasis in this text is on the functions performed by such ICs and how to interconnect them, rather than on the circuitry within the IC. It is not possible to alter the internal connections or repair any components within an IC. Neither is it possible to troubleshoot individual components within an IC, unless they are connected to the external connecting pins.

The main advantages of building digital systems from SSI, MSI, and LSI devices, instead of using discrete components, are greater reliability, lower cost, fewer components to mount, reduction in area, fewer connectors required, reduction in the size of the enclosure for the circuit, less design time required (since the IC devices are already designed), reduction in assembly time (since fewer components need to be mounted), and simplified testing of the system (only a few test points instead of hundreds!).

1.5 DIGITAL INTEGRATED CIRCUITS 15

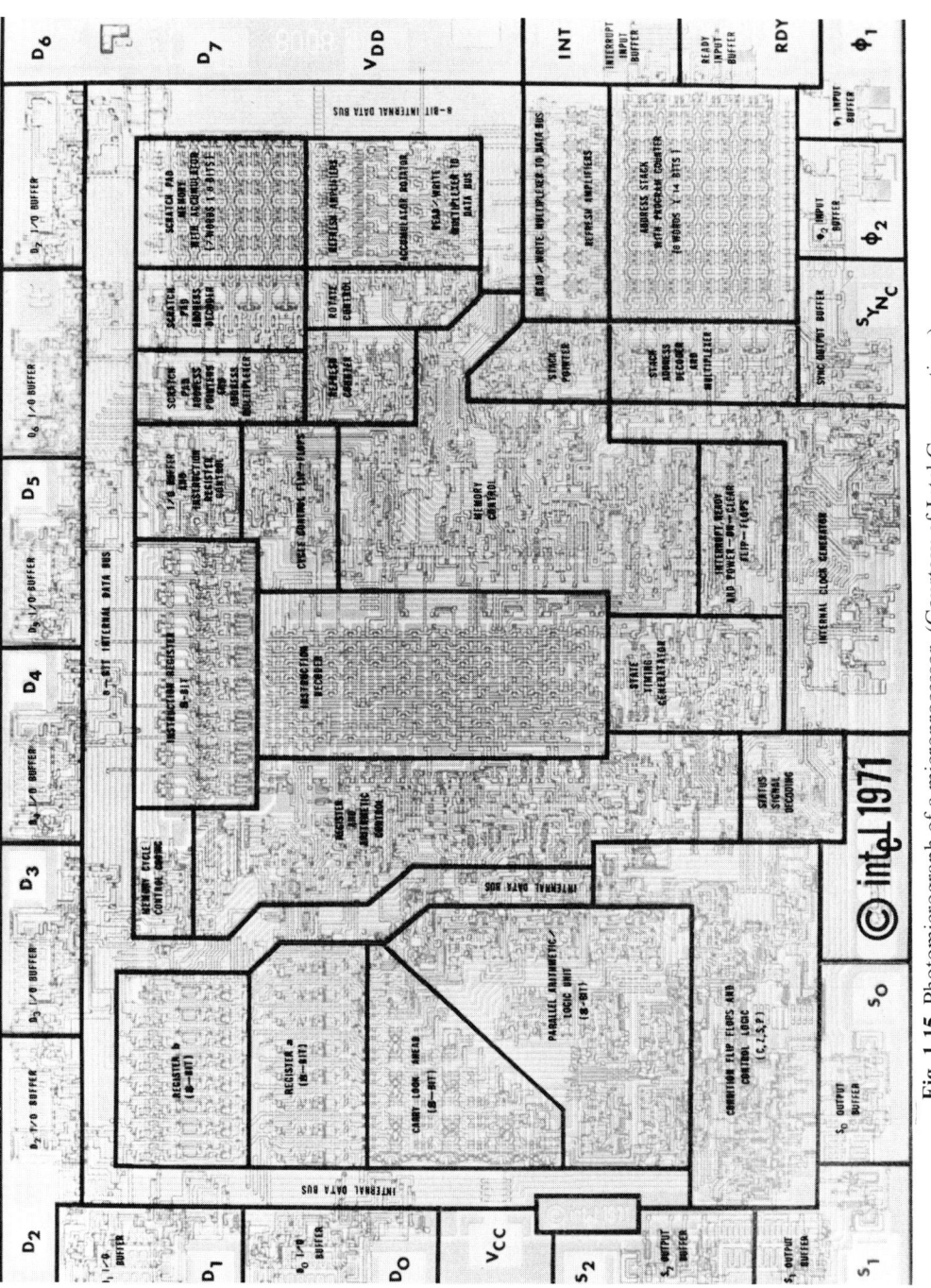

Fig. 1.15 Photomicrograph of a microprocessor. (Courtesy of Intel Corporation.)

Fig. 1.16 The effect of MSI and LSI on digital systems. (Reprinted from *Electronics*, April 18, 1974; copyright © 1974 McGraw-Hill, Inc.)

The trend in most digital systems is toward using more MSI and LSI devices instead of SSI devices. Figure 1.16 illustrates this trend. The microprocessor on the five printed circuit boards in the background contains 451 ICs, most of them SSI devices. Using MSI devices and semiconductor memories requires 114 ICs and one circuit board (middle board, Fig. 1.16). And the single 40-pin LSI package shown in the foreground replaces the entire 114 packages! Clearly digital MSI and LSI devices will have a fantastic effect on the electronics field in the future.

With this discussion we have nearly finished laying the groundwork for material to be discussed in the rest of the book. To complete the task, we must devote some attention to binary arithmetic. Therefore the last two sections of this chapter deal with binary numbers.

1.6 CONVERSION BETWEEN DECIMAL AND BINARY NUMBERS

The representation of numbers in binary form is convenient for the operation of binary devices, but inconvenient for humans, who are used to working with decimal numbers. Thus there is a need to be able to convert from binary to decimal numbers and from decimal to binary numbers.

1.6 CONVERSION BETWEEN DECIMAL AND BINARY NUMBERS

The Decimal Number System The decimal number system is said to have a **base** of ten since it has ten digits, represented by the symbols 0, 1, 2, 3, 4, 5, 6, 7, 8, and 9. Decimal numbers greater than 9 are formed by putting digits in columns to the left of the first digit. Each column has ten times the value (or **weight**) of the column immediately to its right. This method of representing decimal numbers is called **positional notation**. For example, the number 264 contains three decimal digits, 2, 6, and 4, and can be written as $200 + 60 + 4$. Recall the powers of 10: $10^0 = 1$, $10^1 = 10$, $10^2 = 100$, $10^3 = 1000$, etc. Powers of 10 can be used when writing a decimal number in **scientific notation**; thus 264 can be written

$$264 = 2 \times 10^2 + 6 \times 10^1 + 4 \times 10^0.$$

Similarly, the number 4695 can be written in scientific notation as

$$4696 = 4 \times 10^3 + 6 \times 10^2 + 9 \times 10^1 + 6 \times 10^0.$$

The Binary Number System The binary number system has two **binary digits**, called **bits**, which are assigned the symbols 0 and 1. Binary numbers are formed by using positional notational. An example of a binary number is 1011, which is read as one-zero-one-one. Some additional examples of binary numbers are 101, 0111, and 10111. The number 1301 is not a binary number, since 3 is not a binary digit. The binary number system is said to have a **base** of 2, since it has only two symbols (0 and 1).

1.6.1 Conversion of Binary Numbers to Decimal

The binary number 1100 can be converted to its equivalent decimal number by writing it in terms of powers of 2:

$$1100_2 = 1 \times 2^3 + 1 \times 2^2 + 0 \times 2^1 + 0 \times 2^0$$
$$= 1 \times 8 + 1 \times 4 + 0 \times 2 + 0 \times 1 = 12_{10}.$$

The **least significant bit** (LSB) in 1100_2 is the rightmost bit, or 0, and the **most significant bit** (MSB) is the leftmost bit, or 1. The subscript 2 tells us that 1100 is a binary (base 2) number. The binary weights are 8, 4, 2, and 1, which are the powers of 2 (see Appendix B).

Example 1.3 Convert the following binary numbers to decimal numbers: (a) 001_2, (b) 010_2, (c) 10110_2.

Solution

a) $N = 001_2$
$= 0 \times 2^2 + 0 \times 2^1 + 1 \times 2^0 = 1,$ so $001_2 = 1_{10}.$

b) $N = 010_2$
$= 0 \times 2^2 + 1 \times 2^1 + 0 \times 2^0 = 2_{10}.$

c) $N = 10110_2$
$= 1 \times 2^4 + 0 \times 2^3 + 1 \times 2^2 + 1 \times 2^1 + 0 \times 2^0$
$= 16 + 0 + 4 + 2 + 0 = 22_{10}.$

Note from this example that it requires more digits to express a number in binary form than it does in decimal form. ∎

Table 1.1 shows the decimal numbers 0 through 15 and their binary equivalents.

Table 1.1 Decimal Numbers and Binary Equivalents

Decimal	Binary	Decimal	Binary
0	0000	8	1000
1	0001	9	1001
2	0010	10	1010
3	0011	11	1011
4	0100	12	1100
5	0101	13	1101
6	0110	14	1110
7	0111	15	1111

A quicker way of converting from binary to decimal is to write the binary weights $2^0 = 1$, $2^1 = 2$, $2^2 = 4$, $2^3 = 8$, $2^4 = 16$, etc., directly under each bit, starting with the LSB. Wherever a 0 occurs, strike out that power of 2. Then add up the remaining powers of 2. For example:

a) 0 1 0
   $\cancel{4} \, 2 \, \cancel{1} = 2$

b) 1 1 1 0
   $8 \, 4 \, 2 \, \cancel{1} = 8 + 4 + 2 = 14$

c) 1 1 0 0 1
   $16 \, 8 \, \cancel{4} \, \cancel{2} \, 1 = 16 + 8 + 1 = 25$

Some additional examples are: 10111 = 23, 10001 = 17, and 110110 = 54.

1.6.2 Conversion of Decimal Numbers to Binary

The conversion of a decimal number to a binary number is somewhat harder. It can be accomplished by successive division of the decimal number by the base 2, until the quotient decreases to zero. Example 1.4 illustrates this procedure.

Example 1.4 Convert the decimal number 14_{10} to its equivalent binary number.

Solution. Begin by dividing 14 by 2, which gives a quotient of 7 and a remainder of 0. Then divide the 7 by 2, giving a quotient of 3 and a remainder of 1. Continue this process until a quotient of 0 occurs.

$$
\begin{array}{ll}
& \textit{Remainders} \\
14 \div 2 = 7 & 0 \text{ (LSB)} \\
7 \div 2 = 3 & 1 \\
3 \div 2 = 1 & 1 \\
1 \div 2 = 0 & 1 \text{ (MSB)}
\end{array}
$$

The remainders form the binary number. The first remainder is the LSB and the last remainder is the MSB. Thus $14_{10} = 1110_2$. ∎

The following example further illustrates this successive division technique.

Example 1.5 Convert the following decimal numbers to their binary equivalents: (a) 13, (b) 26.

Solution

a)
$$
\begin{array}{ll}
& \textit{Remainders} \\
13 \div 2 = 6 & 1 \text{ (LSB)} \\
6 \div 2 = 3 & 0 \\
3 \div 2 = 1 & 1 \\
1 \div 2 = 0 & 1
\end{array}
$$

So
$$13_{10} = 1101_2,$$

which can be checked by converting 1101 into a decimal number to see if it equals 13.

b)
$$
\begin{array}{ll}
& \textit{Remainders} \\
26 \div 2 = 13 & 0 \text{ (LSB)} \\
13 \div 2 = 6 & 1 \\
6 \div 2 = 3 & 0 \\
3 \div 2 = 1 & 1 \\
1 \div 2 = 0 & 1
\end{array}
$$

Thus
$$26_{10} = 11010_2. \blacksquare$$

1.7 BINARY ADDITION

Binary addition is an important principle on which digital arithmetic operations are based. Details of binary arithmetic are discussed in textbooks that deal with arithmetic applications of digital circuits, such as computer texts. Here we need only discuss the basic principles of binary addition.

The rules for binary addition are:

$$0 + 0 = 0,$$
$$0 + 1 = 1,$$
$$1 + 0 = 1,$$
$$1 + 1 = 0 \text{ with a carry of } 1.$$

The first three addition rules are just the same as for decimal numbers and are therefore easily remembered. When the decimal digits 9 and 1 are added, a 1 is carried over to the next most significant position and a 0 is put in the units position, that is, $9 + 1 = 10$. Similarly, in binary $1 + 1$ is not 2, since 2 is not a binary digit; the sum is simply called 0 and a 1 is carried to the next MSB, that is, $1 + 1 = 10$. Three bits can be added together by adding them two at a time, just as is done in decimal addition. Thus $1 + 1 + 1 = (1 + 1) + 1 = 0 + 1 = 1$, with a carry of 1, written as 11.

Some examples will clarify the binary addition procedure.

```
         1        11       111   ← carry
  10     11      110       111
 +01    +10     +011      +011
 ---    ---     ----      ----
  11    101     1001      1010
```

SUMMARY

The widespread use of digital circuits today makes a basic understanding of digital electronics indispensable to the electronics technician or engineer.

An analog electronic system deals with continuous (analog) voltages whereas a digital system deals with discrete (digital) voltages. Binary systems use only two voltages, V_H and V_L, to represent the two states (HIGH and LOW) of the system.

Electronic symbols are important for understanding the functions of the complex circuits they represent.

A logic gate is an electronic circuit that makes decisions, such as the AND decision. It is one of the basic building blocks of most digital systems. The output of an AND gate is HIGH only when all of its inputs are HIGH, as summarized in its truth table. The flip-flop is a memory device and is the second basic building block for most digital systems.

The main advantages of digital systems over analog systems are, in general: (1) lower susceptibility to noise, (2) lower cost, (3) simpler and easier-to-read digital readouts, (4) ease of duplication, (5) ease of storage and manipulation of information, (6) reliability and ease of maintenance, and (7) ease of troubleshooting.

Almost all modern digital systems use ICs rather than discrete circuits. The trend is toward greater use of MSI and LSI devices in digital systems.

The binary digits, called bits, are denoted by 0 and 1. Binary numbers, such as 1011_2, can be converted to their decimal equivalents by multiplying each bit by ascending powers of 2, beginning with the LSB, and adding. Decimal-to-binary conversion is performed by successive division by 2 until a zero quotient is obtained. The remainders form the binary number.

The rules of binary addition are $0 + 0 = 0$, $0 + 1 = 1$, $1 + 0 = 1$, and $1 + 1 = 0$ with a carry of 1.

PROBLEMS

Section 1.1

1. Why is an understanding of digital electronics important for technicians and engineers in the electronics field?

Section 1.2

2. Define an analog system.
3. Give an example of a digital system that uses ten voltage levels.
4. Name two advantages of a binary system over a digital system with four states.
5. What are the two states of a binary system called?

Section 1.3

6. What is meant by binary logic?
7. Draw the logic diagram for a digital system that uses an AND gate. Include a truth table that uses descriptive terms, such as on, off, true, false, rather than the generalized form using HIGHs and LOWs.
8. Describe the function of an *RS* latch and give its truth table.
9. *Automotive Warning System.* Draw the logic diagram of an automotive warning system that indicates a thermal overload only if two locations in the starter motor overheat. [*Hint*: Refer to Example 1.2 and use a two-input AND gate.]
10. *Lock System.* The business office in a department store must keep large amounts of change available for each department. The money is kept in a drawer that is locked with two key-operated switches. The keys are kept by the office manager and the department supervisor. Design a logic system that will allow access to the money only if both keys are used.

Section 1.4

11. Make up an example of a simple analog circuit and a digital circuit that you are familiar with, and that would be useful in comparing digital and analog circuits.
12. Make a table which summarizes the advantages and disadvantages of digital circuits as compared with analog circuits.

Section 1.5

13. What is an IC chip?
14. Draw a top view of a 14-pin DIP. Label the pin numbers.
15. Define an SSI device.
16. List five advantages of LSI circuits over discrete circuits.

Section 1.6

17. Make a table showing the following powers of 2: $2^0, 2^1, 2^2, \ldots, 2^{10}$.
18. What are the LSB and the MSB in the binary number 10110?
19. Convert the following binary numbers to decimal numbers.
    a) 010   b) 011   c) 110   d) 1011
    e) 1001   f) 01001   g) 10111   h) 10001
    i) 111111   j) 010001   k) 101010   l) 100100
20. Convert the following decimal numbers to binary numbers.
    a) 6   b) 9   c) 12   d) 22
    e) 40   f) 16   g) 140 h) 296
    i) 29   j) 50   k) 320 l) 149

Section 1.7

21. Add the following binary numbers together.
    a) 11 + 01   b) 110 + 011
    c) 111 + 110   d) 1101 + 0111
    e) 10 + 10   f) 1011 + 0100
    g) 11011 + 01100   h) 011000 + 1110001
22. Check your answers to Problem 21 by converting the binary numbers to decimal numbers, adding, and then converting to binary numbers.
23. *Computer Program.* (a) Write a digital computer program that will add any two seven-bit binary numbers. Have your program print out the augend, addend, and sum. Reserve an eighth bit for the possibility of a carry-out from the most significant bit position (called *overflow*). (b) Use your program to compute the answers for Problem 21.

REFERENCES

1. Richard A. Karlin and Gary F. Comisky. "What Can Digital Do?" *Radio-Electronics*, pp. 48–50, February 1975.
2. Robert G. Hibberd. *Integrated Circuits.* New York: McGraw-Hill, 1969, p. 29.

2: LOGIC GATES

2.1 INTRODUCTION

The logic gate is one of the basic building blocks from which most digital systems are built. Its applications appear in many systems—for example, in traffic light controllers, in digital computers, in digital voltmeters.

As you will recall from Chapter 1, a logic gate is an electronic circuit that makes a logic decision. Logic gates are available in IC form in all families of logic. Presently the three most widely used logic families are transistor-transistor logic (TTL or T^2L), emitter-coupled logic (ECL), and complementary metal-oxide-semiconductor logic (CMOSL).

In this chapter we present the principles of logic gates and explain them in terms of mechanical switches. We also give examples of their practical applications. We shall examine logic gate circuits for the TTL, ECL, and CMOSL families in Chapter 3, but thereafter we shall use IC logic gates in the applications with little reference to the circuitry. Additional information about logic gate circuits may be found in the references [1, 2].

The five logic gates that we shall discuss in this chapter are the OR, AND, NOT, NAND, and NOR gates. Let's begin by considering the OR gate.

2.2 THE OR GATE

An **OR gate** is a logic gate having one output and two or more inputs. Its output is LOW (*L*) only when all inputs are LOW; if any input is HIGH (*H*) the output is HIGH.

An OR gate can be visualized as two parallel switches in series with the output lamp *F* as shown in Fig. 2.1(a). The lamp can be any two-state device, for example an incandescent light or a light-emitting diode (LED). Let's see how this simple switching circuit performs OR decisions. If switch *A* is open (LOW) and switch *B* is open, no current will flow through the lamp in Fig. 2.1(a), so the lamp will be OFF. If switch *A* is open and switch *B* is closed (HIGH), as shown in Fig. 2.1(b), then current *I* will flow through the closed switch, so the lamp will be ON. (The use of **conventional current**, i.e., current which flows from positive to negative, is assumed throughout this text.) Similarly, if switch *A* is closed and switch *B*

24 LOGIC GATES 2.2

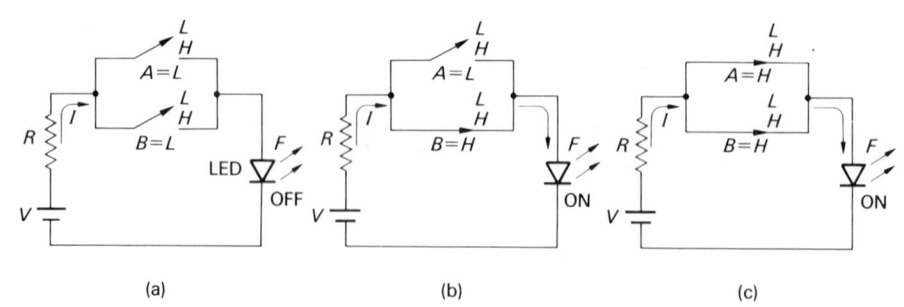

Fig. 2.1 OR gate: (a) $A = L, B = L$; (b) $A = L, B = H$; (c) $A = H, B = H$.

is open, the lamp will be ON. Finally, if both switches A and B are closed as shown in Fig. 2.1(c), current will flow through both switches, so the lamp will be ON.

The output of an OR gate can be represented by the symbol F and the inputs by the symbols A and B, as in Fig. 2.1. The functional relationship between the output F and the inputs A and B is then written

$$F = A + B,$$

where the symbol + means OR, rather than addition. The statement is read, "F equals A or B."

The properties of the logic operation OR, as performed by an OR gate, can be summarized as follows:

Switch A open	and	switch B open:	F is OFF
Switch A open	and	switch B closed:	F is ON
Switch A closed	and	switch B open:	F is ON
Switch A closed	and	switch B closed:	F is ON

These properties can be summarized in tabular form as shown in Fig. 2.2(a). This table is a convenient way of describing, i.e., defining, the logic operator OR.

Inputs		Output
Switch A	Switch B	Lamp F
Open	Open	OFF
Open	Closed	ON
Closed	Open	ON
Closed	Closed	ON

Inputs		Output
A	B	$F=A+B$
L	L	L
L	H	H
H	L	H
H	H	H

(a)   (b)   (c)   (d)

Fig. 2.2 OR gate: (a) truth table using switch terminology; (b) truth table using HIGH and LOW notation; (c) standard logic symbol; (d) gate with detector output.

It is called a **truth table** or a **table of combinations**. Truth tables are extremely useful for understanding logic gates because they show the output of a logic gate for all possible inputs.

Truth tables can be written more compactly as shown in Fig. 2.2(b). An open switch or an OFF detector is represented by a LOW (L). A closed switch or an ON detector is represented by a HIGH (H). Thus when A is LOW (open) and B is HIGH (closed) the detector F is HIGH (ON).

Notice in Fig. 2.2(b) the order in which the inputs are normally written. Both input columns begin with L. Then the rightmost column (B) alternates values in each row, that is, it is an L, then an H, then an L, and finally an H. The leftmost column (A) alternates values in every second row, that is, LL and then HH. This is a standard way of writing truth tables. The values in the output column are determined by the type of logic gate being represented.

When drawing circuits that have many gates, it would be extremely time-consuming to draw the complete schematic of each gate or to draw the switch representation. A standard logic symbol has therefore been adopted for each type of gate [3]. The standard OR gate logic symbol is shown in Fig. 2.2(c). Other symbols are sometimes used in industry, but they will not be used in this text.

Figure 2.2(d) shows a logic circuit that will visually indicate the state of the OR gate inputs. The lamp will turn ON if one or both gate inputs are HIGH.

We have explained the OR gate in terms of mechanical switches only to illustrate the principles of the gate. Actual OR gates are made from transistors for the TTL, ECL, or CMOSL families. OR gates can also be made from diodes or relays.

Now let's look at an example of how an OR gate can be used to solve a practical problem.

Example 2.1 *Warning System.* A small bank has two tellers, each of which has a foot-activated switch connected to an alarm system. When either or both of the switches are depressed a red warning light turns on in the bank president's private office. Draw the logic symbol for the warning system.

Solution. Let T_1 and T_2 be the teller-operated switches. When T_1 or T_2 is depressed the warning light should go ON. An OR gate will provide the correct

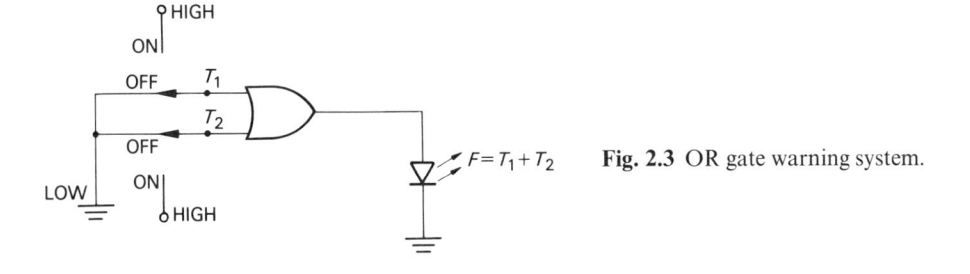

Fig. 2.3 OR gate warning system.

26   LOGIC GATES

logic as shown in Fig. 2.3. When T_1 is ON (HIGH) or T_2 is ON, or both T_1 and T_2 are ON, then the warning light F turns ON as required. ∎

An excellent discussion of logic principles and many examples of logic applications in industrial plants are given in the references [4].

2.3 THE AND GATE

An **AND gate** is a logic gate whose output is HIGH only when all of its inputs are HIGH; if any input is LOW its output is LOW.

An AND gate can be visualized as two switches in series with the output as shown in Fig. 2.4(a). Let's see how an AND gate works. If switch A is open (LOW) and switch B is open, as shown in Fig. 2.4(a), no current will flow through the lamp, so the lamp will be OFF. If switch A is open and switch B is closed (HIGH), no current will flow, so the lamp will be OFF (Fig. 2.4b). Similarly, if A is closed and B is open, F will be OFF. Finally, if both switches A and B are closed (Fig. 2.4c), current I will flow through the lamp, so it will turn ON.

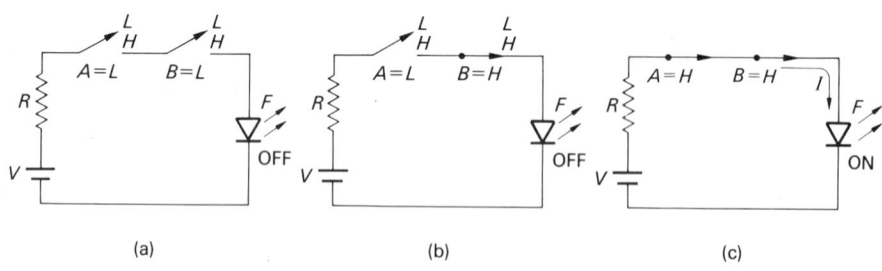

Fig. 2.4 AND gate: (a) $A = L, B = L$; (b) $A = L, B = H$; (c) $A = B = H$.

The functional relationship between the inputs A and B and the output F of an AND gate is written

$$F = A \cdot B,$$

where the symbol · means AND, rather than multiplication. Frequently the dot is omitted between the A and the B, and the relationship is written

$$F = AB.$$

This statement is read, "F equals A and B."

The properties of the logic AND gate can be conveniently summarized in a truth table as shown in Fig. 2.5(a). Note that the output F will be ON only when both switches A and B are closed.

Figure 2.5(b) shows the truth table for a two-input AND gate using the HIGH and LOW notation. In comparing the truth tables for the AND gate and the OR

2.3   THE AND GATE   27

Fig. 2.5 AND gate: (a) truth table using switch terminology; (b) truth table using HIGH and LOW notation; (c) logic symbol; (d) gate with detector output.

(a)

Inputs		Output
Switch A	Switch B	Lamp F
Open	Open	OFF
Open	Closed	OFF
Closed	Open	OFF
Closed	Closed	ON

(b)

Inputs		Output
A	B	F=AB
L	L	L
L	H	L
H	L	L
H	H	H

gate note that the output of the AND gate is HIGH for only one combination of inputs. The OR gate, on the other hand, has a HIGH output for three combinations of inputs.

The logic symbol for an AND gate is shown in Fig. 2.5(c). Figure 2.5(d) illustrates an AND gate connected to a lamp indicator. The lamp will light only when A and B are both HIGH.

Logic gates can have more than two inputs. Figure 2.6(a) shows the switch representation for a three-input AND gate. Just as for the two-input AND gate, the detector will light only if all switches are closed.

The functional relationship between the inputs and the outputs is written

$$F = ABC$$

and is read, "F equals A and B and C."

The truth table for a three-input AND gate is given in Fig. 2.6(b). Any truth table with three inputs has eight rows since there are $2^3 = 8$ possible combinations of the three inputs A, B, and C. Note that this truth table begins with the first

(b)

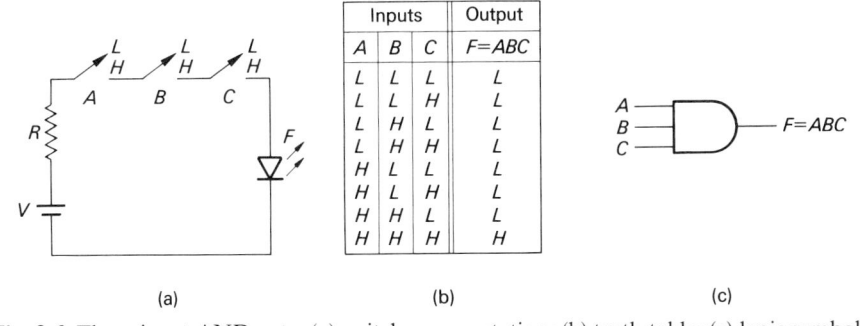

Inputs			Output
A	B	C	F=ABC
L	L	L	L
L	L	H	L
L	H	L	L
L	H	H	L
H	L	L	L
H	L	H	L
H	H	L	L
H	H	H	H

Fig. 2.6 Three-input AND gate: (a) switch representation; (b) truth table; (c) logic symbol.

28 LOGIC GATES 2.3

row of inputs LLL and the last row HHH. The rightmost input column (C) in Fig. 2.6(b) has alternate L's and H's in it. The middle input column (B) alternates two L's and two H's. The leftmost column (A) has four L's followed by four H's in it.

Figure 2.6(c) shows the logic symbol for a three-input AND gate. It is the same as the two-input logic symbol except for the addition of the third input.

The following example illustrates a practical application of an AND gate.

Example 2.2 *Safety System.* A numerically controlled (NC) drill press has three switches on it. The main power switch supplies power to the drill press. A thermal overload circuit breaker switch opens if the drill press motor temperature exceeds a safe limit. A light-operated safety switch opens if the operator's hand is in the immediate area of the drill. Draw the switch representation, truth table, and logic diagram for this system.

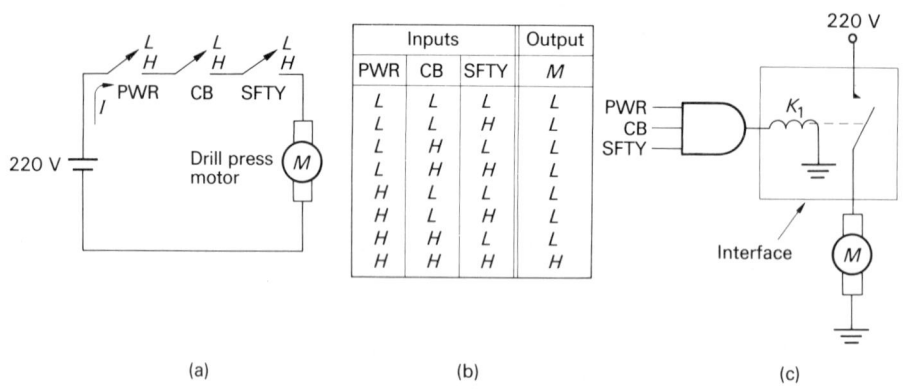

Inputs			Output
PWR	CB	SFTY	M
L	L	L	L
L	L	H	L
L	H	L	L
L	H	H	L
H	L	L	L
H	L	H	L
H	H	L	L
H	H	H	H

(a)   (b)   (c)

Fig. 2.7 Safety system: (a) switch representation; (b) truth table; (c) logic diagram.

Solution. The main power switch (PWR), thermal switch (CB), and safety switch (SFTY) must all be closed for the NC drill press to operate. Figure 2.7(a) shows the switch representation of the system. It consists of a dc power source, three switches, and the drill press motor. All three switches must be closed for the motor to run. Figure 2.7(b) shows the truth table for this system. Since there are three switches the truth table will have three input columns with $2^3 = 8$ rows in it. An L is placed in each row of the motor (M) column except when all three switches are closed (H). The truth table represents a three-input AND gate as shown in Fig. 2.7(c). Logic gates supply only a few milliamperes of current and on the order of ten volts output. Therefore, an interface is required between the logic gate output in Fig. 2.7(c) and the motor to increase the current. A mechanical or solid state relay K_1 might be used for the interface between the logic gate and the motor. ∎

2.4 THE NOT GATE

In systems that are more complicated than those illustrated in the above two examples, several logic gates can be used. A **logic diagram** is a drawing that shows the connections between logic symbols. Pin numbers and power supplies are not shown. The next example illustrates a system that includes an AND gate and an OR gate.

Example 2.3 Draw the logic diagram for a 10¢ coffee vending machine that will take nickels, dimes, or quarters and give change only from quarters. Any money in excess of 10¢, except quarters, is lost.

Solution. Let the inputs to the machine be Q (quarters), D (dimes), and N (nickels). Figure 2.8 shows the logic diagram for the system, which consists of one three-input OR gate and one two-input AND gate. The logic diagram can be arrived at by intuition, or by formal means as shown in Chapter 4.5.

If a quarter is inserted, 15¢ change is returned and coffee is dispensed because one input to the OR gate is HIGH. If a dime is inserted the OR gate output is HIGH, coffee is dispensed, and no change is returned. If two nickels are inserted the output of the AND gate is HIGH, so the output of the OR gate is also HIGH. If only one nickel is inserted the AND gate output is LOW, so the OR gate output is also LOW and no coffee is dispensed.

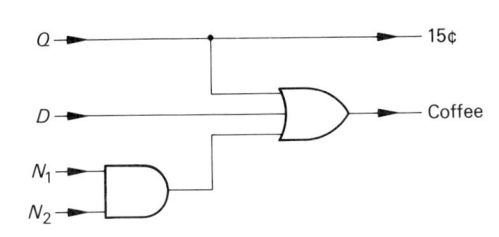

Fig. 2.8 Coffee vending machine.

Note that if a dime and one or two nickels are inserted in the machine no change is returned. We shall postpone developing a foolproof vending machine until the problems at the end of Chapter 4. ∎

The next section discusses the NOT gate.

2.4 THE NOT GATE

A **NOT gate** is a logic gate that has only one input. When its input is HIGH its output is LOW, and when its input is LOW its output is HIGH. Thus the output of a NOT gate is always the **inverse** or the **complement** of its input. The NOT gate is therefore often called an **inverter**.

A NOT gate can be visualized as a switch A connected in parallel with the output F as shown in Fig. 2.9(a). If the switch is open, that is, $A = L$, then

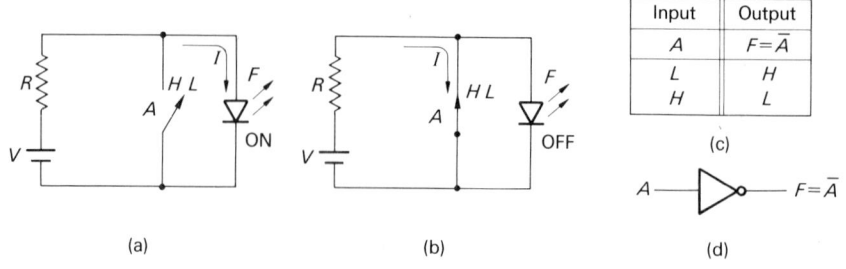

Fig. 2.9 NOT gate: (a) $A = L$; (b) $A = H$; (c) truth table; (d) logic symbol.

current I will flow through the lamp, so the lamp will turn ON (H). If the switch is closed, that is, $A = H$, as in Fig. 2.9(b), the current I will flow through the closed switch and not through the lamp. Hence the lamp will be OFF.

The functional relationship between the input A and the output F is written

$$F = \bar{A},$$

where the superscript bar symbol (¯) means NOT. The statement $F = \bar{A}$ is read, "F equals not A."

The truth table given in Fig. 2.9(c) summarizes the properties of the NOT gate. Since there is only one input variable, there are only $2^1 = 2$ possible truth values for A. Hence the inverter truth table has only two rows.

The logic symbol for a NOT gate is shown in Fig. 2.9(d). The small circle on the output of this symbol is used to denote inversion of the input.

The next example shows how AND, OR, and NOT gates can be combined to perform binary addition.

Example 2.4 Draw the logic diagram for a system that will add two binary numbers together. Include an output for a carry-out.

Solution. The **binary half-adder** performs the operation of binary addition of any two binary digits A and B as shown in the block diagram of Fig. 2.10(a). The arrows on the inputs and outputs show the direction of the flow of the logic signals. The half-adder has two outputs, S for the sum and C for the carry-out. If the inputs are both zero, that is, $A = 0$ and $B = 0$, then the sum S should be $0 + 0 = 0$ and there is no carry-out, that is, $C = 0$. The half-adder inputs and outputs are the binary digits 0 and 1 and not the logic variables logic-1 and logic-0 so the + symbol means addition and not the logic OR operation. If $A = 1$ and $B = 0$, then $S = 1 + 0 = 1$ and $C = 0$. If $A = 0$ and $B = 1$, then $S = 0 + 1 = 1$ and $C = 0$. Finally, if $A = 1$ and $B = 1$, then the sum $S = 1 + 1 = 0$, with a carry-out of 1. The truth table for the half-adder is given in Fig. 2.10(b). For example, $0 + 1 = 01$, $1 + 1 = 10$, etc.

The half-adder can be implemented as follows. The half-adder truth table

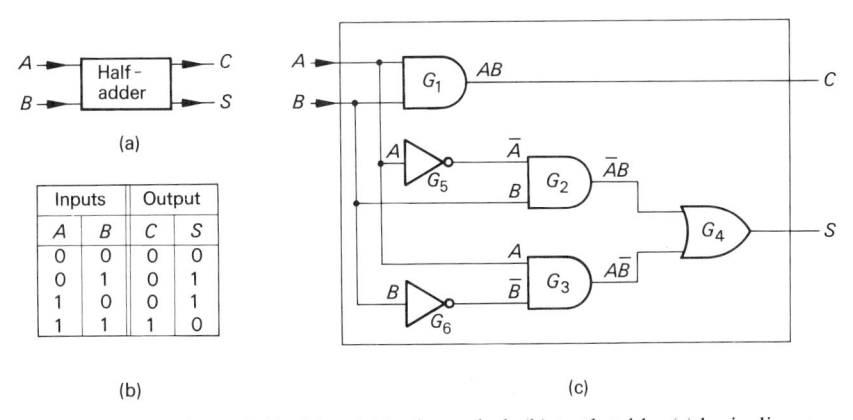

Fig. 2.10 The binary half-adder: (a) logic symbol; (b) truth table; (c) logic diagram.

shows that the carry-out output function is $C = AB$. This function can be implemented with AND gate G_1 as shown in Fig. 2.10(c). The sum output S is somewhat harder to implement. Note that S is 1 when $A = 0$ and $B = 1$; that is, an AND gate G_2 with inputs $\bar{A} = 1$ (i.e., $A = 0$) and $B = 1$ would have a 1 output as shown in Fig. 2.10(c). Also, the S output is 1 when $A = 1$ and $B = 0$, so an AND gate G_3 with inputs $A = 1$ and $\bar{B} = 1$ (i.e., $B = 0$) would also give $S = 1$. Since either of the inputs $\bar{A} \cdot B$ or $A \cdot \bar{B}$ gives an output of $S = 1$, then gates G_2 and G_3 must be ORed together by gate G_4, to give $S = \bar{A}B + A\bar{B}$.

Let's verify the operation of the half-adder in Fig. 2.10(c) for all possible combinations of inputs A and B.

$A = 0, B = 0$. The carry-out output is $C = A \cdot B = 0 \cdot 0 = 0$, which agrees with the half-adder truth table (Fig. 2.10b). If $A = 0$ and $B = 0$, then $\bar{A} = 1$ and $\bar{B} = 1$. Hence the sum output is $S = \bar{A} \cdot B + A \cdot \bar{B} = 1 \cdot 0 + 0 \cdot 1 = 0 + 0 = 0$, which agrees with the first row of the truth table for output S.

$A = 0, B = 1$. $C = A \cdot B = 0 \cdot 1 = 0$. Since $A = 0$ and $B = 1$, then $\bar{A} = 1$ and $\bar{B} = 0$. Hence $S = \bar{A} \cdot B + A \cdot \bar{B} = 1 \cdot 1 + 0 \cdot 0 = 1 + 0 = 1$, as given by the second row in the half-adder truth table.

$A = 1, B = 0$. $C = A \cdot B = 1 \cdot 0 = 0$. $S = \bar{A} \cdot B + A \cdot \bar{B} = 0 \cdot 0 + 1 \cdot 1 = 0 + 1 = 1$.

$A = 1, B = 1$. $C = A \cdot B = 1 \cdot 1 = 1$. $S = \bar{A} \cdot B + A \cdot \bar{B} = 0 \cdot 1 + 1 \cdot 0 = 0 + 0 = 0$.

Since the sum and carry-out outputs have the correct values for all four combinations of inputs, the implementation given in Fig. 2.10(c) is valid. ∎

The circuit of Fig. 2.10(c) is called a half-adder, rather than a full adder, because it does not have an input to handle a carry-out from a *previous* addition. Note that logic gates can perform arithmetic operations such as addition, as well as logic operations such as AND, OR, and NOT.

2.5 THE NAND GATE

A **NAND gate** is a logic gate whose output is HIGH when any or all inputs are LOW. Its output is LOW only when all inputs are HIGH.

A NAND gate can be visualized as series switches connected in parallel with the output as shown in Fig. 2.11(a). If either or both switches are open (L), the current I will flow through the lamp, so the lamp will turn ON (H) as shown in Fig. 2.11(a). If both switches A and B are closed, the lamp is shorted by the switches and all the current I flows through the switches, so the lamp is OFF as shown in Fig. 2.11(b).

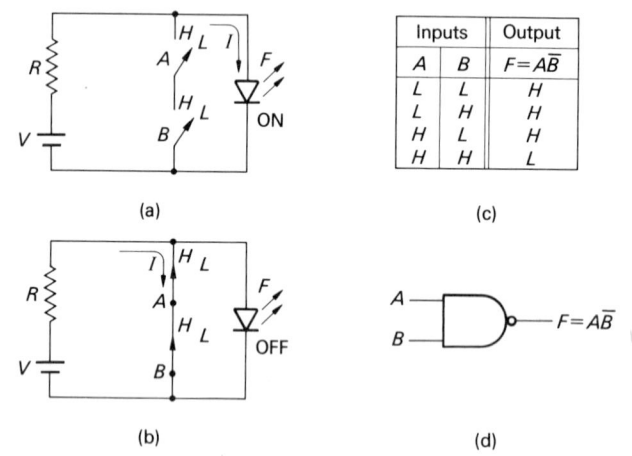

Inputs		Output
A	B	$F = \overline{AB}$
L	L	H
L	H	H
H	L	H
H	H	L

Fig. 2.11 NAND gate: (a) $A = B = L$; (b) $A = B = H$; (c) truth table; (d) logic symbol.

The truth table for a two-input NAND gate is given in Fig. 2.11(c). Note that the output is LOW only when both inputs are HIGH. This is just the inverse (complement) of the AND gate situation. If we were to complement each output in the NAND gate truth table (i.e., change H to L and L to H), an AND gate truth table would result. Thus NAND means NOT(AND). The functional relationship between the inputs A and B and the output F is written

$$F = \overline{AB}.$$

This statement is read, "F equals not A and B."

The logic symbol for a NAND gate is shown in Fig. 2.11(d). The small circle at the output of this logic symbol indicates that the NAND gate is an inverted AND gate. In fact, a NAND gate can be made from an AND gate and a NOT gate as shown in Fig. 2.12(a).

A NOT gate can be made from a NAND gate by connecting all its inputs together as shown in Fig. 2.12(b). When the input is LOW, both inputs to the

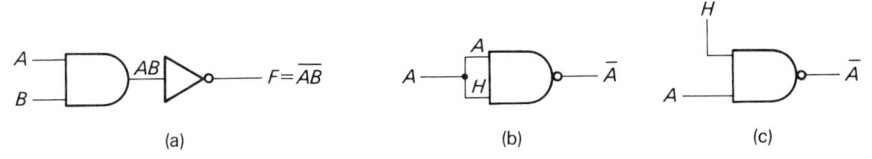

Fig. 2.12 (a) NAND gate made from an AND gate and a NOT gate; (b, c) NOT gate made from a NAND gate.

NAND gate are LOW, so the output is HIGH. When input A is HIGH, both inputs are HIGH, so the output is LOW. A NOT gate can also be made from a NAND gate by connecting all but one input to a HIGH, as shown in Fig. 2.12(c). The advantage of this method for making a NOT gate, as opposed to tying all the inputs together, is that it draws less current from the driving source of the A input.

The following example shows how NAND and NOT gates are used in a decoder.

Example 2.5 *Binary Decoder.* Draw the logic diagram for a two-bit binary-to-decimal decoder and verify that it decodes its inputs.

Solution. A **binary decoder** is a device that converts a binary number into a decimal number. As we pointed out in Chapter 1, humans are accustomed to using decimal numbers while electronic circuits are designed to use binary numbers. Therefore, it is usually desirable to have the outputs of electronic circuits in decimal form. A decoder is therefore required. For example, a decoder would be useful in a digital voltmeter that uses binary gate circuits but displays the output in decimal form.

Figure 2.13 shows the logic diagram of a two-bit binary decoder that decodes any two binary inputs into one of four decimal outputs. Let's follow the logic signals through the decoder for inputs $A = H$, $B = L$ using Fig. 2.13(b). The A input to NAND G_0 is LOW while the B input is HIGH, hence its output is HIGH, $F_0 = H$. The inputs to NAND gate G_1 are both HIGH, so its output is $F_1 = L$. The inputs to NAND gate G_2 are both LOW, so its output is $F_2 = H$. Finally, the inputs to NAND gate G_3 are HIGH and LOW, so $F_3 = H$.

Using the above logic signal tracing approach, you can verify that when $A = B = L$, then only F_0 is LOW. When $A = L$, $B = H$, then only F_2 is LOW, and when $A = B = H$, only F_3 is LOW. The properties of this decoder are summarized in the truth table of Fig. 2.13(c) letting $L = 0$ and $H = 1$. ∎

Note that when a given output for the decoder in Example 2.5 is selected it goes LOW instead of HIGH. This is called an **active LOW output** since the selected output goes LOW when its inputs are activated. Active LOW inputs or outputs are indicated on a logic symbol with a small circle. Active HIGH inputs or outputs are denoted by the absence of the circle. The logic symbol for the decoder of Example 2.5 is shown in Fig. 2.13(d). It has active HIGH (no circle) inputs and active LOW (circle) outputs. Active LOW outputs are useful

34 LOGIC GATES 2.6

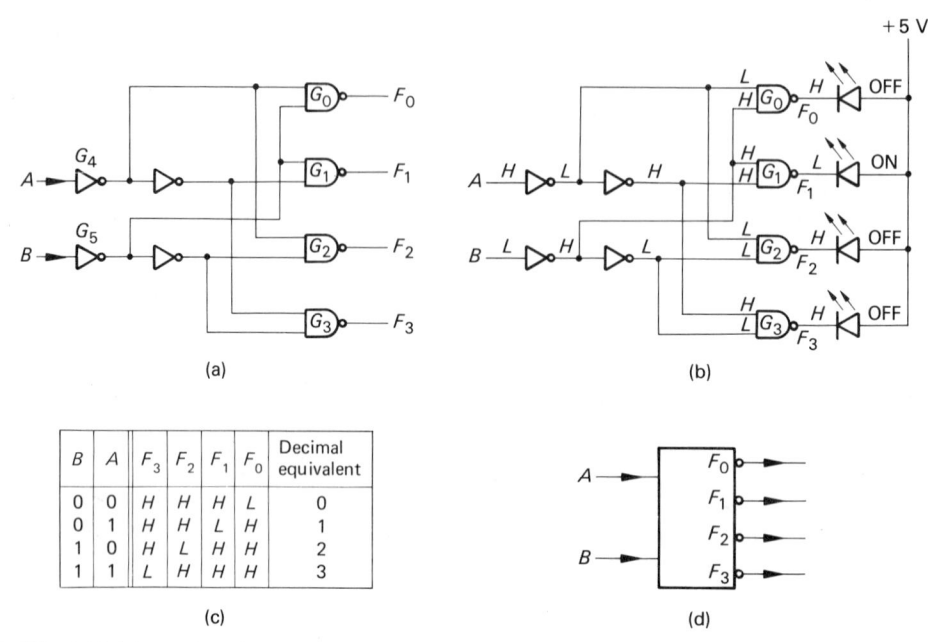

Fig. 2.13 Binary decoder: (a) logic diagram; (b) $A = H, B = L$ inputs; (c) truth table; (d) logic symbol.

B	A	F_3	F_2	F_1	F_0	Decimal equivalent
0	0	H	H	H	L	0
0	1	H	H	L	H	1
1	0	H	L	H	H	2
1	1	L	H	H	H	3

(c)

for driving external devices such as light-emitting diodes, since they can sink (input) more current than they can source (output).

In the decoder shown in Fig. 2.13(a), four input NOT gates are used instead of the minimum of two, in order to electrically separate (i.e., **isolate**) the decoder NAND gates from the outputs of the previous circuit. Isolation, or **buffering**, results in a lower current drain (load) on the driving circuit because it has to drive only the two input gates G_4 and G_5 instead of the four gates G_0–G_3.

2.6 SPECIFICATION SHEETS

A manufacturer's specification sheet is a table listing the important characteristics of a particular device. Specification sheets are furnished by the manufacturer as separate sheets or in a data book. The manufacturer assumes the user has some knowledge of the device when he makes up the specification sheet. It is very important that the user of a device understand the information on a specification sheet so that he may use the device properly and be sure that it performs as it should.

A manufacturer's specification sheet for a TTL NAND gate is shown in Fig. 2.14. There are four NAND gates in one IC package. Each gate has two inputs. This IC is called a quad two-input NAND gate and is designated by the

DIP (TOP VIEW)   FLATPAK (TOP VIEW)

in accordance with
JEDEC (TO-116) outline
14 Lead SSI Dual In-line

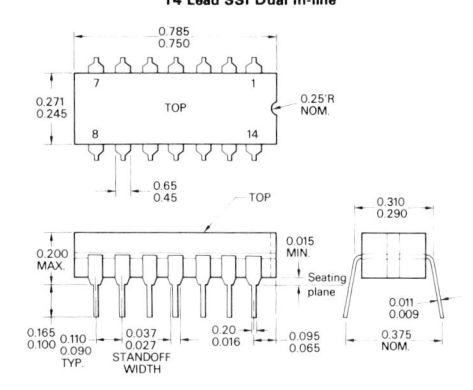

RECOMMENDED OPERATING CONDITIONS

PARAMETER	9N00XM/5400XM MIN.	TYP.	MAX.	9N00XC/7400XC MIN.	TYP.	MAX.	UNITS
Supply Voltage V_{CC}	4.5	5.0	5.5	4.75	5.0	5.25	Volts
Operating Free-Air Temperature Range	−55	25	125	0	25	70	°C
Normalized Fan-Out from Each Output, N			10			10	U.L.

X = package type; F for Flatpak, D for Ceramic Dip, P for Plastic Dip. See Packaging Information Section for packages available on this product.

ELECTRICAL CHARACTERISTICS OVER OPERATING TEMPERATURE RANGE (Unless Otherwise Noted)

SYMBOL	PARAMETER	LIMITS MIN.	TYP. (Note 2)	MAX.	UNITS	TEST CONDITIONS (Note 1)	TEST FIGURE
V_{IH}	Input HIGH Voltage	2.0			Volts	Guaranteed Input HIGH Voltage	1
V_{IL}	Input LOW Voltage			0.8	Volts	Guaranteed Input LOW Voltage	2
V_{OH}	Output HIGH Voltage	2.4	3.3		Volts	V_{CC} = MIN., I_{OH} = 0.4 mA, V_{IN} = 0.8 V	2
V_{OL}	Output LOW Voltage		0.22	0.4	Volts	V_{CC} = MIN., I_{OL} = 16 mA, V_{IN} = 2.0 V	1
I_{IH}	Input HIGH Current			40	µA	V_{CC} = MAX., V_{IN} = 2.4 V, Each Input	4
				1.0	mA	V_{CC} = MAX., V_{IN} = 5.5 V	
I_{IL}	Input LOW Current			−1.6	mA	V_{CC} = MAX., V_{IN} = 0.4 V, Each Input	3
I_{OS}	Output Short Circuit Current (Note 3)	−20		−55	mA	9N00/5400, V_{CC} = MAX.	5
		−18		−55	mA	9N00/7400	
I_{CCH}	Supply Current HIGH		4.0	8.0	mA	V_{CC} = MAX., V_{IN} = 0V	6
I_{CCL}	Supply Current LOW		12	22	mA	V_{CC} = MAX., V_{IN} = 5.0 V	6

SWITCHING CHARACTERISTICS (T_A = 25°C)

SYMBOL	PARAMETER	LIMITS MIN.	TYP.	MAX.	UNITS	TEST CONDITIONS	TEST FIGURE
t_{PLH}	Turn Off Delay Input to Output		11	22	ns	V_{CC} = 5.0 V	A
t_{PHL}	Turn On Delay Input to Output		7.0	15	ns	C_L = 15 pF, R_L = 400Ω	

NOTES:
1) For conditions shown as MIN. or MAX., use the appropriate value specified under recommended operating conditions for the applicable device type.
2) Typical limits are at V_{CC} = 5.0 V, 25°C.
3) Note more than one output should be shorted at a time.

Fig. 2.14 Quad two-input TTL 5400/7400 NAND gate. (Courtesy of Fairchild Camera and Instrument Corporation.)

54 series number, 5400, or the 74 series number, 7400. Note that this gate comes in two types of packages, a dual in-line package (DIP) and a flatpack. The DIP pins are numbered counterclockwise from the notch or dot at one end of the IC package, when viewed from the top of the package. The gate inputs, gate outputs, supply voltage, and ground pins are shown. Note that the two packages are not pin for pin compatible, e.g., the ground pin (number 7) on the DIP is a gate input on the flatpack. The outline of the 14-lead SSI DIP is given in Fig. 2.14.

Let's look now at some of the information supplied in this data sheet.

Recommended Operating Characteristics The 7400 TTL series dc supply voltage V_{CC} can range from 4.75 to 5.25 V, and its temperature from 0 to 70°C. The only difference between the 7400 and the 5400 series is that the 5400 series has a wider operating temperature range and a greater supply voltage range for more demanding operating conditions, such as military applications. The maximum number of loads which each gate output can drive in its logic family, called **fan-out**, is 10.

Electrical Characteristics Logic gate circuits use voltages to represent the HIGH and the LOW logic states. The manufacturer guarantees a HIGH input when the input voltage V_{IH} is 2.0 V or more, as shown in Fig. 2.14. It guarantees a LOW input for any input voltage $V_{IL} = 0.8$ V or less. A HIGH output will have a voltage V_{OH} of 2.4 V or more. A LOW output will have a voltage V_{OL} of 0.4 V or less.

The input HIGH current I_{IH} is the input current that flows into an input that is HIGH for the given input voltage. The input LOW current I_{IL} is the input current that flows out of an input that is LOW for the given input voltage. The output short circuit current I_{OS} is the output current that flows out of a HIGH output that is connected directly to ground. The HIGH supply current I_{CCH} is the total power supply current (per IC) that flows when the output of the gate is HIGH. Similarly, I_{CCL} is the total supply current that flows when the output of the gate is LOW.

A single minimum or maximum value given in a data sheet is a guaranteed value over the entire temperature and supply voltage range given for the device. This guarantee means the device can operate reliably under variable temperatures and supply voltages that occur in a practical system.

The test figures noted in the right-hand column of Fig. 2.14 refer to the circuits used for measuring the parameters listed. These test figure circuits are found in the references [5].

Switching Characteristics In order to discuss the switching characteristics in Fig. 2.14, we must first discuss pulses. A **pulse** is a relatively square, narrow voltage (or current) waveform, as illustrated in Fig. 2.15(a). Pulses are used as inputs to most logic devices. The outputs of most logic devices are also pulses. A voltage pulse waveform is a plot of the voltage amplitude versus time. The pulse shown in Fig. 2.15(a) has an amplitude of +5 V from times 2 sec to 3 sec, and has zero

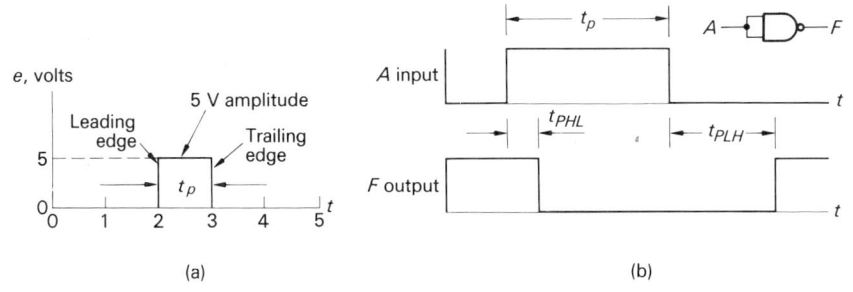

Fig. 2.15 Pulses: (a) single pulse; (b) timing diagram.

amplitude for all other times. The width of the pulse t_p, called the **pulse width**, is 3 sec − 2 sec = 1 sec.

Pulses have a **leading (rising) edge** and a **trailing (falling) edge**, as shown in Fig. 2.15(a). An **ideal pulse** has a vertical leading edge, a vertical trailing edge, and a flat top.

When the input voltage to a logic gate changes, it takes time for the output to change. The **propagation delay time** is the time required for the output of a device to change after receiving an input signal. Delay times in the TTL, ECL, and CMOSL families are on the order of 1 to 30 nanoseconds (1 ns = 10^{-9} sec). Under "Switching Characteristics" in Fig. 2.14, two times are listed. The **turn-off delay** t_{PLH} is the time required for the gate output to switch from a LOW to a HIGH, measured relative to the gate input. The **turn-on delay** t_{PHL} is the time required for the gate output to switch from a HIGH to a LOW.

Figure 2.15(b) shows an input pulse applied to a NAND gate that has been converted into a NOT gate. A figure that compares input or output waveforms is called a **timing diagram**. Timing diagrams are very important in the study of digital systems. The input pulse width applied to the NOT gate is t_p. When the input goes HIGH, the NOT gate output doesn't go LOW until one delay time t_{PHL} later. Figure 2.14 shows that the turn-on delay time for a 7400 gate is typically 7 ns (7 × 10^{-9} sec) and never more than 15 ns. Similarly, when the input goes LOW, the output does not go HIGH until one delay time t_{PLH} later. The 7400 NAND gate has a typical turn-off delay time of 11 ns. Notice that the turn-off and turn-on delay times are not necessarily equal. The delay times are due to the time required for the transistors in the device to turn on (begin conducting) and turn off (stop conducting). (See Ch. 3.5 for a detailed discussion.)

2.7 THE NOR GATE

A **NOR gate** is a logic gate whose output is HIGH only when all inputs are LOW, and whose output is LOW when any input is HIGH.

A NOR gate can be visualized as two parallel switches connected in parallel

38 LOGIC GATES

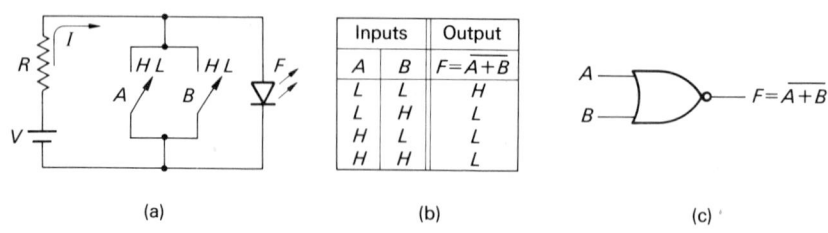

Fig. 2.16 NOR gate: (a) switch representation; (b) truth table; (c) logic symbol.

Inputs		Output
A	B	$F=\overline{A+B}$
L	L	H
L	H	L
H	L	L
H	H	L

with the output as shown in Fig. 2.16(a). If both switches are open (L), then current I will flow through the lamp and turn it ON. If one or both switches are closed, the current flows through the switches, so the lamp does not light.

The truth table for a two-input NOR gate is given in Fig. 2.16(b). The output F is HIGH only when both inputs are LOW; this is just the inverse of what happens with an OR gate. If each output in the NOR gate truth table were complemented, an OR gate would result. Therefore, NOR means NOT(OR). The functional relationship between the inputs A and B and the output F is written

$$F = \overline{A + B}.$$

This statement is read, "F equals not (A or B)."

The logic symbol for a two-input NOR gate is an OR symbol with a small circle at the output to denote NOT, as shown in Fig. 2.16(c). A NOR gate can be made from an OR gate and a NOT gate as shown in Fig. 2.17(a). A NOT gate can be made from a NOR gate by connecting all its inputs together, or by connecting all but one input to a LOW as shown in Fig. 2.17(b) and (c).

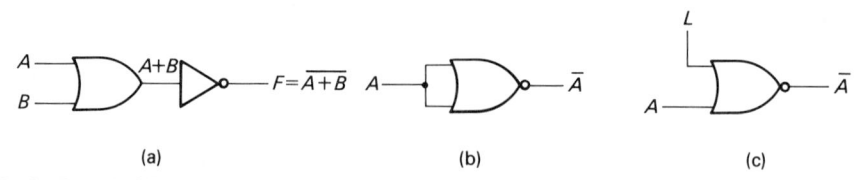

Fig. 2.17 (a) NOR gate made from an OR gate and a NOT gate; (b, c) NOT gate made from a NOR gate.

The next example shows how two NOR gates can be connected together to make an *RS* latch (memory).

Example 2.6 *Memory.* Draw the logic diagram of an *RS* latch made from two NOR gates. Explain how the latch can store one bit of information, i.e., act as a one-bit memory.

Solution. Figure 2.18(a) shows an *RS* latch made by cross-coupling the outputs of two NOR gates back into their inputs. The two outputs are labeled Q and $\bar{Q}$, and the inputs are labeled S (set) and R (reset).

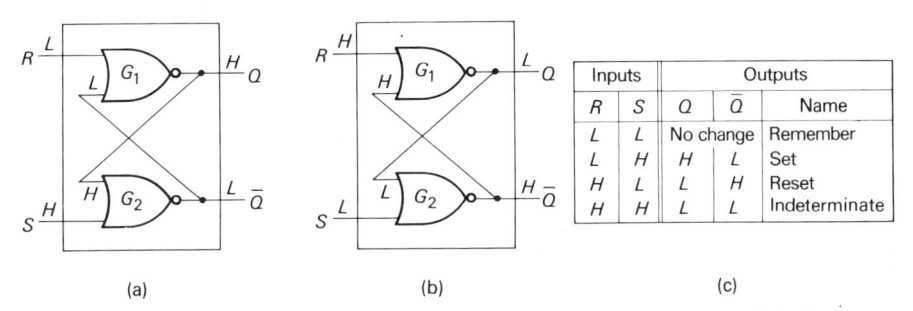

Fig. 2.18 The NOR gate *RS* latch: (a) logic diagram with $R = L$ and $S = H$; (b) logic diagram with $R = H$ and $S = L$; (c) truth table.

Inputs		Outputs		
R	S	Q	$\bar{Q}$	Name
L	L	No change		Remember
L	H	H	L	Set
H	L	L	H	Reset
H	H	L	L	Indeterminate

The latch works as follows. Recall that the output of a NOR gate is LOW when one or more inputs are HIGH. Suppose the reset input is LOW ($R = L$) and the set input is HIGH ($S = H$), as shown in Fig. 2.18(a). Then the $\bar{Q}$ output is forced LOW by the $S = H$ input (regardless of the other input to gate G_2). With $\bar{Q} = L$, both inputs to NOR gate G_1 are LOW, so Q goes HIGH, i.e., sets.

Next suppose $R = H$ and $S = L$, as shown in Fig. 2.18(b). Then G_1 is forced LOW by $R = H$, so Q goes LOW, i.e., resets. With both inputs to G_2 LOW, $\bar{Q}$ becomes HIGH. When $S = R = H$, then both Q and $\bar{Q}$ become LOW. This is called the indeterminate state because if both inputs return from this state to a LOW simultaneously, the output state cannot be determined, i.e., it could be either HIGH or LOW.

Finally, if $S = R = L$ and $Q = H$, then one input to G_2 is HIGH so $\bar{Q} = L$. With $\bar{Q} = L$, both inputs to G_1 are LOW, so $Q = H$ as assumed initially. If $S = R = L$ and $Q = L$, then $\bar{Q} = H$, making Q LOW as assumed. So when $S = R = L$, the outputs do not change state, i.e., they remember their last state. Thus the output of a latch can be made HIGH or LOW by applying appropriate inputs, and when both inputs are returned to LOW the latch will remember its last output state. ∎

A latch can also be made by cross-coupling two NAND gates together. An example of a NAND gate latch was given in Chapter 1; see also Problem 24 at the end of this chapter.

A convenient method for troubleshooting digital ICs, such as the logic gates and latches discussed in this chapter, is illustrated in Fig. 2.19. A logic pulser applies a pulse to one of the inputs of a device. A probe with a LED readout is then used to detect the logic state of the outputs.

40 LOGIC GATES

Fig. 2.19 Troubleshooting digital systems using a logic pulser and logic probe. (Courtesy of Hewlett-Packard Company.)

SUMMARY

In this chapter we discussed five logic gates, the OR, AND, NOT, NAND, and NOR gates. We gave the switch representation of each logic gate to help visualize the logic performed. Table 2.1 summarizes these five gate symbols, their switch representation, and their truth tables. It is important to learn this table, since the logic gate is one of the basic building blocks used in almost all digital systems.

Table 2.1 Logic Gate Summary

Operator	Symbol	Switch Circuit	Truth Table
			A B F
OR	$F = A + B$		L L L L H H H L H H H H
AND	$F = A \cdot B$		L L L L H L H L L H H H
NOT	$F = \overline{A}$		L H H L
NAND	$F = \overline{A \cdot B}$		L L H L H H H L H H H L
NOR	$F = \overline{A + B}$		L L H L H L H L L H H L

PROBLEMS

Section 2.1

1. Give three examples of useful tasks that logic gates can perform.

Section 2.2

2. Explain how an OR gate works, using the switch representation.
3. Draw the OR gate truth table, using switch terminology.

42   LOGIC GATES

4. Draw the OR gate truth table, using HIGH and LOW notation.
5. *Motor Warning System.* A large electric motor has two overload switches that open in case of an overload and turn off the motor. Draw a logic diagram for a warning system that indicates when either or both overload switches are open.

Section 2.3

6. Using the switch representation, explain how an AND gate works.
7. Draw the truth table for an AND gate, using the HIGH and LOW notation.
8. Draw the logic diagram for the logic function $F = AB + C$.
9. Draw the logic diagram for the logic function $G = ABD + CB$.

Section 2.4

10. *Elevator Warning System.* Draw the logic diagram for a warning system that turns a light on only if two of three elevators arrive at the ground floor at the same time. Three-input AND gates, four-input OR gates, and NOT gates are available.
11. Draw the logic diagram for a half-adder. Explain how a half-adder works by giving the output of *each* gate for inputs $A = 0$, $B = 1$.
12. A half-adder has inputs $A = 0$ and $B = 1$ and yet the output $S = 0$. Which gates could be faulty? Use the notation of Fig. 2.10(c).

Section 2.5

13. Draw the switch representation, truth table, and standard symbol for a NAND gate.
14. Verify that a NAND gate can be made from an AND gate and a NOT gate. [*Hint*: Use a truth table with input columns A and B and output columns AB and $\overline{AB}$.]
15. Explain how a NAND gate can be converted into a NOT gate.
16. (a) Draw the logic diagram for a two-bit binary decoder. Use input buffer gates. (b) Show the inputs and output for each gate in the decoder for inputs $A = L$ and $B = H$.

Section 2.6

17. Draw a **connection diagram** (i.e., logic symbol, pin numbers, supply, and ground) for one gate in a 7400 DIP IC package. Show the correct polarity and value of the supply voltage. Show the typical HIGH and LOW output voltage values. Also show the value of $V_{IH(min)}$ and $V_{IL(max)}$.
18. Make a table showing the following specifications for a 7400: maximum supply voltage, temperature range, fan-out, maximum input HIGH current for $V_{in} = 2.4$ V, maximum short circuit current, and turn-off delay time. Use the correct symbol notation for each of these specifications.
19. *Microcomputer.* Refer to sheet 1 of the IMP–16C schematic diagram in Appendix G. Draw the connection diagram for gates 8E, 7F, and 8F in the conditional jump multiplexer and control flag logic circuit. Write the functional relationship between pins 9 and 11 on 8E, and pin 6 on 7F. Label the inputs and outputs with the symbols used in the schematic.

20. Draw and carefully label the propagation delay times on the input and output timing diagram for a 7400 NAND gate whose inputs are tied together.
21. *Digital Automobile Ignition Lock.* An automobile ignition lock consists of five separate pushbuttons (spring-loaded open) connected to a logic gate circuit that will start the engine if the proper buttons are pushed simultaneously. The code to start the engine is $A = C = D = L, B = E = H$. Draw the logic diagram for the circuit. Five-input logic gates are available.

Section 2.7

22. Draw the switch representation, truth table, and symbol for a NOR gate.
23. *NOR Latch.* Explain how a NOR gate latch works. Draw a separate logic diagram for each possible combination of inputs and show the logic levels (H or L) at each gate input and output.
24. *NAND Latch.* (a) Explain how the NAND gate latch shown in Fig. P2.1 works. Draw a logic diagram for each possible combination of inputs and show the logic levels (H or L) at each gate input and output. (b) Complete a truth table.

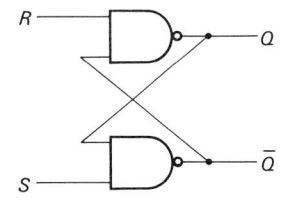

Fig. P2.1 The NAND gate latch.

25. *Signal Generator.* Refer to the phase detector circuit in the SL2 Phase Detector schematic in Appendix F. (a) Draw the connection diagram for gates U3A, U3B, U3C, and U3D. (b) What is the output of gate U3B (pin 6) if pin 12 and pin 13 of U3D are HIGH and pin 1 of U3A is LOW?
26. *Troubleshooting.* Study Fig. C.2 in Appendix C. (a) How is a test point shown? (b) How are a plug-in connector and a pin number shown? (c) How can you find a signal line that continues on another drawing?

REFERENCES

1. Jacob Millman and Christos C. Halkias. *Integrated Electronics: Analog and Digital Circuits and Systems.* New York: McGraw-Hill, 1972.
2. Arpad Barna and Dan I. Porat. *Integrated Circuits in Digital Electronics.* John Wiley and Sons, 1973.
3. "IEEE Standard Graphic Symbols for Logic Diagrams (Two-State Devices)." IEEE Std 91–1973, ANSI Y32. 14–1973, April 6, 1973, IEEE, 345 East 47th Street, New York, N.Y. 10017.
4. Robert G. Nelson. *Logic Level Design.* Winchester, Mass.: University Press, 1969.
5. *TTL Data Book.* Fairchild Semiconductor, June 1972.

3: Digital Circuits

3.1 INTRODUCTION

The functional behavior of a digital system can be understood in terms of logic symbols. However, the practical implementation of a working system, the understanding of its capabilities and limitations, and the ability to troubleshoot digital systems comes from some knowledge of the circuits which make up the system. The purpose of this chapter is to give an understanding of the digital circuits used in three logic families: (1) the transistor-transistor logic (TTL or T^2L) family, (2) the emitter-coupled logic (ECL) family, and (3) the complementary metal-oxide semiconductor logic (CMOSL) family.

A **logic family** refers to a general structure of the circuits such as TTL, ECL, or CMOSL. Within each logic family are one or more **logic series** that have distinctive characteristics relative to other series within a given family. Not all these series can be discussed, so we shall consider only the following representative series: TTL (the 54/74 standard, Schottky, and low-power Schottky series), ECL (10K series), and CMOSL (14000 series). Semiconductor manufacturers produce digital IC devices in one, and often more, logic families and generally produce devices in several series.

It may be desirable to omit the sections in this chapter that are preceded by an asterisk, and refer to them as needed at a later time. We begin with a discussion of semiconductor diodes.

3.2 SEMICONDUCTOR DIODES

In this section we shall discuss only those basic properties of the junction diode and the Schottky diode that are useful in digital circuits. Additional details may be found in the references [1].

Junction Diodes A semiconductor **junction diode** consists of two different semiconductor materials separated by a junction as shown in Fig. 3.1(a). The two materials are called the *n*-type and the *p*-type materials, and are made from specially prepared silicon or germanium semiconductors. The terminal connected to the *p* material is called the **anode** and the terminal connected to the *n* material is called the **cathode**. Figure 3.1(b) shows the schematic symbol for a diode.

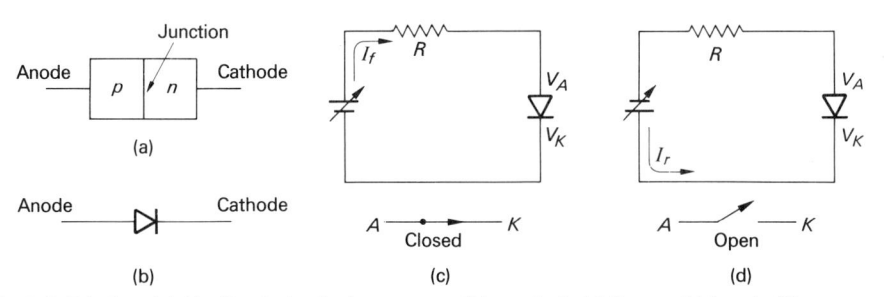

Fig. 3.1 Diodes: (a) idealized physical structure; (b) symbol; (c) forward-biased; (d) reverse-biased.

A diode is said to be **forward-biased** if the voltage at its anode V_A is more positive than the voltage at its cathode V_K; that is, $V_f = V_A - V_K > 0$ (Fig. 3.1c). The current that flows from anode to cathode in a forward-biased diode is called the **forward current** I_f. A diode is said to be **reverse-biased** if its anode voltage is less positive than its cathode voltage; that is, $V_r = V_K - V_A > 0$ (Fig. 3.1d). The current that flows from cathode to anode in a reverse-biased diode is called the **reverse current** I_r.

Let's forward-bias a diode and make a plot of the current through the diode versus the forward voltage V_f across the diode. As the voltage increases from 0 V, the forward current I_f increases very little. At a forward voltage V_D, called the **cut-in voltage**, the current increases very rapidly, as shown in Fig. 3.2(a). The cut-in voltage is about 0.3 V for germanium (Ge) diodes and about 0.7 V for silicon (Si) diodes. For forward voltages greater than the cut-in voltage, that is, for $V_f > V_D$, the diode has a very low resistance (on the order of 50 Ω) and acts like a closed switch. The current through the diode is then limited primarily by the external resistance R and is on the order of milliamperes.

For forward voltages between 0 V and V_D, and for reverse voltages V_r, the diode has a very high resistance, on the order of megohms, and acts like an open

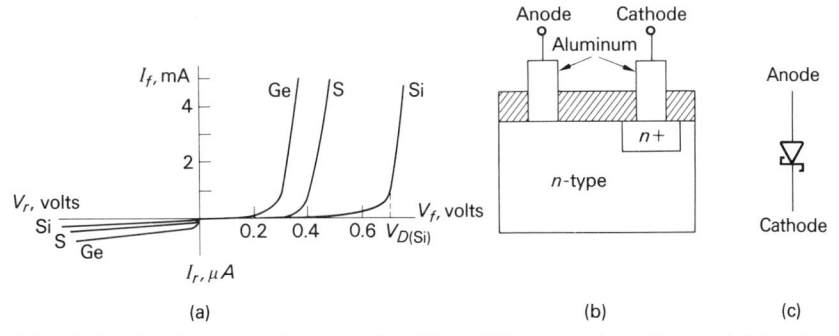

Fig. 3.2 Diodes: (a) characteristic curves for silicon (Si), germanium (Ge), and Schottky (S); (b) Schottky diode physical structure; (c) Schottky diode symbol.

switch. The reverse current I_r in a diode, also called the **leakage current**, is very small (on the order of nanoamperes for Si). A curve of the current through a diode versus the voltage across it, such as the one shown in Fig. 3.2(a), is called a **characteristic curve**. The almost vertical portions of the characteristic curve represent a low diode resistance. The horizontal portions of the curve represent a very high diode resistance.

Schottky* (Barrier) Diodes A **Schottky (barrier) diode** consists of a semiconductor and a metal in contact with each other as shown in Fig. 3.2(b) [1]. The metal is usually gold or aluminum. The semiconductor is usually of the *n* type. The symbol for the Schottky diode is shown in Fig. 3.2(c).

When a junction diode is forward-biased some excess charge carriers (electrons or holes) are stored in the region of the junction; this is called **charge storage** [1]. When the diode is reverse-biased, forward current continues to flow for a short time, called the **reverse recovery time** t_{rr}, until these excess charges have been eliminated. In a Schottky diode, charge storage is negligible; therefore, when the bias is reversed, conduction of current ceases much more rapidly than in a junction diode. For this reason Schottky diodes are very useful in high-speed switching circuits. The reverse recovery time for a Schottky diode can be as short as 10 picoseconds (1 ps = 10^{-12} sec), compared to 750 ps for a 1N4244 high-speed computer junction diode. The cut-in voltage V_D for a Schottky diode is about 0.4 V.

3.3 BIPOLAR TRANSISTORS [1]

A bipolar transistor consists of three semiconductor regions separated by two junctions, as shown in Fig. 3.3(a). The three regions are called the **emitter** (*E*), the **base** (*B*), and the **collector** (*C*). In an *n-p-n* transistor, the *p*-type material is between the two *n*-type materials as shown in Fig. 3.3(a). The symbol for an

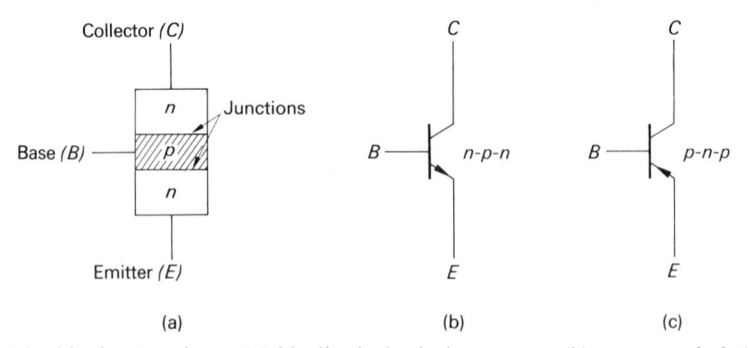

Fig. 3.3 The bipolar transistor: (a) idealized physical structure; (b) *n-p-n* symbol, (c) *p-n-p* symbol.

---

* Walter Schottky (1886–), German physicist born in Zurich. In 1940 he proposed the theory of the Schottky barrier diode.

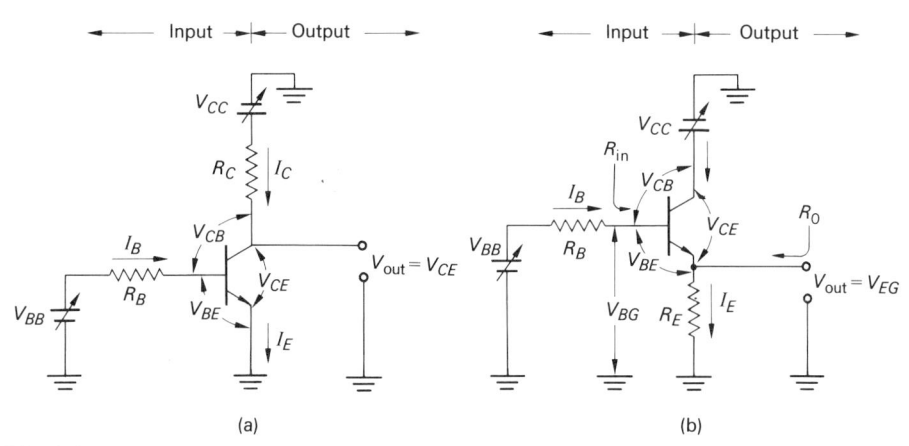

Fig. 3.4 The *n-p-n* transistor switch: (a) common emitter configuration; (b) emitter follower configuration.

n-p-n transistor is shown in Fig. 3.3(b) and for a *p-n-p* transistor in Fig. 3.3(c). The direction of the arrow on the emitter terminal distinguishes the two types and shows the direction of positive current. The *n-p-n* transistor generally has a faster switching time and is therefore usually used in digital circuits.

Figure 3.4(a) shows an *n-p-n* transistor connected in the **common emitter** configuration, i.e., with the emitter terminal common to the input and the output circuit. The directions of conventional current are shown for the base–emitter junction forward-biased and the collector–base junction reverse-biased (that is, $V_{CE} > V_{BE}$).

Figure 3.4(b) shows an *n-p-n* transistor connected in the **emitter follower** configuration. In this configuration the input is applied to the base and the output is taken across the emitter resistor R_E.

First let's study the properties of the common emitter configuration (Fig. 3.4a). Suppose V_{BB} is adjusted until the base current I_B equals a constant 0.02 mA. When the collector supply voltage V_{CC} is zero, and hence the collector–emitter voltage V_{CE} is zero, the collector current I_C is also zero. As V_{CC} increases from zero to a few tenths of a volt, the collector current rises very sharply, as shown in Fig. 3.5(a). A further increase in V_{CC} results in a very small increase in I_C, that is, the collector current is almost constant. For most transistors the collector current is on the order of 100 times greater than the base current. The ratio of the collector current I_C to the base current I_B is called the **current gain** h_{FE}:

$$h_{FE} = \frac{I_C}{I_B}.$$

For example, if $h_{FE} = 100$ and V_{BB} is increased so that the base current is 0.04 mA, then $I_C = h_{FE}I_B = 100 \times 0.04 = 4$ mA. Similarly, if $I_B = 0.06$ mA, then $I_C = 6$ mA.

If we draw a graph of I_C versus V_{CE} for several values of base current, we

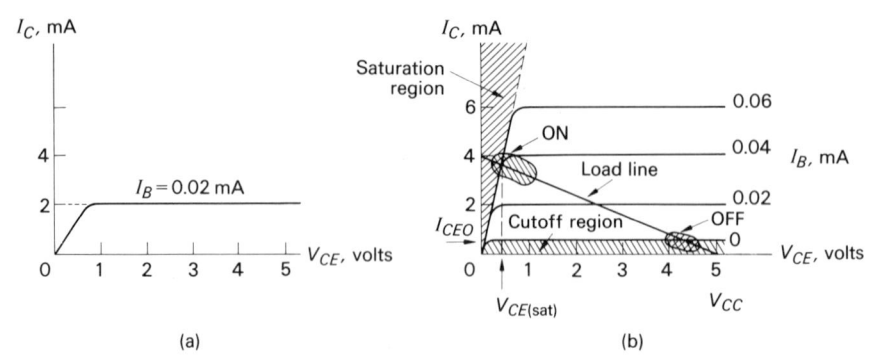

Fig. 3.5 Characteristic curves for the common emitter configuration.

obtain the *transistor characteristic curve* of Fig. 3.5(b). Note that the collector current becomes almost constant for collector–emitter voltages above a few tenths of a volt. Also, the base current controls the collector current.

The vertical shaded region in Fig. 3.5(b) is called the **saturation region**. In the saturation region $V_{BE} > V_{CE}$, so the collector–base junction is forward-biased. Since $V_{BE} > 0$ the base–emitter junction is also forward-biased. In the saturation region the collector current is no longer controlled or limited by the base current, but is limited only by the external resistance in the collector and emitter circuit. The saturated collector–emitter voltage, denoted $V_{CE(sat)}$, is on the order of a few tenths of a volt for most digital IC transistors. Note that the saturation voltage depends on the collector current. When a transistor is at or near saturation it has a low resistance and thus acts like a closed (on) switch.

The horizontal shaded region in Fig. 3.5(b) is called the **cutoff region**. In the cutoff region the base–emitter voltage is zero or negative, that is, $V_{BE} \leq 0$. Hence the base–emitter junction is reverse-biased. If $V_{BE} = 0$ no base current flows. If $V_{BE} < 0$ a reverse base current flows out of the base. Since $V_{CE} > V_{BE}$, the collector–base junction is also reverse-biased. For $I_B = 0$, the corresponding collector current I_{CEO}, called the **leakage current**, is almost zero. A transistor at or near cutoff has a large resistance and thus acts like an open (off) switch.

In summary, a transistor can be operated like a switch by controlling the base current. When the base–emitter junction is sufficiently forward-biased, the switch between the collector and the emitter is closed. When the base-emitter junction is not biased, or is reverse-biased, the switch is open.

The **load line** in Fig. 3.5(b) shows all the possible values of the collector current and the collector–emitter voltage that a transistor can have for a given collector resistance R_C and a given base current. For example, in Fig. 3.5(b), if $V_{CC} = 5$ V and $R_C = 5 \text{ V}/4 \text{ mA} = 1.25 \text{ k}\Omega$ then for $I_B = 0.03$ mA, $V_{CE} = 1.2$ V and $I_C = 3$ mA.

The input resistance of a transistor in the common emitter configuration is approximately 1 kΩ. Its output resistance is approximately 100 kΩ.

The above discussion applies to *p-n-p* transistors if *n* is replaced by *p*, *p* is replaced by *n*, and the voltage and current polarities are reversed.

3.4 THE EMITTER FOLLOWER CONFIGURATION

Now let's look at the important switching properties of the emitter follower configuration shown in Fig. 3.4(b).

The output voltage is one diode voltage drop V_{BE} below its base voltage V_{BG}, that is,

$$V_{out} = V_{BG} - V_{BE}.$$

For example, if $V_{BG} = +5$ V and $V_{BE} = 0.7$ V, then $V_{out} = 5$ V $- 0.7$ V $= 4.3$ V.

The output resistance R_0 of an emitter follower is

$$R_0 \approx R_B/h_{FE}.$$

The input resistance R_{in} is approximately

$$R_{in} \approx h_{FE} R_E.$$

Thus for $h_{FE} = 100$ and $R_B = 1$ kΩ, the input and output resistances are $R_{in} = 100$ kΩ and $R_0 = 10$ Ω.

In the next section we discuss why switching transistors cannot switch on and off instantaneously. A grasp of this principle is essential to an understanding of logic circuits.

3.5 SWITCHING TIMES FOR BIPOLAR TRANSISTORS

The base current can be used to control the collector current and hence turn a transistor on or off. Ideally the transistor would become a short circuit between its emitter and collector the instant the base current reached its saturated level, and V_{CE} would drop to $V_{CE(sat)} \approx 0$ V. Actually it takes time for the collector voltage to change. Consider the common emitter circuit shown in Fig. 3.6(a). The input and output voltage waveforms are shown in Fig. 3.6(b). When V_{in} equals

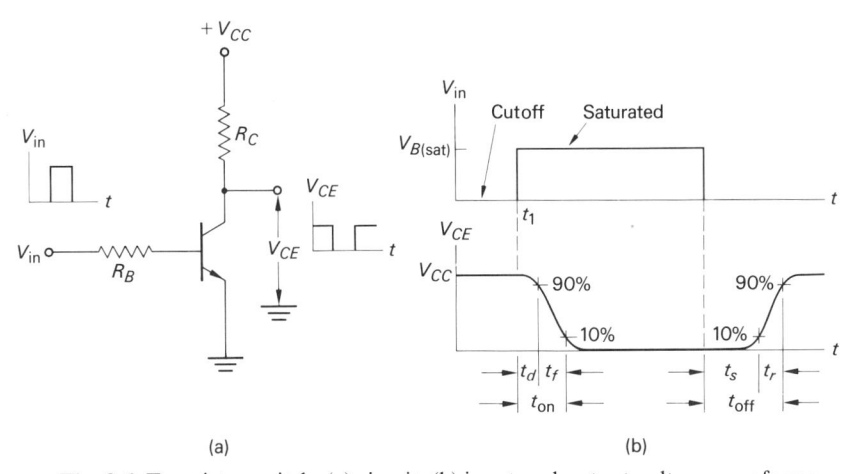

Fig. 3.6 Transistor switch: (a) circuit; (b) input and output voltage waveforms.

0 V, the transistor is cut off, and when V_{in} equals the saturation voltage $V_{B(sat)}$, the transistor saturates. Let's consider the four times involved.

Delay Time (t_d)   This is the time interval between the instant the input voltage rises (t_1) and the instant the collector voltage V_{CE} decrease to the 90% level. The delay time is due to (1) the time required for the charge carriers to travel from the collector to the emitter, (2) the time needed for charging up the base–emitter capacitance of the transistor, and (3) the time it takes for the collector voltage to drop by 10%.

Fall Time (t_f)   This is the time required for the collector voltage to drop from 90% to 10% of V_{CC}. It reflects the time required for the transistor collector current to traverse the region of the characteristic curve from cutoff to saturation and is due to the collector capacitance of the transistor. The total ON time t_{on} is the sum of the delay and fall times, that is, $t_{on} = t_d + t_f$.

Storage Time (t_s)   This is the time interval between the instant the input voltage drops to zero and the instant the collector voltage reaches the 10% level. Storage time is due to the time needed for the removal of excess charges that are stored at the base junction when the transistor is saturated.

Rise Time (t_r)   This is the time required for the collector voltage to rise from the 10% level to the 90% level. It reflects the time required for the transistor collector current to traverse the region from saturation to cutoff. The OFF time t_{off} is defined as $t_{off} = t_s + t_r$. For example, a 2N 2222 has $t_d = 5$ ns, $t_f = 15$ ns, $t_s = 190$ ns, and $t_r = 23$ ns.

3.6 THE TTL FAMILY [2]

The TTL (transistor-transistor logic) family is a widely used, relatively high-speed logic family. In this section we shall consider the TI (Texas Instruments Inc.) 54/74 standard, the Schottky, and the low-power Schottky series. Each logic family has one type of logic gate that is basic to that family. The basic gate in the TTL family is the NAND gate and we shall study its circuit in some detail. The circuits and characteristics of most other gates in the TTL family can be understood from a knowledge of the basic gate. Some NOR and AND-OR-INVERT gates (see Chapter 4) use a differential amplifier circuit, which is discussed in Section 3.9.1. We shall not discuss flip-flop circuits, since they can be understood in terms of logic gate circuits and characteristics.

This section will serve as a basic introduction to the TTL family, but does not cover all topics in detail. The references should be consulted for additional details [3].

3.6.1 Introduction

In this section we shall discuss three topics: voltage levels, multiple-emitter transistors, and current conventions for devices.

Voltage Levels We have assumed that the HIGH and LOW voltage levels, V_H and V_L, that represent the two logic states are exact voltages such as $V_H = +5$ V and $V_L = 0$ V. However, the parameters of a device, such as the saturation voltage or the capacitance of a transistor, vary from sample to sample and vary with temperature. Also noise, due to power supply ripple or external sources, may be present in the circuit. For these reasons binary voltage levels cannot be exact, but are specified over a range of voltages as illustrated in Fig. 3.7(a).

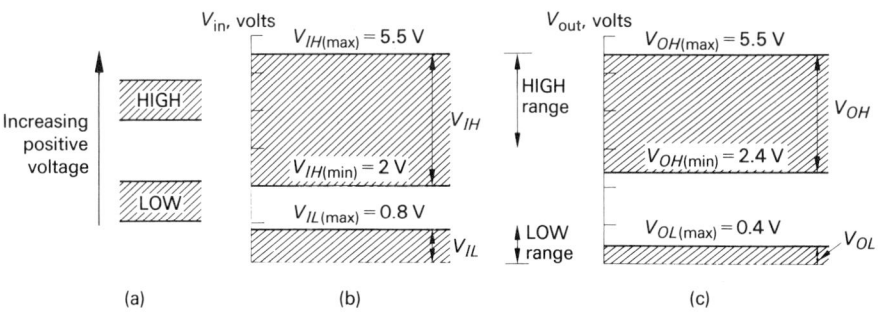

Fig. 3.7 Voltage levels: (a) general case; (b) TTL input range; (c) TTL output range.

The most positive voltage level is always called a HIGH. The least positive voltage level is always called a LOW. Two symbols often used in mathematical logic are the logic-1 (TRUE) and the logic-0 (FALSE) symbols (0 and 1 for short). There are two conventions for defining logic-1 and logic-0 in terms of H and L voltage levels. For **positive logic**, logic-1 is defined as the HIGH state and logic-0 as the LOW state. For **negative logic**, logic-1 is defined as the LOW state and logic-0 as the HIGH state. The convention chosen is arbitrary [3]. This text will always assume the more widely used positive logic convention. In arithmetic devices the binary digits, 0 and 1, are interpreted as L and H, respectively, when using positive logic.

Recall from the discussion of manufacturer's specifications (Section 2.6) that the range of voltages is not the same for input voltages and output voltages. We shall discuss this further under noise immunity (Section 3.6.4). For TTL, the maximum input voltage, $V_{IL(max)}$, that the device will consider a LOW is $+0.8$ V and the minimum, $V_{IL(min)}$, is usually 0 V, as shown in Fig. 3.7(b). The minimum input voltage $V_{IH(min)}$ that the device will consider a HIGH is $+2.0$ V and the maximum, $V_{IH(max)}$, is generally $+5.5$ V. The maximum output LOW voltage $V_{OL(max)}$ is $+0.4$ V, and the minimum output HIGH voltage $V_{OH(min)}$ is $+2.4$ V, as shown in Fig. 3.7(c).

Multiple-Emitter Transistors Many TTL gates with more than one logic input use transistors with more than one emitter. The schematic symbol for a transistor with two emitters is shown in Fig. 3.8(a). A two-emitter transistor is functionally

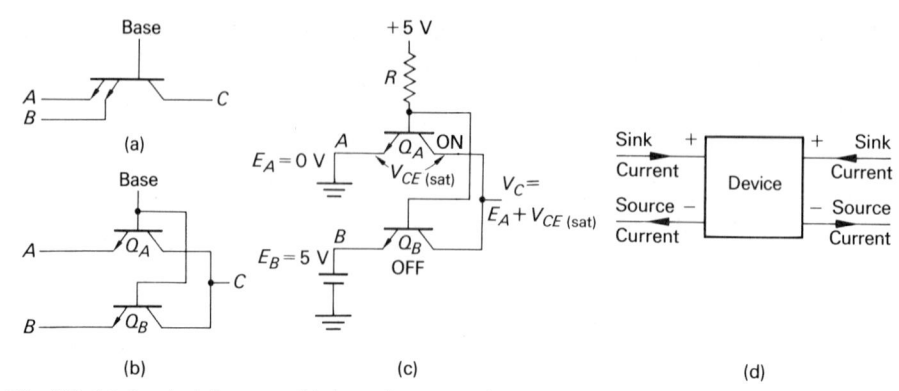

Fig. 3.8 (a) Symbol for a multiple-emitter transistor; (b) equivalent circuit for a multiple-emitter transistor; (c) two-emitter transistor with $E_A = 0$ V and $E_B = +5$ V; (d) device current conventions.

equivalent to two transistors with their bases connected together and their collectors connected together as shown in Fig. 3.8(b). Multiple-emitter transistors typically have 2, 3, 4, 8, to as many as 13 emitters.

The two-emitter transistor in Fig. 3.8(c) works as follows. Suppose input A is LOW (say 0 V), input B is HIGH (say $+5$ V), and the base is at $+5$ V. Then the BE junction of Q_B is reverse-biased. The BE junction of Q_A is forward-biased, so Q_A looks like a closed switch between its emitter and collector. The collector voltage is then

$$V_C = E_A + V_{CE(\text{sat})Q_A} \approx 0 \text{ V} + 0.1 \text{ V}.$$

Current Conventions Recall that in this text we assume use of conventional current, i.e., current that flows from positive to negative. Current flowing into a device is called **sink current** and is given a positive sign as shown in Fig. 3.8(d). Thus a current of 10 mA flowing into the output of a device is a sink current and is designated as $+10$ mA. A current flowing out of a device is called a **source current** and is given a negative sign.

3.6.2 The Standard 54/74 Series NAND Gate

The schematic of a two-input NAND gate is shown in Fig. 3.9(a). It consists of a two-emitter input transistor Q_1, transistors Q_2, Q_3, and Q_4, diodes D_1, D_2, and D_3, and resistors R_1, R_2, R_3, and R_4. There are two logic inputs A and B, and one logic output Y. Finally, there is a ground and a power supply input V_{CC}. The circuit can be divided into three stages: the input stage, the driver stage, and the output stage.

For the normal positive input voltages, input clamp diodes D_1 and D_2 are reverse-biased. They therefore represent a very high resistance and will be considered the equivalent of open circuits. Oscillations, due to high-frequency

3.6   THE TTL FAMILY   53

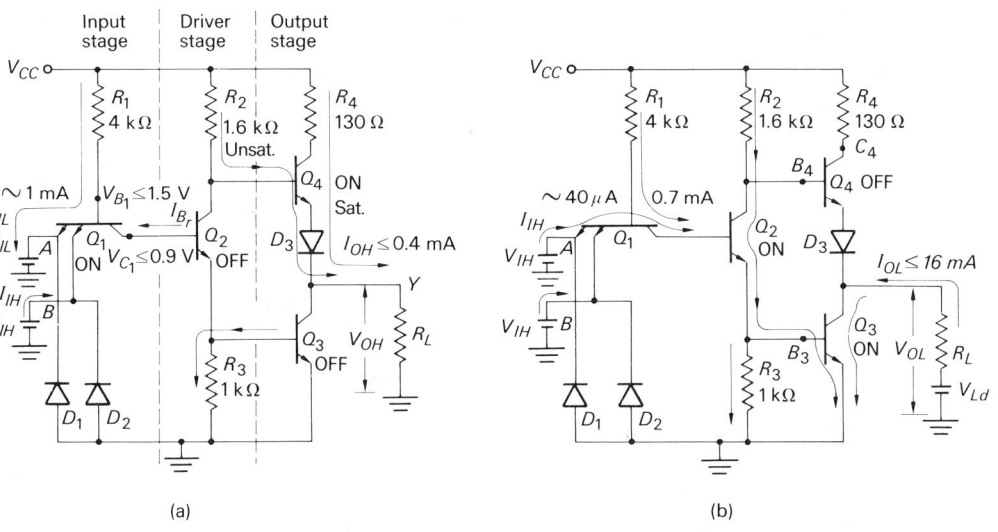

Fig. 3.9 The standard 7400 series NAND gate: (a) at least one input LOW; (b) all inputs HIGH.

effects, can cause input voltages to become negative. Any negative input voltage greater than the diodes' cut-in voltage V_D will cause the diodes to become forward-biased. This shorts to ground the negative portion of the input voltage, thereby protecting the input transistor from damage due to negative voltages. The maximum input source current through these diodes must be limited to 20 mA [3, p. 71]. Now let's see how the circuit in Fig. 3.9 works as a NAND gate.

LOW Inputs If one or more inputs to a NAND gate are LOW the output will be HIGH. Suppose input A is LOW and input B is HIGH (Fig. 3.9a). The BE junction for input A of Q_1 will be forward-biased because the base of Q_1 is more positive than emitter A. Resistor R_1 is so chosen that sufficient base current I_{B_1} flows into Q_1 to saturate it when at least one input is LOW. This current is approximately

$$I_{IL} = \frac{V_{CC} - (V_{BE_1} + V_{IL(\text{typ})})}{R_1} = \frac{5 \text{ V} - 0.7 \text{ V} - 0.4 \text{ V}}{4 \text{ k}\Omega} = -0.97 \text{ mA}.$$

The current in R_1 flows through emitter A, out of the input lead through the V_{IL} supply into ground.

The base–emitter junction for input B is reverse-biased for the following reasons. Input B is HIGH, so its minimum voltage is 2 V. The voltage at the base of Q_1 is one diode drop (V_{BE_1}) above the LOW input voltage V_{IL} since Q_1 is saturated, i.e.,

$$V_{B_1} = V_{IL(\text{max})} + V_{BE_1} \leqslant 0.8 \text{ V} + 0.7 \text{ V} = 1.5 \text{ V}.$$

Since the voltage at the base of Q_1 is no more than 1.5 V and the emitter is at least at 2 V the base–emitter junction for input B is reverse-biased. The leakage current I_{IH} flows from input B into Q_1. The specification sheet shows it to be a maximum of $+40\ \mu A$ for $V_{IH} = 2.4$ V and $+1$ mA for $V_{IH} = 5.5$ V.

Transistor Q_2 is off for the following reasons. If Q_2 were on, then Q_3 would also be on and the voltage at the base of Q_2 would be $V_{BE_3} + V_{BE_2} = 0.7\text{ V} + 0.7\text{ V} = 1.4$ V. Q_1 is saturated, so its collector voltage must be the input LOW voltage plus Q_1's saturation voltage, i.e.,

$$V_{C_1} = V_{B_2} = V_{IL(\text{max})} + V_{CE_1(\text{sat})} \leqslant 0.8\text{ V} + 0.1\text{ V} = +0.9\text{ V}.$$

Since V_{B_2} would have to be at least 1.4 V for Q_2 to be on, then Q_2 is off.

With Q_2 turned off, the current flow to the base of Q_3 is nil, so Q_3 is off. Resistor R_2 is so chosen that with Q_2 off, sufficient current flows into the base of Q_4 to turn it on. With Q_4 on, source current I_{OH} flows through R_4, Q_4, D_3, and into the load resistor R_L and the output is HIGH.

When the output switches from a LOW to a HIGH, the output voltage rises exponentially from approximately 0 V toward V_{CC} as it charges up internal and external capacitances. Initial charging of capacitances draws considerable base current through Q_4, so Q_4 may saturate. As the capacitances are charged up and the output voltage nears its typical value $V_{OH(\text{typ})}$, the base current I_{B4} in Q_4 decreases and Q_4 is no longer saturated, although it stays turned on. For low base currents I_{B4}, Q_4 acts like an emitter follower, so the output voltage for the open circuit case, $R_L = \infty$, is

$$V_{OH} = V_{B_4} - V_{BE_4} - V_{D_3} \approx V_{CC} - V_{BE_4} - V_{D_3}.$$

For example, if $R_L = \infty$, $V_{CC} = 4.75$ V, and $V_{BE_4} = V_{D_3} = 0.7$ V, then $V_{OH} = 4.75\text{ V} - 0.7\text{ V} - 0.7\text{ V} = 3.35$ V. As R_L decreases, it draws more current, so V_{OH} decreases almost linearly with I_{OH} (see Appendix D1 curves). The purpose of R_4 is to limit the current through Q_4.

HIGH Inputs Suppose both inputs of the NAND gate in Fig. 3.9(b) are HIGH. Then both base–emitter junctions of transistor Q_1 are reverse-biased and only leakage current I_{IH} flows into each emitter. The collector–base junction of Q_1 is forward-biased, so current

$$I_{R_1} = \frac{V_{CC} - V_{BC_1} - V_{BE_2} - V_{BE_3}}{R_1} = \frac{5\text{ V} - 2.1\text{ V}}{4\text{ k}\Omega} = 0.72\text{ mA}$$

flows through R_1 and the collector of Q_1 into the base of Q_2, turning Q_2 on. With Q_2 on, current flows from V_{CC} through R_2, Q_2, R_3, and into Q_3, turning Q_3 on. With Q_3 on, the output is LOW, i.e.,

$$V_{OL(\text{typ})} = V_{CE_3(\text{sat})} \approx 0.2\text{ V}.$$

Current flows from the load supply V_{LD} (if there is one) through R_L and Q_3, to ground. The maximum LOW current which will ensure that the output voltage does not rise above $V_{OL(\text{max})} = 0.4$ V is $I_{OL(\text{max})} = 16$ mA. Currents greater than this may cause Q_3 to come out of saturation, thereby increasing V_{OL} above the maximum LOW voltage.

Transistor Q_4 is off for the following reasons. Since Q_3 is on, its base is $V_D = 0.7$ V above its emitter, or 0.7 V. Transistor Q_2 is on, so its collector is at least 0.1 V above its emitter; therefore $V_{B_4} > 0.1$ V $+ 0.7$ V $= 0.8$ V. Since Q_3 is on, the voltage at the emitter of Q_4 is $V_{E_4} = V_{OL(min)} + V_{D_3} \geq 0$ V $+ 0.7$ V $= 0.7$ V. Hence the base–emitter junction of Q_4 is not forward-biased because its emitter is at 0.7 V and its base is at 0.8 V, so Q_4 is off. The purpose of D_3 is to ensure that Q_4 will be off by producing a 0.7 V drop.

Thus with one or more inputs in Fig. 3.9 LOW the output is HIGH, and with all inputs HIGH the output is LOW, so the circuit in Fig. 3.9 functions as a NAND gate.

The advantages of multiple-emitter transistors are their small size and the ease and economy of their fabrication in IC form. Smaller size reduces the parasitic capacitance of the transistor, resulting in faster switching.

Different digital IC manufacturers may have different resistor values, or even a different schematic for a given device such as a gate, flip-flop, or counter. However, if two devices have the same device number they will perform the same function (i.e. they are functionally equivalent) and they will be pin-for-pin compatible. For example, the TI and Fairchild versions of the 7400 NAND gate are functionally equivalent and pin-for-pin compatible.

The next section illustrates another way of describing the NAND gate.

*3.6.3 Transfer Curves

A **voltage transfer curve** for a device is a graph of its input voltage versus its output voltage, like that of Fig. 3.10(a). The main value of transfer curves is that they simultaneously show V_{OH}, V_{OL}, and transition region of a device as a function

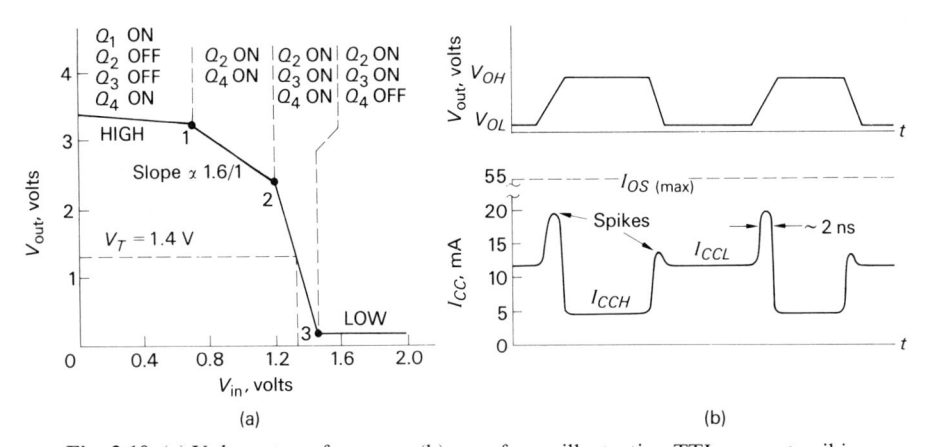

Fig. 3.10 (a) Voltage transfer curve; (b) waveforms illustrating TTL current spiking.

of its input voltage, i.e., they completely describe the dc voltage parameters. Assume that the inputs of a NAND gate are tied together so that it becomes a NOT gate. When the input voltage V_{in} is 0 V the output is HIGH, as shown in Fig. 3.10(a). As the input increases from 0 to 0.7 V the output stays HIGH (that is, Q_4 in Fig. 3.9 is ON), and remains at a relatively constant voltage. As V_{in} increases, the base current in Q_1 is gradually diverted from the emitter of Q_1 to the collector of Q_1. When $V_{in} \approx 0.7$ V, the base current flowing out of the collector of Q_1 and into the base of Q_2 is sufficient to cause Q_2 to conduct. Between points 1 and 2 in Fig. 3.10(a), Q_2 is conducting and acts like a linear amplifier whose gain (base to collector) is determined by the ratio $R_2/R_3 = 1.6 \text{ k}\Omega/1 \text{ k}\Omega = 1.6$. Since Q_4 is ON, it has the same gain as Q_2, so its output decreases with a slope of 1.6. At point 2, sufficient current flows from V_{CC} through R_2, Q_2, and R_3 to cause Q_3 to conduct. At point 2 the effective emitter impedance of Q_2 is decreased, due to increased conduction, which causes an increase in the gain of Q_2. This increase in gain causes the steep slope in the transfer curve between points 2 and 3. Operation of a digital device in this region may cause the output to oscillate. At point 3, Q_4 turns off and the output is LOW.

The region between points 1 and 3 is called the **transition region** because in this region the transistor is in transit between its HIGH and LOW states. The **average threshold voltage** V_T for a device is that voltage at which the output voltage equals the input voltage. The input threshold voltage is determined by

$$V_T = V_{BE_3} + V_{BE_2} + (V_{CB_1} - V_{BE_1}).$$

For example, if $V_{BE_3} = V_{BE_2} = 0.65$ V and $V_{CB_1} - V_{BE_1} = 0.1$ V, then $V_T = 1.4$ V. If all devices had exactly the same threshold voltage, then the HIGH and LOW voltage ranges could be defined as any voltage greater than or less than V_T, respectively. In actual devices, V_T varies from sample to sample and with variations in temperature and supply voltage. For example, the threshold voltage decreases by about -4.2 mV for every 1°C rise in temperature. This rate is about twice that of a single junction (-2.5 mV/°C). Thus it is necessary to define the HIGH and LOW voltage ranges with a gap in them around V_T. The transfer curves for different temperatures are given in Appendix D.

The next section discusses some important characteristics of TTL devices.

3.6.4 Characteristics of TTL Devices

The output stage in Fig. 3.9 is called a **totem pole** output because of the way transistor Q_4 sits on top of Q_3. The output stage is also called an **active pull-up** because an active device, transistor Q_4, pulls the output up toward V_{CC}. A **passive pull-up** would use a resistor in place of Q_4. Variations in the output stage circuitry exist, and they are treated in the references [4].

Output Short Circuit Current The output short circuit current I_{OS} is the HIGH current that can flow out of the output of a device whose output is connected directly to ground. This current is meant as a measure of the device's ability, in the HIGH state, to rapidly charge external capacitances; it should not be used as an indication of its dc drive capability. For the 7400 NAND gate, the short circuit current is measured with all gate inputs LOW; its maximum value is $I_{OS} = -55$ mA. Only one output in an IC package should be shorted at one time.

Current Spiking A problem associated with totem-pole outputs is current spiking. When the output is HIGH, a constant amount of supply current I_{CCH} is drawn from the power supply by the IC, as shown in Fig. 3.10(b). Similarly, when the output of the gate is LOW, a constant supply current I_{CCL} is drawn from the power supply. Typically, for the 7400 NAND gate $I_{CCH} = 4$ mA and $I_{CCL} = 12$ mA per IC, measured with the output open. The supply current for a LOW output is greater than that for a HIGH output (see Problem 9). When the gate output is changing states, transistors Q_3 and Q_4 are both on for a short period of time. This results in a larger current drain from the power supply during switching because the power supply is then connected to ground through a small resistance R_4, Q_4, and Q_3. The result is **current spikes** (narrow pulses) in the power supply line, as shown in Fig. 3.10(b). The largest current spike occurs in the LOW to HIGH change of state. This is because Q_4 in Fig. 3.9 does not saturate (with normal loads), so it turns on faster than Q_3 can turn off. Also, in going from the LOW to the HIGH state, external load capacitances must be charged, causing charging currents up to I_{OS}. To prevent these current spikes from propagating through the power supply and ground system as noise, one decoupling capacitor (0.01 μF to 0.1 μF) for each five to ten IC packages should be connected from the supply line to ground, near the IC pins.

Power Dissipation One of the important characteristics of a digital IC is its power dissipation P_D. As the power dissipation in a system increases more heat must be dissipated from the system and larger, more costly power supplies are required. The static power dissipation P_D of an IC is the product of the supply voltage V_{CC} and the static supply current I_{CC}, that is, $P_D = V_{CC}I_{CC}$. The supply current includes internal operating currents for the device, plus currents supplied (sourced) to the input or output loads, and is measured under dc conditions only. The supply current is usually measured with the outputs open, so that the only source current is the input LOW current. If the output of a device is HIGH half of the time and LOW half of the time, then the average supply current is $I_{avg} = (I_{CCH} + I_{CCL})/2$.

Example 3.1 Calculate the typical average power dissipation for a single 7400 NAND gate.

Solution. The data book gives typical currents of $I_{CCH} = 4$ mA, $I_{CCL} = 12$ mA for a typical V_{CC} of 5 V. Hence

$$P_{D,\text{avg(typ)}} = V_{CC(\text{typ})}I_{CC,\text{avg(typ)}}$$
$$= 5\text{ V} \times \frac{4\text{ mA} + 12\text{ mA}}{2} = 40\text{ mW/package.}$$

Since there are four gates in one package, the power dissipation per gate is 10 mW. ∎

When a gate changes logic states frequently, current spikes increase the power consumption by approximately 0.3 mW/MHz [2]. Power dissipation is essentially constant for capacitive loads of less than 20 pF up to about 1 MHz (Appendix D). The increased power dissipation with frequency is due to the added energy required to charge and discharge load capacitances C_L and internal circuit capacitances.

Impedance The output impedance (Z_0) of TTL gates is quite low due to the totem-pole configuration of the output transistors Q_3 and Q_4. In the HIGH state, transistor Q_4 acts as an emitter follower, which results in a low impedance. In the LOW state, Q_3 is saturated, which also results in a low impedance. Typically, in the low state $Z_{OL} = 10\,\Omega$ and in the HIGH state $Z_{OH} = 70\,\Omega$. The approximate input impedances are $Z_{IH} = 400\,\text{k}\Omega$ and $Z_{IL} = 4\,\text{k}\Omega$ [2]. The advantage of a high input impedance is that less current is drawn from the source that drives these inputs. The advantage of a low output impedance is increased switching speed when driving capacitive loads, due to a lower RC time constant.

Propagation Delay Times As discussed in Section 3.5, it takes a finite amount of time for a single switching transistor to change its output voltage in response to an input voltage. Similarly, in a digital device that contains switching transistors, a time lag exists between the input and output response. The ON and OFF times discussed in Section 3.5 are not satisfactory for specifying logic circuit delay times because (1) the input pulses are **nonideal pulses**, i.e., they have nonzero rise and fall times, and (2) the input voltage only has to reach the threshold level before the device begins to change state. For these reasons the delay time is measured from a **reference voltage level** V_{ref} ($V_{\text{ref}} = 1.5$ V for standard series TTL) near the threshold voltage.

Recall from Section 2.6 that the **(propagation) delay time** for a digital device is the time interval required for the output to respond to an input, measured relative to the reference level. Consider the NAND gate shown in Fig. 3.11(a), connected as a NOT gate. The input waveform, shown in Fig. 3.11(b), is a nonideal

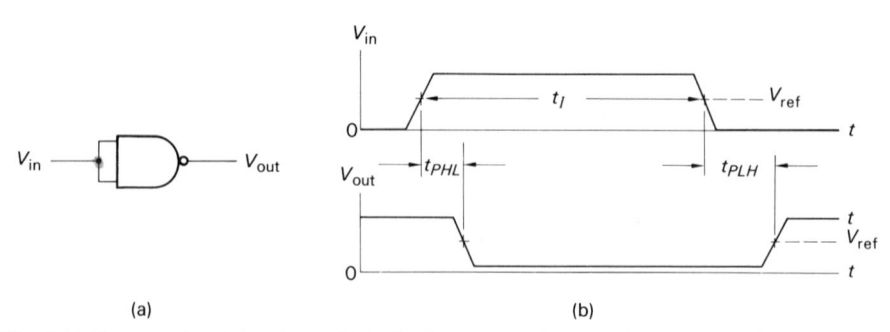

Fig. 3.11 Propagation delay times: (a) logic diagram; (b) input and output waveforms showing t_{PLH} and t_{PHL}.

pulse. Its pulse width t_I is measured at the reference level between the leading and trailing edge. When the inputs go HIGH the output will go LOW after one propagation delay time, called the **turn-on delay time** t_{PHL}. Figure 3.11(b) illustrates the turn-on delay time for a nonideal output pulse. The typical turn-on delay time for a 7400 NAND gate is 7 ns.

When the inputs go LOW, the output of the NAND gate goes HIGH after one **turn-off delay time** t_{PLH}. The typical turn-off delay time for a 7400 gate is 11 ns. The unequal delay times are due to the fact that Q_3 saturates, so its reverse recovery time is greater than that for Q_4, which does not saturate (refer to Fig. 3.9). The average delay time is (7 ns + 11 ns)/2 = 9 ns. Since other gates have slightly longer delay times, the standard TTL series delay time is usually given as 10 ns.

A transistor in a standard series TTL device is turned on by applying sufficient base current to saturate it for the lowest expected current gain ($h_{FE(min)}$). Hence the average transistor receives much more base current than necessary, driving it deeply into saturation. To turn off a saturated transistor, the excess charge stored in the base region of the transistor must first be removed. This results in an appreciable inherent delay in standard series devices.

Propagation delay times depend on three parameters: (1) capacitive loading (Section 3.6.5), (2) supply voltage, and (3) temperature. The delay time decreases slightly with increasing supply voltage, as shown in the references [3, p. 37]. In TTL devices an increase in temperature causes an increase in transistor current gain, increase in resistor values, and decrease in the forward voltage drop in semiconductor junctions. The result is an almost constant average delay time with variations in temperature. The variation of delay time with temperature, together with other curves, is shown in Appendix D.

The maximum pulse frequency at which a gate can be operated is approximately

$$f_{max} = \frac{1}{t_{PLH} + t_{PHL}}.$$

Thus if $t_{PLH} = 11$ ns and $t_{PHL} = 7$ ns, then $f_{max} = 1/(11 + 7)$ ns = 55 MHz.

Noise Immunity Noise immunity is a measure of the ability of a digital circuit to prevent noise voltages from changing a given logic voltage level in the circuit. Consider a digital system consisting of two NAND gates as shown in Fig. 3.12(a). Suppose noise from an external source, such as an arcing motor, or internal noise, such as current spiking, produces a ±200-mV change in the output voltage level of gate G_1. Will this noise voltage affect the operation of the system?

Let's compare the HIGH and LOW voltage ranges for the driving device (G_1) and the driven device (G_2). Figure 3.12(b) shows the relations between output and input voltage ranges. If the output of G_1 is LOW, a positive noise voltage will have to be at least 400 mV before the maximum input LOW voltage of G_2 is exceeded. Even if the output of G_1 exceeded $V_{IL(max)}$, gate G_2 would not change

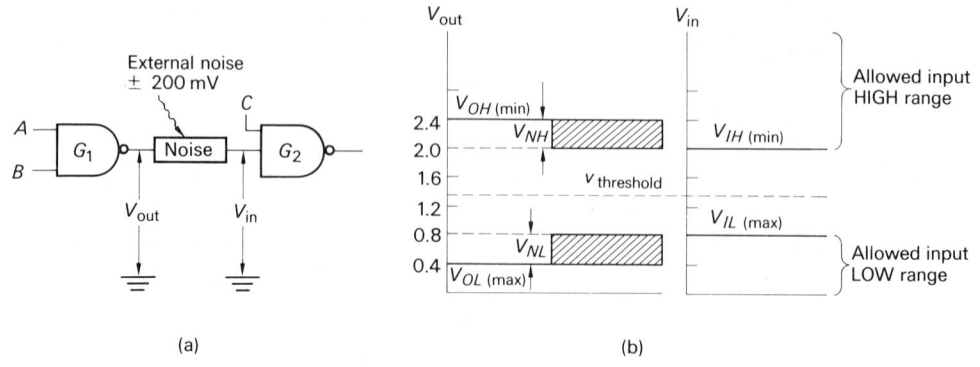

Fig. 3.12 Noise in a digital system: (a) logic diagram; (b) input and output TTL voltage levels, illustrating dc noise margins.

state until its input voltage reached the threshold voltage of about 1.4 V. Similarly, if the output of gate G_1 is HIGH, a negative-going noise voltage will have to be at least -400 mV before the input voltage of G_2 falls below the minimum, $V_{IH(\text{min})}$.

The dc HIGH and LOW **noise margins** are defined as

$$V_{NH} = V_{OH(\text{min})} - V_{IH(\text{min})}$$

and

$$V_{NL} = V_{IL(\text{max})} - V_{OL(\text{max})}.$$

For standard TTL, $V_{NH} = V_{NL} = 400$ mV (refer to Fig. 2.14). This represents worst-case conditions; more typical noise margins of more than 1 V result if the threshold voltage $V_T = 1.4$ V is used in place of V_{IH} and V_{IL}, and typical values are used for V_{OH} and V_{OL} in the above formulas. For a discussion of ac noise margins see the references [3, p. 41].

Unused Gates and Inputs Referring to Fig. 3.9, note that an open input, called a **floating input**, floats at the threshold voltage for the device. A floating input usually has the same effect on the circuit operation as a HIGH input. However, open inputs allow noise voltages to enter an input via the package leads, which act like antennas for noise. A few hundred millivolts of negative noise voltage is sufficient to drive a floating input to the LOW state. Therefore, it is advisable to tie all unused inputs of any device to a HIGH or a LOW.

Unused inputs of AND gates (with multiple-emitter inputs) and NAND gates should be tied to used inputs of the same gate, provided the HIGH level output current drive capability of the circuit is not exceeded. If output drive current is limited, tie unused inputs to a source of a permanent HIGH; for example, (1) tie to V_{CC}, if V_{CC} is less than 5.5 V ($V_{\text{in(max)}}$), (2) tie to a separate power supply of 2.4 V to 5.5 V, (3) tie to an unused gate whose output is permanently HIGH, or (4) tie to V_{CC} through a 1-kΩ current-limiting resistor ($V_{CC(\text{max})} = +7.0$ V, $I_{\text{in(max)}} \approx 1.0$ mA at $+5.5$ V, so $R = (7 - 5.5)/1.0$ mA $= 1$ kΩ).

Unused inputs of NOR or OR gates should also be connected to used inputs of the same gate, provided the current capability of the driving circuit is not exceeded. Each OR and NOR gate input has a separate single-emitter input transistor, so this connection increases the input current requirements for both the HIGH and the LOW states. If output drive current is limited, unused inputs should be tied to ground.

The outputs of unused gates in an IC package should be forced HIGH by appropriate connection of the inputs. This provides a convenient HIGH, reduces the power dissipation (since $I_{CCH} < I_{CCL}$), and prevents oscillation of the gate, which causes increased power dissipation.

3.6.5 Loading Effects

In this section we shall consider the effects of resistive loads and capacitive loads on the performance of gates.

Capacitive Loading If the load impedance connected to the output of a gate has a capacitive component, the output switching time can be increased considerably. Consider the output circuit of a NAND gate, whose inputs are all tied together, which is connected to an external purely capacitive load C_L as shown in Fig. 3.13(a). This external load capacitance can be due to loads such as another logic device, an oscilloscope, or stray wiring capacitance.

The propagation delay times given on data sheets are due to internal delay mechanisms (already discussed for transistors) and for a given maximum external capacitive load C_L (15 pF for most TTL gates) and resistive load R_L (400 Ω). Figure 3.13(b) illustrates the delay time for $C_L = 15$ pF.

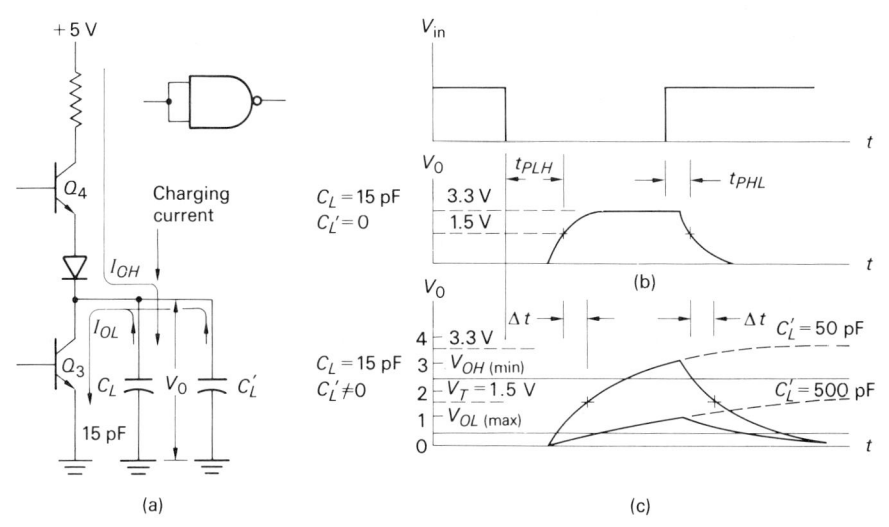

Fig. 3.13 Effect of capacitive loading on delay times: (a) totem pole output circuit; (b) $C'_1 = 0$; (c) $C'_1 \neq 0$.

Suppose the output of the NAND gate in Fig. 3.13(a) is LOW and an external load capacitor C'_L is added to the output of the NAND gate (in addition to $C_L = 15$ pF). Since Q_3 is essentially a short circuit to ground, when the gate output is LOW the capacitor will be completely discharged, so V_0 will be 0 V as shown in Fig. 3.13(c). Next suppose V_{in} goes LOW, so V_0 would become HIGH one delay time t_{PLH} later if C'_L were 0 (Fig. 3.13b). However, when $C'_L \neq 0$ the output voltage rises exponentially due to the $R_L C'_L$ time constant, where R_L is the total output resistance of the gate in the HIGH state and of the load. The delay time Δt, in addition to t_{PLH}, is $\Delta t \approx R_L C'_L$.

Example 3.2 (a) Calculate the total delay time in a 7400 NAND gate due to a load capacitance of 50 pF, assuming $R_L = 100 \ \Omega$. (b) Repeat for $C'_L = 150$ pF.

Solution

a)
$$\Delta t = 100 \ \Omega \times 50 \times 10^{-12} \ F = 5 \ ns$$
$$t_{total} = t_{PLH} + \Delta t = 11 \ ns + 5 \ ns = 16 \ ns$$

b)
$$\Delta t = 100 \ \Omega \times 150 \times 10^{-12} \ F = 15 \ ns$$
$$t_{total} = 11 \ ns + 15 \ ns = 26 \ ns \ \blacksquare$$

When specifying the propagation delay times for a device, reasonable values for the load resistance and external load capacitance should be used. For example, 15 pF is too small for most applications.

Next consider the effect of a large capacitance load in Fig. 3.13(a) say $C'_L = 500$ pF. Then Δt will be 10×5 ns $= 50$ ns. Suppose the capacitor is discharged and the output becomes HIGH. Then the capacitor begins charging slowly as shown in Fig. 3.13(c). Now if the input goes HIGH again, before the output voltage reaches the threshold voltage V_T, the output has ignored the input signal, thereby causing a logic error in the system. Thus the effect of capacitive loads is to increase the delay times and reduce the maximum speed at which a device can be operated.

Resistive Loading Suppose the output of a logic gate is connected to resistive load as shown in Fig. 3.14(a). In the HIGH state a maximum continuous source current $I_{OH(max)}$ can flow through R_4, Q_4, and into the load R_L. If the load resistance is too small it will try to draw more current than is available. As a result the output voltage will be reduced, perhaps to less than $V_{OH(min)}$, causing a logic error in the output.

If the output of a logic gate is LOW, as shown in Fig. 3.14(b), then sink current I_{OL} flows into the output and through Q_3 to ground. If this sink current is large enough, sufficient voltage will be applied to the collector of Q_3 to bring it out of saturation, in which case the output voltage could exceed $V_{OL(max)}$. Thus excessive resistive loading causes changes in the output voltage, which can reduce the noise margin, while capacitive loading reduces the speed of the device.

The **dc fan-out** (F.O.) is the maximum number of inputs that can be con-

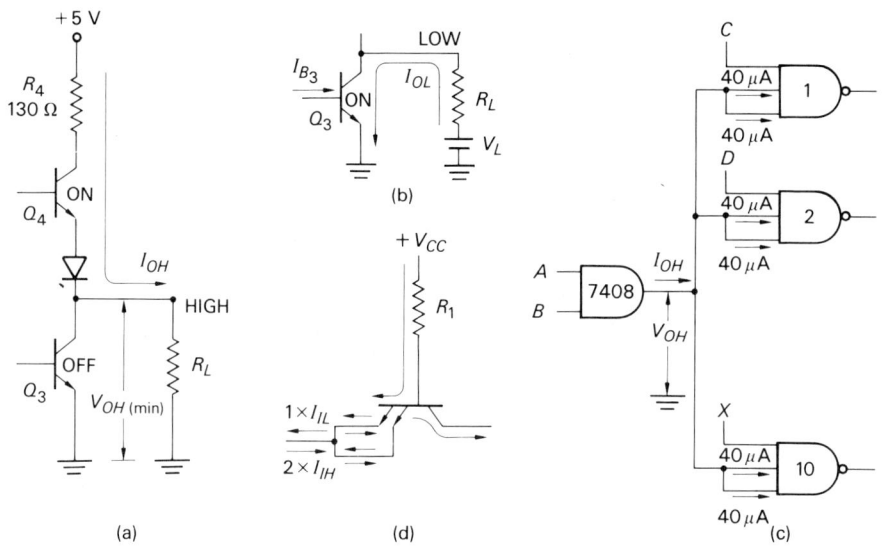

Fig. 3.14 Resistive loading: (a) HIGH current; (b) LOW current; (c) circuit with 20 loads; (d) input currents in a multiple-emitter transistor.

nected to the output of the driving gate in its HIGH or LOW state, without the output voltage of the driving device dropping below $V_{OH(min)}$ or exceeding $V_{OL(max)}$, respectively:

$$\text{F.O.(HIGH)} = \frac{I_{OH(max)}}{I_{IH(max)}}, \quad \text{F.O.(LOW)} = \frac{I_{OL(max)}}{I_{IL(max)}}.$$

The currents $I_{OH(max)}$ and $I_{OL(max)}$ are the maximum HIGH and LOW currents, respectively, of the driving gate, and $I_{IH(max)}$ and $I_{IL(max)}$ are the maximum input currents for each input of the devices being driven. A TTL **unit load** (U.L.) is defined as an input HIGH current of 40 μA and an input LOW current of 1.6 mA. The unit load concept is useful when comparing the current capabilities between different series of devices within the same logic family. The next example illustrates the ideas of fan-out and unit loads.

Example 3.3 Calculate the fan-out for a 7408 AND gate for a unit load.

Solution. The 7408 will source 800 μA. If each load on the output of the 7408 is a unit load (that is, 40 μA/1.6 mA), such as a 7400, then

$$\text{F.O.(H)} = \frac{800 \ \mu\text{A}}{40 \ \mu} = 20 \quad \text{40-}\mu\text{A loads, i.e., 20 unit loads;}$$

$$\text{F.O.(L)} = \frac{16 \ \text{mA}}{1.6 \ \text{mA}} = 10 \quad \text{1.6-mA unit loads, i.e., 10 unit loads.}$$

64 DIGITAL CIRCUITS 3.6

Thus when the output of the 7408 is HIGH, 20 unit loads can be connected to its output without drawing so much current that the output voltage will decrease below $V_{OH(min)}$, as illustrated in Fig. 3.14(c). Similarly, 10 U.L. can be connected to the output of the 7408 when it is LOW without danger of exceeding $V_{OL(max)}$. ∎

The reason for a different HIGH and LOW fan-out for some devices is as follows. Suppose two emitters of a multiple-emitter transistor in a load gate are connected together as shown in Fig. 3.14(d). When the input to these two emitters is HIGH they will draw two times the input HIGH current, or $2I_{IH}$. When these two emitters are LOW, they will source the same amount of current as one emitter because both emitters share one common pull-up resistor R_1. Thus a HIGH fan-out of 20 makes it possible to tie as many as 10 unused inputs to used inputs on the same gate without exceeding a fan-out of 10 for either the HIGH or the LOW state.

The **input load factor**, also called the **fan-in**, is a number which tells how much input current an input requires relative to a unit load (40 µA/1.6 mA for TTL). For example, the input load factor for a 7400 NAND gate is 40 µA/40 µA = 1 U.L. A device whose input draws 50 µA of current, such as a Schottky 74S00 NAND gate (Section 3.6.8), has an input load factor of 50 µA/40 µA = 1.25 U.L., in other words, it draws 1.25 times as much current as a standard series 7400 gate. Thus a 7400 gate can drive only 10/1.25 = 8 74S00 gates, that is, its fan-out is 8.

Current requirements for SSI devices other than logic gates, such as flip-flops, and for MSI and LSI devices vary greatly. Individual data sheets should be consulted for these devices.

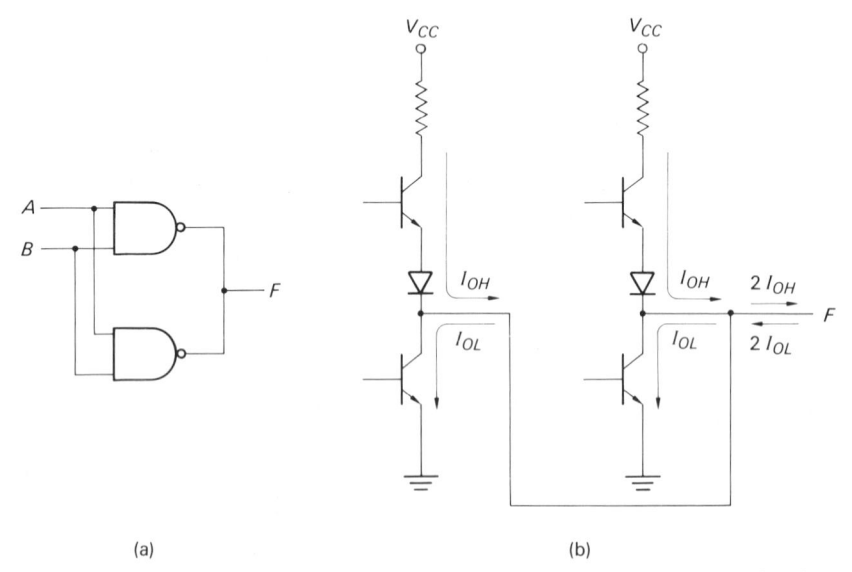

Fig. 3.15 Paralleling gates to increase fan-out: (a) logic diagram; (b) schematic of output circuit.

The fan-out of a device can be increased by connecting the inputs and the outputs of two or more devices in the same IC package in parallel. Figure 3.15(a) shows the logic diagram for two parallel NAND gates. Figure 3.15(b) shows that the HIGH and LOW output currents for two parallel gates are twice those for one gate. It is preferable to limit the gates being paralleled to one IC package to avoid large power supply transient pulses due to different switching times of gates.

*3.6.6 Open Collector Devices

It is often desirable to connect together the outputs of several devices, such as logic gates, in order (1) to increase the fan-in, (2) to perform some desired logic function, or (3) to connect several devices to a common line (called a bus). Suppose that the outputs of two gates having totem pole outputs were connected together. If the output of one gate were going to the HIGH state while the other gate output was LOW, the LOW gate would have a low resistance to ground. Then the HIGH gate could sink as much as $I_{OS(max)} = 55$ mA into the LOW gate, and since the LOW gate can safely sink a maximum of only 16 mA, the transistor would probably be damaged.

Open collector gates are specially designed gates whose outputs can be connected together. Figure 3.16(a) shows the schematic of an open collector NAND gate. It is the same as an ordinary NAND gate (Fig. 3.9) except that transistor Q_4, resistor R_4, and diode D_3 have been omitted.

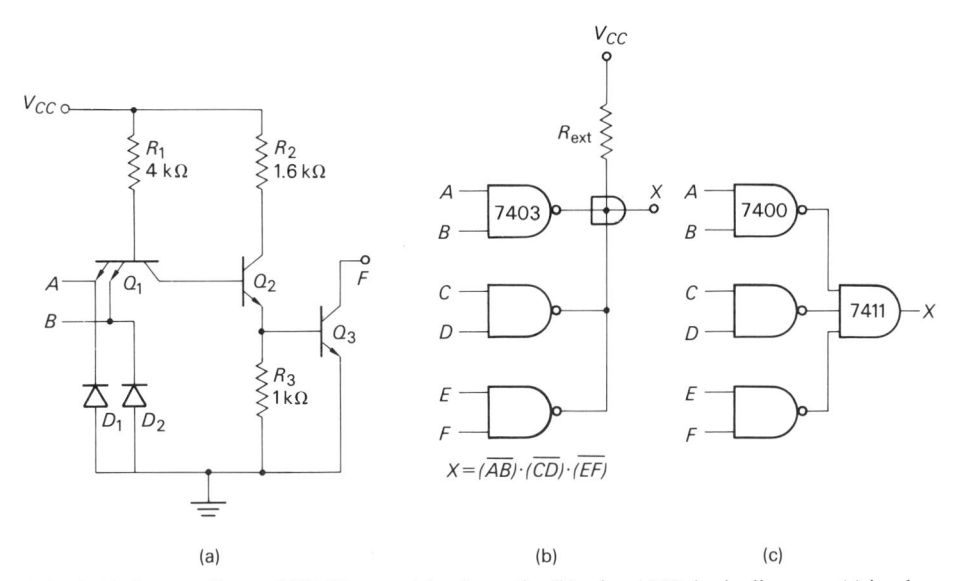

Fig. 3.16 Open collector NAND gate: (a) schematic; (b) wire-AND logic diagram; (c) implementation of (b) using totem pole gates.

Suppose three open collector gates in a quad 7403 NAND gate package are connected (wired) together as shown in Fig. 3.16(b). The external **pull-up resistor** R_{ext} acts as the collector load resistor for each of the NAND gates. Its value can be calculated from standard formulas (see Problem 16). The AND gate symbol in Fig. 3.16(b) is not an actual logic gate; rather, it shows the type of logic performed. The output X is HIGH only if the outputs of each gate are HIGH. If the output of one gate is LOW, Q_3 will be ON, and it will pull all the other gate outputs LOW. Thus the circuit performs the AND logic operation so $X = (\overline{AB}) \cdot (\overline{CD}) \cdot (\overline{EF})$. Note that the ANDing is achieved by making a wiring connection without the need for an additional gate. Hence the name **wire-AND logic**, which means that the gate outputs are connected together and perform the AND operation. The terms **wire-OR**, **dot-OR**, and **collector-dotting** are also used because, as will be shown in Chapter 4, X can also be written in OR form as $X = \overline{AB + CD + EF}$.

Figure 3.16(c) shows a method of implementing $X = (\overline{AB})(\overline{CD})(\overline{EF})$ using three two-input 7400s and one three-input 7411 AND gate. The wire-AND circuit in Fig. 3.16(b) requires only three gates and one package, thus saving gate count and package count compared to the totem-pole implementation in Fig. 3.16(c).

The advantage of the wire-AND configuration in general is that one or more levels of gating may be saved. This can result in reducing the overall propagation delay time and the total power dissipation. Since the minimum value of R_{ext} is larger than the corresponding active totem-pole HIGH resistance, single open collector devices are slower than single totem-pole devices. Hence the disadvantages of open collector devices, as compared with totem-pole devices, are (1) discrete resistor required, (2) slower, (3) less ac noise immunity, and (4) less able to drive large capacitive loads.

A wide variety of logic gates, and other devices such as decoders and memories, are available with open collector outputs.

A **buffer/driver gate** is a gate with a higher than normal output current and/or voltage rating for the output transistor. Buffer/driver gates are available with totem-pole or open collector outputs. For example, the open collector 7407 buffer/driver has a maximum HIGH output voltage of 30 V and will sink 40 mA of current. Buffer/drivers which invert the output are called **inverter buffer/drivers**.

The next section shows another method for connecting the outputs of several devices together.

*3.6.7 Three-State Devices

In digital systems, such as computers and data communication systems, data is often transferred along a common wire called a data bus. A **data bus** is a signal line which connects two or more outputs and one or more inputs to the same line, as illustrated in Fig. 3.17(c).

3.6 THE TTL FAMILY 67

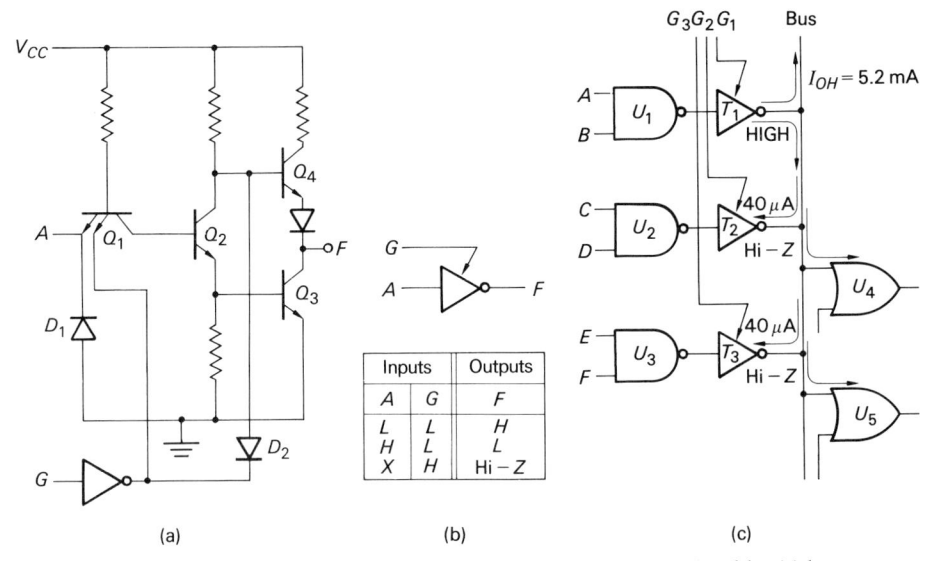

Fig. 3.17 Three-state gate: (a) inverter schematic; (b) symbol and truth table; (c) bus system.

A **three-state** (**tristate** or **disabled**) **device** is a special device which has three states: the normal low-impedance LOW, the normal low-impedance HIGH, and a high-impedance state. Its output circuit is especially designed for driving capacitive loads. The schematic for a tristate inverter is shown in Fig. 3.17(a). When the control input G is HIGH, the base voltage for Q_4 is LOW, so Q_4 is turned off. Also, the collector to Q_2 is LOW, so Q_2 is off, and therefore Q_3 is off. Hence *both* output transistors (Q_3 and Q_4) are off and therefore the output impedance is high (the Hi-Z state). When the control input is LOW the circuit operates like a normal inverter (see Problem 19).

The logic symbol and truth table for this three-state gate are given in Fig. 3.17(b). Figure 3.17(c) shows a system with three logic gates, with totem-pole outputs connected to a bus through three (National Semiconductor) DM8093 tri-state buffers. Suppose we wish to connect NAND gate U_1 to the bus. Then the control inputs should be $G_1 = L, G_2 = G_3 = H$, which will connect U_1 to the bus and disconnect U_2 and U_3 from the bus. Buffer T_1 can source up to 5.2 mA of current. Buffers T_2 and T_3, in their Hi-Z state, sink only 40 µA of leakage current. Therefore, 50 devices could be connected to this bus, since 50×40 µA $=$ 2.0 mA, with 3.2 mA of current left over for driving the bus and devices.

Since the output stage of a three-state device is a totem-pole, only one driving device on the same bus should be in the low-impedance state; all other devices must be in their Hi-Z state. Thus, the circuit in Fig. 3.17(c) does not perform the wire-AND operation. The advantage of a system using three-state devices, such as the one shown in Fig. 3.17(c), is the large number of devices that can be

68   DIGITAL CIRCUITS   3.6

connected to one bus while maintaining high speed. It is also possible to transmit data in two directions along a bus, called a **bidirectional bus**, by use of three-state devices [5, Ch. 10, p. 59]. The main disadvantage of three-state devices, as compared with open collector devices, is that the logic for the control inputs to the three-state devices must ensure that no more than one device is in the low-impedance state at one time. Secondly, rather large current demands on the power supply occur during switching (especially for the LOW to HIGH transition). Usually three-state gates are designed so that the enable time is longer than the disable time. Thus when the control inputs to several gates are changed simultaneously the output of the enabled gate will not become active until all other gates become disabled.

*3.6.8 The Schottky Series

High-speed logic devices are necessary in some applications such as in digital computers and digital data communication systems. In these applications, where 10-ns delay times (typical of the standard TTL series) are too slow, there are currently two meaningful choices, namely, Schottky TTL or emitter-coupled logic (ECL). The Schottky TTL series is covered in this section and ECL is covered in Section 3.9.

The Schottky (S) series is the highest-speed TTL series, with a typical/average gate delay time of 3 ns and a power dissipation of 19 mW per gate. Thus the Schottky series is over three times faster than the standard series, but has almost twice the power dissipation. However, above about 25 MHz, Schottky devices dissipate less power than the standard series.

The Schottky series is pin for pin compatible with the equivalent standard series devices, so converting to Schottky devices is easily accomplished. Also, Schottky devices can be readily mixed with non-Schottky devices to achieve high speed in those parts of the circuit that require it.

A wide variety of Schottky series SSI and MSI devices are available. Open collector and three-state devices are also available. Some examples of systems that use Schottky devices are minicomputers, high-speed instruments, high-speed intelligent terminals, and some communication systems.

Figure 3.18(a) shows the schematic of the 74S00 NAND gate. This circuit differs from the standard series 7400 NAND gate in several ways.

First, the input clamp diodes D_1 and D_2 are Schottky diodes (Section 3.2), which are more effective in suppressing negative input voltages than silicon junction diodes, because their cut-in voltage is only 0.4 V instead of 0.7 V.

Second, all transistors that normally saturate (i.e., all but Q_4) are Schottky transistors. A **Schottky transistor** is a transistor with a Schottky barrier diode connected between its base and collector as shown in Fig. 3.18(b). As a Schottky transistor begins to saturate, its base current is diverted through the Schottky diode, instead of through the collector–base junction of the transistor. This is due to the lower cut-in voltage of the Schottky diode as compared with the cut-in

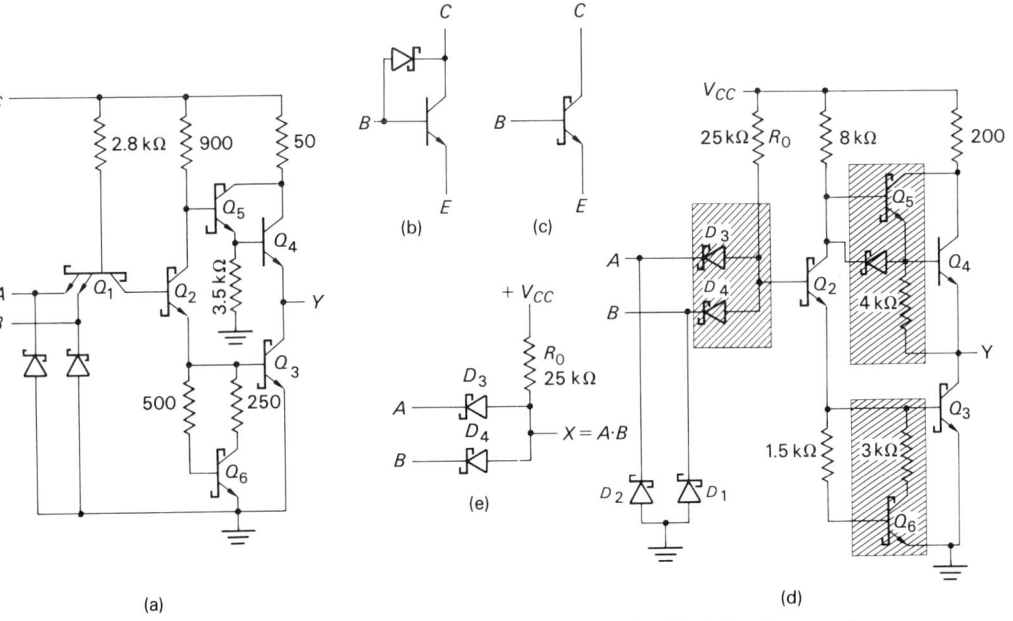

Fig. 3.18 (a) The 54/74S00 Schottky NAND gate schematic; (b) Schottky transistor construction; (c) Schottky transistor symbol; (d) the 54/74S00 low-power Schottky NAND gate schematic; (e) diode logic AND gate.

voltage of the collector–base junction diode. As a result, the Schottky transistor never becomes fully saturated and few excess charge carriers are stored in the base of the transistor, so the reverse recovery time is greatly reduced. The symbol for a Schottky transistor is shown in Fig. 3.18(c). The disadvantage of the Schottky transistor is that the Schottky diode increases the input capacitance of the transistor, causing an RC time constant limitation on its speed.

Third, note that resistor R_3 in Fig. 3.9 has been replaced with an active pull-down, transistor Q_6, in Fig. 3.18(a). This allows Q_3 to turn off more rapidly because of the low collector–emitter resistance of Q_6 and results in a squarer voltage transfer curve (see Appendix D).

The final difference between the Schottky and standard series gate is that the upper totem-pole transistor is in the Darlington configuration for the Schottky gate. A **Darlington configuration** consists of a pair of transistors (Q_4 and Q_5) with the emitter of the first transistor connected to the base of the second transistor and the collectors of both transistors connected together. The Darlington configuration provides a high current drive capability, resulting in faster switching speeds. The open circuit (i.e., no load) output HIGH voltage is $V_{OH} = V_{CC} - 2V_{BE}$, which results in higher-speed switching.

Some maximum properties of the 74S00 NAND gate are: $I_{OH} = -1$ mA, $I_{OL} = 20$ mA, $I_{IH} = 50$ µA, $I_{IL} = -2$ mA, and $I_{OS} = -100$ mA. The HIGH level dc noise margin is $V_{NH} = V_{OH(min)} - V_{IH(min)} = 2.7$ V $- 2.0$ V $= 700$ mV and $V_{NL} = V_{IL(max)} - V_{OL(max)} = 0.8$ V $- 0.5$ V $= 300$ mV. The HIGH and LOW level fan-outs are 20 and 10, respectively. Additional Schottky series device characteristics are given in Appendix D.

It is more difficult to design systems using Schottky devices than other TTL series devices because Schottky devices have very fast rise and fall times ($t_{r,f} \sim 1$ V/ns), which cause reflection and cross-talk problems [6]. Open wire lengths should not exceed 25 cm. Current spikes are of greater amplitude and narrower than for the standard series, which means that better power supply decoupling is required. One 0.01 to 0.1 µF RF (ceramic disk) decoupling capacitor for each five IC packages should be used. Signal path lengths should be minimized if speed is important, because each 15 cm of conductor on a PC board adds about 1 ns to the delay time. A 150-Ω wirewrap connection in the backplane adds about 1 ns delay for every 25 cm of wire length [6]. A ground plane on the PC board and low-impedance ground lines should be used.

*3.6.9 The Low-Power Schottky Series [26]

For many applications, such as a vending machine, the high speed of the standard or Schottky series is not needed. The low-power Schottky (LS) series has a power dissipation of only 2 mW per gate and an average gate delay of 5 ns. Thus the LS series is twice as fast as the standard series and dissipates only one-fifth as much power. The LS series achieves high speed by using Schottky transistors and low power dissipation by using larger resistor values.

The LS series is pin for pin compatible with the equivalent standard and Schottky series devices. Thus a standard series system can be easily converted to LS series devices. A wide variety of SSI and MSI LS series devices are available. Open collector and three-state devices are also available.

Figure 3.18(d) shows the schematic for the 74LS00 NAND gate. This circuit differs from the Schottky series schematic (Fig. 3.18a), discussed in Section 3.6.8, in the following ways.

First, the two-emitter input transistor has been replaced by two Schottky diodes D_3 and D_4 and resistor R_0. These diodes form a two-input **diode logic** (DL) AND gate as shown in Fig. 3.18(e). If any input in Fig. 3.18(e) is LOW, say input A, then the output X is shorted to ground through D_3, so $X = L$. If both inputs A and B are HIGH, both diodes act as open switches, so the output is connected to V_{CC} through R_0, hence $X = H$. The advantages of DL inputs using Schottky diodes are higher speed and lower input currents than those for multiple-emitter transistors. When the diode inputs are followed by a transistor, such as Q_2 in Fig. 3.18(d), this is called **diode-transistor logic** (DTL). Some LS series MSI devices, such as the 74LS83 full adder, do not have DL inputs.

The second difference between the LS and S series NAND gate is that the LS series ties the emitter resistor for Q_5 to the output, instead of to ground.

The final difference is that the LS series uses larger resistor values than the S series. This results in lower power dissipation but longer delay times, due to larger RC time constants.

Some maximum properties of the 74LS00 gate are: $I_{OH} = -400\ \mu A$, $I_{OL} = 8$ mA, $I_{IH} = 20\ \mu A$, $I_{IL} = -0.36$ mA, and $I_{OS} = -42$ mA. The HIGH level dc noise margin is $V_{NH} = V_{OH(\min)} - V_{IH(\min)} = 2.7\ V - 2.0\ V = 700$ mV and $V_{NL} = V_{IL(\max)} - V_{OL(\max)} = 0.8\ V - 0.5\ V = 300$ mV. The HIGH and LOW level fan-outs are 20 and 22, respectively. Additional LS series device characteristics are given in Appendix D. The reference voltage V_{ref} for the LS series is 1.3 V, instead of 1.5 V as it is in the standard and Schottky series.

Let's consider the main advantages of the LS series. Low power dissipation by the device means smaller size, weight, and cost of the power supply, which in turn results in lower heat dissipation and operating costs. Lower heat dissipation means greater device density on one PC board and reduction in size, or elimination, of cooling fans. Lower power dissipation also allows a greater density of components within an IC of a given size without exceeding the power dissipation capabilities of the IC package. It also means lower junction temperatures for the transistors and diodes, which increases the longevity of the device. The LS series switches approximately 25% less current than the standard series, resulting in lower-amplitude current spikes, i.e., less internal noise generation. Consequently fewer and lower-valued decoupling capacitors are required. The low input current requirements of the LS series make it an ideal interface between TTL-compatible MOS devices and TTL. The high output drive current allows the LS series to drive capacitive loads, other TTL series, and CMOSL devices.

Unused inputs of the LS series devices, with diode inputs, may be connected directly to $V_{CC} \leqslant 7$ V, i.e., no resistor is required to drop the input voltage to 5.5 V. This is due to the higher input breakdown voltage V_{CBO} of the input collector–base junction of input transistor Q_2—about 25 V, as compared with only ~ 6.5 V in the standard series. The high parasitic capacitance of the collector of the standard series input transistor is also reduced, which results in higher switching speeds.

In the next section we shall briefly discuss MOS transistors before undertaking a description of the CMOSL family in the following section.

*3.7 MOS TRANSISTORS [12, 13, 14]

A **MOS (metal-oxide semiconductor) transistor** consists of four regions called the **source** (*S*), the **gate** (*G*), the **drain** (*D*), and the **substrate** (*B*), as shown in Fig. 3.19. A MOS transistor controls the current between its source and drain with an applied gate-to-source voltage, i.e., it is a voltage controlled switch. By contrast, a bipolar transistor acts as a current controlled switch.

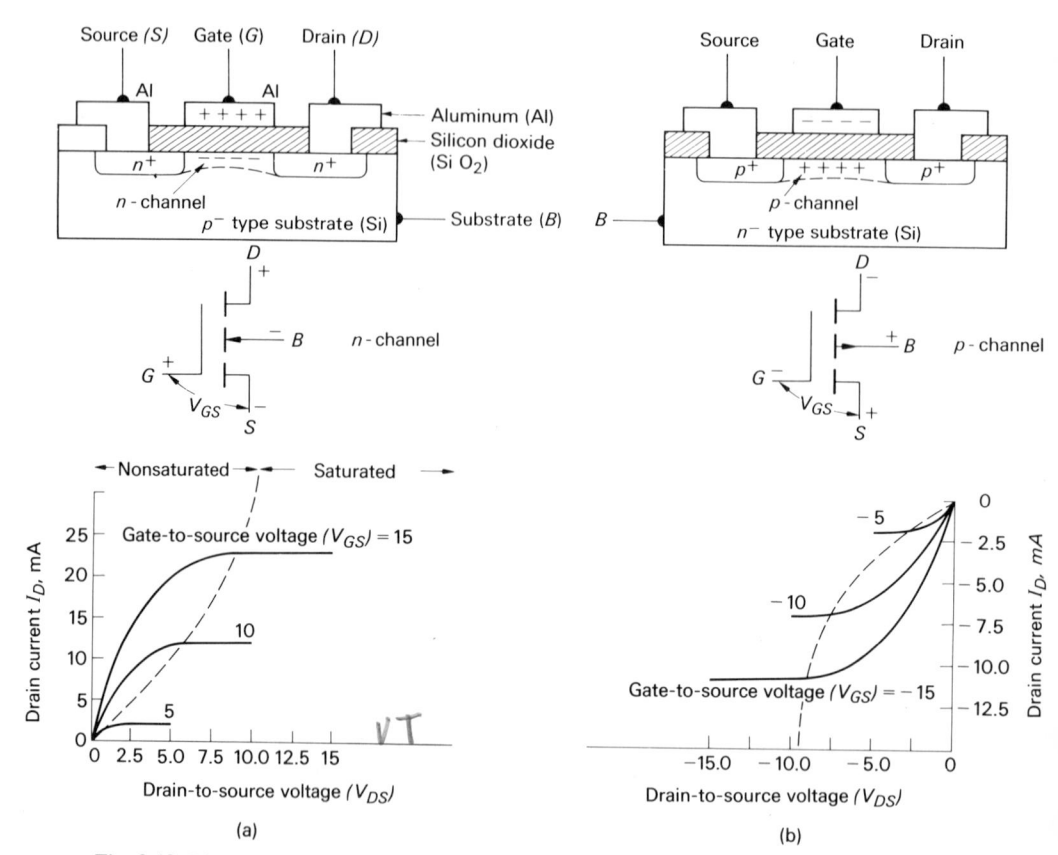

Fig. 3.19 Physical structure, schematic symbol, and characteristic curve for the MOS metal gate enhancement mode transistor: (a) n-channel; (b) p-channel.

There are two modes of operation for a MOS transistor, the **enhancement mode** and the **depletion mode**. A depletion mode transistor conducts, i.e., it is on, when zero voltage is applied to its gate. An enhancement mode transistor does not conduct, i.e., it is off, until the gate voltage reaches a certain level (~ 2 V). Thus the enhancement mode transistor has an inherent noise immunity and for this reason is used more extensively in digital devices than the depletion mode transistor. The following paragraphs therefore discuss only the enhancement mode MOS transistor.

There are two types of enhancement mode transistors, the **p-channel transistor** (PMOS) and the **n-channel transistor** (NMOS). The (channel) **threshold voltage** V_T is the gate-to-source voltage at which an n-channel or a p-channel transistor begins to conduct. Consider the n-channel MOS transistor shown in Fig. 3.19(a). The symbol n^+ is used to denote a heavy concentration of n-type impurities, while the symbol p^- denotes a light concentration of p-type impurities. The metal gate electrode is separated from the p-type semiconductor substrate by a thin ($\sim 10^{-5}$ cm) silicon dioxide (SiO_2) insulating layer; in other words, the gate forms a capacitor. If a positive voltage is applied to the gate, negative charges are induced on the semiconductor side of this capacitor. As the positive charge on the gate is increased, the negative charge induced in the semiconductor increases until at the threshold voltage the region beneath the gate effectively becomes an n-type semiconductor. If a positive voltage (with respect to the source) is applied to the drain of this transistor then current will flow between the source and drain through this induced channel. Thus a conducting n-channel is created, or enhanced, by applying a positive gate voltage.

A MOS transistor is **saturated** when its channel is carrying as much drain current as it can for a given gate-to-source voltage. The characteristic curve for an n-channel MOS transistor is shown in Fig. 3.19(a). Note that for example, if V_{GS} is 10 V the drain current increases almost linearly as the drain-to-source voltage V_{DS} is increased. At about 12 mA of drain current the channel becomes saturated and thereafter the curve becomes essentially flat. For $V_{GS} = 15$ V the transistor saturates for a drain current of about 23 mA. The substrate (B) for an n-channel transistor must be connected to the most negative power supply potential.

The p-channel MOS transistor is shown in Fig. 3.19(b). A p-channel is enhanced (induced) by applying a negative voltage to the gate. The characteristic curve in Fig. 3.19(b) shows that this transistor saturates at $I_D = 7$ mA for $V_{GS} = -10$ V. The substrate for a p-channel transistor must be connected to the most positive power supply potential.

In the saturation region a MOS transistor acts like a constant current source since the drain current is constant, independent of the drain-to-source voltage. When a MOS transistor is not saturated it acts like a resistor since its V–I characteristic curve is approximately linear there. Thus a MOS transistor can be used as a load resistor by connecting its gate to a constant supply voltage. The drain-to-source resistance is on the order of a thousand ohms. The input impedance of a MOS transistor is primarily capacitive and is on the order of 10^{14} Ω.

For a given channel width-to-length ratio, n-channel transistors have a higher operation frequency than p-channel transistors because the n-channel charge carriers (electrons) have about 2.5 times greater mobility μ (measured in cm^2/V-s) than the p-channel holes ($\mu \approx 200$ cm^2/V-s). However, n-channel transistors are more difficult to fabricate than p-channel transistors due to contaminants in the semiconductor.

The threshold voltage V_T for a p-channel transistor is about 2 V and for an n-channel about 1 V. An n-channel transistor conducts when the gate voltage is more positive than the source voltage by an amount equal to V_T, that is, when $V_{GS} \geq V_S + V_T$. For example, if the source voltage is $V_S = 1$ V and $V_T = 2$ V the n-channel transistor will conduct if $V_{GS} \geq 1$ V $+ 2$ V $= 3$ V. A p-channel transistor conducts when the gate voltage is more negative than the source voltage by an amount equal to V_T, that is, when $V_{GS} \leq V_S - V_T$.

The threshold voltage decreases about -2 mV for every 1°C increase in temperature. Thus the threshold voltage of a MOS transistor is not very sensitive to temperature variations. The drain current decreases with increasing temperature because the charge carriers experience more collisions in the channel as the temperature is increased. A low threshold voltage is advantageous because (1) lower voltage power supplies can be used, (2) the logic voltages are more compatible with bipolar logic families, and (3) the transistor switches faster, due to the reduced voltage swing during switching. The threshold voltage can be reduced by using a heavily doped silicon layer for the gate instead of the metal gate shown in Fig. 3.19. Silicon gate devices also have a higher operating speed than metal gate devices [13].

The advantages of making logic devices from MOS transistors as compared with bipolar transistors are (1) smaller size, (2) lower power consumption, (3) ease and economy of fabrication in IC form, and (4) high input impedance. Their main disadvantages are (1) lower speed, and (2) susceptibility to rupturing the gate oxide layer due to high electric fields on device inputs or outputs.

*3.8 THE CMOSL FAMILY [7, 15, 16, 17, 19–23]

The CMOSL (complementary MOS logic) family has a very low dc power dissipation of only 6 nW per gate and a very high dc noise margin. CMOSL devices consist of pairs of *n*-channel and *p*-channel enhancement mode transistors. CMOSL is used in applications requiring low power consumption, such as portable test equipment and wristwatches, and in noisy environments, such as industrial plants. Low power dissipation is achieved because, under static conditions, one transistor in each pair is off when the other one is on.

The Motorola 14000 and 14500 CMOSL series are discussed in this section. The basic gate in these series is the NOR or NAND gate. A wide variety of SSI, MSI, and some LSI CMOSL devices are available in both series. Most LSI devices, such as memories (flip-flops), use NMOS or PMOS transistors because they are smaller in size than CMOS transistors.

The typical CMOSL supply voltages are V_{DD} from $+3$ V $+ 16$ V, with the V_{SS} supply usually grounded. The input and output logic voltages depend on the supply voltage:

$$V_H \cong V_{DD} \qquad (V_{SS} = 0),$$

$$V_L \cong 0 \text{ V}.$$

Let's see how the CMOSL NOT and NOR gates work.

3.8.1 Basic Circuits

Figure 3.20(a) shows the circuit for a NOT gate. It consists of one *p*-channel and one *n*-channel enhancement mode transistor. The source and substrate of the *p*-channel are connected directly to the positive supply voltage V_{DD}. The *p*-channel

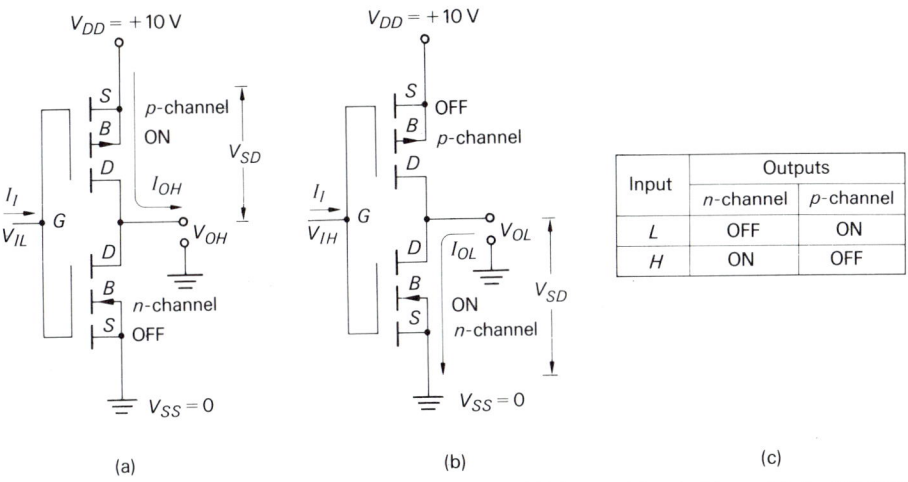

Fig. 3.20 CMOSL NOT gate: (a) LOW input; (b) HIGH input; (c) switching table for NMOS and PMOS transistors.

drain is connected to the drain of the *n*-channel transistor. The substrate and source of the *n*-channel are connected to ground. The gates of the two transistors are connected together and form the input of the NOT gate. The output is taken from the common drain terminals.

Now let's apply a LOW (0 V) to the input of the NOT gate (Fig. 3.20.a) with $V_{DD} = +10$ V and $V_{SS} = 0$ V. The *n*-channel transistor is off because V_{GS} is less than V_T. The *p*-channel transistor is turned on because its gate is more negative than V_T. Thus the output is directly connected to V_{DD} through the *p*-channel transistor. An on MOS transistor has about a 10-mV voltage drop between its source and drain, i.e., $V_{SD} \approx 10$ mV. The output of the NOT gate in Fig. 3.20(a) is HIGH, since $V_{OH} = V_{DD} - V_{SD} = 10$ V $- 10$ mV ≈ 10 V. The output source current is labeled I_{OH}.

When the input is HIGH (+10 V), the *n*-channel transistor is on because $V_{GS} > V_T$ as shown in Fig. 3.20(b). The *p*-channel transistor is turned off because $V_{GS} < V_T$. Therefore the output is connected to V_{SS} (ground), so the output is LOW, since $V_{OL} = V_{SS} - V_{SD} = 0 - (10$ mV$) \approx 0$ V. The output sink current is I_{OL}. Thus the circuit in Fig. 3.20 performs the NOT logic operation.

Figure 3.20(c) summarizes the conditions for a *p*-channel and an *n*-channel switching transistor.

The schematic for a CMOSL 14001 NOR gate is shown in Fig. 3.21(a). It consists of two *p*-channel transistors in series and two *n*-channel transistors in parallel. Each input is connected to one *n*-channel and one *p*-channel transistor. If input *A* is HIGH its *p*-channel transistor is turned off, which disconnects V_{DD} from the output. The *n*-channel transistor for *A* is turned on, so the output is connected to 0 V (LOW). When both inputs are LOW, both *p*-channel transistors

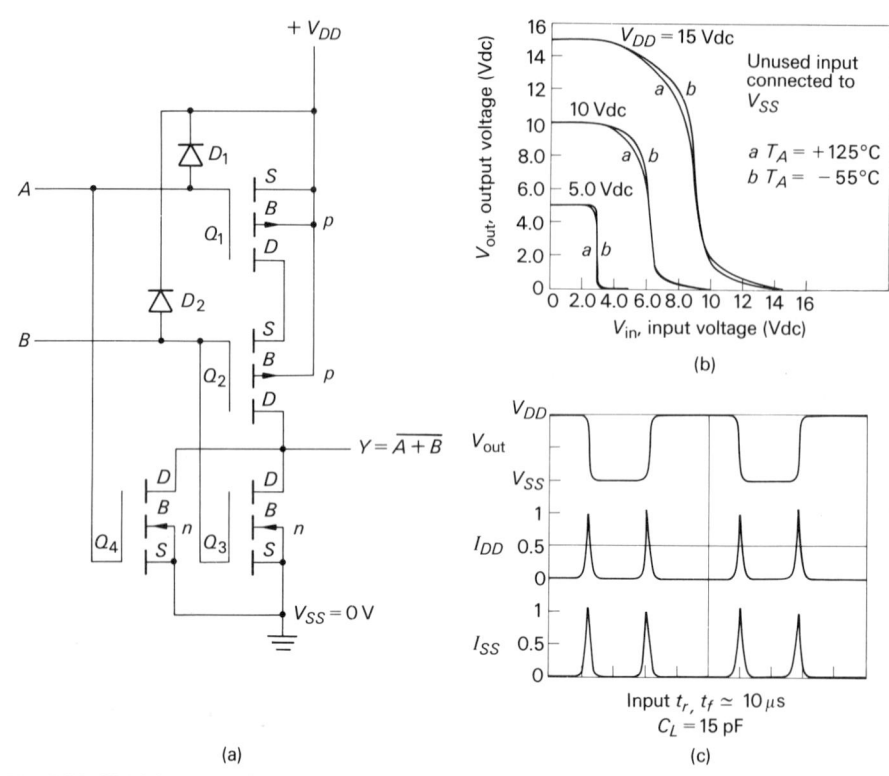

Fig. 3.21 CMOSL 14001 NOR gate: (a) schematic; (b) typical voltage transfer curves; (c) current spikes in the supply (I_{DD}) and ground (I_{SS}) lines. (Parts b and c, courtesy of Motorola Corporation.)

are turned on and both *n*-channel transistors are turned off. In this case V_{DD} is connected to the output, so the output Y is HIGH. Thus this circuit performs the NOR logic operation.

Diodes D_1 and D_2 in the input of the NOR gate in Fig. 3.21(a) illustrate one method for protecting the transistors against high electric fields. The diode-resistor method is discussed in the references [16]. These diodes can operate in the forward-biased region as long as currents are kept under 10 mA. Their reverse breakdown voltage is about 30 V.

The advantages of CMOSL devices over devices which use only *n*-channel or only *p*-channel transistors are higher speeds, lower power dissipation, higher noise immunity, greater fan-out, and less power supply regulation needed.

3.8.2 Characteristics of CMOSL Devices

The brief discussion of device characteristics in this section assumes familiarity with the logic circuit terminology covered in Sections 3.6.3 through 3.6.7.

3.8   THE CMOSL FAMILY   77

Transfer Curves The voltage and current transfer curves for a 14001 NOR gate are shown in Fig. 3.21(b) for supply voltages of 5, 10, and 15 volts. Note that the curves in the transition region are almost vertical. This is due to the fact that as one transistor is turning on, the other is increasing in impedance. This speeds up the transition process, resulting in a narrow transition region. A narrow transition region results in a high noise immunity because not much voltage range is wasted in going from one state to the other. By contrast with TTL devices, the input switching threshold voltage for a CMOSL gate depends on the supply voltage. This allows CMOSL devices to operate over a wide range of supply voltages. The input switching threshold voltage V_T is approximately 50% of the supply voltage. For example, for $V_{DD} = 5$ V, V_T is about 2.5 V.

Output Short Circuit Current Shorting the outputs of a CMOSL device to V_{SS} or V_{DD} can cause the power dissipation of the device to exceed the safe limit of 200 mW for special high current devices, such as the 14007 inverter. The typical low current device outputs can be connected directly to $V_{DD} - V_{SS} \leq 5$ V.

Current Spiking Current spiking does occur during switching, as shown in Fig. 3.21(c), because both an *n*-channel and a *p*-channel are on during the transition between logic levels [14]. At high switching speeds bypass capacitors should be used. Generally, unregulated power supplies can be used because of the high noise immunity.

Power Dissipation The quiescent power dissipation P_{QD} is the power dissipation of a device that is not changing logic states. It is due to transistor leakage currents. For the 14001 at $V_{DD} = 5$ V, $I_{\text{leak}} = 5$ nA and P_{QD} is about 25 nW per IC. As the frequency of switching increases, the dynamic power dissipation P_{DD} becomes important, as shown in Fig. 3.22(a). Above 1 MHz, the dynamic power dissipation

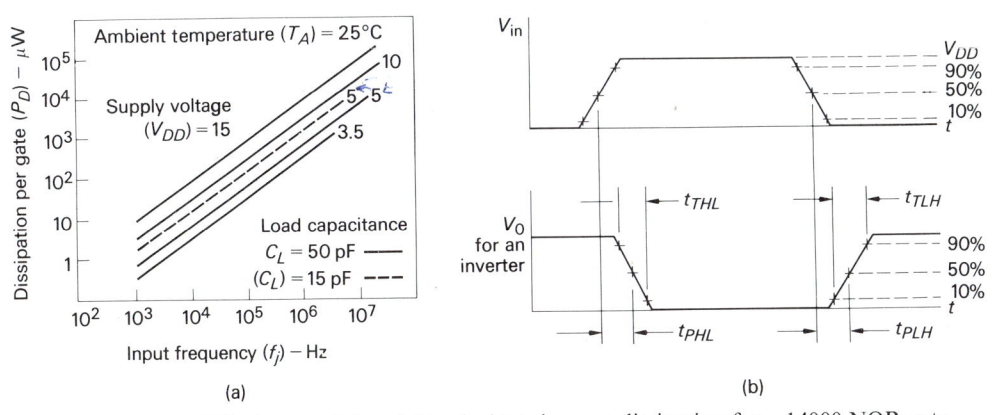

Fig. 3.22 CMOSL characteristics: (a) typical total power dissipation for a 14000 NOR gate (courtesy of Motorola Corporation); (b) voltage waveforms showing propagation delay times and transition times.

predominates, and can exceed TTL and ECL power dissipations. The dynamic power dissipation is due primarily to charging and discharging of the external load capacitance C_L and the internal circuit capacitance. Power dissipation due to capacitive loads is given by $P_{DD} = VI = VQ/T$, and since $Q = VC_L$ and $f = 1/T$,

$$P_{DD} = C_L V^2 f.$$

For example, if $V_{DD} = 10$ V, $C_L = 10$ pF, and $f = 10$ MHz, then $P_{DD} = 10$ mW.

The output of a CMOSL device is equally isolated from both V_{DD} and V_{SS}. This means that CMOSL devices can operate with negative or positive supplies. The only requirement is that V_{DD} be more positive than V_{SS} by ~ 3 V or more.

Impedance The input impedance, in either state, is typically 10^{12} Ω. The input capacitance is about 5 pF. The output impedances depend on the particular device but are often nearly equal in both states, and on the order of 1 kΩ or less.

Propagation Delay Times The propagation delay times for CMOSL devices are relatively long due to their relatively high output impedance. Typical delay times for the 14001 gate, for $V_{DD} = 5$ V, are $t_{PLH} = t_{PHL} = 60$ ns, and 25 ns for $V_{DD} = 10$ V for a 15 pF load, as shown in Fig. 3.23. Thus doubling the supply voltage more than doubles the speed of a CMOS gate. Transition times for the 14001 are $t_{TLH}(=t_r) = t_{THL}(=t_f) = 70$ ns for $V_{DD} = 5$ V, measured at the 10% and 90% levels of V_{DD}. The voltage waveform reference level for measuring delay times is 50% of V_{DD} as shown in Fig. 3.22(b). The use of **silicon-on-sapphire** (SOS) CMOS devices increases their operating frequencies to the 30-MHz region and decreases the threshold voltage to about 0.6 V [18].

Noise Immunity The HIGH and LOW noise margins for $V_{DD} = 5$ V, specified at the 70% and 30% supply voltage levels, are $V_{NH} = V_{OH(min)} - V_{IH(min)} = 4.99$ V $- 0.7 \times 5$ V ≈ 1.5 V and $V_{NL} = V_{IL(max)} - V_{OL(max)} = 0.3 \times 5$ V $- 0.01$ V ≈ 1.5 V. In general, noise immunity is a minimum of 30% of V_{DD}, and typically 45% of V_{DD}, so that CMOSL devices are good in noisy environments such as automobiles and industrial plants. The HIGH and LOW noise margins are essentially equal because the output impedance, output voltage, and input threshold voltages are symmetrical with respect to the supply voltage. However, CMOSL devices are much more sensitive to capacitively coupled noise than TTL devices.

Unused Inputs Unused inputs should be terminated in V_{DD} or V_{SS}, as appropriate for the logic. Floating (unused) inputs guarantee neither a LOW nor a HIGH at the output of the device and cause increased susceptibility to noise, as well as excessive power dissipation.

Loading Effect The dc fan-out of a CMOSL device is not limited by the current drive available, because the input current requirement of CMOSL loads is nil ($\sim$ pA). The dc fan-out is usually listed as greater than 50. Fan-out then must be determined in terms of the reduction in propagation delay time due to capacitive loading. For example, the typical propagation delay time for the 14001 NOR gate,

ELECTRICAL CHARACTERISTICS

				MC14001AL							MC14001CL/CP							
				-55°C		+25°C			+125°C		-40°C		+25°C			+85°C		
Characteristic	Figure	Symbol	V_{DD} Vdc	Min	Max	Min	Typ	Max	Min	Max	Min	Max	Min	Typ	Max	Min	Max	Unit
Output Voltage "0" Level	1,2,3	V_{out}	5.0	–	0.01	–	0	0.01	–	0.05	–	0.01	–	0	0.01	–	0.05	Vdc
			10	–	0.01	–	0	0.01	–	0.05	–	0.01	–	0	0.01	–	0.05	
			15	–	–	–	0	–	–	–	–	–	–	0	–	–	–	
"1" Level			5.0	4.99	–	4.99	5.0	–	4.95	–	4.99	–	4.99	5.0	–	4.95	–	Vdc
			10	9.99	–	9.99	10	–	9.95	–	9.99	–	9.99	10	–	9.95	–	
			15	–	–	–	15	–	–	–	–	–	–	15	–	–	–	
Noise Immunity* ($V_{out} \geq 3.5$ Vdc)	–	V_{NL}	5.0	1.5	–	1.5	2.25	–	1.4	–	1.5	–	1.5	2.25	–	1.4	–	Vdc
($V_{out} \geq 7.0$ Vdc)			10	3.0	–	3.0	4.50	–	2.9	–	3.0	–	3.0	4.50	–	2.9	–	
($V_{out} \geq 10.5$ Vdc)			15	–	–	–	6.75	–	–	–	–	–	–	6.75	–	–	–	
($V_{out} \leq 1.5$ Vdc)		V_{NH}	5.0	1.4	–	1.5	2.25	–	1.5	–	1.4	–	1.5	2.25	–	1.5	–	Vdc
($V_{out} \leq 3.0$ Vdc)			10	2.9	–	3.0	4.50	–	3.0	–	2.9	–	3.0	4.50	–	3.0	–	
($V_{out} \leq 4.5$ Vdc)			15	–	–	–	6.75	–	–	–	–	–	–	6.75	–	–	–	
Output Drive Current (V_{OH} = 2.5 Vdc) Source	–	I_{OH}	5.0	-0.62	–	-0.50	-1.7	–	-0.35	–	-0.23	–	-0.20	-1.7	–	-0.16	–	mAdc
(V_{OH} = 9.5 Vdc)			10	-0.62	–	-0.50	-0.9	–	-0.35	–	-0.23	–	-0.20	-0.9	–	-0.16	–	
(V_{OH} = 13.5 Vdc)			15	–	–	–	-3.5	–	–	–	–	–	–	-3.5	–	–	–	
(V_{OL} = 0.4 Vdc) Sink		I_{OL}	5.0	0.50	–	0.40	0.78	–	0.28	–	0.23	–	0.20	0.78	–	0.16	–	mAdc
(V_{OL} = 0.5 Vdc)			10	1.1	–	0.90	2.0	–	0.65	–	0.60	–	0.50	2.0	–	0.40	–	
(V_{OL} = 1.5 Vdc)			15	–	–	–	7.8	–	–	–	–	–	–	7.8	–	–	–	
Input Current	–	I_{in}	–	–	–	–	10	–	–	–	–	–	–	10	–	–	–	pAdc
Input Capacitance (V_{in} = 0 Vdc)	–	C_{in}	–	–	–	–	5.0	–	–	–	–	–	–	5.0	–	–	–	
Quiescent Dissipation ** Per Package	4	P_Q	5.0	–	0.00025	–	0.000025	0.00025	–	0.015	–	0.0025	–	0.000025	0.0025	–	0.075	mW
			10	–	0.001	–	0.0001	0.001	–	0.060	–	0.01	–	0.0001	0.01	–	0.30	
			15	–	–	–	0.00023	–	–	–	–	–	–	0.00023	–	–	–	
Total Power Dissipation (Dynamic Plus Quiescent) (C_L = 15 pF) Per Package	4	P_D	5.0	P_D = (2.0 mW/MHz) f + 0.000025 mW														mW
			10	P_D = (8.0 mW/MHz) f + 0.00010 mW														
			15	P_D = (18 mW/MHz) f + 0.00023 mW														
Output Rise Time** (C_L = 15 pF) t_r = (3.0 ns/pF) C_L + 25 ns	5	t_r	5.0	–	–	–	70	175	–	–	–	–	–	70	200	–	–	ns
t_r = (1.5 ns/pF) C_L +12 ns			10	–	–	–	35	75	–	–	–	–	–	35	110	–	–	
t_r = (1.1 ns/pF) C_L + 8.0 ns			15	–	–	–	25	–	–	–	–	–	–	25	–	–	–	
Output Fall Time** (C_L = 15 pF) t_f = (1.5 ns/pF) C_L + 47 ns	5	t_f	5.0	–	–	–	70	175	–	–	–	–	–	70	200	–	–	ns
t_f = (0.75 ns/pF) C_L + 24 ns			10	–	–	–	35	75	–	–	–	–	–	35	110	–	–	
t_f = (0.55 ns/pF) C_L + 17 ns			15	–	–	–	25	–	–	–	–	–	–	25	–	–	–	
Turn-Off, Turn-On Delay Time (C_L = 15 pF) t_{PLH}, t_{PHL} = (1.8 ns/pF) C_L + 33 ns	5	t_{PLH}, t_{PHL}	5.0	–	–	–	60	75	–	–	–	–	–	60	100	–	–	ns
t_{PLH}, t_{PHL} = (0.73 ns/pF) C_L + 14 ns			10	–	–	–	25	50	–	–	–	–	–	25	60	–	–	
t_{PLH}, t_{PHL} = (0.60 ns/pF) C_L + 10 ns			15	–	–	–	19	–	–	–	–	–	–	19	–	–	–	

*DC Noise Margin (V_{NH}, V_{NL}) is defined as the maximum voltage change from an ideal "1" or "0" input level before producing an output state change.
**The formula given is for the typical characteristics only.
†For dissipation at different external load capacitances refer to corresponding formula: $P_T(C_L) = P_D + 4 \times 10^{-3} (C_L - 15 \text{ pF}) V_{DD}^2 f$
Where: P_D in mW, C_L in pF, V_{DD} in Vdc, and f in MHz

Fig. 3.23 CMOSL 14001 quad two-input NOR gate specifications. (Courtesy of Motorola Corporation.)

with $V_{DD} = 5$ V, goes from 60 ns for a 15-pF load to 150 ns for a 65-pF load, since $t_{PLH} = t_{PHL} = 1.8\ C_L + 33 = 1.8 \times 65 + 33 = 150$ ns (Fig. 3.23). Since each CMOSL input typically represents a 5-pF load, then 65 pF − 15 pF = 50 pF represents a load of 50 pF/5 pF = 10 gates. Curves of delay time versus capacitive loads, as well as other CMOSL curves, are shown in Appendix D.

CMOSL devices are about 10 times more sensitive to capacitive loading than TTL devices because their output impedance is about 10 times larger.

Wire-ORing and Three-State Devices Wire-ORing is not permitted with CMOSL devices because excessive current may flow in the low-impedance output transis-

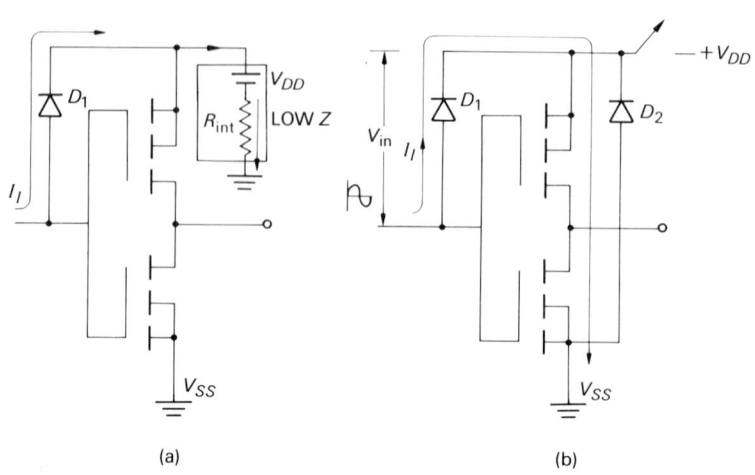

Fig. 3.24 CMOSL single-diode protection circuit.

tors. The equivalent result can be obtained with transmission gates and/or functional logic [15, p. 72]. Some three-state CMOSL devices, such as the 14034 register (a set of flip-flops) and the 14506 AOI gate, are available [19].

The power supply should be turned on before applying any logic signals to a CMOSL device. Also, logic signals should be removed before turning off the power supply, thereby keeping protection diode D_1 in Fig. 3.24(a) reverse-biased by voltage $+V_{DD}$. This prevents excessive input currents I_I from flowing through D_1 into the low-impedance power supply, which could damage the diode. When the power supply is off, current can flow through the protection diode and the CMOS transistors into V_{SS} (Fig. 3.24b). In this case the device may not be damaged; however, ac signals can be rectified by D_1 and turn the device on. Most diode protected devices can safely handle 10 mA of input current. Most devices also have an output protection diode D_2, shown in Fig. 3.24(b). Frequently CMOSL schematics will not show protection diodes since they have no effect under normal operating conditions, except to increase the input capacitance.

All CMOSL devices have parasitic *n-p-n* and *p-n-p* bipolar transistors in their structure, as shown in Fig. 3.25(a). When current flows through the bases of these transistors they can turn on, causing excessive heating and device damage. Another effect of these parasitic bipolar transistors is the latch configuration shown in Fig. 3.25(b). If sufficient current flows to activate this latch, the V_{DD} supply shorts to V_{SS}, which destroys the IC. This condition is called the **latch-up** or **SCR condition**. Latch-up is avoided by ensuring that the voltage at any pin is less than $V_{DD} + 0.3$ V and greater than $V_{SS} - 0.3$ V.

Large valued current-limiting resistors in series with the V_{DD} or V_{SS} line can cause the input protection diodes to turn on, resulting in faulty logic operation.

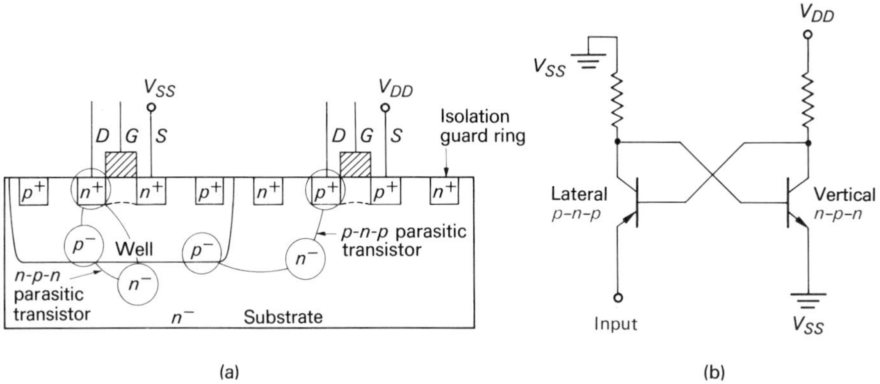

Fig. 3.25 (a) Physical structure of a CMOS transistor pair; (b) latch-up condition.

Input signals that exceed V_{DD} or drop below V_{SS} can cause the input diode to conduct; therefore this condition should be avoided.

In some CMOSL devices, such as high current buffers, input signal rise and fall times should be kept less than about 15 μs. This avoids biasing the device in the linear region for long periods of time, which can cause excessive device dissipation.

Gates within the same IC package can be paralleled to increase the source or sink current.

Another advantage of the CMOSL family is the relative ease with which it can be interfaced, using a single power supply, with other logic families such as TTL, NMOS, and PMOS.

*3.9 THE ECL FAMILY [7–11]

The ECL (emitter-coupled logic) family is a very high-speed (~ 1 ns) logic family which uses bipolar transistors. ECL is used in applications such as computers, high-speed memories, electronic instruments, and in digital data communications where very high operating speeds are required. The name emitter-coupled logic comes from the fact that the emitters of the input transistors are all coupled together.

High switching speeds are achieved in ECL by designing the circuits so that the transistors do not saturate, thereby avoiding storage delay times. Also small, low-capacitance device geometries are used to minimize RC time constants. The Motorola 10K series is discussed in this section. The basic gate in this series is a combined OR and NOR gate. A wide variety of SSI and MSI ECL devices are available.

82 DIGITAL CIRCUITS 3.9

The typical ECL supply voltage is $V_{EE} - V_{CC} = -5.2$ V, with the V_{CC} supply normally grounded (that is, $V_{CC} = 0$ V). The approximate input and output logic voltages are

$$V_H = -0.9 \text{ V}, \quad V_L = -1.75 \text{ V}.$$

Note that the logic voltages are negative, in contrast to those for the TTL and CMOSL families. In the next section we turn our attention to the differential amplifier, which is basic to understanding ECL device schematics.

3.9.1 The Differential Amplifier [1]

A **differential amplifier** is an amplifier with two inputs, whose output voltage is proportional to the difference between its input voltages. Consider the differential amplifier shown in Fig. 3.26(a). It consists of two n-p-n transistors Q_1 and Q_2 in the common emitter configuration with a common emitter resistor R_1. It has two inputs A and D, where the D input is a fixed reference voltage V_{BB}. The outputs, taken at the collectors, are labeled V_{O1} and V_{O2}. The power supply voltage V_{EE} is -5.2 V, V_{BB} is -1.29 V, and the collectors are grounded through load resistors R_{C_1} and R_{C_2}. Let's see how this amplifier operates as a switch.

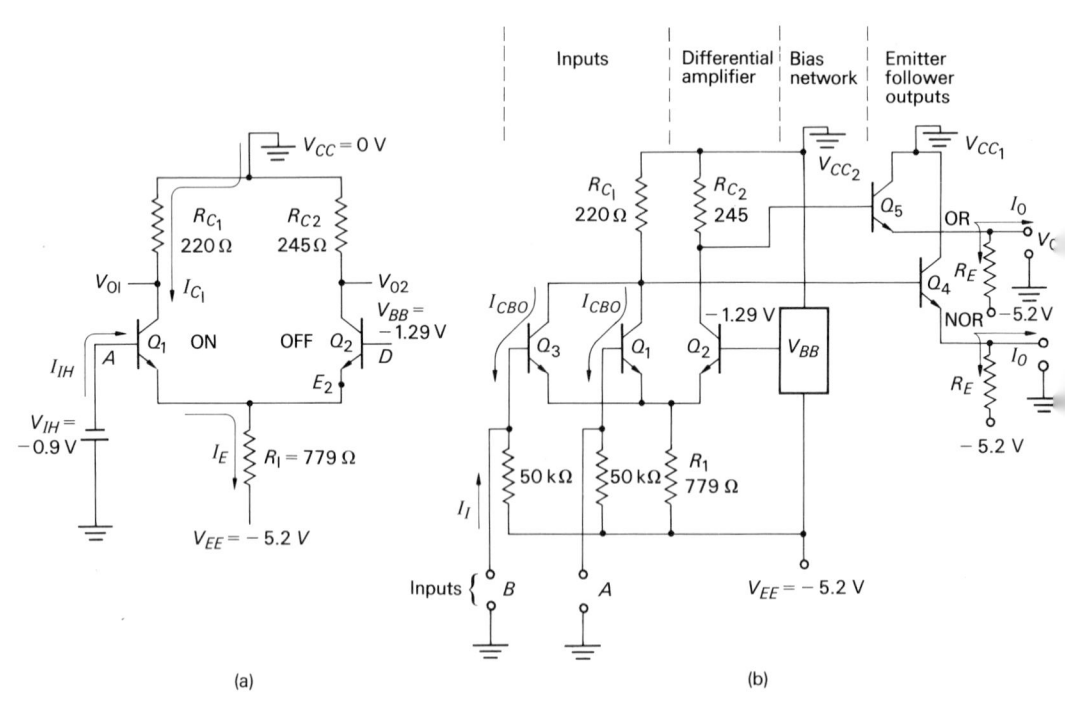

Fig. 3.26 ECL: (a) differential amplifier; (b) schematic for a 10105 OR/NOR gate.

Suppose a voltage V_{IH} of -0.9 V is applied to input A as shown in Fig. 3.26(a). The voltage drop across the base–emitter junction for this type of transistor is 0.8 V. Since Q_1 has the most positive base it offers less resistance to current from V_{CC} than Q_2. The voltage at the emitter of Q_2 is $V_{E_2} = V_{IH} + V_{BE_1} = -0.9$ V $+ (-0.8$ V$) = -1.7$ V, which is not sufficiently negative to forward-bias Q_2. Hence, Q_2 is "off" and Q_1 is "on".* If a voltage of -1.75 V is applied to input A, then the input voltage is more negative than V_{BB}. Transistor Q_2 has the most positive base voltage, so Q_2 turns on and Q_1 turns off. Hence, in a differential amplifier, the transistor with the most positive base turns on, while the other transistor turns off; in other words, the circuit can operate like a switch. By changing the voltage levels, other logic levels could be used. For example, shifting all voltages by $+2$ V would give $V_{CC} = +2.0$ V, $V_{EE} = -3.2$ V, $V_{BB} = +0.71$ V, $V_H = +1.1$ V, and $V_L = +0.25$ V.

By proper choice of resistors R_1, R_{C_1} and R_{C_2}, the "on" transistor is prevented from saturating. The output logic voltage swing is correspondingly reduced. This type of switching is called **nonsaturated switching**. The collector resistors are also chosen to allow rapid charging and discharging of parasitic capacitances. The power supply current I_E is very nearly constant, even during switching, because of the diversion of current from one transistor to the other. The input impedance of a differential amplifier is high for LOW or HIGH inputs [1, p. 589].

3.9.2 The 10K Series OR/NOR Gate

The ECL 10K series provides typical delay times of 2 ns and a power dissipation of 25 mW per gate. The trend is toward LSI ECL devices where the power dissipation per functional gate can be as low as 5 mW.

The schematic for a typical 10105 gate is shown in Fig. 3.26(b). The circuit can be divided into four parts: (1) inputs, (2) differential amplifier, (3) bias voltage network, and (4) outputs. The inputs A and B are fed into the bases of Q_1 and Q_3, respectively. The 779-Ω common emitter resistor R_1 is tied to the power supply V_{EE}. The differential amplifier consists of transistor pairs $Q_2 Q_1$ and $Q_2 Q_3$. The bias network supplies the constant voltage $V_{BB} = -1.29$ V [10]. Two outputs are available, OR and NOR. The outputs are in the emitter follower configuration in order to provide low output impedance (~ 7 Ω) and to shift the output voltage levels from the differential amplifier levels back to ECL logic levels (this is called **level shifting**). All 10K series devices have open emitters and therefore require that the user supply external emitter resistors, called **output pull-down resistors** R_E. The values of R_E range from 270 Ω to 410 Ω, for $V_{EE} = -5.2$ V. Let's see how this gate works [11].

LOW Inputs If both inputs are LOW (-1.75 V) then Q_2 has the most positive base ($V_{BB} = -1.29$ V), so Q_2 turns on while Q_1 and Q_3 turn off. Current then

---

* For ECL the term "on" means a conducting but unsaturated transistor. The term "off" means a transistor which is cut off.

flows from ground through R_{C_2}, Q_2, and R_1. Sufficient voltage (-1.0 V) is developed across R_{C_2} to turn Q_5 on. Transistor Q_5 is an emitter follower amplifier, so its emitter voltage is one diode drop below its base voltage, i.e.,

$$V_{OR} = -1.0 \text{ V} + (-0.8 \text{ V}) = -1.8 \text{ V},$$

which is a LOW.

Since Q_1 and Q_3 are off, the base of Q_4, due to leakage current, is at approximately -0.05 V, so the emitter of Q_4 is one diode drop below its base, i.e.,

$$V_{NOR} = -0.05 \text{ V} + (-0.8 \text{ V}) = -0.85 \text{ V},$$

which is a HIGH.

HIGH Inputs If one or more inputs are HIGH (-0.9 V), then the corresponding transistors Q_1 or Q_3 are turned on (Fig. 3.26b) and Q_2 is off. The voltage at the base of Q_5 is approximately -0.05 V; therefore

$$V_{OR} = -0.05 \text{ V} + (-0.8 \text{ V}) = -0.85 \text{ V},$$

which is a HIGH.

Current flows from ground through the on transistor Q_1 or Q_3, R_1 to V_{EE}. This produces approximately -1.0 V at the base of Q_4, so the emitter is one diode drop (-0.8 V) below this, i.e.,

$$V_{NOR} = -1.0 \text{ V} + (-0.8 \text{ V}) = -1.8 \text{ V},$$

which is a LOW.

Note that inputs always sink current. This current is divided between the internal pull-down resistor and the base of the input transistor. Outputs always source current from V_{CC_1}.

The next section discusses some of the important characteristics of 10K devices.

3.9.3 Characteristics of ECL Devices

The brief discussion of device characteristics in this section assumes a familiarity with the logic circuit terminology covered in Sections 3.6.3 through 3.6.6.

Transfer Curves The voltage transfer curve for a typical ECL OR gate is shown in Fig. 3.27(a). The curve is centered about the voltage $V_{BB} = -1.29$ V. Note that the curve is considerably steeper in the transition region than the TTL transfer curve (Fig. 3.10a). Several specification voltages are shown in Fig. 3.27(a) [8]. An *A* subscript refers to the value of voltage or current closest to positive infinity and a *B* subscript represents the value closest to negative infinity. Thus V_{IHA} represents the most positive (i.e., the maximum) input HIGH voltage, while V_{IHB} represents the most negative (i.e., the minimum) input HIGH voltage.

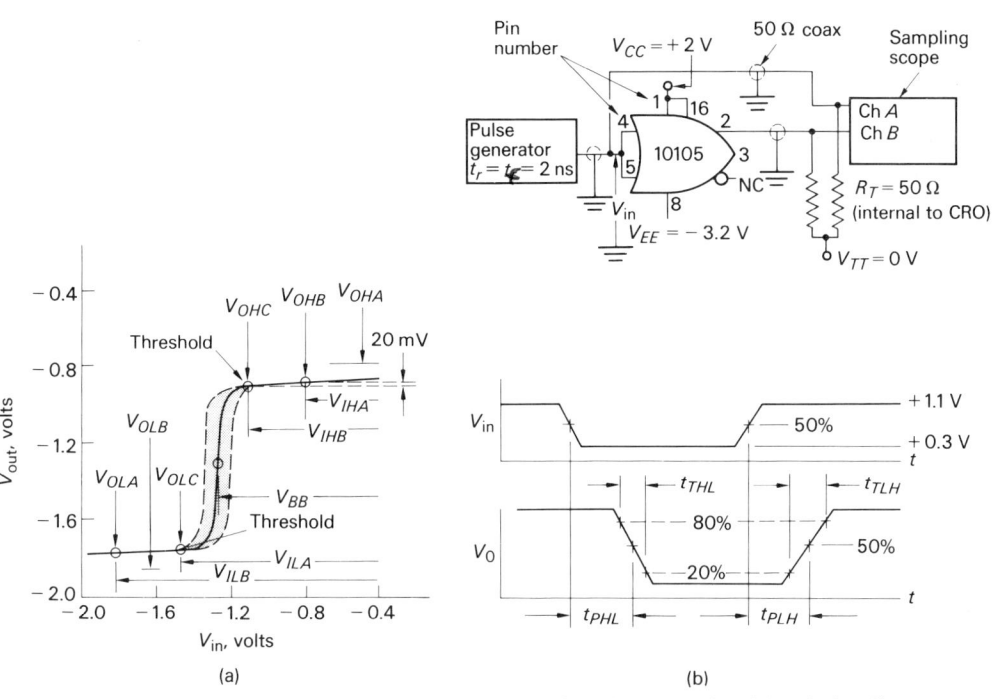

Fig. 3.27 ECL: (a) voltage transfer curve; (b) test circuit and propagation delay timing diagram.

V_{IHB} also represents the guaranteed input **HIGH threshold voltage**. The corresponding input LOW voltages are denoted V_{ILA} (maximum) and V_{ILB} (minimum), and V_{ILA} also represents the guaranteed input **LOW threshold voltage**.

Output voltages V_{OHA} and V_{OLA} represent the most positive HIGH and LOW output voltages respectively under specified input loading conditions. Voltages V_{OHB} and V_{OLB} represent the most negative output HIGH and LOW voltages respectively under specified input loading conditions. Voltage V_{OHC} is the guaranteed output HIGH threshold voltage with the inputs at their respective threshold levels. Voltage V_{IHB}, the guaranteed input HIGH threshold voltage, is the input HIGH voltage that will cause the output voltage to decrease from V_{OHB} to V_{OHC}. This change in V_{OH} is 20 mV for the 10K series. Similarly, V_{OLC} is the guaranteed output LOW threshold voltage with the inputs set at their respective threshold levels. The shaded transition region marks the sample-to-sample variations in V_{BB} of ± 65 mV for the 10K series.

Note that for ECL devices we define both input and output threshold voltages, for the HIGH and the LOW state, and not the average threshold voltage commonly used in TTL. Transfer curves and other characteristics of the 10105 gate are found in Appendix D. The data sheet for several 10K series gates is given in Fig. 3.28.

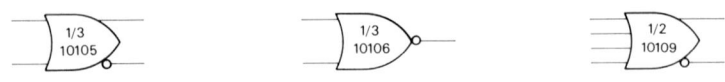

DC CHARACTERISTICS: V_{EE} = -5.2 V, V_{CC} = GND F10101 • F10102 • F10103 • F10105 • F10106
F10107 • F10109 • F10110 • F10111

SYMBOL	CHARACTERISTIC	LIMITS B	TYP	LIMITS A	UNITS	T_A	CONDITIONS	
V_{OH}	Output Voltage HIGH	-1000 -960 -900		-840 -810 -720	mV	0°C 25°C 75°C	$V_{IN} = V_{IHA}$ or V_{ILB} per Logic Function	
V_{OL}	Output Voltage LOW	-1870 -1850 -1830		-1665 -1650 -1625	mV	0°C 25°C 75°C		Loading is 50Ω to -2.0 V
V_{OHC}	Output Voltage HIGH	-1020 -980 -920			mV	0°C 25°C 75°C	$V_{IN} = V_{IHB}$ or V_{ILA} per Logic Function	
V_{OLC}	Output Voltage LOW			-1645 -1630 -1605	mV	0°C 25°C 75°C		
V_{IH}	Input Voltage HIGH	-1145 -1105 -1045		-840 -810 -720	mV	0°C 25°C 75°C	Guaranteed Input Voltage HIGH for All Inputs	
V_{IL}	Input Voltage LOW	-1870 -1850 -1830		-1490 -1475 -1450	mV	0°C 25°C 75°C	Guaranteed Input Voltage LOW for All Inputs	
I_{IH}	Input Current HIGH F10101 Lead 12 F10107 Leads 5,7,15 F10110 All Inputs F10111 All Inputs			265 550 220 435 435	µA	25°C	$V_{IN} = V_{IHA}$	
I_{IL}	Input Current LOW	0.5			µA	25°C	$V_{IN} = V_{ILB}$	

I_{EE}	Power Supply F10101
	Current F10102
	                       F10103
	                       F10105
	                       F10106
	                       F10107
	                       F10109
	                       F10110
	                       F10111
		-26
		-26
		-26
		-21
		-21
		-28
		-14
		-38
		-38
			-20
			-20
			-20
			-15
			-15
			-22
			-10
			-28
			-28
					mA	25°C	Inputs and Outputs Open	

SWITCHING CHARACTERISTICS: V_{EE} = -5.2 V, T_A = 25°C F10101 • F10102 • F10103 • F10105 • F10106 • F10109

SYMBOL	CHARACTERISTIC	B	TYP	A	UNITS	CONDITIONS
t_{PLH}	Propagation Delay, LOW to HIGH	1.0	2.0	2.9	ns	
t_{PHL}	Propagation Delay, HIGH to LOW	1.0	2.0	2.9	ns	See Figure 1
t_{TLH}	Output Transition Time LOW to HIGH (20% to 80%)	1.5	2.2	3.3	ns	
t_{THL}	Output Transition Time HIGH to LOW (80% to 20%)	1.5	2.2	3.3	ns	

Fig. 3.28 ECL 10K logic gate specifications. (Courtesy of Fairchild Camera and Instrument Corporation.)

3.9 THE ECL FAMILY

Output Short Circuit Current The outputs of ECL gates should not generally be connected directly to ground, to V_{EE}, to a logic HIGH, or to a logic LOW. Such connections could cause excessive current to flow through the output transistors. The maximum rated output current is -50 mA. The specified output current is usually for a termination voltage of $V_{TT} = -2.0$ V with a 50-Ω load (Fig. 3.27b), that is, $I_0 = (-2.0 \text{ V} - 0.9 \text{ V})/50 \text{ } \Omega = -22$ mA.

Current Spiking The use of differential amplifiers in ECL makes the supply current I_{EE} essentially constant even during the switching transition. However, one 0.01–0.1 µF decoupling capacitor should be used with every four to six IC packages to reduce currents through stray capacitances. Ceramic, mica, or other *RF* capacitors should be used.

Power Dissipation The power dissipation P_D is given by $I_{EE}(V_{CC} - V_{EE})$ and is essentially independent of the frequency at which the device is switched, since the supply current is constant. Typically, for the 10105 OR/NOR gate, $I_{EE} = 15$ mA (measured with the inputs and the outputs open) and $V_{EE} \approx 5$ V, so $P_D = 75$ mW total or 75 mW/3 = 25 mW per gate.

The V_{CC} supply inputs are connected to the most positive supply voltage. Two V_{CC} inputs are used to ensure that any changes in supply current during switching do not cause a change in V_{CC_2} due to the inductance of the V_{CC_1} IC bond wire and the package lead. Externally, V_{CC_1} and V_{CC_2} are connected to a common V_{CC} line.

Impedance The input impedance of a differential amplifier, as already discussed, is very high in either the HIGH or LOW state. Notice in Fig. 3.26(b) the 50-kΩ input resistors, called **input pull-down resistors**, connected between the logic inputs and the negative supply V_{EE}. These resistors result in a net input resistance of approximately 50 kΩ in either the HIGH or LOW state. The emitter follower outputs have a very low resistance, typically 7 Ω in either the HIGH or LOW state. The input impedance of 10K devices is predominately capacitive and is typically 3 pF [8, p. 5–6].

Propagation Delay Times The input test voltage waveform for a 10105 OR/NOR gate and the OR output waveform are shown in Fig. 3.27(b). The pulse generator must produce a pulse with 2.0 ± 0.2 ns rise and fall times. The high speed of ECL devices requires a sampling oscilloscope for accurate measurements. These scopes have a 50-Ω emitter input impedance. To avoid having two different termination resistances, R_E and R_T, the supply voltage levels are shifted so that a 50-Ω load resistor can be used. The V_{CC} inputs are connected to $+2.0$ V and V_{EE} to -3.2 V (net -5.2 V on V_{EE}). Thus all the voltages are raised by $+2.0$ V. The input HIGH voltage will then be $V_H = +1.1$ V, $V_L = +0.31$ V, and $V_{BB} = +0.71$ V. The open emitter outputs can then be directly connected to a 50-Ω coaxial cable terminated with a 50-Ω termination resistor R_T in the sampling oscilloscope. No additional pull-down resistor R_E is therefore used.

The 50% level, $(1.1 + 0.3)\text{V}/2 = 0.7 \text{ V} = V_{BB}$, is the reference level from which the delay times are measured. On the output waveform, the transition times (rise and fall times) are measured between the 20% and 80% voltage levels. The typical propagation delay times for the 10105 are $t_{PLH} = t_{PHL} = 2.0$ ns. The typical output transition times are $t_{TLH} = t_{THL} = 2.2$ ns.

Noise Immunity The HIGH state noise margin is $V_{NH} = V_{OHB} - V_{IHB} = -0.960 \text{ V} - (-1.105 \text{ V}) = 145 \text{ mV}$. The LOW state noise margin is $V_{NL} = V_{ILA} - V_{OLA} = -1.475 \text{ V} - (-1.650 \text{ V}) = 175 \text{ mV}$. Slightly more conservative values are obtained when V_{OHC} is used in place of V_{OHB} and when V_{OLC} is used in place of V_{OLA}. One disadvantage of ECL is its low noise margin. However, its internal noise due to current spiking is much less than for TTL. Their low noise margin means that ECL circuits must be well shielded from noisy environments.

Unused Inputs The purpose of the input pull-down resistors is to provide a path to V_{EE} for the transistor leakage currents I_{CBO} (Fig. 3.26b). They also prevent the buildup of charge on input capacitances. Hence most nonessential ECL inputs are designed to be active LOW. The presence of these pull-down resistors eliminates the need to externally wire unused inputs to a LOW in order to make the inputs active.

Loading Effects The low output impedance and high input impedance give ECL devices a large dc fan-out of 100. However, noise margins and ac loading reduce this number considerably in a practical system. The type of termination also affects fan-out [8, Ch. 5].

Capacitive loading causes reduced device speed due to the increase in the rise time of the output waveform. For example, a 60-pF load can add 2 ns to the nonloaded delay time [8, p. 4–8]. Each gate input adds about 3 pF [11, p. 141], so if the output is loaded with just 20 gate loads it more than doubles the propagation delay time.

Wire-ORing Since 10K series devices have open emitters, their outputs can be connected together, which results in the wire-OR operation. Thus if the outputs

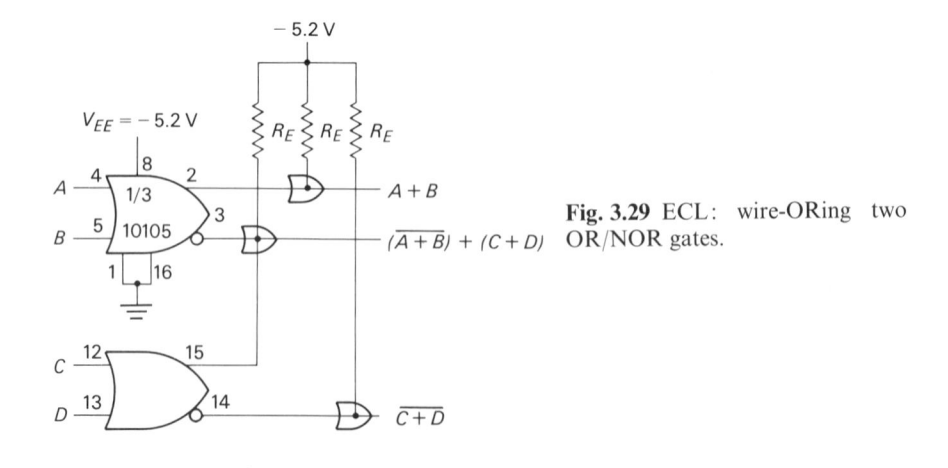

Fig. 3.29 ECL: wire-ORing two OR/NOR gates.

of two two-input 10105 OR/NOR gates are wired together, as illustrated in Fig. 3.29, the output functions shown result. Other functions can be obtained by wiring different outputs together. Using the wire-OR feature often saves gate and package count, eliminating the need for inverters for example.

Internally, ECL circuits can use the techniques of **wired collectors** and **series gating** to provide high performance and low part count [8, p. 2–4]. These internal connections cannot be changed by external connections.

The design of high-speed ECL systems is a complex subject involving transmission line concepts and is treated in the references [8, 11].

SUMMARY

In this chapter we discussed the main switching properties of junction diodes, Schottky diodes, and bipolar and MOS transistors. These devices we essentially treated as conducting (on) or nonconducting (off) devices.

We discussed the TTL, ECL, and CMOSL families in some detail. Their main characteristics are summarized in Table 3.1. The values given must be interpreted in terms of the factors which affect them, as discussed in this chapter. The last entry in the table, the clock rate, refers to the maximum frequency at which a flip-flop can operate.

Table 3.1 Typical Average Logic Family Electrical Characteristics

Parameter	Units	TTL(74) Standard	TTL(74) Schottky	TTL(74) Schottky Low Power	ECL 10K	CMOSL 14000
Basic gate		NAND	NAND	NAND	OR/NOR	NOR/NAND
Gate delay time	ns	10	3	5	2	60
P_D (per gate)	mW	10	19	2	25	6 nW
dc noise margin[a]	mV	400	300	300	145	1500
dc fan-out[b]		10	10	20	100	>50
R_{IH}/R_{IL}	kΩ	400/4	100/3	400/4	50/50	~$10^9/10^9$
R_{OH}/R_{OL}	Ω	70/10	50/8	50/30	7/7	1000/1000
Supply voltage	V	5	5	5	−5.2	5–15
V_{IH}/V_{IL}	V	3/0.4	3/0.4	3/0.4	−1.10/−1.66	3/0.7
V_{OH}/V_{OL}	V	3.4/0.2	3.2/0.2	3.4/0.35	−0.88/−1.75	5/0
I_{IH}/I_{IL}[c]	μA	40/1600	50/2000	20/0.36	265/0.5[a]	~pA
I_{OH}/I_{OL}[c]	mA	0.4/16	1/20	0.4/8	50/50	2[d]/1[d]
Clock rate	MHz	35	125	50	200	10

[a] Minimum values.

[b] Minimum number of loads a device can drive in its own series.

[c] Maximum values

[d] Typical values.

90   DIGITAL CIRCUITS

The main advantages of TTL are fairly high speed, moderate power dissipation, variety of series available, and large variety of SSI and MSI devices available. The advantages of the ECL family are its very high speed, wire-OR capability, complementary outputs, and low noise generation. The CMOSL family offers very low power dissipation (at low frequencies), high noise immunity, and tolerance to wide variations in supply voltage.

Other logic families such as diode transistor logic (DTL) [1] and integrated injection logic (I^2L) [24, 25] are treated in the references.

PROBLEMS

Section 3.2

1. A diode has +3 V at its anode and +5 V at its cathode. (a) Draw the circuit. (b) Is the diode forward- or reverse-biased?

2. What is the advantage of a Schottky diode over a junction diode for high-speed devices?

Section 3.3

3. Draw an *n-p-n* transistor in the common emitter configuration. Label all voltages and currents.

Section 3.5

4. Draw the input and output voltage waveforms for a switching transistor. Label delay, rise, fall, and storage times. Use an ideal input pulse.

Section 3.6.1

5. List two reasons why binary voltage levels have a range of voltage values instead of only two specific values.

Section 3.6.2

6. Briefly explain how a 7400 NAND gate works in terms of each transistor being either on or off.

7. Using a TTL data book, explain how a 7408 AND gate works in terms of each transistor being either on or off.

Section 3.6.3

8. Explain how a 7400 NAND gate works in terms of its voltage transfer curve.

Section 3.6.4

9. (a) What causes current spiking in totem-pole outputs? (b) How are they prevented from propagating through a digital system? (c) Calculate the HIGH and LOW supply currents for a 7400 device. (Remember that each gate draws one-fourth of the total supply current.)

PROBLEMS 91

10. (a) Draw a timing diagram that illustrates propagation delay times. (b) Give typical values for t_{PLH} and t_{PHL} for a 7400 gate.
11. *Signal Generator.* Refer to Appendix F, the SL2 Phase Detector schematic diagram. Calculate the maximum average delay time from pin 1 in IC U4A to pin 8 in IC U5C.
12. Calculate the noise margins for a 7400 gate if excessive loading causes the minimum output HIGH voltage to decrease to 2.3 V and the maximum input LOW voltage to increase to 0.6 V.
13. *Microcomputer.* Refer to sheet 1 of the IMP-16C schematic diagram in Appendix G. Using a data book, calculate the maximum average current and power dissipation required by all ICs in the "Conditional Jump Multiplexer and Control Flag Logic" circuit.

Section 3.6.5

14. With the aid of a timing diagram, explain how a capacitive load decreases the maximum speed at which a gate can operate.
15. Explain why the input currents for a gate with N emitters are $N \times I_{IH}$ and $1 \times I_{IL}$.

Section 3.6.6

16. A control circuit requires the implementation of the AND function $X = ABCDEF$. Two-input open collector AND gates (7409) are available. Calculate the external pull-up resistor maximum value $R_{L(max)}$ and minimum value $R_{L(min)}$ when two loads are being driven. The formulas are

$$R_{L(max)} = \frac{V_{CC(min)} - V_{OH(min)}}{N_W I_{OH(max)} + N_L I_{IH(max)}}$$

and

$$R_{L(min)} = \frac{V_{CC(max)} - V_{OL(max)}}{I_{OL(max)} - N_L I_{IL(max)}},$$

where N_W is the total number of gates wired together and N_L is the number of loads being driven. [The data book currents are used without regard to their sign in these formulas.] Figure P3.1 illustrates NAND gates for the case $N_W = 2$ and $N_L = 3$.

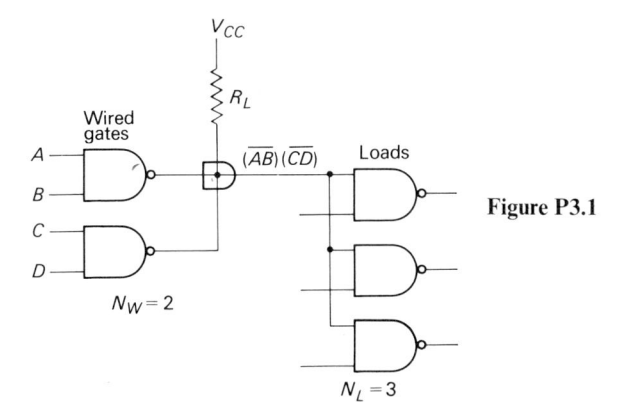

Figure P3.1

92   DIGITAL CIRCUITS

17. What is the advantage of implementing a function using wire-ANDing as compared with cascading totem-pole gates?

Section 3.6.7

18. Explain why only one output of three-state devices connected to a data bus can be in the low-impedance state at one time.
19. Explain how the three-state NOT gate shown in Fig. 3.17(a) operates like an ordinary NOT gate when the control input is LOW.

Section 3.6.8

20. Briefly explain how a 74S00 gate schematic differs from a 7400 gate and why this results in a faster gate.
21. List three factors that are important in the construction of digital systems that use Schottky devices.

Section 3.6.9

22. Briefly explain the differences between the schematic of a 74LS00 and a 74S00 gate.
23. (a) List five advantages of the LS series as compared with the standard series. (b) How many 7400 gates can one 74LS00 drive? (c) Referring to Appendix D, what is the power dissipation for a 74LS00 for a load capacitance of 15 pF and 50 pF at 1 MHz and at 30 MHz?

Section 3.7

24. (a) Draw a circuit for biasing a *p*-channel MOS transistor using a +10 V supply. (b) Make a table showing whether the transistor conducts (on) or does not conduct (off) when 0 V and +10 V are applied to the gate. (c) Briefly explain how an *n*-channel MOS transistor works with the aid of a drawing of its physical structure and characteristic curves.
25. Make a table that compares the advantages and disadvantages of MOS transistors versus bipolar transistors.

Section 3.8

26. (a) List the typical supply voltage range for a CMOSL gate. (b) List the input and output logic voltages for a 10-V supply.
27. List two general types of applications where CMOSL is especially useful.

Section 3.8.1

28. Explain how a CMOSL NOT gate works.
29. Using a CMOSL data book, draw the schematic of a single 14012 NAND gate and explain how it works.

Section 3.8.2

30. Make a table that shows the power dissipation per gate for a CMOSL 14001 operated at 1 kHz, 1 MHz, and 5 MHz for capacitive loads of 25 pF and 50 pF. The supply voltage is 10 V. [*Hint*: Use Fig. 3.23.]

31. How many TTL standard and low-power Schottky series gates can one 14001 CMOSL gate drive with $V_{DD} = 5$ V? (Use typical values.)

32. Referring to the CMOSL curves in Appendix D, determine the typical average propagation delay time for a 14501 NOR gate driving one 74LS00 gate when the load capacitance is 100 pF and $V_{DD} = +5$ V.

33. (a) Briefly explain why the power supply should be turned on before applying a logic signal to a CMOSL device. (b) Explain why logic signals should be removed before turning off the power supply. Use schematics to aid your explanation. (c) Draw the physical structure of a CMOS transistor and explain the SCR condition.

Section 3.9

34. How do the ECL devices achieve their high speed?

Section 3.9.2

35. (a) Explain how the ECL OR/NOR gate works for a HIGH input. (b) Show the direction of all currents. (c) Show how the HIGH OR and NOR output voltages are arrived at.

Section 3.9.3

36. Referring to the typical 10K curves in Appendix D, determine the output HIGH and LOW currents for typical output voltages, when $V_{EE} = -2$ V and the load is 50 Ω.

37. (a) Draw a logic diagram that shows the best implementation of the functions $X = A + B + C + D$ and $Y = \overline{(A + B)} + \overline{(C + D)}$ using 10105 gates. (b) Draw the implementation of X and Y using totem-pole two-input TTL OR and NOR gates. (c) Compare the package count and gate count for TTL versus ECL.

38. The **speed power product** in picojoules (pJ) is the product of the typical average propagation delay time in nanoseconds and the average power dissipation in milliwatts. A low value for the speed-power product is desirable because this means the device has a high speed and a low power dissipation. Calculate the speed-power product for a typical logic gate in each of the following families: standard TTL, Schottky TTL, low-power Schottky, 10K ECL, and 14000 CMOSL. (a) Plot these speed-power products on a graph that shows delay time on the vertical axis and power dissipation on the horizontal axis. Use log-log graph paper. (b) Which logic family has the best speed-power product?

Summary

39. Make a table that gives the main advantages of the TTL, ECL, and CMOSL families.

REFERENCES

1. Jacob Millman and Christos C. Halkias. *Integrated Electronics: Analog and Digital Circuits and Systems*. New York: McGraw-Hill, 1972.

2. Lane S. Garrett. "Integrated-Circuit Digital Logic Families, Part II: TTL Devices." *IEEE Spectrum*, pp. 63–72, November 1970.

3. Robert L. Morris and John R. Miller (eds.). *Designing with TTL Integrated Circuits*. New York: McGraw-Hill, 1971.

4. Peter Alfke and Ib Larsen (eds.). *The TTL Applications Handbook.* Fairchild Semiconductor, August 1973.
5. *Interface Integrated Circuits.* National Semiconductor Corp., 1974.
6. Gene Cavinaugh. "Using Schottky to Upgrade Performance of Your TTL Systems." *EDN*, pp. 38–42, September 20, 1973.
7. Tom Ormand. "IC Logic Families: They Keep Growing in Number and Size." *EDN*, pp. 30–35, October 5, 1974.
8. Charles H. Alford. *The ECL Handbook.* Fairchild Semiconductor, July 1974.
9. Lane S. Garrett. "Integrated-Circuit Digital Logic Families, Part III: ECL and MOS Devices." *IEEE Spectrum*, pp. 30–42, December 1970.
10. H.H. Muller, *et al.* "Fully Compensated Emitter-Coupled Logic." *Progress* (Fairchild Semiconductor), July 1973.
11. William R. Blood, Jr. *MECL System Design Handbook.* Motorola Inc., December 1972.
12. Joe Kroeger. "Review the Basics of MOS Logic." *Electronic Design*, pp. 98–105, March 15, 1974.
13. Jack McCullen. *Motorola Complementary MOS I/Cs.* AN-538A, Motorola Inc., 1972.
14. *MOS Users Guide.* Signetics Applications Handbook, pp. 7–1 to 7–33, 1974.
15. *COS/MOS Integrated Circuits Manual.* RCA Corp., June 1972.
16. *McMOS Data Book.* Motorola Corp., December 1975.
17. *McMOS Handbook,* Motorola Corp., 1974.
18. Edward A. Torrero. "Focus on CMOS." *Electronic Design*, pp. 86–95, March 15, 1974.
19. Jerry Tonn. *Introduction to CMOS ICs with 3-State Outputs.* AN-715, Motorola Inc., 1974.
20. Dick Funk. "Handling CMOS: Some Guidelines." *Electronic Products Magazine*, pp. 59–61, August 19, 1974.
21. Laurence Altman (ed.). "Special Report: CMOS Enlarges Its Territory." *Electronics*, pp. 77–88, May 15, 1975.
22. Ken Stephenson. "CMOS: Opportunities Are Limited—If You Ignore the Failure Modes." *EDN*, pp. 50–54, June 5, 1975.
23. Ken Stephenson. "CMOS: Opportunities Unlimited—If You Know the Basic Ground Rules." *EDN*, pp. 41–25, February 5, 1975.
24. C.M. Hart, *et al.* "Bipolar LSI Takes a New Direction with Integrated Injection Logic." *Electronics*, pp. 111–118, October 3, 1974.
25. Horst H. Berger and Siegfried K. Wiedmann. "The Bipolar LSI Breakthrough, Part I: Rethinking the Problem." *Electronics*, pp. 89–95, September 4, 1975.
26. Peter Alfke and Charles Alford. "Low Power Schottky TTL." *Progress* (Fairchild Semiconductor), pp. 3–9, September/October, 1975.

4: Logic Gate Applications

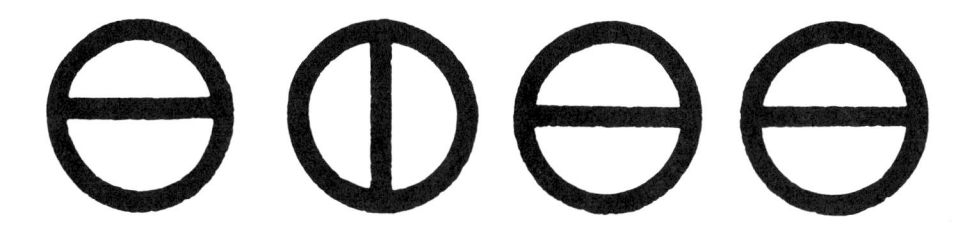

4.1 INTRODUCTION

In this chapter we shall give some practical examples of logic gates and introduce three additional logic gates; the AND-OR-INVERT gate, the EXCLUSIVE-OR gate, and the EXCLUSIVE-NOR gate. Next we shall discuss drawing logic diagrams from truth tables and rules of logic. We shall give a simple method of drawing logic diagrams using only NAND or NOR gates. This is important because some logic families have predominately NAND or NOR gates available. In the last section we shall introduce the *JK* flip-flop and give several important additional applications.

4.2 APPLICATIONS USING AND, OR, AND NOT GATES

Our first example illustrates the development of the control logic for a digital door lock.

Example 4.1 *Digital Door Lock Control.* Develop the control logic for an electronic door lock that will open the lock only when the correct one of 32 combinations is selected. If an incorrect combination is tried an alarm will sound.

Solution. We can use toggle switches for setting the correct combination. Since there are to be 32 combinations, we shall need five switches ($2^5 = 32$). Next we need to choose a specific combination of five switches, labeled *A*, *B*, *C*, *D*, and *E*, that will open the lock. Let's select $A = B = H$, $C = L$, $D = H$, and $E = L$ for the correct combination. The output of the five-input AND gate, shown in Fig. 4.1(a), will go HIGH when the correct combination is chosen. This HIGH output can then be used to open the lock. Note that any of the other 31 combinations of inputs will result in a LOW output from the AND gate, so the lock will not open.

Next an alarm system must be added. The alarm should not go on while the five switches are being set. A pushbutton switch *F* is added so that when it is pushed, either the alarm will sound, or the door will open if the correct combination has been set. The complete door lock control system is shown in Fig. 4.1(b). If the correct combination is set with toggle switches *A* through *E*, then the output of

96   LOGIC GATE APPLICATIONS   4.2

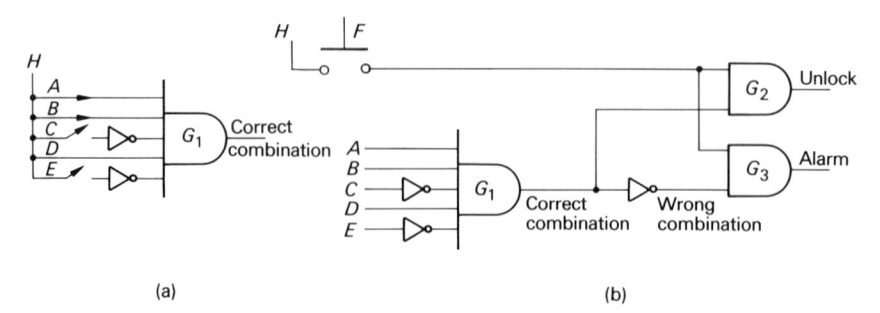

Fig. 4.1 Digital door lock control: (a) lock combination; (b) complete system.

gate G_1 is HIGH, so when F is made HIGH the output of gate G_2 is HIGH and the door unlocks. If the wrong combination is used, the output of gate G_1 will be LOW, so when switch F is pushed the door will not unlock and G_3 will go HIGH, sounding the alarm.

We could also add an *RS* latch to the alarm output so that if one incorrect combination is tried, the alarm will continue to sound even if the correct combination is eventually set; see Problem 2.

The probability of guessing the correct combination can be reduced by increasing the number of switch inputs. For example, six switches give $2^6 = 64$ combinations, seven switches give $2^7 = 128$ combinations, etc. ■

Fig. 4.2 Optical card reader. (Courtesy of Hewlett-Packard Company.)

4.2   APPLICATIONS USING AND, OR, AND NOT GATES

The next example shows how logic gates can be used in an automatic punched card reader, such as the one shown in Fig. 4.2. This reader can read 300 punched cards per minute. Each card can have up to 80 characters punched in it.

Example 4.2   *Optical Punched Card Reader.* The students in an electronics class are required to attend a group of five special lectures in order to complete a course. Each student is given a card that will be punched for each lecture he attends. Devise a logic system that will automatically detect any student who has missed one or more lectures.

Solution. A card with a specific location for punching a hole for each lecture attended is shown in Fig. 4.3(a). A hole corresponds to attending a lecture, the absence of a hole indicates a missed lecture. Figure 4.3(b) shows an optical system for reading cards. When a card is inserted in the reader, the photodetectors will receive light from their corresponding lamps only if a hole is punched in the card. Light shining through the hole onto the photodetector produces a HIGH output. Thus a missed lecture will result in a LOW output from the photodetector. The card in Fig. 4.3(a) shows that lectures 1 and 3 were missed.

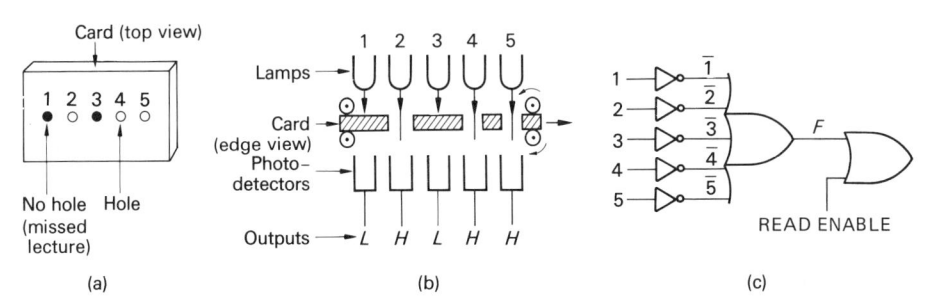

Fig. 4.3 Punched card sorter: (a) punched card, top view; (b) optical sorter, edge view; (c) detector logic diagram.

The automatic detector must have a HIGH output if one or more of its inputs from the card sorter are LOW. Figure 4.3(c) shows the solution. The five inverters ensure that *F* goes HIGH only if one or more inputs are LOW. ■

The last example in this section illustrates how logic gates can be used in an automobile warning system.

Example 4.3   *Automobile Warning System.* Develop an alarm system that turns on a buzzer alarm if the ignition switch is on and the shift is not in park and the seat is occupied and the seat belt is not buckled and the door is open.

Solution. A five-input AND gate is required. The logic diagram is shown in Fig. 4.4(a). The alarm sounds if all five conditions in the statement of the problem

98   LOGIC GATE APPLICATIONS   4.3

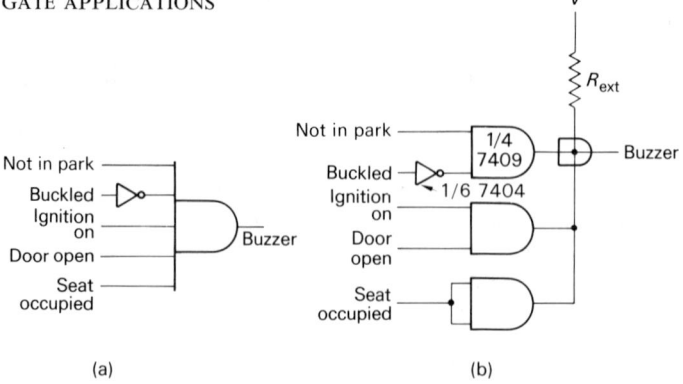

Fig. 4.4 Automobile warning system implementation using (a) one five-input AND gate, (b) a TTL 7409 wire-AND gate, (c) cascaded two-input AND gates.

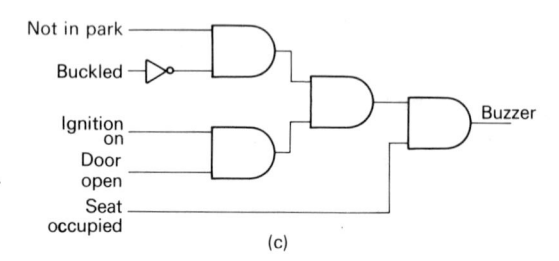

are met. For example, if the driver is in his seat (H) and the ignition switch is on (H) and the shift is not in park (H) and the seat belt is unbuckled (L) and the door is open (H), the buzzer will sound. This problem can be modified so that the buzzer sounds only when certain combinations of inputs occur.

The next section shows how to cascade gates and introduces a new gate called an AND-OR-INVERT (AOI) gate.

4.3 CASCADED AND AOI GATES

Suppose a five-input AND gate were not available to implement the automobile warning system of Example 4.3. In general, three alternatives exist for obtaining larger fan-ins: (1) use wired gates (wire-AND or wire-OR), (2) cascade gates together, or (3) use AOI gates.

Figure 4.4(b) shows the TTL wire-AND implementation of the automobile warning system using open collector gates which we discussed in Chapter 3. This implementation requires one 7404 NOT gate package, a 7409 AND gate package, and an external pull-up resistor R_{ext}.

4.3.1 Cascaded Gates

The connection of several levels of devices together is called **cascading**. Figure 4.4(c) shows the implementation of the automobile warning system by cascading three levels of two-input AND gates to produce one five-input AND gate.

4.3 CASCADED AND AOI GATES

The disadvantage of cascading devices is the additional delay time incurred for each level added; also, the power dissipation will be greater than for the five-input gate implementation. Thus the cascaded implementation, shown in Fig. 4.4(c), has two more delay times than the wire-AND or five-input gate implementation. For the automobile warning system application, the reduced speed is of no consequence. The wire-AND implementation has the disadvantage of requiring an external resistor, and it is slower than the five-input gate implementation. The disadvantage of the five-input gate implementation is that a five-input AND gate may not be available in the logic family used, or if it is manufactured, it may not be readily available to the user.

A gate with any number of inputs can be obtained by cascading a sufficient number of gates together as long as the current drive capabilities of the devices are not exceeded. Figure 4.5(a) shows a six-input OR gate made by cascading five two-input OR gates together.

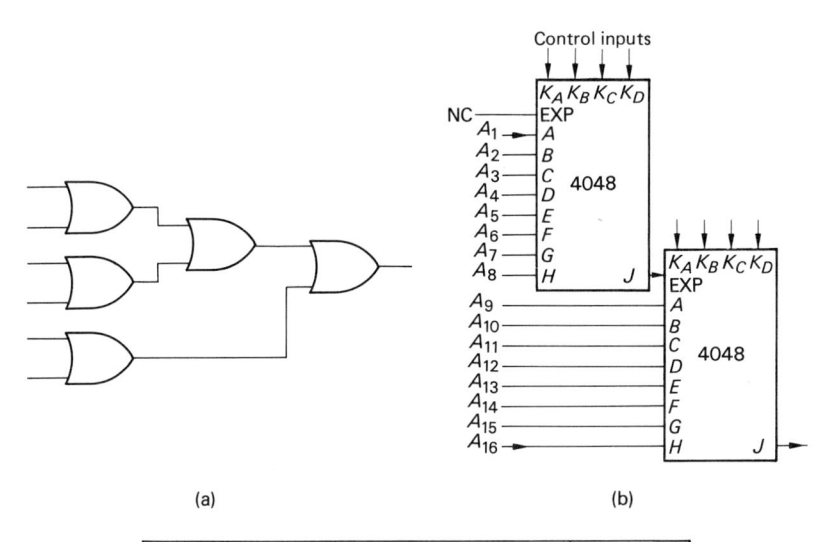

Output function	Boolean expression	K_A	K_B	K_C	K_D
NOR	$J=\overline{A+B+C+D+E+F+G+H}$	0	0	0	0
OR	$J=A+B+C+D+E+F+G+H$	0	0	1	0
OR/AND	$J=(A+B+C+D)\cdot(E+F+G+H)$	0	1	0	0
OR/NAND	$J=\overline{(A+B+C+D)\cdot(E+F+G+H)}$	0	1	1	0
AND	$J=ABCDEFGH$	1	0	0	0
NAND	$J=\overline{ABCDEFGH}$	1	0	1	0
AND/NOR	$J=\overline{ABCD+EFGH}$	1	1	0	0
AND/OR	$J=ABCD+EFGH$	1	1	1	0

Fig. 4.5 Cascading gates: (a) six-input OR gate; (b) 16-input CMOSL 4048 multifunction gate; (c) function table for the 4048 gate.

100   LOGIC GATE APPLICATIONS   4.3

Some devices have special inputs, called **expand inputs**, which make cascading easy. Figure 4.5(b) shows a CMOSL three-state 4048 multifunction gate with eight inputs. The expand input (EXP) allows two or more devices to be cascaded by connecting the J output of one device to the EXP input of the next device. The control inputs K_A, K_B, and K_C determine the output function. For example, if $K_A = K_B = L$ and $K_C = H$, the output J is the OR function, as shown in Fig. 4.5(c). The input K_D controls the three-state output. When $K_D = H$, the output is in its high-impedance state; when $K_D = L$, the output is in its low-impedance state.

4.3.2 AOI Gates

An **AND-OR-INVERT (AOI) gate** consists of several AND gates connected to one OR gate followed by an INVERTER gate, as shown in Fig. 4.6(a). Most AOI gates also have an expand input that allows expander gates to be connected to it in order to further increase the fan-in.

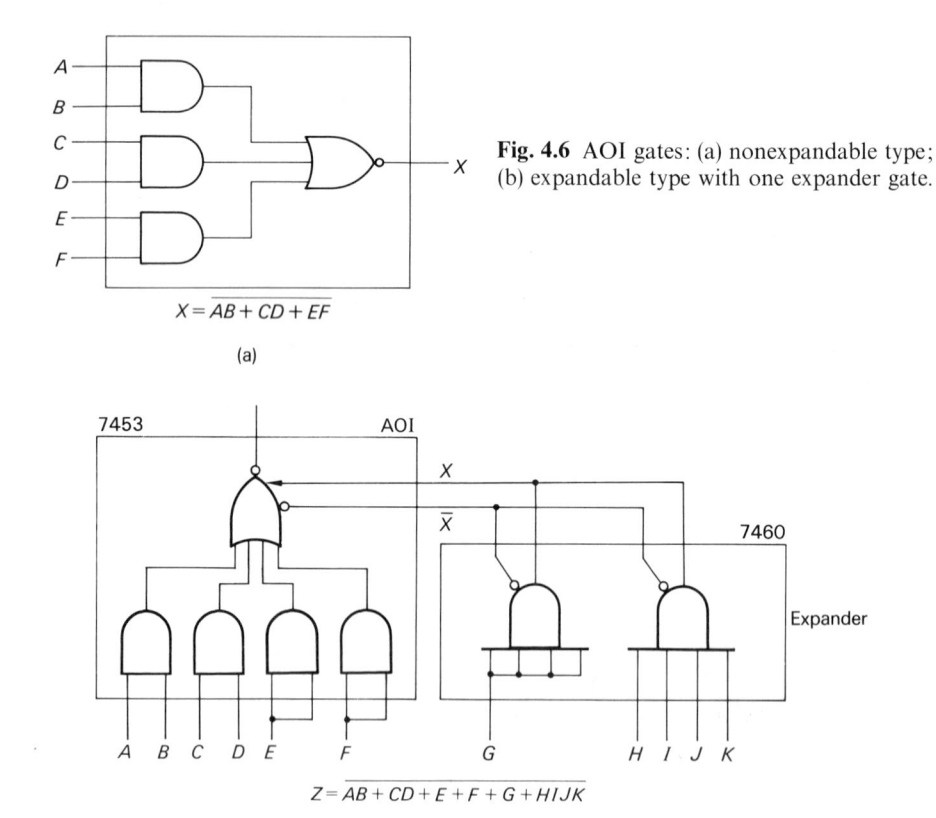

Fig. 4.6 AOI gates: (a) nonexpandable type; (b) expandable type with one expander gate.

$X = \overline{AB + CD + EF}$

(a)

$Z = \overline{AB + CD + E + F + G + HIJK}$

(b)

ELECTRICAL CHARACTERISTICS

Characteristic	Symbol	V_{DD} Vdc	T_{low}* Min	T_{low}* Max	25°C Min	25°C Typ	25°C Max	T_{high}* Min	T_{high}* Max	Unit
Output Voltage "0" Level	V_{out}	5.0	–	0.01	–	0	0.01	–	0.05	Vdc
		10	–	0.01	–	0	0.01	–	0.05	
		15	–	0.05	–	0	0.05	–	0.25	
"1" Level		5.0	4.99	–	4.99	5.0	–	4.95	–	Vdc
		10	9.99	–	9.99	10	–	9.95	–	
		15	14.95	–	14.95	15	–	14.75	–	
Noise Immunity #	V_{NL}									Vdc
($\Delta V_{out} \leq$ 0.8 Vdc)		5.0	1.5	–	1.5	2.25	–	1.4	–	
($\Delta V_{out} \leq$ 1.0 Vdc)		10	3.0	–	3.0	4.50	–	2.9	–	
($\Delta V_{out} \leq$ 1.5 Vdc)		15	4.5	–	4.5	6.75	–	4.4	–	
($\Delta V_{out} \leq$ 0.8 Vdc)	V_{NH}	5.0	1.4	–	1.5	2.25	–	1.5	–	Vdc
($\Delta V_{out} \leq$ 1.0 Vdc)		10	2.9	–	3.0	4.50	–	3.0	–	
($\Delta V_{out} \leq$ 1.5 Vdc)		15	4.4	–	4.5	6.75	–	4.5	–	
Output Drive Current (AL Device)	I_{OH}									mAdc
(V_{OH} = 2.5 Vdc) Source		5.0	-0.62	–	-0.50	-1.7	–	-0.35	–	
(V_{OH} = 9.5 Vdc)		10	-0.62	–	-0.50	-0.9	–	-0.35	–	
(V_{OH} = 13.5 Vdc)		15	-1.8	–	-1.5	-3.5	–	-1.1		
(V_{OL} = 0.4 Vdc) Sink	I_{OL}	5.0	0.50	–	0.40	0.78	–	0.28	–	mAdc
(V_{OL} = 0.5 Vdc)		10	1.1	–	0.90	2.0	–	0.65	–	
(V_{OL} = 1.5 Vdc)		15	4.2	–	3.4	7.8	–	2.4	–	
Output Drive Current (CL/CP Device)	I_{OH}									mAdc
(V_{OH} = 2.5 Vdc) Source		5.0	-0.23	–	-0.20	-1.7	–	-0.16	–	
(V_{OH} = 9.5 Vdc)		10	-0.23	–	-0.20	-0.9	–	-0.16	–	
(V_{OH} = 13.5 Vdc)		15	-0.69	–	-0.60	-3.5	–	-0.48	–	
(V_{OL} = 0.4 Vdc) Sink	I_{OL}	5.0	0.23	–	0.20	0.78	–	0.16	–	mAdc
(V_{OL} = 0.5 Vdc)		10	0.60	–	0.50	2.0	–	0.40	–	
(V_{OL} = 1.5 Vdc)		15	1.8	–	1.5	7.8	–	1.2	–	
3-State Output Leakage Current	I_{TL}	5.0	–	±0.05	–	±0.0001	±0.05	–	±3.0	μAdc
(AL Device)		10	–	±0.1	–	±0.0002	±0.1	–	±6.0	
(C_L = 15 pF)		15	–	±0.2	–	±0.0003	±0.2	–	±12	
3-State Output Leakage Current	I_{TL}	5.0	–	±0.5	–	±0.0001	±0.5	–	±7.0	μAdc
(CL/CP Device)		10	–	±1.0	–	±0.0002	±1.0	–	±14	
(C_L = 15 pF)		15	–	±2.0	–	±0.0003	±2.0	–	±28	
Input Current	I_{in}	–	–	–	–	10	–	–	–	pAdc
Input Capacitance (V_{in} = 0)	C_{in}	–	–	–	–	5.0	–	–	–	pF
Quiescent Dissipation (AL Device)	P_Q	5.0	–	0.0025	–	0.000025	0.0025	–	0.15	mW
		10	–	0.01	–	0.00010	0.01	–	0.60	
		15	–	0.03	–	0.00023	0.03	–	1.8	
Quiescent Dissipation (CL/CP Device)	P_Q	5.0	–	0.025	–	0.000025	0.025	–	0.75	mW
		10	–	0.10	–	0.00010	0.10	–	3.0	
		15	–	0.30	–	0.00023	0.30	–	9.0	
Power Dissipation **†	P_D	5.0	P_D = (1.0 mW/MHz) f + P_Q							mW
(Dynamic plus Quiescent		10	P_D = (4.0 mW/MHz) f + P_Q							
(C_L = 15 pF)		15	P_D = (10 mW/MHz) f + P_Q							

SWITCHING CHARACTERISTICS* (C_L = 15 pF, T_A = 25°C)

Characteristic	Symbol	V_{DD}	Typ All Types	Max AL Device	Max CL/CP Device	Unit
Output Rise Time	t_r					ns
t_r = (3.0 ns/pF) C_L + 25 ns		5.0	70	175	200	
t_r = (1.5 ns/pF) C_L + 12 ns		10	35	75	110	
t_r = (1.1 ns/pF) C_L + 8 ns		15	25	55	80	
Output Fall Time	t_f					ns
t_f = (1.5 ns/pF) C_L + 47 ns		5.0	70	175	200	
t_f = (0.75 ns/pF) C_L + 24 ns		10	35	75	110	
t_f = (0.55 ns/pF) C_L + 17 ns		15	25	55	80	
Data Propagation Delay Time	t_{PLH}					ns
t_{PLH} = (1.0 ns/pF) C_L + 225 ns		5.0	240	360	480	
t_{PLH} = (0.53 ns/pF) C_L + 72 ns		10	80	135	175	
t_{PLH} = (0.36 ns/pF) C_L + 50 ns		15	55	110	140	
t_{PHL} = (1.1 ns/pF) C_L + 225 ns	t_{PHL}	5.0	240	360	480	ns
t_{PHL} = (0.41 ns/pF) C_L + 74 ns		10	80	135	175	
t_{PHL} = (0.26 ns/pF) C_L + 51 ns		15	55	110	140	
Expand Propagation Delay Time	t_{PLH}					ns
		5.0	125	270	330	
		10	45	90	110	
		15	30	60	90	
	t_{PHL}	5.0	170	270	330	ns
		10	65	90	110	
		15	45	60	90	
Inhibit Propagation Delay Time	t_{PLH}					ns
		5.0	165	360	400	
		10	70	135	175	
		15	45	100	150	
	t_{PHL}	5.0	200	360	400	ns
		10	80	135	175	
		15	50	100	150	
3-State Propagation Delay Time						
"1" to High Impedance	$t_{"1"H}$	5.0	46	125	150	ns
		10	30	75	100	
		15	23	60	90	
"0" to High Impedance	$t_{"0"H}$	5.0	77	125	150	ns
		10	39	75	100	
		15	30	60	90	
High Impedance to "1"	$t_{H"1"}$	5.0	93	250	300	ns
		10	36	100	120	
		15	30	60	90	
High Impedance to "0"	$t_{H"0"}$	5.0	158	250	300	ns
		10	55	100	120	
		15	40	60	90	

*The formula given is for the typical characteristics only.

Fig. 4.7 Specifications for the CMOSL 14506 AOI gate. (Courtesy of Motorola Corporation.)

By connecting various AOI gate inputs together any logic function can be generated. For example, to generate the function

$$Z = \overline{AB + CD + E + F + G + HIJK},$$

using TTL gates, one four-wide two-input (i.e., four gates at the input, with two inputs per gate) expandable 7453 AOI gate and one 7460 expander gate can be used, as shown in Fig. 4.6(b). The outputs of the gates in the expander can be connected together because they have open collector outputs. A total of four 7460 expanders (eight gates total) can be connected to the expand inputs, X and $\bar{X}$, of the 7453.

Without the expander connected, the 7453 has an average typical supply current of 4.55 mA and an average typical delay time of 10.5 ns. The average typical delay time of the expander, through the 7453, is 12.5 ns.

Figure 4.7 shows the specifications for the CMOSL 14506 dual two-wide two-input expandable AOI gate. This device has a three-state output, controlled by a disable input, for connection to a common bus. Notice the rapid decrease in delay time with increasing supply voltage V_{DD}. Formulas are given, under delay time, for calculating the delay time for any capacitive load C_L. The logic diagram for the 14506 is given in the problems.

The advantage of AOI gates over open-output or cascaded gates is their higher speed and lower power dissipation. Their disadvantage is that they are harder to troubleshoot because each gate input and output is not accessible.

4.4 EXCLUSIVE-OR AND EXCLUSIVE-NOR GATES

The **EXCLUSIVE-OR (EX-OR) gate** is a logic gate with just two inputs and one output. Its output is HIGH only when one of its two inputs is HIGH; its output is LOW when both inputs are LOW or when both inputs are HIGH.

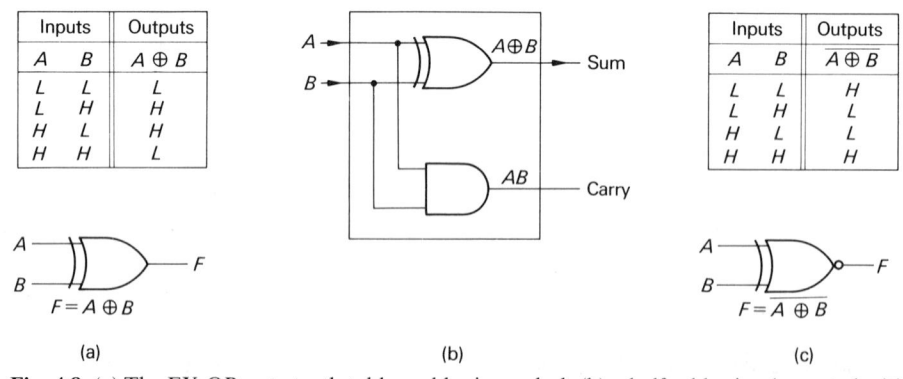

Fig. 4.8 (a) The EX-OR gate truth table and logic symbol; (b) a half-adder implemented with an EX-OR gate; (c) the EX-NOR gate truth table and logic symbol.

4.4 EXCLUSIVE-OR AND EXCLUSIVE-NOR GATES

The truth table for the EX-OR gate is given in Fig. 4.8(a). Note that it is identical to the ordinary OR gate, sometimes called the **INCLUSIVE-OR gate**, except for inputs $A = B = H$. Note also that the output of an EX-OR gate is HIGH only when an odd number of inputs are HIGH. When an even number of inputs are HIGH or LOW, its output is LOW. Thus an EX-OR gate can be used as an **odd-bit detector**.

The logic symbol for the EX-OR gate is the OR gate symbol with an additional curved line, as shown in Fig. 4.8(a). The symbol for the EX-OR operator is the OR operator with a circle around it. The functional relationship between inputs A and B and the output F is written

$$F = A \oplus B$$

and is read, "F equals A exclusive-or B."

An EX-OR gate can be converted into an inverter by connecting one input to a permanent HIGH. It can be made into a noninverting gate by connecting one input to a permanent LOW.

The following example shows how the EX-OR gate can be used to build the half-adder discussed in Chapter 2.

Example 4.4 *Two-Bit Half-Adder.* Draw the logic diagram for a two-bit half-adder using an EX-OR gate and an AND gate.

Solution. Referring to Fig. 2.10, note that the sum S output of a half-adder is the same as the output of an EX-OR gate. Thus the logic diagram for a half-adder becomes as shown in Fig. 4.8(b). A full adder is described in the problems. ∎

The **EXCLUSIVE-NOR (EX-NOR) gate** is an EX-OR gate whose output is inverted (complemented). Figure 4.8(c) shows the truth table for an EX-NOR gate. Note that its output is HIGH only when both inputs are the same, i.e., it is an **equality detector**. The functional relation between the output F and the inputs A and B is written

$$F = \overline{A \oplus B}$$

and is read, "F equals the complement of A exclusive-or B." The logic symbol for an EX-NOR gate is the EX-OR symbol with a small circle at the output to indicate inversion, as shown in Fig. 4.8(c).

EX-OR or EX-NOR gates are available in IC form in the TTL, ECL, and CMOSL families. The ECL 10107 triple EX-OR/NOR gate has a delay time of 2.4 ns. The CMOSL 4030 quad EX-OR gate has a delay time of 100 ns for $V_{DD} = 5$ V and 40 ns for $V_{DD} = 10$ V. Figure 4.9 shows the specifications for the TTL 74LS86 quadruple two-input low-power Schottky EX-OR gate. Note that the typical average power dissipation per gate is only 6.1 mA × 5 V/4 = 7.5 mW, and thus is less than that of the standard series TTL EX-OR gate, which draws a total

104 LOGIC GATE APPLICATIONS 4.5

recommended operating conditions

	SN54LS86 MIN	SN54LS86 NOM	SN54LS86 MAX	SN74LS86 MIN	SN74LS86 NOM	SN74LS86 MAX	UNIT
Supply voltage, V_{CC}	4.5	5	5.5	4.75	5	5.25	V
High-level output current, I_{OH}			−400			−400	µA
Low-level output current, I_{OL}			4			8	mA
Operating free-air temperature, T_A	−55		125	0		70	°C

electrical characteristics over recommended operating free-air temperature range (unless otherwise noted)

PARAMETER		TEST CONDITIONS†	SN54LS86 MIN	SN54LS86 TYP‡	SN54LS86 MAX	SN74LS86 MIN	SN74LS86 TYP‡	SN74LS86 MAX	UNIT
V_{IH}	High-level input voltage		2			2			V
V_{IL}	Low-level input voltage				0.7			0.8	V
V_I	Input clamp voltage	V_{CC} = MIN, I_I = −18 mA			−1.5			−1.5	V
V_{OH}	High-level output voltage	V_{CC} = MIN, V_{IH} = 2 V, V_{IL} = V_{IL} max, I_{OH} = −400 µA	2.5	3.4		2.7	3.4		V
V_{OL}	Low-level output voltage	V_{CC} = MIN, V_{IH} = 2 V, V_{IL} = V_{IL} max, I_{OL} = MAX		0.25	0.4		0.35	0.5	V
I_I	Input current at maximum input voltage	V_{CC} = MAX, V_I = 5.5 V			0.2			0.2	mA
I_{IH}	High-level input current	V_{CC} = MAX, V_I = 2.7 V			40			40	µA
I_{IL}	Low-level input current	V_{CC} = MAX, V_I = 0.4 V			−0.6			−0.6	mA
I_{OS}	Short-circuit output current§	V_{CC} = MAX	−6		−40	−5		−42	mA
I_{CC}	Supply current	V_{CC} = MAX, See Note 2		6.1	10		6.1	10	mA

†For conditions shown as MIN or MAX, use the appropriate value specified under recommended operating conditions for the applicable type.
‡All typical values are at V_{CC} = 5 V, T_A = 25°C.
§Not more than one output should be shorted at a time.
NOTE 2: I_{CC} is measured with the inputs grounded and the outputs open.

switching characteristics, V_{CC} = 5 V, T_A = 25°C

PARAMETER¶	FROM (INPUT)	TEST CONDITIONS		MIN	TYP	MAX	UNIT
t_{PLH}	A or B	Other input low	C_L = 15 pF, R_L = 2 kΩ, See Note 7		10	17	ns
t_{PHL}					10	17	
t_{PLH}	A or B	Other input high			10	17	ns
t_{PHL}					10	17	

¶t_{PLH} ≡ propagation delay time, low-to-high-level output
t_{PHL} ≡ propagation delay time, high-to-low-level output
NOTE 7: Load circuit and voltage waveforms are shown on page 149.

Fig. 4.9 Specifications for the TTL quad 74LS86 EX-OR gate. (Courtesy of Texas Instruments Incorporated.)

supply current of 30 mA. The 74LS86 has a typical average delay time of 10 ns, as compared with 14 ns for the standard series.

Now let's look at a method for obtaining a logic diagram from a verbal statement of a problem.

4.5 FROM TRUTH TABLE TO LOGIC DIAGRAM

A nuclear power generation station has in it a critical digital control circuit that must have very high reliability against failure. To increase the reliability, three

4.5 FROM TRUTH TABLE TO LOGIC DIAGRAM

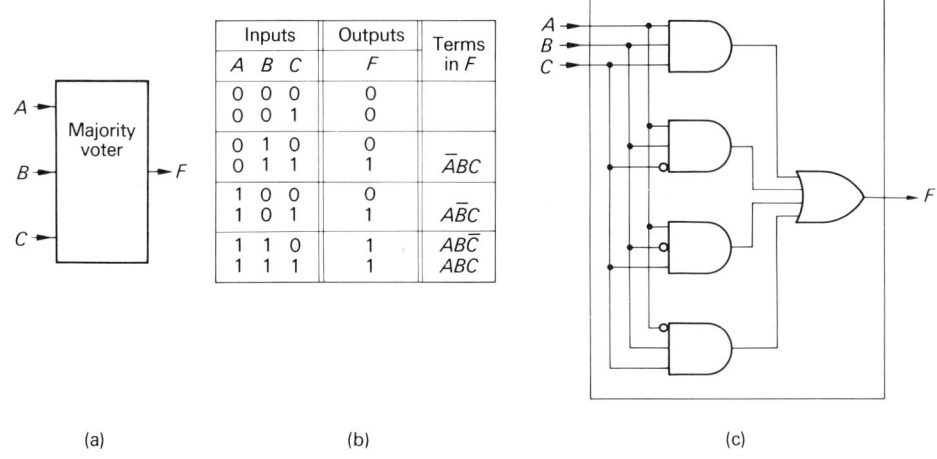

Fig. 4.10 Majority voter circuit: (a) logic symbol; (b) truth table; (c) logic diagram.

identical circuits are used. This is called **triple modular redundancy** (TMR). The outputs of these three circuits are fed into a majority voter circuit as shown in Fig. 4.10(a). A **majority voter circuit** is a device whose output always agrees with the majority of its inputs. Thus, if one of the TMR circuits fails, the output of the majority voter circuit will agree with the other two correct inputs. Let's develop the logic diagram for a majority voter circuit.

Example 4.5 *Majority Voter Circuit.* Using a truth table, develop the logic function and draw the logic diagram for a majority voter circuit.

Solution. The verbal statement that describes the majority voter circuit is, "The output must be HIGH when the majority of the inputs are HIGH, i.e., when two or three inputs are HIGH." An alternative and equivalent statement is, "The output must be LOW when the majority of inputs are LOW."

The first step is to complete a truth table from the verbal description. We shall use the logic-0 and logic-1 notation, 0 and 1 for short, in the truth table because we are dealing with the logic value of the binary variables, rather than their circuit HIGH and LOW nature. Figure 4.10(b) shows the truth table; note that in the output F column a 0 appears when the majority of inputs are 0 and a 1 appears when the majority of inputs are 1.

Let's adopt a convenient notation for developing the output function F. A logic variable, such as A, can have only one of two values, 0 or 1. Let's say that A represents the logic-1 value of A and that $\bar{A}$ represents the logic-0 value of A. Similarly, let B represent the logic-1 value of B and let $\bar{B}$ represent the logic-0 value of B.

Referring to the majority voter truth table, we see that the output F is a 1 for any of the following four cases:

$A = 0$ and $B = 1$ and $C = 1$ written $\bar{A}BC$,

or

$A = 1$ and $B = 0$ and $C = 1$ written $A\bar{B}C$,

or

$A = 1$ and $B = 1$ and $C = 0$ written $AB\bar{C}$,

or

$A = 1$ and $B = 1$ and $C = 1$ written ABC.

These four terms can be conveniently written in an additional output column, labeled "terms in F" in Fig. 4.10(b). Since any one of these terms produces a 1 output in F, the function F is the ORed combination of these four terms, i.e.,

$$F = \bar{A}BC + A\bar{B}C + AB\bar{C} + ABC.$$

For example, if $A = 0$, $B = 1$, and $C = 1$, then $\bar{A} = 1$, $\bar{B} = 0$, and $\bar{C} = 0$. Hence F becomes

$$F = 1 \cdot 1 \cdot 1 + 0 \cdot 0 \cdot 1 + 0 \cdot 1 \cdot 0 + 0 \cdot 1 \cdot 1 = 1 + 0 + 0 + 0 = 1.$$

The majority voter can be implemented by ORing the four terms together. Each AND term is generated from one three-input AND gate, as shown in Fig. 4.10(c). In the next section we shall show how to simplify this implementation. ∎

Example 4.6 *Digital Comparator.* A **digital comparator** compares the magnitude of two N-bit binary numbers $A = A_1 A_2 \cdots A_n$ and $B = B_1 B_2 \cdots B_n$ as shown in Fig. 4.11(a). Draw the logic diagram for a one-bit comparator (that is, $A = A_1$ and $B = B_1$).

(a)

(b)

A	B	A=B	A<B	A>B
0	0	1	0	0
0	1	0	1	0
1	0	0	0	1
1	1	1	0	0

(c)

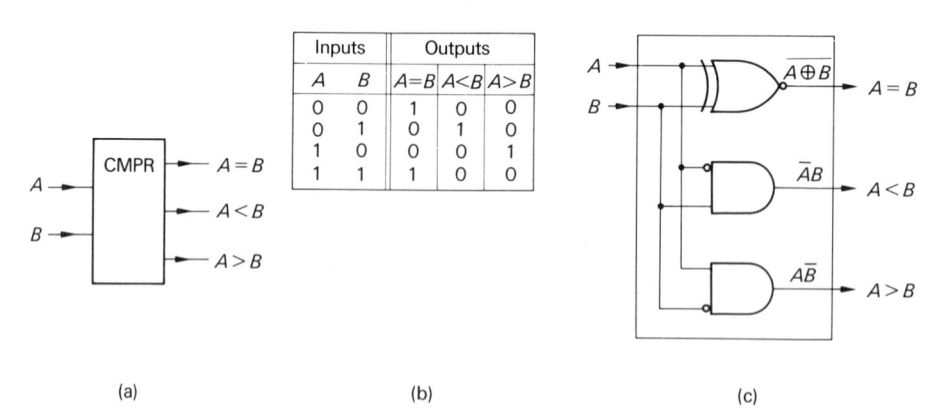

Fig. 4.11 A one-bit comparator: (a) logic symbol; (b) truth table; (c) logic diagram.

4.6　　　　　　　　　　　　　　　　　　　　　　　　　　LOGIC RULES　　107

Solution. The first step is to state the problem precisely. The $A = B$ output of the comparator must be 1 when $A = B$, the $A > B$ output must be 1 when A is greater than B, and the $A < B$ output must be 1 when A is less than B. The truth table will have three output columns, one for each of the three comparator outputs, as shown in Fig. 4.11(b).

The $A = B$ output column will have a 1 in it when $A = B$, that is, when $A = B = 0$ or when $A = B = 1$. The $A < B$ output column will have a 1 in it when A is less than B, that is, when $A = 0$ and $B = 1$. The $A > B$ column will have a 1 in it when $A = 1$ and $B = 0$. The three output functions will be

$$F_1 = \bar{A}\bar{B} + AB, \qquad F_2 = \bar{A}B, \qquad \text{and} \qquad F_3 = A\bar{B}.$$

The $A = B$ column in the truth table is the same as an EX-NOR output, that is, $\bar{A}\bar{B} + AB = \overline{A \oplus B}$, so this output can be implemented with an EX-NOR gate. The logic diagram for the comparator is given in Fig. 4.11(c). ■

A comparator can also be used as a **coincidence detector**, as shown in the problems.

The next section discusses some of the logic rules that are useful for simplifying logic functions. The rules are also useful for implementing a function using NAND or NOR gates.

4.6　LOGIC RULES [1, 2, 3]

First let's consider the **double negation** (DN) rule

$$(\bar{\bar{A}}) = A, \qquad \text{(DN)}$$

which says that two NOT gates connected in series is logically the same as not having a gate. Figure 4.12(a) illustrates this idea. The double negation rule can be proved using a truth table as shown in Fig. 4.12(a). Each row in the A column and $\bar{\bar{A}}$ column has identical values, so A and $\bar{\bar{A}}$ are the same, i.e., they are equal.

Another useful rule is

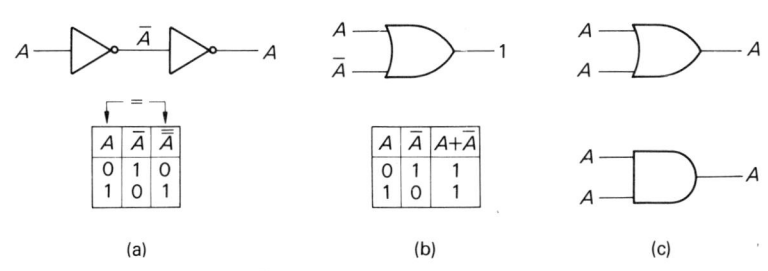

$$A + \bar{A} = 1,$$

Fig. 4.12 Logic rules: (a) $\bar{\bar{A}} = A$; (b) $A + \bar{A} = 1$; (c) $A + A = A$, $A \cdot A = A$.

108   LOGIC GATE APPLICATIONS   4.6

which says that when A and its complement $\bar{A}$ are ORed, the result is always a logic-1, as illustrated in Fig. 4.12(b). The truth table proof, also given in Fig. 4.12(b), shows that for all possible values of A, $A + \bar{A}$ is always equal to 1.

Two other rules of importance, called the **idempotent rules**, are

$$A + A = A \tag{ID.1}$$

and

$$AA = A. \tag{ID.2}$$

These rules are illustrated in Fig. 4.12(c) and proved in the problems. They say that when A is ORed or ANDed with itself, the result is just A.

Logic variables obey the distributive rule just as ordinary algebraic variables do. Let's show that

$$A(B + C) = AB + AC. \tag{Distributive}$$

We set up a truth table for three variables as shown in Fig. 4.13(a), with columns for A, B, C, $B + C$, $A(B + C)$, AB, AC, and $AB + AC$. Next we place the appropriate 0 or 1 in each row of each output column. Now note that for every possible combination of inputs A, B, and C, $A(B + C)$ has the same value as $AB + AC$, hence they are equal. Figure 4.13(b) shows the implementation of $F = A(B + C)$ and $F = AB + AC$. The two implementations perform identical logic; however, the second method requires one more gate than the first method.

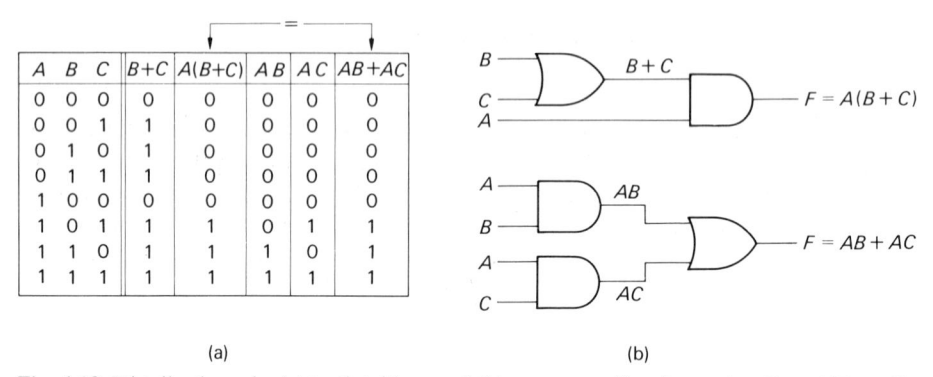

A	B	C	B+C	A(B+C)	AB	AC	AB+AC
0	0	0	0	0	0	0	0
0	0	1	1	0	0	0	0
0	1	0	1	0	0	0	0
0	1	1	1	0	0	0	0
1	0	0	0	0	0	0	0
1	0	1	1	1	0	1	1
1	1	0	1	1	1	0	1
1	1	1	1	1	1	1	1

(a)   (b)

Fig. 4.13 Distributive rule: (a) truth table proof; (b) two ways of implementing $F = A(B + C)$.

Note that when the distributive rule is read from right to left, it shows us how to factor a Boolean* (logic) expression, and when read from left to right, how to expand a Boolean expression.

Additional rules of logic are given in Appendix A.

---

* George Boole (1815–1864), English mathematician and logician. Developed the algebra for manipulating logic variables, called **Boolean algebra**.

4.6   LOGIC RULES

The next example shows how the majority voter function can be simplified, which results in a simpler implementation of the circuit.

Example 4.7   Simplify the majority voter function
$$F = \bar{A}BC + A\bar{B}C + AB\bar{C} + ABC.$$

Solution. Since $A + A = A$ (rule ID.1) it follows that $ABC + (ABC + ABC) = ABC$. Let's OR the term ABC twice with the term ABC in F. This gives

$$F = \bar{A}BC + A\bar{B}C + AB\bar{C} + ABC + (ABC + ABC). \quad \text{(ID.1)}$$

Next regroup the terms as follows:

$$F = (\bar{A}BC + ABC) + (A\bar{B}C + ABC) + (AB\bar{C} + ABC). \quad \text{(Regroup)}$$

Now factor out the common factors in each group, i.e.,

$$F = BC(\bar{A} + A) + AC(\bar{B} + B) + AB(\bar{C} + C). \quad \text{(Distributive)}$$

Finally, since $\bar{A} + A = 1$, $\bar{B} + B = 1$, and $\bar{C} + C = 1$, this simplifies to

$$F = BC + AC + AB. \quad (A + \bar{A} = 1)$$

The implementation of this simplified majority voter circuit function requires only three two-input AND gates and one three-input OR gate. ∎

With the increased availability of MSI and LSI devices, and their continual decrease in cost, the need for SSI gates in most digital systems has been greatly reduced. Hence the need for developing truth tables, logic functions, and simplification [4, 5] of logic functions has been greatly reduced. Therefore, the simplification of logic functions will not be pursued in detail in this text (see Ch. 6 problems).

The last two rules we shall discuss are DeMorgan's rules. These rules are useful for (1) simplifying Boolean expressions [by removing the complement (bar) symbol over groups of terms] and (2) implementing functions using NAND or NOR logic.

DeMorgan's Rules   The first DeMorgan rule is

$$\overline{(\bar{A}) \cdot (\bar{B})} = A + B. \quad \text{(DM.1a)}$$

This rule says that an OR gate can be made by inverting the inputs to a NAND gate, as Fig. 4.14(a) illustrates. The rule can be easily remembered as follows: remove the bar, replace the · by a +, and invert each separated term; thus,

$$\overline{(\bar{A}) \cdot (\bar{B})} = (\bar{\bar{A}}) + (\bar{\bar{B}}) = A + B.$$

Figure 4.14(b) shows the truth table proof of DeMorgan's first rule. The two columns $A + B$ and $\overline{\bar{A} \cdot \bar{B}}$ have identical values for every combination of inputs, so the expressions $A + B$ and $\overline{\bar{A} \cdot \bar{B}}$ are equal.

Another form of DeMorgan's first rule is

$$\overline{A + B} = \bar{A} \cdot \bar{B}. \quad \text{(DM.1b)}$$

110 LOGIC GATE APPLICATIONS 4.6

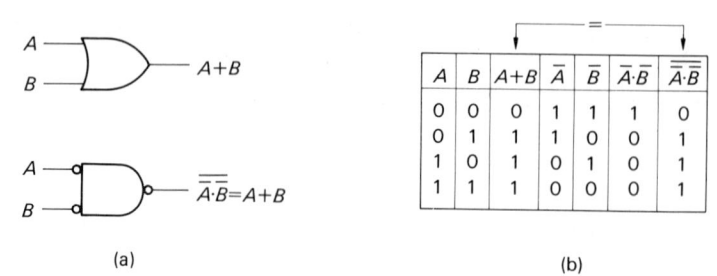

Fig. 4.14 DeMorgan's rule, $A + B = \overline{\overline{A} \cdot \overline{B}}$: (a) logic symbol; (b) truth table proof.

A	B	A+B	$\overline{A}$	$\overline{B}$	$\overline{A} \cdot \overline{B}$	$\overline{\overline{A} \cdot \overline{B}}$
0	0	0	1	1	1	0
0	1	1	1	0	0	1
1	0	1	0	1	0	1
1	1	1	0	0	0	1

This rule can be proved using a truth table, as shown in the problems, or by taking the complement of both sides of DM.1a and recalling that

$$\overline{(\overline{A} \cdot \overline{B})} = \overline{A} \cdot \overline{B} \qquad \text{(double negation rule)}.$$

Rule DM.1b says that a NOR gate is the same as an AND gate whose inputs are inverted. Both forms of the rule are illustrated in Table 4.1.

The second DeMorgan rule is

$$\overline{\overline{A} + \overline{B}} = A \cdot B. \qquad \text{(DM.2a)}$$

This rule says that an AND gate can be made from a NOR gate by inverting its

Table 4.1 Alternate Implementations of OR, NOR, AND, NAND, and NOT Gates

Function	Standard symbol	Equivalent implementation	De Morgan implementation	Rule
OR				DM-1(a)
NOR				DM-1(b)
AND				DM-2(a)
NAND				DM-2(b)
NOT				

inputs (see Table 4.1). Another form of DeMorgan's second rule is

$$\overline{A \cdot B} = \overline{A} + \overline{B}, \qquad \text{(DM.2b)}$$

which says that a NAND gate can be made by inverting the inputs of an OR gate, as illustrated in Table 4.1.

Note that all the DeMorgan rules can be remembered as: remove the bar, interchange $\cdot$ and $+$, and then invert each separated term.

DeMorgan's rules can be used to simplify the expression $\overline{AB\overline{C}(A + B)}$ as follows:

$$\begin{aligned}
\overline{AB\overline{C}(A + B)} &= \overline{(AB\overline{C})} + \overline{(A + B)} & \text{(DM.2b)} \\
&= (\overline{A} + \overline{B} + C) + (\overline{A} \cdot \overline{B}) & \text{(DM.2b, 1b)} \\
&= (\overline{A} + \overline{A}\overline{B}) + \overline{B} + C & \text{(Regroup)} \\
&= \overline{A}(1 + \overline{B}) + \overline{B} + C & \text{(Distributive)} \\
&= \overline{A} + \overline{B} + C. & (1 + B = 1)
\end{aligned}$$

For reference, Table 4.1 summarizes the DeMorgan and additional methods for implementing each of the five logic gates.

NAND and NOR gates are called **universal gates**, or **functionally complete**, because they can each be used to generate any of the other four gates.

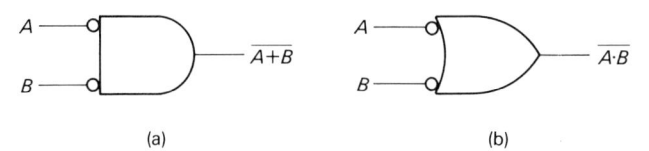

Fig. 4.15 (a) NOR gate represented by AND logic; (b) NAND gate represented by OR logic.

Frequently logic diagrams will show an AND gate with inverted inputs instead of its equivalent, the NOR gate. Figure 4.15(a) shows a NOR gate represented as an AND gate with inverted inputs, as given by DeMorgan's first rule (DM.1b). Similarly, a NAND gate can be thought of as an OR gate with inverted inputs as shown in Fig. 4.15(b). The use of AND and OR gates makes logic diagrams easier to read since, in their everyday experience, humans are more accustomed to AND and OR logic than to NAND and NOR logic [6].

Now let's see how to use the results of this section to implement logic functions using NAND and NOR gates.

4.7 NAND/NOR IMPLEMENTATION

In many logic families, such as TTL, ECL, and CMOSL, the basic gate is either a NAND gate or a NOR gate. In these families a *much* smaller selection of AND and OR gates exists and their cost is usually higher. Further, the AND/OR gates

112 LOGIC GATE APPLICATIONS 4.7

in these families generally have a higher power dissipation and longer typical delay times due to the added stages required in the IC. For these reasons it is necessary to learn how to implement logic functions using only NAND, NOR, and NOT gates. The procedures for implementing a function using NAND or NOR logic are:

1. Draw the logic diagram using only AND, OR, and NOT gates.
2. Replace each AND, OR, and NOT gate by a NAND (NOR) gate using Table 4.1.
3. Simplify the resulting logic diagram.

Example 4.8 (a) Implement $F = AB + \bar{C}$ using only NAND gates. (b) Repeat using only NOR gates.

Solution. (a) Figure 4.16(a) shows F implemented by using AND, OR, and NOT gates. Referring to Table 4.1, we see that each AND gate can be replaced by a NAND-NOT gate combination. Each OR gate can be replaced by a NAND gate with inverted inputs as illustrated in Fig. 4.16(b). When two inverters occur in series, they can be canceled since $(\bar{\bar{A}}) = A$. This results in the simplified logic diagram shown in Fig. 4.16(c). The small circles at the output of a logic symbol cannot be canceled since they indicate the type of gate (NAND or NOR).

(b) Figure 4.16(d) shows the implementation of F using only NOR gates. Referring to Table 4.1, we see that the AND gate can be replaced by a NOR gate with inverted inputs and the OR gate by a NOR-NOT combination. Note that

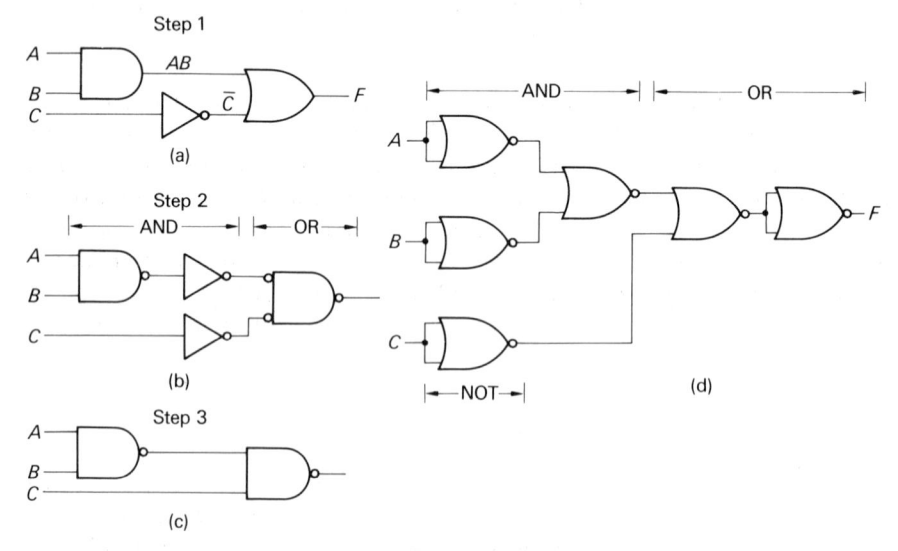

Fig. 4.16 Implementation of $F = AB + \bar{C}$ using (a) AND/OR/NOT gates, (b) NAND/NOT gate, (c) simplified NAND implementation, (d) NOR gate implementation.

the NOR implementation requires more gates than the NAND implementation. Hence, if both NAND and NOR gates were available in the chosen logic family, the simpler NAND implementation would be preferred. ■

Some functions can be implemented most economically using both NAND and NOR gates.

Example 4.9 *Automobile Seat Belt Warning System.* Implement the seat belt warning system of Example 4.3 using only three-input NOR gates.

Solution. Referring to Table 4.1, we replace each AND gate in Fig. 4.4(a) by a NOR gate with inverted inputs. The result is shown in Fig. 4.17(a). ■

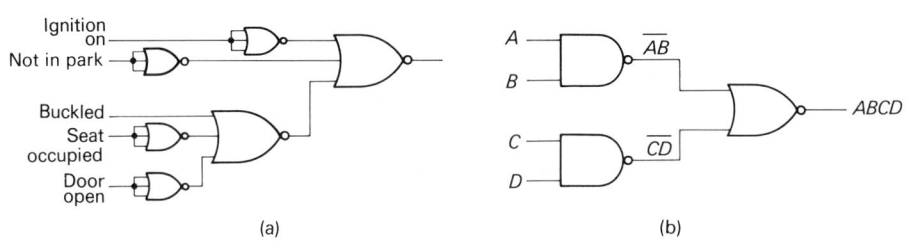

Fig. 4.17 (a) NOR implementation of automobile seat belt warning system; (b) NAND/NOR implementation of a four-input AND gate.

The next example shows how to increase the fan-in of NAND gates by cascading them.

Example 4.10 Draw the logic diagram of a four-input AND gate using only two-input NAND gates and two-input NOR gates.

Solution. If two two-input NAND gates are fed into a two-input NOR gate, as shown in Fig. 4.17(b), the NOR output will be $\overline{(\overline{AB}) + (\overline{BC})}$. Using DeMorgan's rule (remove the bar, replace + by ·, and complement each term) gives

$$\overline{(\overline{AB}) + (\overline{CD})} = (\overline{\overline{AB}}) \cdot (\overline{\overline{CD}}) = ABCD,$$

a four-input AND gate. ■

4.8 MISCELLANEOUS APPLICATIONS

In this section we consider several applications using logic gates and flip-flops.

4.8.1 Frequency Dividers

Many digital systems use a pulse generator, called a **clock**, to control the timing of events within the system. Often pulse generators having different frequencies are required to control devices at different rates or times. A frequency divider made from flip-flops can be used to generate accurate lower-frequency pulse waveforms from the system clock.

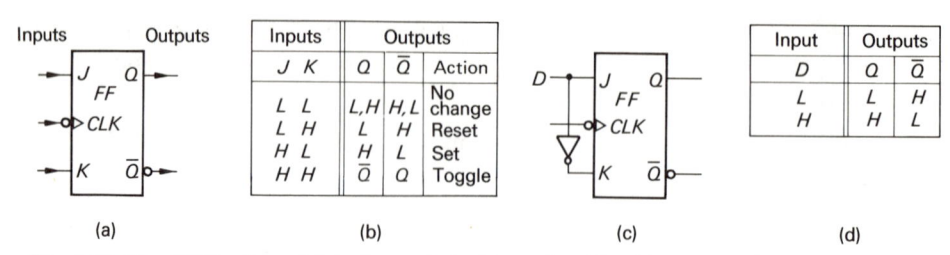

Fig. 4.18 The JK flip-flop: (a) logic symbol: (b) truth table: (c) converted to a D flip-flop; (d) D flip-flop truth table.

(a) Logic symbol: JK FF with inputs J, K, CLK and outputs Q, $\bar{Q}$.

(b) Truth table:

J	K	Q	$\bar{Q}$	Action
L	L	L,H	H,L	No change
L	H	L	H	Reset
H	L	H	L	Set
H	H	$\bar{Q}$	Q	Toggle

(c) D flip-flop from JK FF: D input connected to J, and through inverter to K.

(d) D flip-flop truth table:

D	Q	$\bar{Q}$
L	L	H
H	H	L

A **JK flip-flop** (FF) is a flip-flop that has three inputs J, K, and CLOCK (CLK) and two complementary outputs Q and $\bar{Q}$, as shown in Fig. 4.18(a). The J and K inputs determine whether the Q and $\bar{Q}$ outputs will be HIGH or LOW. The CLOCK input allows the outputs to change only at the HIGH to LOW ($H \rightarrow L$) transition of any clock pulse. A steady HIGH or LOW at the clock input will not allow the outputs to change. The **dynamic indicator** symbol

at any input, such as the clock input, is used to indicate that only a transition in the voltage level at that input will activate that input. The absence of the dynamic indicator symbol at any input means that input is static, that is, it is activated by a constant voltage level. A small circle in front of the dynamic indicator symbol

indicates that the input is activated by a HIGH to LOW ($H \rightarrow L$) transition. The absence of a small circle in front of the dynamic indicator symbol means that input is activated by a $L \rightarrow H$ transition.

The truth table for the JK flip-flop is given in Fig. 4.18(b). When $J = L$ and $K = L$, the output does not change from its previous state, i.e., if Q was LOW (or HIGH) before the inputs $J = L$, $K = L$ were applied, then Q stays LOW (or HIGH) until after the next clock pulse. When $J = L$ and $K = H$, the output resets (i.e., Q becomes LOW) at the $H \rightarrow L$ transition of the next clock pulse. When $J = H$ and $K = L$, the output sets (i.e., Q becomes HIGH) at the next $H \rightarrow L$ clock pulse. When $J = H$ and $K = H$, the Q and $\bar{Q}$ outputs change state (i.e., toggle) at each $H \rightarrow L$ clock transition. Note that the JK flip-flop, unlike the RS latch, has no indeterminate state.

A D flip-flop can be made from a JK flip-flop by connecting the J input through an inverter to the K input as shown in Fig. 4.18(c). The output of a D flip-flop becomes the same as its input at the $H \rightarrow L$ transition of each clock pulse, as shown in the truth table of Fig. 4.18(d).

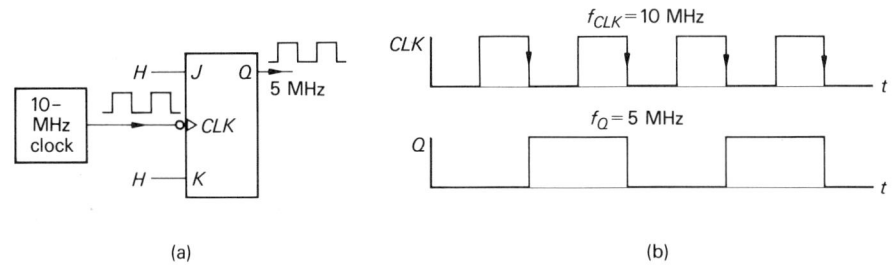

Fig. 4.19 Frequency divider: (a) logic diagram; (b) timing diagram.

The next example shows how the toggle mode of a *JK* flip-flop can be used as a frequency divider.

Example 4.11 *Frequency Divider.* Draw the logic diagram for a circuit that will reduce the frequency of the clock in a digital system from 10 MHz to 5 MHz.

Solution. The logic diagram of a divide-by-two ($\div 2$) frequency divider using a *JK* flip-flop in the toggle mode is shown in Fig. 4.19(a). Let's construct the timing diagram for this circuit. In the toggle mode the *Q* output changes state at the trailing edge ($H \rightarrow L$ transition) of each clock pulse as shown in Fig. 4.19(b). Note that the frequency of the *Q* output is exactly one-half of the frequency of the *CLK* input, i.e., 5 MHz. ∎

If the *Q* output of one flip-flop is connected to the *CLK* input of a second flip-flop, the output frequency of the second flip-flop will be $(\frac{1}{2})(\frac{1}{2}) = \frac{1}{4}$ of the input frequency of the first flip-flop, as shown in the problems.

4.8.2 Pulse-Shaping Circuits

It is often desirable in a digital system to delay one pulse with respect to another. The logic symbol for a time delay element is shown in Fig. 4.20(a). The two vertical lines indicate the input terminal of the delay element. The time t_p is the total delay from the input to the output.

Short delays, on the order of nanoseconds, can be produced using noninverting logic gates. In the standard TTL family, two 7404 NOT gates will produce an average total typical time delay of about 20 ns, as shown in Fig. 4.20(b).

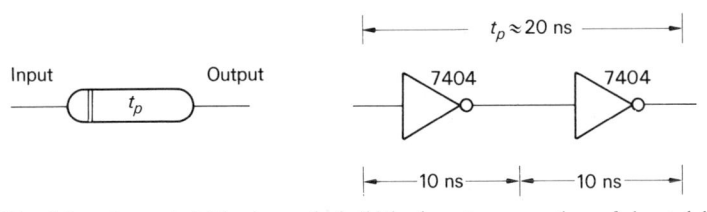

Fig. 4.20 The delay element: (a) logic symbol; (b) logic gate generation of short delay times.

An **RC integrator circuit** is a series circuit that consists of a resistor R and a capacitor C, with the output taken across the capacitor as shown in Fig. 4.21(a). This circuit can be used to increase the width of the input pulse as follows. At time t_0 almost all the input voltage appears across the resistor because the capacitor has not had time to charge, so $e_{out} = 0$, where $E_0 = 0$ V if the capacitor is not charged at time t_0. As the capacitor charges, the output voltage increases exponentially to 63% of its maximum value E_2 in one time constant $\tau(\sec) = R(\Omega) \times C(F)$. The voltage across a discharging capacitor decreases to 63% below its charged voltage in time $\tau = RC$. For medium time constants, i.e., $\frac{1}{10}\tau < T < 10\tau$, and long time constants ($T > 10\tau$) the output pulse width T at some voltage level E_1 will be greater than the input pulse width t_{in}. Thus an RC integrator can be used to increase the width of the input pulse; this circuit is therefore sometimes called a **pulse stretcher**. A pulse stretcher circuit that doesn't use RC components is discussed in the problems.

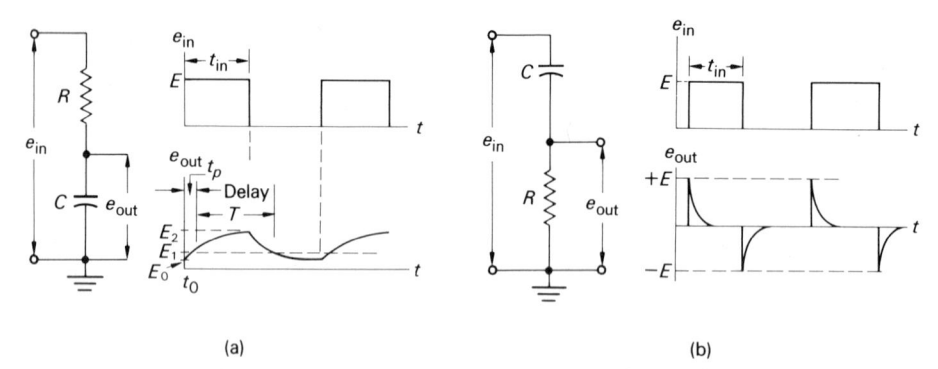

Fig. 4.21 (a) RC integrator circuit and waveforms; (b) RC differentiator and waveforms.

An integrator can also be used as a delay element because its output doesn't reach voltage level E_1 until after a time delay t_p (Fig. 4.21a).

Sometimes it is desirable to produce a narrow pulse from a wider pulse. An **RC differentiator circuit** is a series circuit consisting of a resistor R and a capacitor C, with the output taken across the resistor as shown in Fig. 4.21(b). A narrow output pulse will result if $RC \leq \frac{1}{10} t_{in}$. Initially the capacitor is uncharged, that is, $e_C = 0$ V, so all the input voltage appears across R. The capacitor charges rapidly (since the RC time constant is small) so the voltage across R decreases rapidly to 0 V. Now the polarity of the capacitor is negative with reference to ground, i.e., it appears as a voltage source of $-E$ volts. Therefore when e_{in} becomes zero, the capacitor quickly discharges from $-E$ volts to 0 V. If pulses of only one polarity are desired, a diode can be placed in the output line to eliminate either the positive or the negative pulses.

4.8 MISCELLANEOUS APPLICATIONS 117

*4.8.3 Applications Using *RC* Elements

Let's see how time delay elements can be used in some practical applications.

Example 4.12 *Frequency Doubler (Pulse Edge Detector).* Draw the logic diagram for a circuit whose output frequency is twice the input frequency.

Solution. Consider a two-input EX-OR gate with a time delay t_p introduced at the *B* input as shown in Fig. 4.22(a). The *B* input to the EX-OR gate is the same as the *A* input except that it occurs t_p seconds later. Figure 4.22(b) shows the timing diagram for inputs *A* and *B* and output *F*. Input *B* is delayed, relative to input *A*, by a total time t_p. The output *F* of the EX-OR gate will be HIGH only during the time *A* is HIGH and *B* is LOW (leading edge of *A*) or during the time *A* is LOW and *B* is HIGH (trailing edge of *A*). Thus an output occurs at each edge of the input pulse, which results in a doubling of the input frequency. It is also a circuit that detects each edge of an input pulse, i.e., a pulse edge detector. The width of the output pulse equals delay time t_p. If $t_p = t_{in}$, the output will always be HIGH. A 2^N multiplier can be constructed by cascading *N* frequency doublers. ∎

Short delay times t_p can be generated by using one or more noninverting EX-OR gates as shown in Fig. 4.22(c). Longer delay times can be generated using an *RC* integrator as shown in Fig. 4.22(d). The time delay is approximately *RC*, plus the delays of the two EX-OR gates. The purpose of EX-OR gate G_1 is to act as a buffer between the source of signal *A* and the *RC* integrator. Gate G_2 is used to square up the integrated waveform. The rise time of the integrated waveform must be fast enough to avoid gate oscillations for the logic family used. Long delays can be generated with monostables as discussed in Chapter 7.

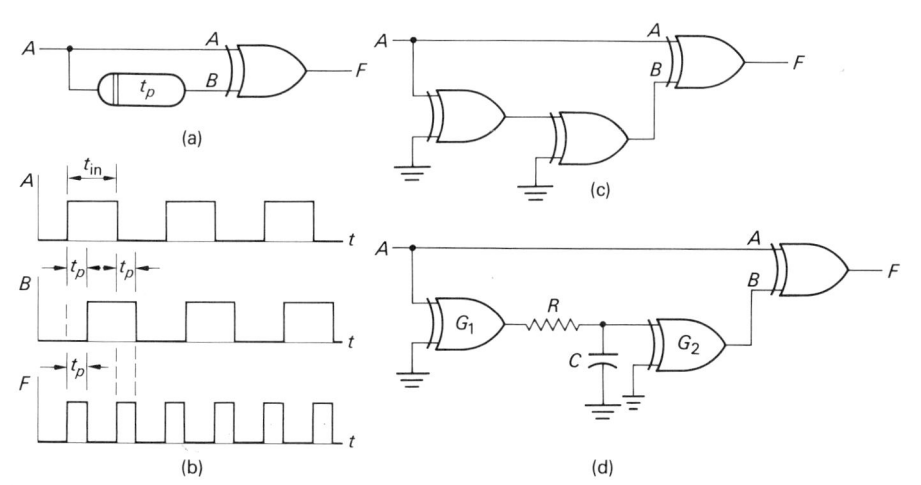

Fig. 4.22 The frequency doubler (pulse edge detector): (a) conceptual logic diagram; (b) timing diagram; (c) EX-OR implementation; (d) *RC* integrator delay implementation.

The next example illustrates the use of RC integrators and logic gates to build a circuit that will ignore noise pulses.

Example 4.13 *Noise Discriminator.* Draw the logic diagram for a circuit that discriminates against, i.e., ignores, noise pulses whose time duration T_n is less than a specified time t_1.

Solution. A noise discriminator circuit is shown in Fig. 4.23. It consists of two NOR gates, one inverter, and two RC integrators R_1C_1 and R_2C_2. The circuit discriminates against noise pulses whose duration T_n is less than $t_1 = R_1C_1$. The time constant $t_2 = R_2C_2$ determines the width of the output pulse. Diode CR_1 allows C_1 to charge rapidly. Diode CR_2 allows C_2 to discharge rapidly. The detailed timing diagram for this circuit is developed in the problems. A monostable (Chapter 7) can also be used as a noise discriminator. ∎

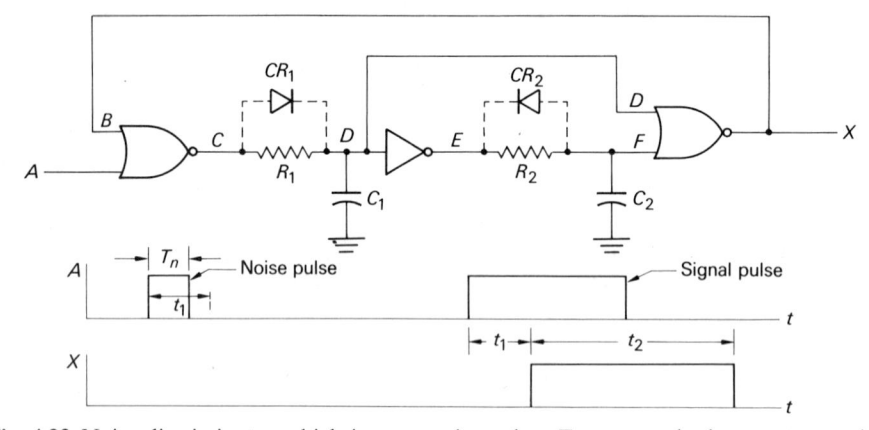

Fig. 4.23 Noise discriminator which ignores noise pulses $T_n < t_1$, and whose output pulse width is t_2.

*4.8.4 The Strobed RS Latch

An RS latch with an enable input, as shown in Fig. 4.24, is called a **strobed latch**. It is useful for applications such as the phase detector in a phase lock loop (PLL) signal generator, like the one described in Appendix F.

Let's develop the truth table for a strobed RS latch. When the enable input E is LOW the outputs of gates 3A and 3D are HIGH, independent of whether S and R are HIGH or LOW. A HIGH input to gates 3B and 3C produces no change in the outputs of the latch. Thus, the latch is disabled when $E =$ LOW.

When $E = H$, $S = L$, and $R = L$, the outputs of gates 3A and 3D are HIGH,

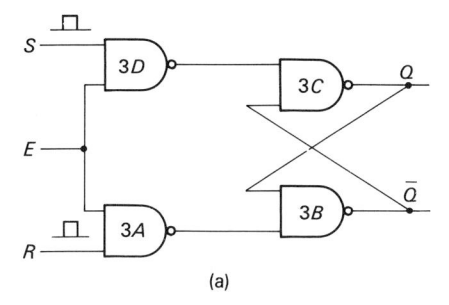

Fig. 4.24 Strobed *RS* latch: (a) logic diagram; (b) truth table.

so the latch outputs (Q and $\bar{Q}$) do not change. When $E = H$, $S = L$, and $R = H$, the output of 3A is LOW, forcing $\bar{Q}$ HIGH and Q LOW. When $E = H$, $S = H$, and $R = L$, the output of gate 3D is LOW, forcing Q HIGH and $\bar{Q}$ LOW. Finally, when $E = H$, $S = H$, and $R = H$, the outputs of 3A and 3D are both LOW, forcing Q and $\bar{Q}$ HIGH.

SUMMARY

In this chapter we gave several practical applications using gates and flip-flops. We discussed three methods for increasing the fan-in of gates: (1) wired gates (open collector or open emitter), (2) cascaded gates, and (3) AOI gates. Wired gates require an external load resistor and generally have longer delays than single gates having a sufficient number of inputs. Cascading of gates results in more levels of gating than exist with wired gates or AOI gates, which in turn results in longer delay times and greater power dissipation. The main disadvantage of AOI gates is that they are more difficult to troubleshoot.

Expandable AOI gates and expander gates are capable of generating any desired function.

The output of an EX-OR gate is HIGH if only one input is HIGH. An EX-NOR gate has a HIGH output when both of its inputs are the same.

To draw logic diagrams from truth tables, we OR each term in the output column of the truth table that has a logic-1 in it.

Logic rules are useful for simplifying Boolean functions and for NAND/NOR implementations. NAND/NOR implementations involve replacing each AND and each OR gate in the original logic diagram by its NAND (or NOR) equivalent, as summarized in Table 4.1.

In the last section we described the *JK* and *D* flip-flop, the strobed *RS* latch and several applications, such as the frequency divider, delay devices, frequency doubler, and noise discriminator.

PROBLEMS

Section 4.2

1. *Digital Door Lock.* Draw the logic diagram for the control of a digital door lock that has 128 possible input combinations. Include an alarm circuit.

2. Draw a logic diagram for a latch that is connected to the alarm output of the digital door lock in Fig. 4.1(b) in such a way that the alarm will continue to sound once it has been activated, even if the correct combination is eventually found. Include a provision on the latch for turning the alarm off if the correct combination is set.

Section 4.3

3. Draw a logic diagram that generates the function
$$Z = \overline{AB + C + D + EF + G + H + J + K}$$
using TTL 7453 and 7460 gates.

4. The logic diagram for a dual two-wide two-input CMOSL 14506 AOI gate is shown in Fig. P4.1. (a) Write the expression for the output Z in terms of inputs A, B, C, D, and E. (b) Draw the logic diagram for a four-wide two-input AOI gate using one 14506 IC.

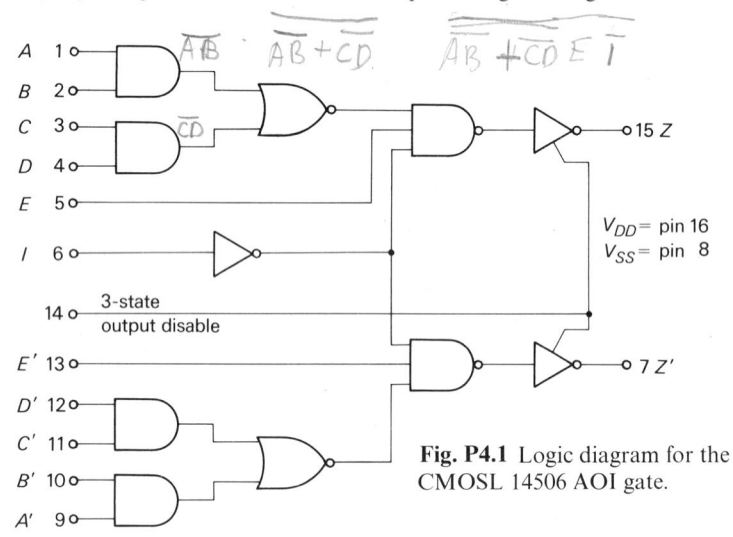

Fig. P4.1 Logic diagram for the CMOSL 14506 AOI gate.

5. Using a TTL data book, compare the average typical delay time and average typical power dissipation of a six-input AND gate made from (a) cascaded two-input 7408 AND gates, (b) wire-ANDed 7409 AND gates, and (c) AOI 7453 and 7460 expander gates [Hint: $\overline{\overline{A} + \overline{B} + \overline{C} + \cdots} = A \cdot B \cdot C \ldots$]

Section 4.4

6. Draw the logic diagram for an EX-NOR gate made from two EX-OR gates.

7. *Full Adder.* A full adder has a carry-in C_i input to handle the carry-out bit from previous additions as shown in Fig. P4.2(a). The truth table for a full adder is given in Fig. 4.2(b).

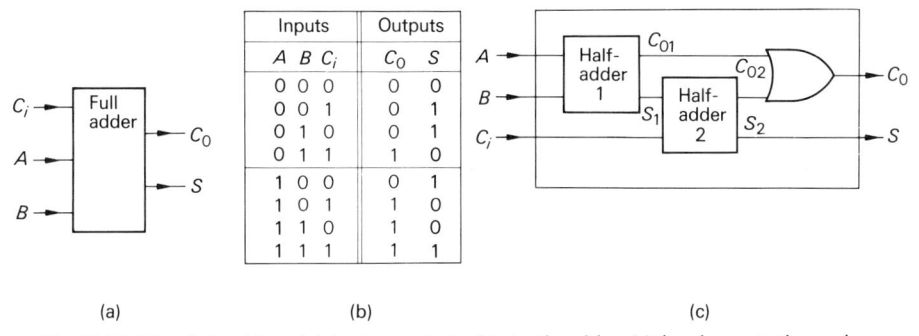

Fig. P4.2 The full adder: (a) logic symbol; (b) truth table; (c) implementation using half-adders.

The full adder adds three bits at a time. Thus, if $A = 1$, $B = 1$, and $C_i = 0$, then the sum $S = 1 + 1 + 0 = 0$ and the carry-out is $C_0 = 1$. The full adder can be implemented from two half-adders and an OR gate as shown in Fig. P4.2(c). (a) Draw the logic diagram of a full adder using EX-OR, AND, and OR gates. (b) Show the 0 and 1 values at the input and output of each gate for $A = 1$, $B = 0$, and $C_i = 1$.

Section 4.5

8. *Optical Card Reader.* For the optical card reader detector of Example 4.2, show the implementation that will detect if three or more out of five lectures have been missed.
9. *Vending Machine.* Derive the Boolean equation for the vending machine given in Chapter 2, Example 2.3.
10. *Foolproof Vending Machine.* Derive the Boolean equations for a vending machine that is similar to the one described in Example 2.3 but that will give the correct change and dispense coffee for any combination of one dime, one or two nickels, or a single quarter.
11. Show the 0 and 1 values at the input and output of each gate in the digital comparator of Example 4.6 for the cases $A = 1$, $B = 0$ and $A = 1$, $B = 1$.
12. *Coincidence Detector.* Refer to the SL2 Phase Detector phase lock loop schematic in Appendix F, and explain in detail how the coincidence gate detector works. Use a timing diagram showing the input and output waveforms to NAND gates U7B, U4D, U4A, U7D, U1B, and U7C.

Section 4.6

13. Using a truth table prove the following.
    a) $\bar{\bar{A}} = A$ b) $A + A = A$
    c) $AA = A$ d) $\overline{A + B} = \bar{A} \cdot \bar{B}$
14. Use the rules of logic to evaluate the following EX-OR expressions.
    a) $A \oplus A =$ b) $A \oplus 1 =$
    c) $A \oplus A \oplus A =$ d) $A \oplus 0 =$
    [*Hint*: Since $A \oplus B = A\bar{B} + \bar{A}B$, therefore $A \oplus A = A\bar{A} + \bar{A}A$.]

122 LOGIC GATE APPLICATIONS

15. Draw the DeMorgan implementation of the following gates.
    a) OR   b) NOR   c) NAND
16. Use DeMorgan's rules to simplify (i.e., remove the bar from) the Boolean functions.
    a) $F = \overline{A + B + C}$ b) $G = \overline{AB + \overline{C}}$ c) $X = \overline{A(B + C)}$

Section 4.7

17. *Digital Door Lock.* Implement the digital door lock in Fig. 4.1(b) using 2-input (a) NAND gates, (b) NAND and/or NOR gates.
18. Implement the function $X = A\overline{B} + \overline{C}$ using only NAND gates.
19. Implement the function $Z = \overline{A}(B + C) + \overline{B}$ using only NOR gates.

Section 4.8

20. *JK Flip-Flop.* Draw a timing diagram for the clock, with a 50% duty cycle (square wave), the J and K inputs, and Q and $\overline{Q}$ outputs of a trailing edge ($H \rightarrow L$) JK flip-flop. Use pulses to represent HIGH and LOW J and K inputs on your timing diagram. Illustrate all combinations of J and K inputs, such as $J = L, K = L, J = L, K = H$, etc. Use ideal pulses. Initially Q is LOW. {Omit delay times.}
21. *Frequency Divider.* Draw the logic diagram and timing diagram for a device that will reduce the input frequency from 4 MHz to 1 MHz.
22. *Delay Element.* Draw the circuit for an RC delay element that will produce 5-μs delays in one RC time constant. Show component values.
23. *Frequency Quadrupler.* Draw the logic diagram and timing diagram for a device that uses EX-OR gates to increase the input frequency from 10 MHz to 40 MHz. The output pulse width is not specified, so use the general symbol for any delay elements needed.
24. *Noise Discriminator.* (a) Explain how the noise discriminator in Example 4.23 works. Show a timing diagram for points A through F in Fig. 4.23. Assume gates change state at the 50% voltage level. [*Hint*: First show that if $A = L$ then $X = L$ by assuming $X = H$ and showing this cannot occur.] (b) Can the output pulse width be less than, equal to, or greater than the input signal pulse width? Explain.

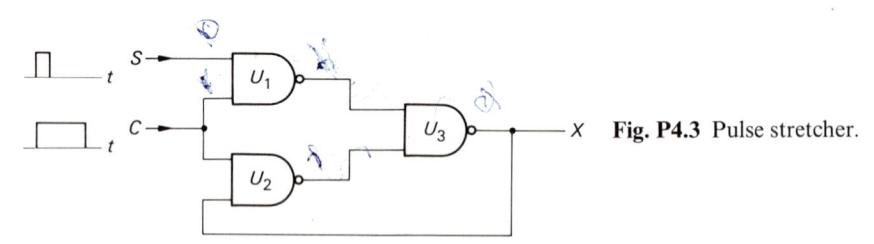

Fig. P4.3 Pulse stretcher.

25. *Pulse Stretcher.* With the aid of a timing diagram, explain how the pulse stretcher shown in Fig. P4.3 works. [*Hint*: Each narrower pulse at the S input produces an output pulse the width of the C pulses. Let C be a square wave input and let S be a narrow pulse that occurs once for every three C pulses.]
26. Draw the timing diagram for the logic diagram shown in Fig. P4.4 (a and b).

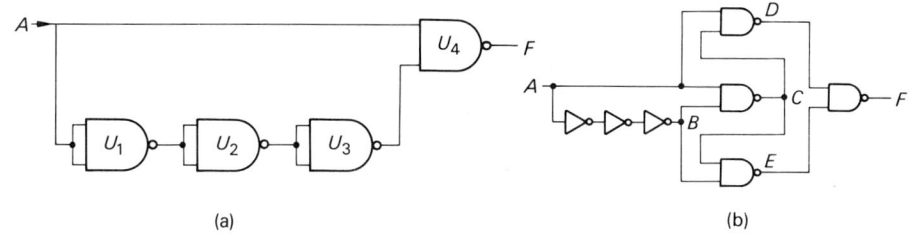

Fig. P4.4 Pulse edge detectors.

27. (a) Draw the timing diagram for points A through D in Fig. P4.5. Input A is a pulse train. (b) Describe what this circuit does.
28. *Clock Generator.* Explain in detail how the clock generator shown in the SL2 Phase Detector schematic in Appendix F works. Use a timing diagram to help in your explanation.
29. *Sense Circuit.* Referring to the SL2 Phase Detector schematic in Appendix F, explain in detail how the sense circuit works. Use a timing diagram.
30. *Phase Detector.* Explain how the phase detector circuit shown in the SL2 Phase Detector schematic in Appendix F works. Use a timing diagram. [*Hint*: Refer to Section 4.8.4.]

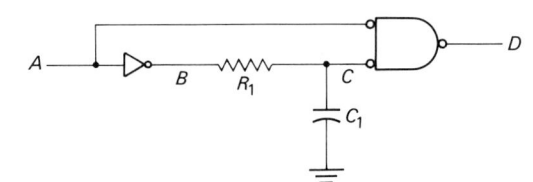

Fig. P4.5 Pulse edge detector.

REFERENCES

1. George Boole. *An Investigation of the Laws of Thought on Which Are Founded the Mathematical Theories of Logic and Probabilities.* New York: MacMillan and Co., 1854. Reprinted by Dover Publications, Inc., 1954.

2. W. Kneal. *Boole and the Algebra of Logic. Royal Society of London Notes and Records*, vol. 12, No. 1, pp. 53–63, August 1956. A short historical development of Boolean algebra.

3. Claude E. Shannon. "A Symbolic Analysis of Relay and Switching Circuits." *AIEE Transactions* **57**, 713–723, 1938.

4. Robert L. Morris and John R. Miller (eds.). *Designing with TTL Integrated Circuits.* New York: McGraw-Hill, 1971.

5. E. Mendelson. *Boolean Algebra and Switching Circuits.* Schaum's Outline Series. McGraw-Hill, 1970. Many examples of Boolean algebra are given.

6. Stanley H. Dinsmore. "Clean Up Four Logic Schematics." *Electronic Design*, pp. 80–84, June 21, 1974.

5: MSI DEVICES AND APPLICATIONS

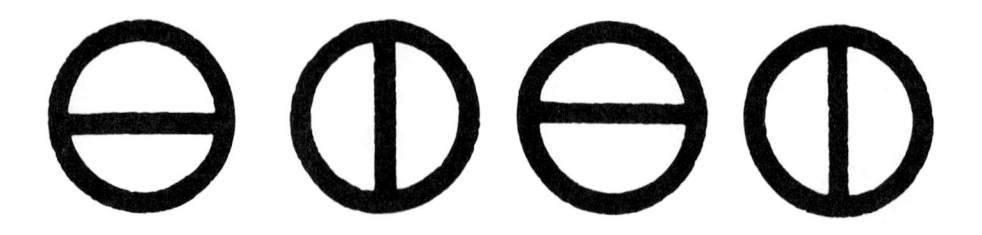

5.1 INTRODUCTION

In Chapter 4 we discussed logic gate applications using ICs with fewer than 12 gates, i.e., small scale integration (SSI). This chapter deals primarily with medium scale integration (MSI) devices and applications. MSI devices contain the equivalent of from 13 to 99 gates in each IC.

MSI devices perform functions such as comparison, addition, multiplexing, and decoding. These functions are sufficiently complex that they can be considered a major part of an entire logic system; therefore, in this text we call them **subsystems**. The TTL, ECL, and CMOSL families have a large and growing selection of MSI devices.

Since we shall need counters for use in the applications that follow, let us first take a look at them. A **binary counter** is an MSI device, made from flip-flops, that counts the number of input pulses in binary. A two-bit binary counter made from two JK flip-flops is shown in Fig. 5.1(a). Both flip-flops (FF) have their J and K inputs connected to a permanent HIGH, which causes them to toggle at the HIGH to LOW ($H \rightarrow L$) transition of their clock input. Flip-flop A is clocked by an external clock (CLK); FF-B is clocked by the Q output of FF-A.

The counter works as follows. Initially the Q outputs of FF-A and FF-B are reset (LOW) by momentarily putting the clear input LOW. Flip-flop A changes state (i.e., toggles) at the $H \rightarrow L$ transition of each clock (CLK) pulse as shown in the timing diagram of Fig. 5.1(b). The Q output of FF-B toggles at the $H \rightarrow L$ transition of each Q_A output, since it is clocked by Q_A. Suppose the HIGHs and LOWs of outputs Q_A and Q_B are interpreted as 1-bits and 0-bits. Then the count table given in Fig. 5.1(c) shows that the counter counts up from $00 = 0_{10}$ through $11 = 3_{10}$. After the count of 3, the counter repeatedly counts 0, 1, 2, 3, 0, 1, 2, 3, etc. as long as a clock input is applied.

A counter with N flip-flops can have a maximum of 2^N different outputs, called **count states**. The two-bit counter just described has $2^2 = 4$ count states. If the Q outputs in Fig. 5.1(a) are used as the counter outputs and $\bar{Q}_A$ is connected to the clock input of FF-B then the counter will count down 3, 2, 1, 0, 3, 2, 1, 0, etc. as shown in the problems.

The next section discusses the digital comparator and shows an example which uses a four-bit binary counter.

5.2 THE DIGITAL COMPARATOR   125

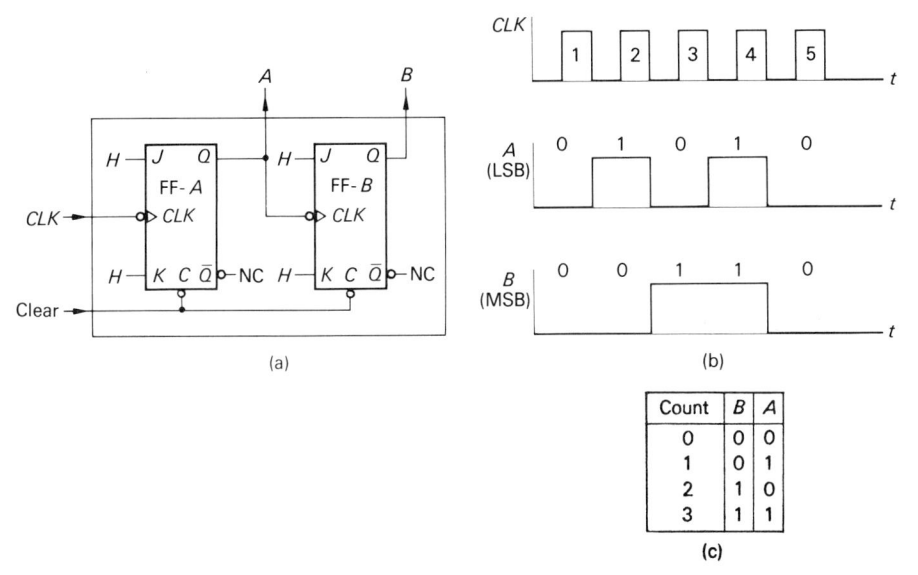

Fig. 5.1 The two-bit binary up-counter: (a) logic diagram; (b) timing diagram; (c) count table.

*5.2 THE DIGITAL COMPARATOR

Many applications require us to compare the magnitudes of two numbers. For example, in control systems it may be desirable to turn off a piece of equipment when the temperature exceeds a safe limit. In doing numerical computations it is often desirable to know when an answer is within a certain specified error.

A digital comparator compares a binary number A with a binary number B and determines whether $A = B$, $A < B$, or $A > B$. The binary numbers A and B may contain any number of bits. We discussed one-bit comparators built from logic gates in the previous chapter. The following example illustrates an application of a four-bit digital comparator.

Example 5.1 *Assembly Line Monitor.* Empty pill bottles are automatically filled on an assembly line. An operator determines how many pills will be put in various batches of bottles. Devise a detector system that indicates when a bottle is being filled, signals when it is full, and issues an alarm if too many pills are put in the bottle.

Solution. A thumbwheel switch is a useful device for setting the number of pills to be put into each bottle. A **thumbwheel switch** is a mechanical switch often having ten positions. Each switch position is mechanically converted into a binary output (Fig. 5.2a). For example, when the thumbwheel switch shown in Fig. 5.2(b) is set

126 MSI DEVICES AND APPLICATIONS							5.2

(a)

Fig. 5.2 Thumbwheel switch: (a) photograph (courtesy of Electronic Engineering Company of California); (b) functional representation for 5_{10}.

(b)

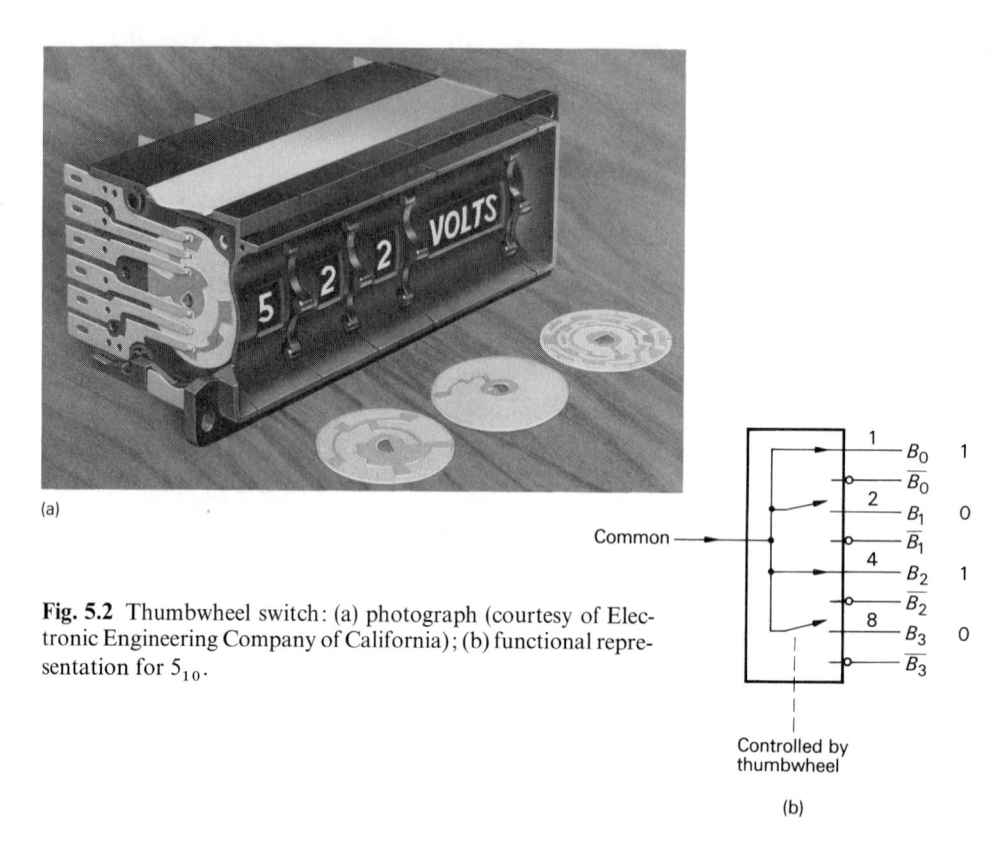

Controlled by thumbwheel

to 5, the binary output on the four output lines is $B_3 = 0, B_2 = 1, B_1 = 0, B_0 = 1$, that is, $0101 = 5_{10}$. Complementary outputs $\bar{B}_3, \bar{B}_2, \bar{B}_1,$ and $\bar{B}_0$ are also usually available.

A counter can be used to count the number of pills entering each bottle and a comparator can detect when each bottle is full. The complete system is shown in Fig. 5.3. Initially the operator sets the thumbwheel switch to a number of pills

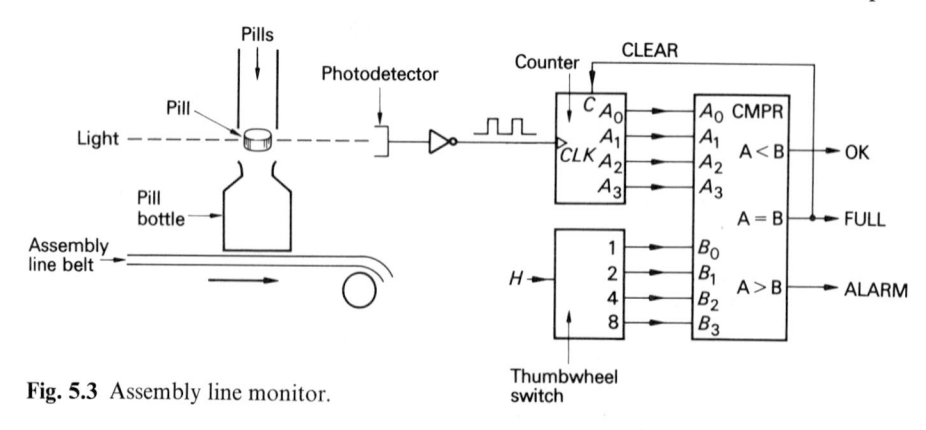

Fig. 5.3 Assembly line monitor.

5.2 THE DIGITAL COMPARATOR

to be put in each pill bottle in a given batch, say six pills. The counter is reset to 0 and the pill bottle is empty. Light illuminates the photodetector, producing a HIGH output that is inverted by the NOT gate to produce a LOW clock input to the counter.

As each pill falls into the pill bottle, the NOT gate output goes HIGH momentarily, advancing the counter output by one count. After the first pill is in the bottle, the comparator inputs are $A \equiv A_3A_2A_1A_0 = 0001$ and $B \equiv B_3B_2B_1B_0 = 0110$, so the $A < B$ output of the comparator is HIGH. After the sixth pill is in the bottle the $A = B$ output of the comparator goes HIGH ($A < B$ goes LOW) and a switch cuts off the further flow of pills. If an extra pill accidentally goes into the bottle the $A > B$ comparator output will go HIGH, sounding an alarm.

The advantage of using a comparator in this example over just a counter (which could be set to count to the desired number) is that it conveniently provides control signals. These control signals can be used to stop the flow of pills, reset the counter, advance the assembly line, and warn of an excess number of pills. ∎

The TTL, ECL, and CMOSL families have MSI digital comparators available. The logic symbol of an MSI four-bit magnitude comparator is shown in Fig. 5.4(a)

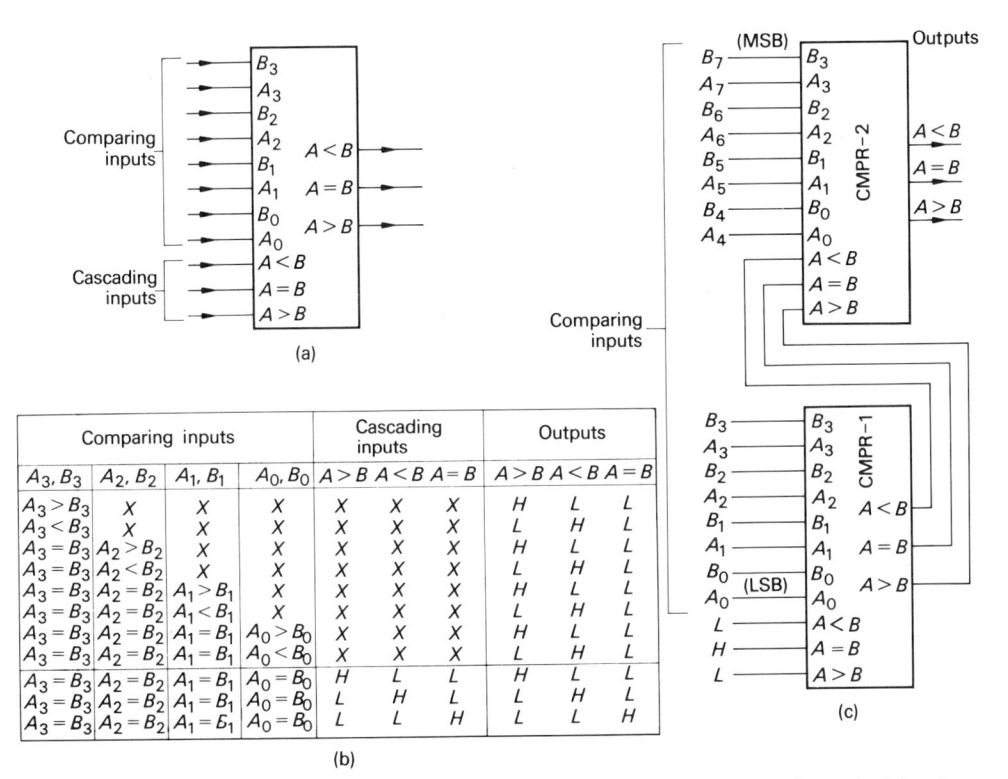

Fig. 5.4 Four-bit magnitude comparator: (a) logic symbol and (b) truth table; (c) eight-bit cascaded magnitude comparator.

and its truth table in Fig. 5.4(b). To begin with, ignore the cascading inputs of the truth table. The LSB of the two four-bit numbers is A_0B_0, and the MSB is A_3B_3. Note from the truth table that the MSB takes priority over the less significant bits, as indicated by the don't cares (X). For example, if $A_3 > B_3$ then the $A > B$ output will be HIGH no matter what the other inputs are. If the MSBs are equal then the next MSB takes priority in determining the output. For example, if $A \equiv A_3A_2A_1A_0 = 1010$ and $B \equiv B_3B_2B_1B_0 = 1001$, then $A_3 = B_3 = 1$, $A_2 = B_2 = 0$, and only the first nonequal bits $A_1 = 1$ and $B_1 = 0$ are compared (it doesn't matter what A_0 and B_0 are). Since $A_1 = 1 > B_1 = 0$, then $A > B$, so the $A > B$ output becomes HIGH.

It is often desirable to compare two eight-bit numbers. This can be done by cascading two four-bit comparators together, using the cascading inputs. Figure 5.4(c) shows the logic diagram for an eight-bit magnitude comparator made from two cascaded four-bit comparators. Note in the truth table of Fig. 5.4(b) that it doesn't matter (X) what the cascading inputs are except when all the comparing inputs are equal. When the comparing inputs of comparator CMPR-1 in Fig. 5.4(c) are equal, and the $A = B$ cascading input CMPR-1 is HIGH, then the $A = B$ output of CMPR-1 will be HIGH, and the $A < B$ and $A > B$ outputs will be LOW. For example, if $A = 1100\ 1001$ and $B = 0111\ 1001$, then the outputs of CMPR-1 are $A > B(L)$, $A < B(L)$, and $A = B(H)$. The outputs of CMPR-2 are $A > B(H)$, $A < B(L)$, and $A = B(L)$. Numbers with more than eight bits can be compared by cascading additional stages.

The typical propagation time for any four-bit comparison for the TTL 7485 comparator is 23 ns, and for the higher-speed 74S85 Schottky TTL comparator the worst case is 12 ns, as shown in Fig. 5.5.

switching characteristics, $V_{CC} = 5$ V, $T_A = 25°$ C

PARAMETER	FROM INPUT	TO OUTPUT	NUMBER OF GATE LEVELS	TEST CONDITIONS	MIN	TYP	MAX	UNIT
t_{PLH}	Any A or B data input	$A < B, A > B$	1			5		ns
			2			7.5		
			3			10.5	16	
		$A = B$	4			12	18	
t_{PHL}	Any A or B data input	$A < B, A > B$	1	$C_L = 15$ pF, $R_L = 280\ \Omega$, See Note 5		5.5		ns
			2			7		
			3			11	16.5	
		$A = B$	4			11	16.5	
t_{PLH}	$A < B$ or $A = B$	$A > B$	1			5	7.5	ns
t_{PHL}	$A < B$ or $A = B$	$A > B$	1			5.5	8.5	ns
t_{PLH}	$A = B$	$A = B$	2			7	10.5	ns
t_{PHL}	$A = B$	$A = B$	2			5	7.5	ns
t_{PLH}	$A > B$ or $A = B$	$A < B$	1			5	7.5	ns
t_{PHL}	$A > B$ or $A = B$	$A < B$	1			5.5	8.5	ns

Fig. 5.5 Propagation delay times for a Schottky TTL 74S85 four-bit MSI magnitude comparator. (Courtesy of Texas Instruments Incorporated.)

Before we can discuss the parity checker and multiplexer devices, we must examine the concept of information transmission.

5.3 WAVEFORM REPRESENTATION OF INFORMATION

Information is a measurable physical quantity such as sound, heat, an electric current, or any other form of energy. **Information theory** is the study of information. Let's see how information can be represented using binary voltages. If switch S_1 in Fig. 5.6(a) is open from time $t_0 = 0$ to $t_1 = 5$ sec, the current I through R is zero, so the output voltage E_0 will be 0 V, as shown in Fig. 5.6(b). If S_1 is closed from $t_1 = 5$ sec to $t_2 = 10$ sec, the output voltage E_0 is $+10$ V, as shown in Fig. 5.6(c). If the switch is open from 10 sec to 20 sec, $E_0 = 0$ V, and Fig. 5.6(d) results. If the "history" of the output voltage from $t = 0$ to $t = 20$ sec is put on one curve, then the single-pulse waveform of Fig. 5.6(e) results.

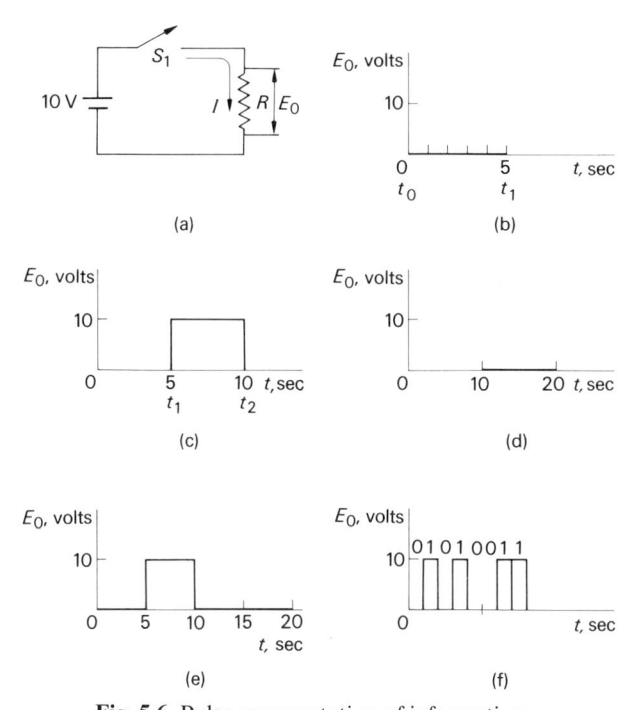

Fig. 5.6 Pulse representation of information.

If the switch is opened and closed at various time intervals, a series of repeated pulses, called a **pulse train**, results. A pulse train representing the binary number 01010011 is illustrated in Fig. 5.6(f), where a logic-1 is represented by $+10$ V and a logic-0 by 0 V. Note that each pulse in this pulse train has the same width and that the bits are written above each pulse.

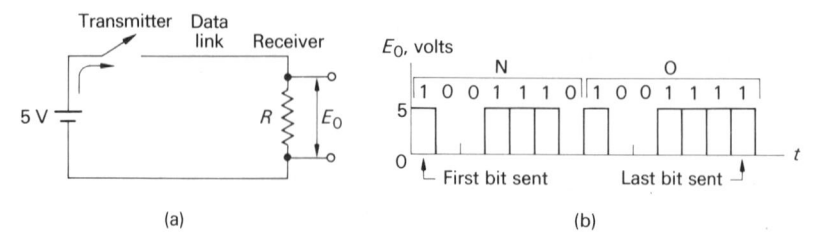

Fig. 5.7 Transmission system: (a) schematic; (b) transmitted waveform.

Suppose a message is to be set along a pair of wires, called a **transmission line** or a **data link**, from a sender to a receiver as illustrated in Fig. 5.7(a). Consider the simple message "NO." A binary representation, called a **code**, is needed for the letters "N" and "O." One standard code, which uses seven bits to represent each **character** (alphabetic, numeric, or special symbol), is the ASCII (pronounced ask'-ee) code [1]. The ASCII code is called an **alpha-numeric** (or **alphameric**) code because its characters represent alphabetic and numeric information as well as special symbols such as +, =, ″, etc. In the ASCII code "N" is represented by the bits 1001110 and "O" by 1001111. If +5 V represents the 1-bit and 0 V the 0-bit, then the message "NO" will appear as the voltage waveform of Fig. 5.7(b).

Note that the letters (information) are sent in **serial** order, i.e., the bits in the letter "N" are sent in the following sequence: a 1-bit, two 0-bits followed by three 1-bits, and a 0-bit. The bits of the letter "O" are then sent serially along the transmission line. The message is also received in serial fashion, with the first 1-bit of the letter "N" being received first. Parallel transmission, where all seven bits of each letter are sent over seven transmission lines simultaneously, is discussed in the next section.

*5.4 THE PARITY CHECKER/GENERATOR

In data communication systems, binary digits are transmitted over wires or through space. If noise enters the system or if any component malfunctions, an error in the received bits may occur. If a bank account is credited with $100,000 instead of $100, this is a serious error.

A parity checker can be used to detect errors in the received data. An additional bit, called a **parity bit**, can be added to each transmitted character to help detect errors. Two parity systems are used. In an **even-parity** system, the total number of 1-bits in any character (including the parity bit) is even. In an **odd-parity** system, the total number of 1-bits in any character (including the parity bit) is odd.

In an even-parity system, the letter "A" in the seven-bit ASCII code (1000001) would be coded as the eight-bit character 01000001, where the 0 parity bit is added to the high-order (i.e., leftmost) position so that the total number of 1-bits is even (2). The letter "C" (1000011), in even parity, becomes 11000011 so that the total

5.4 THE PARITY CHECKER/GENERATOR 131

Character	ASCII code	Even parity	Odd parity
A	1000001	01000001	11000001
C	1000011	11000011	01000011
5	0110101	00110101	10110101
+	0101011	00101011	10101011

(a)

Transmitted character	Received character	Error	Error detected
A	01000001	No	
A	01001001	Yes	Yes
A	01001101	Yes	No

(b)

Fig. 5.8 Parity: (a) ASCII code with even and odd parity; (b) example of parity errors in an even-parity system.

number of 1-bits is even (4). Figure 5.8(a) shows several ASCII characters with parity bits added to give even parity and odd parity.

If a transmitter uses even parity, any odd-parity character received has an error in it. For example, Fig. 5.8(b) shows how the letter "A" might be received from an even-parity transmitter. In the first case no error was received. The second row shows an error in the fourth bit, resulting in odd parity for "A," which is detected as an error. The third row in Fig. 5.8(b) shows errors in two bits, so that the received character still has even parity and no error is detected. Higher-level error codes can detect, and also correct, an odd and an even number of errors [1].

A parity checker is a device that checks the parity of any binary coded character at its inputs. The EXCLUSIVE-OR gate diagram shown in Fig. 5.9(a) is an eight-bit even-parity checker, also called a **parity tree**. Let's see how it works. If the inputs are $A = 1$ and $B = 1$, $C = D = E = F = G = 0$, and $H = 1$, the

$$PE = A \oplus B \oplus C \oplus D \oplus E \oplus F \oplus G \oplus H$$

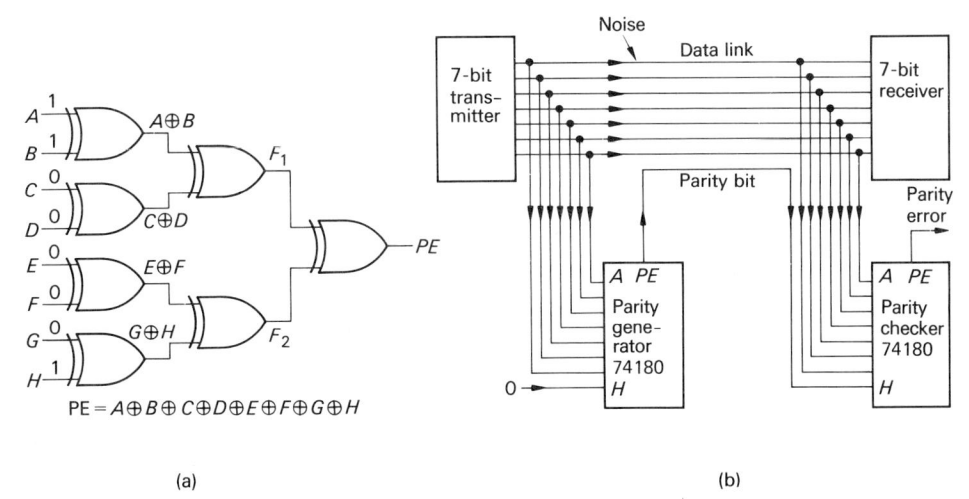

(a)   (b)

Fig. 5.9 Parity: (a) eight-bit even-parity checker logic diagram; (b) eight-bit digital transmission system with parity bit.

parity is odd. The output is determined by

$$F_1 \equiv (A \oplus B) \oplus (C \oplus D) = (1 \oplus 1) \oplus (0 \oplus 0) = 0 \oplus 0 = 0,$$
$$F_2 \equiv (E \oplus F) \oplus (G \oplus H) = (0 \oplus 0) \oplus (0 \oplus 1) = 0 \oplus 1 = 1,$$

and hence the output PE (even parity) is

$$PE = F_1 \oplus F_2 = 0 \oplus 1 = 1.$$

In general, if the number of 1-bits is odd then $PE = 1$, and if the number of 1-bits is even then $PE = 0$. Thus the parity tree in Fig. 5.9(a) checks for odd parity by giving a 1 output if its input has odd parity and provides a 0 output if its input has even parity. When using even parity, if $PE = 1$ is noted at the receiver, an error has been detected and a message can be sent to the transmitter to retransmit that particular character.

The circuit of Fig. 5.9(a) is also an even **parity generator**, since it can be used to add a 0-bit or a 1-bit to each transmitted seven-bit character by connecting the H input to a permanent LOW. If the character has an odd number of 1-bits, such as 0100011, then the output of the generator is

$$F_1 = (0 \oplus 1) \oplus (0 \oplus 0) = 1 \oplus 0 = 1,$$
$$F_2 = (0 \oplus 1) \oplus (1 \oplus 0) = 1 \oplus 1 = 0,$$
$$PE = 1 \oplus 0 = 1.$$

Thus, if the $PE = 1$ bit is sent along with the seven-bit character, an even-parity character results.

MSI parity checker/generators are available in the TTL, ECL, and CMOSL families.

The next example illustrates the use of a parity checker/generator in a data transmission system.

Example 5.2 *Digital Transmission System with Parity.* Draw the logic diagram for a data communication system that uses the seven-bit ASCII code and includes an even-parity bit.

Solution. An eight-bit even-parity generator like the one shown in Fig. 5.9(a) is used at the transmitter end of the communication system. An eight-bit parity checker is used at the receiver as shown in Fig. 5.9(b). The data link consists of eight wires, one of which carries parity information.

As each seven-bit character is transmitted, that character enters the parity generator, where a 0-bit or a 1-bit is generated so as to give the eight bits (character plus parity bit) even parity. At the receiver, the seven-bit character and the parity bit are sent to the parity checker. If the parity checker detects odd parity it produces a 1 output to indicate that an error has been received. This error signal can then be used to request that the transmitter retransmit that character. ∎

Fig. 5.10 Cathode ray tube terminal. (Courtesy of Hewlett-Packard Company.)

In Fig. 5.9(b), note that all eight bits are transmitted and received simultaneously; this is called **parallel transmission**. If the eight bits are sent one after another along a single transmission line, this is called **serial transmission**. Parallel transmission is faster than serial transmission, but it requires more wire for the data link. Generally, for distances longer than a few hundred feet, data is sent serially in order to reduce the number of transmission lines. Parallel-to-serial and serial-to-parallel converters are available to convert from one method of transmission to the other. It is also possible to send messages from several different sources using only one transmission line and multiplexers, as we shall see in the next section. Figure 5.10 shows a CRT terminal and keyboard that uses the ASCII code.

5.5 THE DIGITAL MULTIPLEXER/DATA SELECTOR

Suppose that a computer receives data in serial form from several different sources. It can handle only one source at a time, so a selector switch, called a **data selector**, can be used. Data may come from sources such as teleprinters (TRPs), cathode ray tube (CRT) terminals with keyboards, card readers, frequency counters, or temperature sensors.

The process of switching data from several channels onto one channel is also called **multiplexing**. The terms *multiplexer* and *data selector* are used synonymously in this text.

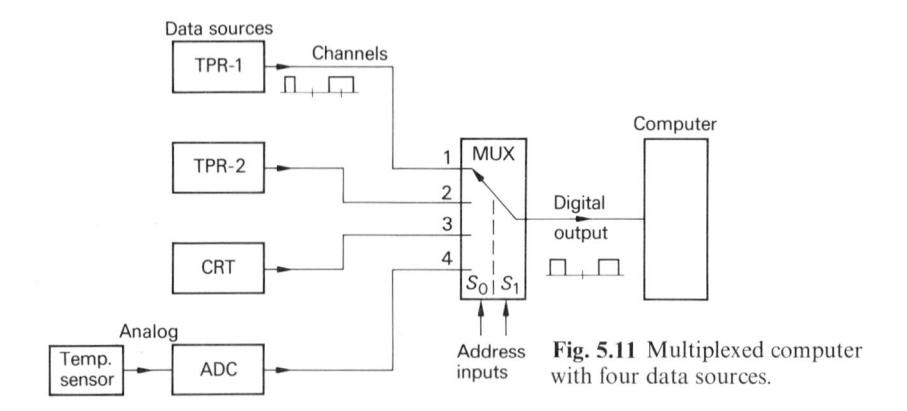

Fig. 5.11 Multiplexed computer with four data sources.

Figure 5.11 shows a multiplexed system consisting of an electronic digital computer and four information sources: two teleprinters, a CRT terminal, and a temperature sensor. Since the output of the temperature sensor is generally an analog (continuous) signal, it is converted to a digital signal with an analog-to-digital converter (ADC).

The data selector (MUX) is connected between the four serial data sources and the computer, and serves to connect any one of four channels to the computer. Thus the data selector acts like a switch. The address inputs (S_0, S_1) are digital inputs that determine the position of the switch. The data selector address inputs could be controlled by the computer.

Data selectors are frequently used in microcomputers to route data from one of several sources to the computer's memory. Appendix G gives the logic diagram for a microcomputer, and the problems at the end of this chapter discuss the data selector logic it uses.

Now let's see in greater detail how a multiplexer works. Suppose three separate messages are to be sent serially along a single transmission line in a data communication system. Instead of using three separate transmission lines, as required in a parallel transmission system, a data selector/multiplexer can be used. This will save two transmission lines but at a reduced net transmission speed. Serial transmission can be accomplished by sending the first character of each message, then the second character of each message, etc., until each character of each message is sent.

For example, suppose three messages "NO," "WE," and "MY" are coded in the ASCII code. Figure 5.12(a) shows a mechanical switch representation of a multiplexer with three input channels y_0, y_1, and y_2, which are selected by addresses A and B. First, channel y_0 is addressed long enough for the seven bits (no parity bit is being sent) of the letter "N" to be transmitted to the output Z. Then channel y_1 is addressed long enough for the "W" to be transmitted. Finally, channel y_2 is addressed and the letter "M" is transmitted. This sequence is then repeated so that the second letters of the messages, "O," "E," and "Y," are sent in sequence. The output of the multiplexer then appears as a bit stream for the letters

5.5 THE DIGITAL MULTIPLEXER/DATA SELECTOR 135

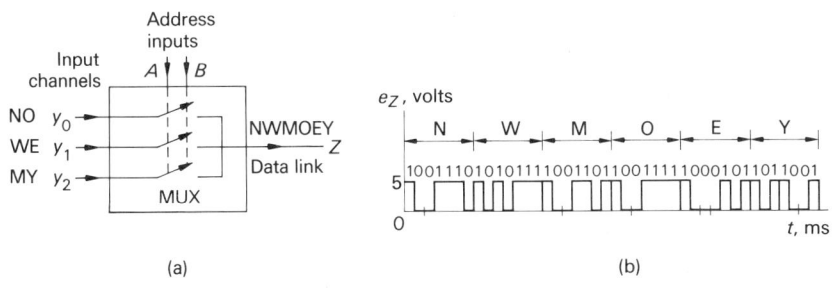

Fig. 5.12 Multiplexer: (a) time division multiplexed (TDM) data transmission; (b) multiplexed output bit stream.

"NWMOEY," as shown in Fig. 5.12(b). This type of multiplexing is called **time division multiplexing**(TDM) because it divides the time during which each message is transmitted along the data link into time intervals. At the receiver end, a demultiplexer (Section 5.8) separates the three messages onto three output channels.

MSI digital data selector/multiplexers are available in the TTL, ECL, and CMOSL families. Multiplexers frequently have from four to sixteen inputs. Figure 5.13(a) shows the logic diagram for a dual four-input data selector. Since each multiplexer has four input lines it is also called a **4-line-to-1-line** or a **1-of-4 multiplexer**. Note that each multiplexer uses a four-wide AND-OR gate.

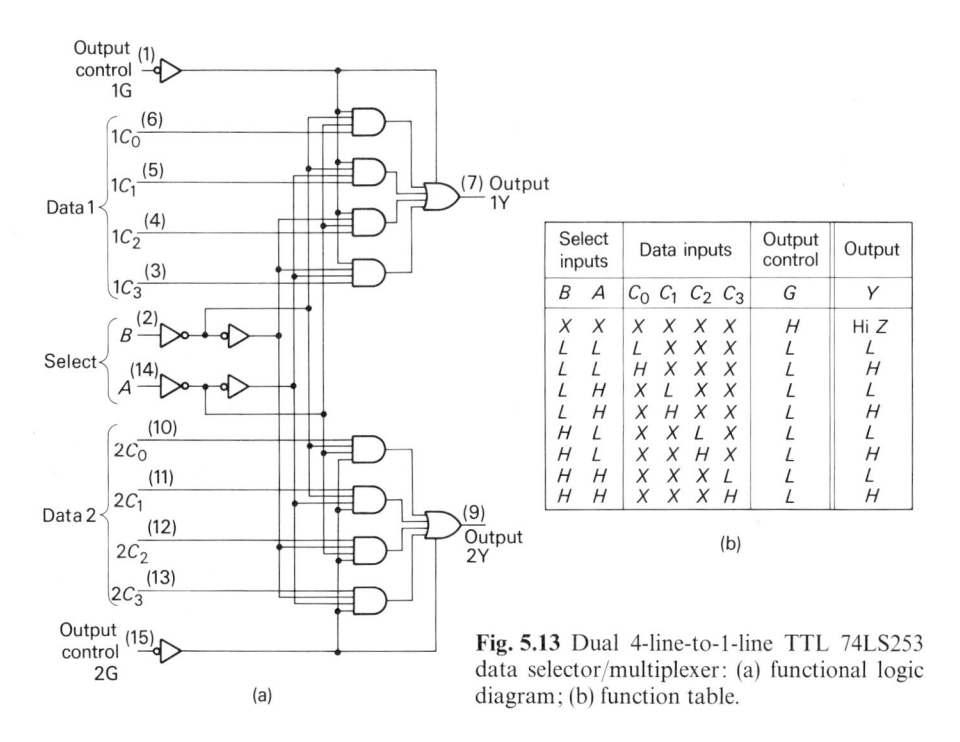

Select inputs		Data inputs				Output control	Output
B	A	C_0	C_1	C_2	C_3	G	Y
X	X	X	X	X	X	H	Hi Z
L	L	L	X	X	X	L	L
L	L	H	X	X	X	L	H
L	H	X	L	X	X	L	L
L	H	X	H	X	X	L	H
H	L	X	X	L	X	L	L
H	L	X	X	H	X	L	H
H	H	X	X	X	L	L	L
H	H	X	X	X	H	L	H

Fig. 5.13 Dual 4-line-to-1-line TTL 74LS253 data selector/multiplexer: (a) functional logic diagram; (b) function table.

recommended operating conditions

	SN54LS253 MIN	NOM	MAX	SN74LS253 MIN	NOM	MAX	UNIT
Supply voltage, V_{CC}	4.5	5	5.5	4.75	5	5.25	V
High-level output current, I_{OH}			−1			−2.6	mA
Low-level output current, I_{OL}			4			8	mA
Operating free-air temperature, T_A	−55		125	0		70	°C

electrical characteristics over recommended operating free-air temperature range (unless otherwise noted)

PARAMETER	TEST CONDITIONS†		SN54LS253 MIN	TYP‡	MAX	SN74LS253 MIN	TYP‡	MAX	UNIT
V_{IH} High-level input voltage			2			2			V
V_{IL} Low-level input voltage					0.6			0.8	V
V_I Input clamp voltage	V_{CC} = MIN,	I_I = −18 mA			−1.5			−1.5	V
V_{OH} High-level output voltage	V_{CC} = MIN, V_{IH} = 2 V, V_{IL} = V_{IL} max, I_{OH} = MAX		2.4	3.4		2.4	3.1		V
V_{OL} Low-level output voltage	V_{CC} = MIN, V_{IH} = 2 V, V_{IL} = V_{IL} max	I_{OL} = 4 mA			0.4				V
		I_{OL} = 8 mA						0.5	
$I_{O(off)}$ Off-State (high-impedance state) output current	V_{CC} = MAX, V_{IH} = 2 V	V_O = 2.7 V			20			20	μA
		V_O = 0.4 V			−20			−20	
I_I Input current at maximum input voltage	V_{CC} = MAX,	V_I = 5.5 V			0.1			0.1	mA
I_{IH} High-level input current	V_{CC} = MAX,	V_I = 2.7 V			20			20	μA
I_{IL} Low-level input current	V_{CC} = MAX,	V_I = 0.4 V			−0.36			−0.36	mA
I_{OS} Short-circuit output current §	V_{CC} = MAX		−6		−40	−5		−42	mA
I_{CC} Supply current	V_{CC} = MAX, See Note 2	Condition A		7	12		7	12	mA
		Condition B		8.5	14		8.5	14	

switching characteristics, V_{CC} = 5 V, T_A = 25°C

PARAMETER¶	FROM (INPUT)	TO (OUTPUT)	TEST CONDITIONS	MIN	TYP	MAX	UNIT
t_{PLH}	Data	Y			11	18	ns
t_{PHL}					13	20	
t_{PLH}	Select	Y	C_L = 15 pF, R_L = 2 kΩ, See Note 3		20	30	ns
t_{PHL}					21	32	
t_{ZH}	Output Control	Y			11	18	ns
t_{ZL}					15	23	
t_{HZ}	Output Control	Y	C_L = 5 pF, R_L = 2 kΩ, See Note 3		27	41	ns
t_{LZ}					12	19	

¶ t_{PLH} ≡ Propagation delay time, low-to-high-level output
t_{PHL} ≡ Propagation delay time, high-to-low-level output
t_{ZH} ≡ Output enable time to high level
t_{ZL} ≡ Output enable time to low level
t_{HZ} ≡ Output disable time from high level
t_{LZ} ≡ Output disable time from low level

Fig. 5.14 TTL 74LS253 multiplexer specifications. (Courtesy of Texas Instruments Incorporated.)

The truth table for this 4-line-to-1-line multiplexer is given in Fig. 5.13(b). For example, if the control (also called the enable) input is LOW and the address inputs are $A = L$ and $B = L$, then input C_0 is connected to the output Y. Most data selectors have an enable input that allows the multiplexer to be activated at any specific time.

5.5   THE DIGITAL MULTIPLEXER/DATA SELECTOR   137

Figure 5.14 shows the specifications for a TTL 74LS253 4-line-to-1-line multiplexer. This multiplexer has a three-state output for bus-organized applications. The output HIGH current drive I_{OH} capability is 2.6 mA, considerably greater than the 0.4 mA maximum for the 74LS00 logic gates. Note that the off-state leakage current is only 20 μA for a LOW or HIGH output (bus) voltage of 0.4 V or 2.7 V, respectively. Its power dissipation is only 35 mW. Delay times for data input to output, select, and control to output are shown in Fig. 5.14.

The ECL 10174 4-line-to-1-line multiplexer has a typical data-to-output delay time of 3.5 ns, and address-to-output delay time of 5 ns.

5.5.1 Cascading Multiplexers

Multiplexers can be cascaded to increase the number of channels that can be multiplexed. Figure 5.15(a) shows a 16-line multiplexer, made from three dual four-input multiplexers whose truth table is given in Fig. 5.13(b). Addresses S_0 and S_1 select four out of the sixteen input channels, which are routed to MUX-3. Addresses S_2 and S_3 select one out of these four channels and route it to the output F. For example, if we want to route channel 2 to the output F, then the address must be $S_0 = L$, $S_1 = H$, $S_2 = S_3 = L$. To route channel 13 to the output F, the address must be $S_0 = H$, $S_1 = L$, $S_2 = S_3 = H$.

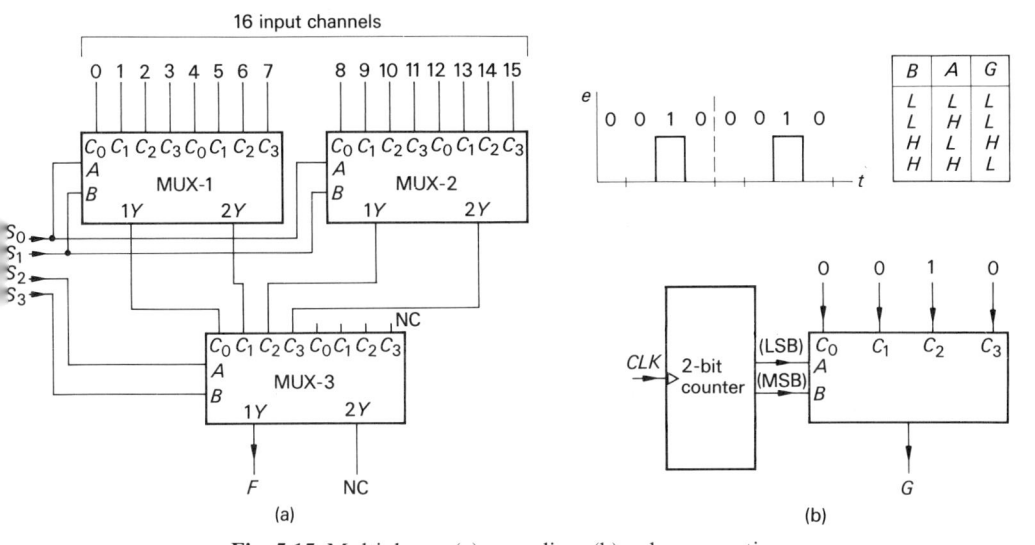

Fig. 5.15 Multiplexer: (a) cascading; (b) pulse generation.

5.5.2 Pulse Generation [2]

In many digital systems it is necessary to generate special logic signals, such as those used to control different parts of the system. These logic signals may be rather irregular and thus difficult to generate with logic gates. Multiplexers can be used to generate arbitrary logic signals easily.

Suppose we want to generate the pulse waveform shown in Fig. 5.15(b). The truth table that represents this function is also given in Fig. 5.15(b). A four-input data selector can be used to generate this function. The four inputs are connected to HIGH or LOW logic levels according to the truth table. The data selector inputs A and B are addressed with a counter in the sequence given in the truth table, and the output is the desired function G. For example, when $A = B = L$, the input C_0 is connected to the output G. Since the input to C_0 is LOW, then the output will be LOW. If the addressing is done in sequence, the waveform of Fig. 5.15(b) is generated. The clock frequency determines the width of the pulses.

More complicated pulse waveforms can be generated by using multiplexers with more input lines, or by using pulse inputs to the multiplexer as shown in the problems.

5.5.3 Function Generation

A multiplexer can be thought of as a function generator instead of a pulse generator. Consider the generation of the Boolean function F given by the truth table in Fig. 5.16(a).

Let's show the logic gate implementation of F first. Using the method developed in Chapter 4 yields the Boolean function $F = \bar{A}\bar{B}C + \bar{A}B\bar{C} + A\bar{B}\bar{C} + A\bar{B}C$. Factoring $A\bar{B}$ out of the last two terms and recalling that $\bar{C} + C = 1$ gives $F = \bar{A}\bar{B}C + \bar{A}B\bar{C} + A\bar{B}$, which is implemented with NAND gates in Fig. 5.16(b). Four NAND gates and three inverters are required, or a total of two TTL IC packages.

The function F can be generated with one 8-line-to-1-line multiplexer as shown in Fig. 5.16(c). For example, when the address inputs are $A = L$, $B = C = H$, input I_4 is connected to the output, so F is LOW as required. Note that the multiplexer implementation requires only one 8-line-to-1-line multiplexer IC to generate *any* function of three variables, and requires no function simplification prior to implementation. A more efficient method for generating functions using multiplexers is given in the references [2].

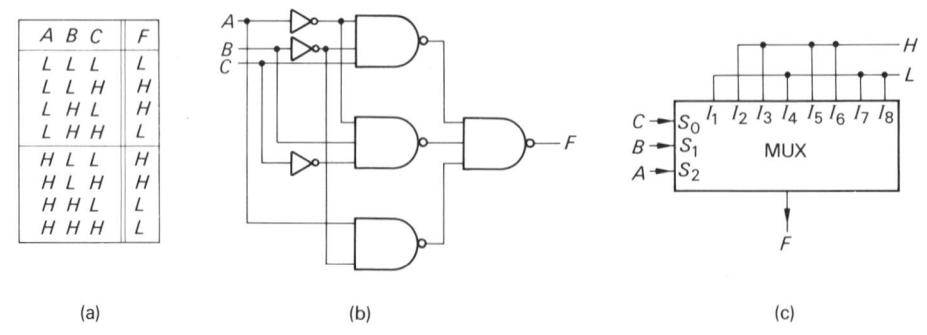

A B C	F
L L L	L
L L H	H
L H L	H
L H H	L
H L L	H
H L H	H
H H L	L
H H H	L

(a)   (b)   (c)

Fig. 5.16 Generation of a Boolean function F: (a) truth table for F; (b) logic gate implementation; (c) multiplexer implementation.

5.6 INTRODUCTION TO BINARY CODES

Before we can discuss decoders, we need to consider some basic concepts of binary codes.

Electronic circuits often process, store, and convey informatin as binary numbers. This information is most conveniently displayed for human consumption as decimal numbers, alphabetic letters, or special symbols. Therefore, devices are required that convert information from one form to another.

A **character** is a symbol used to convey information. Characters may be alphabetic symbols ($A, B, \ldots, Z; \alpha, \beta, \ldots, \eta$), numerical digits (0, 4, 296; 001, 010, ..., 111) or special symbols ($+, =, ;, ", @, <$). A **binary code** is a unique representation of one set of symbols by a set of binary symbols. A code is often put in the form of a table. There are many binary codes, each having its own advantages for particular applications.

A frequently used numeric binary code, which relates decimal numbers and binary numbers, is the **8421 code**. In the 8421 code each decimal digit is represented by a group of four bits. For example, 15_{10} is written in the 8421 code as 0001 0101,

Table 5.1 Binary Codes

Decimal	Natural binary	8421 BCD	Excess-3 BCD	1's complement	ASCII	EBCDIC
0	0000	0000	0011	1111	0110000	11110000
1	0001	0001	0100	1110	0110001	11110001
2	0010	0010	0101	1101	0110010	11110010
3	0011	0011	0110	1100	0110011	11110011
4	0100	0100	0111	1011	0110100	11110100
5	0101	0101	1000	1010	0110101	11110101
6	0110	0110	1001	1001	0110110	11110110
7	0111	0111	1010	1000	0110111	11110111
8	1000	1000	1011	0111	0111000	11111000
9	1001	1001	1100	0110	0111001	11111001
10	1010	0001 0000	0100 0011	0101		
11	1011	0001 0001	0100 0100	0100		
12	1100	0001 0010	0100 0101	0011		
13	1101	0001 0011	0100 0110	0010		
14	1110	0001 0100	0100 0111	0001		
15	1111	0001 0101	0100 1000	0000		
16	10000	0001 0110	0100 1001	01111		
26	11010	0010 0110	0101 1001	00101		
A					1000001	11000001
B					1000010	11000010
Z					1011010	11101001
=					0111101	01111110

since $1_{10} = 0001_2$ and $5_{10} = 0101_2$. The 8421 code gets its name from the fact that in each group of four bits, the LSB has a value (called a **weight**) of 1, the next MSB has a weight of 2, the next MSB has a weight of 4, and the MSB a weight of 8. The decimal number 428 in 8421 is 0100 0010 1000. Table 5.1 shows the natural binary code and the 8421 code for several decimal numbers. Note that the natural binary numbers and the 8421 code are identical for the decimal numbers 0 through 9. The 8421 code is easy to use for humans, but not very useful for binary arithmetic operations in digital devices. Note that the 8421 code is only a numeric code and is not used for alphabetic characters or special symbols. The 8421 code is one example of a class of codes called **binary coded decimal** (BCD) codes. BCD codes are used to code each decimal digit separately into a group of four or more bits.

Example 5.3 Encode the decimal number 268 into (a) its 8421 BCD equivalent and (b) its natural binary code equivalent.

Solution.
a)
$$268_{10} = \underbrace{0010}_{2}\,\underbrace{0110}_{6}\,\underbrace{1000}_{8}$$

b) $268_{10} = 1 \times 2^8 + 0 \times 2^7 + 0 \times 2^6 + 0 \times 2^5 + 0 \times 2^4$
$+ 1 \times 2^3 + 1 \times 2^2 + 0 \times 2^1 + 0 \times 2^0 = 100001100_2$

Clearly, the 8421 coded numbers are easier to code, read, write, and remember. ■

A second numeric binary code shown in Table 5.1 is the **excess-3 BCD code**. It is useful for doing binary arithmetic. The excess-3 code is an **unweighted code**, i.e., the position of the bits cannot be used to calculate their decimal value. The excess-3 code is obtained by adding 3_{10} to each decimal digit in the given decimal number, and then converting the new decimal number to binary using the 8421 code. For example, 4_{10} becomes $4 + 3 = 7$ or 0111 in excess-3, as shown in Table 5.1. Similarly, decimal 15 becomes $1 + 3 = 4$ and $5 + 3 = 8$, or 0100 1000.

The **1's complement code** is useful for doing binary subtraction. The 1's complement of a natural binary number is formed by complementing each bit. For example the 1's complement of 011 is 100. The 1's complement of 11001 is 00110. Table 5.1 shows some additional examples of the 1's complement code.

The **2's complement code** is also a numeric binary code, and is formed by adding 1 to the 1's complement of a number. That is,

$$\text{2's complement} = \text{1's complement} + 1$$

For example, the 2's complement of 0110 is $1001 + 1 = 1010$. The 2's complement of 01100 is $10011 + 1 = 10100$. The 1's and 2's complement codes are used in some computers to do subtraction. For example, to subtract $9 - 4$, form the 2's complement of 4, add this to 9, and discard any carry. Thus $9 - 4$ is $1001 + 1100$ (2's complement of 4) = $\cancel{1}0101 = 5_{10}$. The carry of 1 has been discarded.

The **ASCII code** is shown in Table 5.1 for comparison. Note that the ASCII code is an alphameric binary code. Finally, the eight-bit alphameric binary code used for transmission of data in IBM equipment, called the **EBCDIC** (pronounced ib'-see-dick) **code**, is also illustrated in Table 5.1. Both the seven-bit ASCII and the eight-bit EBCDIC codes have upper and lower case alphabetic characters. If a parity bit (Section 5.4) is added, as it often is, the two codes have a total of eight bits and nine bits, respectively, in each character.

5.7 THE DECODER

A **decoder** is a device that converts simple symbols, like binary 0 and 1, into symbols that contain more information, such as A, $=$, 9, etc. A **binary decoder** decodes binary numbers into characters.

Decoders are useful in applications such as input decoders for digital readout systems (Chapter 6), address decoders for semiconductor memories (Chapter 10), and control circuits (Example 5.4).

$\bar{E}$	A_1	A_0	$\bar{0}$	$\bar{1}$	$\bar{2}$	$\bar{3}$
L	L	L	L	H	H	H
L	L	H	H	L	H	H
L	H	L	H	H	L	H
L	H	H	H	H	H	L
H	X	X	H	H	H	H

(b)

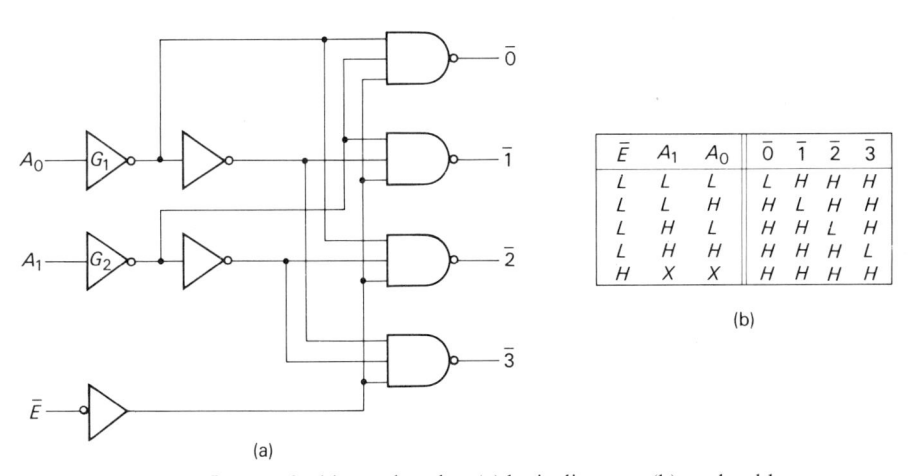

Fig. 5.17 1-of-4 binary decoder: (a) logic diagram; (b) truth table.

Figure 5.17(a) shows the logic diagram for a binary decoder that decodes two binary inputs into one of four decimal outputs. It is called a **1-of-4 decoder**, or a **2-line-to-4-line decoder**. The enable input $\bar{E}$ is used when it is desired to decode inputs only at certain specific times. The decoder will not decode the inputs until the enable input, also called a **strobe input**, goes LOW (active LOW). Buffer gates G_1 and G_2 reduce the input loading. Internal buffers, such as G_1 and G_2, can be designed to drive a known load with a reduced voltage swing, thereby saving power, increasing the speed, and minimizing the number of external SSI gates required.

Note from the truth table of Fig. 5.17(b) that the decoder has active LOW outputs, in order to reduce the power dissipation. For example, if the inputs are $A_0 = L$, $A_1 = H$ (binary 10_2), and $\bar{E} = L$, then the $\bar{2}$ (decimal 2) output will be

142   MSI DEVICES AND APPLICATIONS   5.7

LOW; all other inputs remain HIGH. This decoder is a binary-to-decimal decoder that decodes the decimal digits 0 through 3.

The next example shows how a decoder can be used in a TV digital channel selector [3, 4, 5].

Example 5.4 *TV Digital Channel Selector.* Develop a system that will digitally select one of sixteen TV channels.

Solution. Figure 5.18 shows a TV channel control system. The system consists of two normally open (N.O.) channel selector switches (UP and DOWN), a 2-Hz clock, two NAND gate *RS* latches, a four-bit counter, a 1-of-16 decoder, and 16 transistor-diode switches.

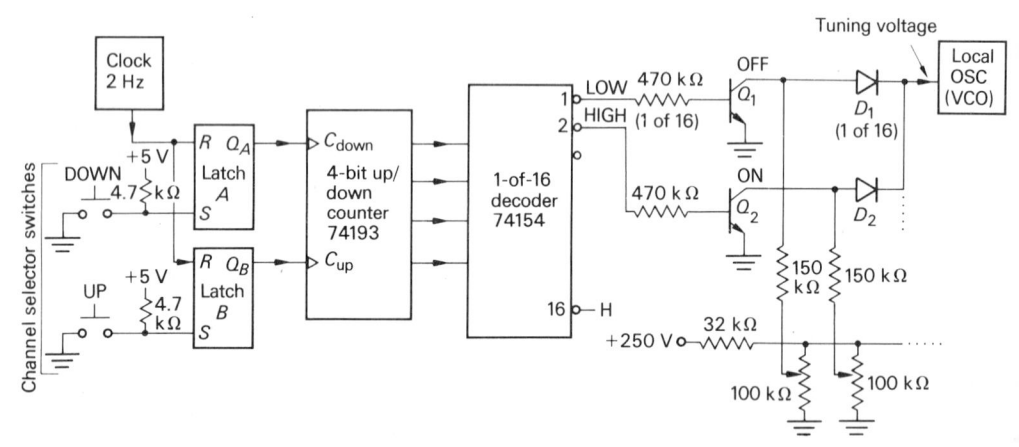

Fig. 5.18 TV digital channel selector. (Courtesy of Heath Company.)

The counter counts up (or down) when the C_{up} (or C_{down}) input to the counter goes from a LOW to a HIGH. The counter output does not change when the C_{down} or C_{up} inputs are a constant HIGH or LOW. When the channel selector switches are not depressed (are open) the set inputs to both latches are HIGH, so neither the Q_A nor the Q_B outputs change with the clock pulse inputs to the *R* inputs, as verified in the problems. If the UP channel selector switch is depressed, the reset input of latch *B* becomes LOW, and the *Q* output of latch *B* toggles with each clock pulse. With each $L \to H$ Q_B transition, the counter counts up by one, until the UP switch is released. If the counter output is *LLLL*, the 1 output of the 1-of-16 decoder will be LOW; all other decoder outputs will be HIGH as shown in Fig. 5.18. With the 1 output of the decoder LOW, transistor Q_1 will be OFF. The channel 1 tuning voltage will then be applied through D_1 to the voltage controlled local oscillator in the TV tuner, which selects channel 1. Transistors Q_2 through Q_{16} are turned on by the HIGH decoder outputs, shorting these tuning voltages to ground so that diodes D_2 through D_{16} act like open switches.

The schematic for this channel selector with the digital channel and time readout is given in the problems at the end of this chapter. ∎

A wide variety of MSI decoders are available in the TTL, ECL, and CMOSL families. One-of-four, 1-of-8, 1-of-10, and 1-of-16 decoders are commonly available. A 1-of-10 decoder has four inputs with a total of $2^4 = 16$ different combinations, only ten of which are used. The remaining six input codes are called **invalid inputs** because they do not correspond to any of the ten valid outputs and therefore should not be used.

In order to obtain more than 16 outputs, decoders can be cascaded. The main disadvantage of cascading decoders is the increased delay time. Figure 5.19(a) illustrates the principle with a 1-of-16 decoder made from one 1-of-4 and two 1-of-10 decoders. The 2^3 address input of the 1-of-4 decoder enables one of the 1-of-10 decoders. The 2^2, 2^1, and 2^0 address inputs then select one of the ten outputs of the enabled decoder. Two more 1-of-10 decoders could be added to the 1-of-4 decoder in Fig. 5.19(a) to make a 1-of-32 decoder. Larger decoders, such as a 1-of-1024, are also possible [2, p. 2–17].

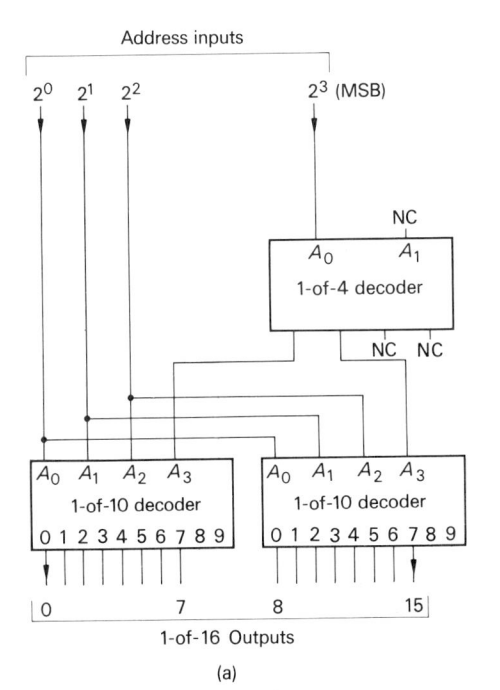

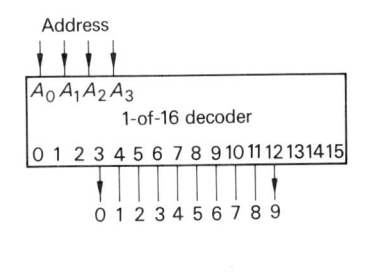

Fig. 5.19 Decoders: (a) cascaded 1-of-16 decoder; (b) decoding table for a 1-of-16 decoder used to decode three codes; (c) 1-of-10 excess-3 decoder logic diagram.

144   MSI DEVICES AND APPLICATIONS   5.8

Some decoders are available with open collectors so that the outputs can be connected together to perform the wired-AND function (Chapter 3) and for driving higher current loads.

A 1-of-16 decoder can be used to decode any four-bit weighted or unweighted code by selecting the decoder outputs in the desired sequence. Figure 5.19(b) shows the decoding table for three different codes. For example, the decimal 6 is decoded in the 8421 code at decoder output 6. For the 5421 code, whose weights are 5, 4, 2, and 1 (5_{10} = 1000, 6_{10} = 1001, etc.), and the excess-3 codes, decimal 6 is decoded at decoder output 9. Thus, to decode the excess-3 code, the decoder outputs would be labeled as shown in Fig. 5.19(c).

Decoders can also be used to generate Boolean functions as described in the references [2, p. 2–7].

5.8 THE DEMULTIPLEXER

The purpose of a **demultiplexer** is to separate a multiplexed signal, containing N channels of information, into N separate output channels. Figure 5.20, which shows a switch representation of a three-channel demultiplexer, illustrates how the demultiplexer input is connected to the outputs.

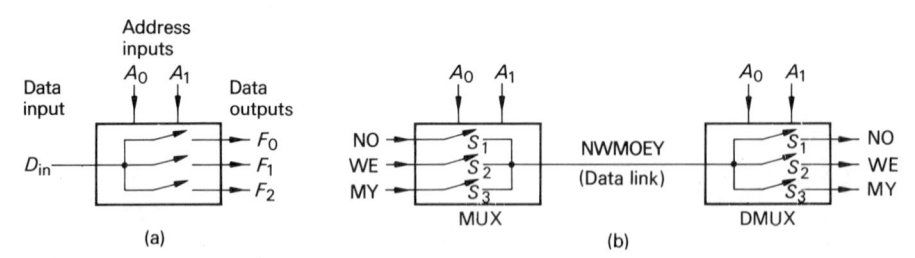

Fig. 5.20 Demultiplexers: (a) switch representation; (b) switch representation of a multiplexed-demultiplexed communication system.

Referring to Fig. 5.12(a), we can see that a demultiplexer is just the reverse of a multiplexer. A demultiplexer has one input and N outputs, while a multiplexer has N inputs and one output.

Let's see how a basic multiplexed data communication system would look using switch representations.

Example 5.5 *Basic Digital Communication System.* Draw a three-channel multiplexed data communication system. Use the switch representation for the multiplexer and demultiplexer.

Solution. The multiplexer and the three messages shown in Fig. 5.12(a), combined with the three-channel demultiplexer of Fig. 5.20(a), are shown in Fig. 5.20(b).

If the multiplexer and demultiplexer inputs (A_0 and A_1) are addressed synchronously and sequentially, the switches S_1, S_2, and S_3 will open and close in sequence. The multiplexed letters "NWMOEY" will be sent in serial form over the data link and demultiplexed into the three separate messages "NO," "WE," and "MY." ∎

The logic diagram for the ECL 10171 demultiplexer is shown in Fig. 5.21(a). There are two address inputs A_0 and A_1, one active LOW data input $\overline{D}_{in}$, one active LOW enable input $\overline{E}$, and four outputs $\overline{F}_0, \overline{F}_1, \overline{F}_2$, and $\overline{F}_3$. The wire-ORing connections of gate outputs G_1 through G_4 are done within the decoder IC and cannot be changed by the user. Since one line is demultiplexed into four lines it is called a **1-of-4** or a **1-line-to-4-line** demultiplexer.

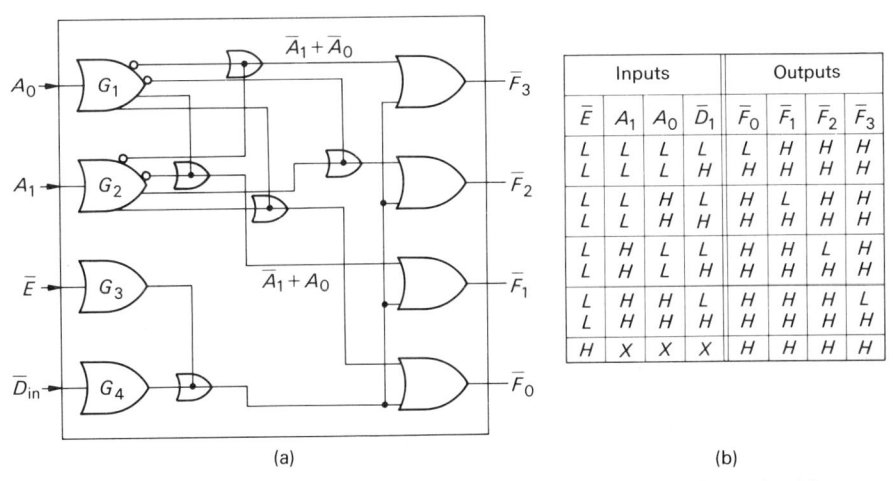

Fig. 5.21 ECL 10171 1-of-4 demultiplexer: (a) logic diagram; (b) truth table.

The active LOW enable input $\overline{E}$ must be LOW in order to enable the outputs. When the address inputs are $A_0 = A_1 = L$, then the data input is connected to the $\overline{F}_0$ output as shown by the truth table in Fig. 5.21(b).

A wide variety of demultiplexers are available in the TTL, ECL, and CMOSL families, including 1-of-4, 1-of-8, 1-of-10, and 1-of-16 demultiplexers.

Demultiplexers can be cascaded, just like decoders, in order to increase the number of output lines.

The essential difference between a decoder and a demultiplexer is that for a demultiplexer the address inputs are used to perform the switching. In a decoder, data to be decoded enters the address inputs, and the enable input, if available, is used to enable the decoder. For example, the decoder of Fig. 5.17(a) becomes a four-line demultiplexer if the data enters the enable input $\overline{E}$, and A_0 and A_1 are used to select the desired output channel.

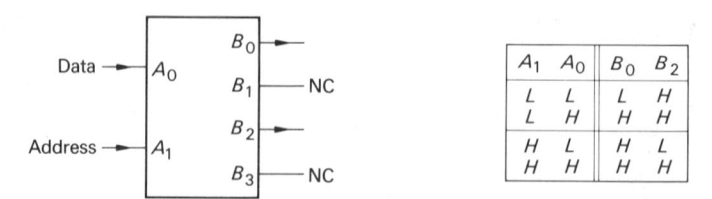

Fig. 5.22 Demultiplexer made from a decoder having no enable input.

A_1	A_0	B_0	B_2
L	L	L	H
L	H	H	H
H	L	H	L
H	H	H	H

A decoder without an enable input can be converted into a demultiplexer by using one of the addresses for the data input. Figure 5.22 shows a 1-of-4 decoder converted into a 1-of-2 demultiplexer by using the A_0 address input for the data input. Address A_1 switches the data between outputs B_0 and B_2. When $A_1 = L$ the data input is connected to B_0, and when $A_1 = H$ the data input is connected to output B_2, as shown by the truth table.

Figure 5.23 shows the antenna and ground station for a satellite communication system that uses multiplexing.

Fig. 5.23 Ground station for a satellite communication system, located at Etam, West Virginia. (Courtesy of Comsat.)

Example 5.6 *Eight-Channel Digital Communication System.* Draw the logic diagram for an eight-channel multiplexed digital communication system that connects six teleprinters, one CRT terminal, and one analog temperature sensor and recorder together. Include an automatic addressing scheme, and a method for synchronizing the transmitter with the receiver.

Solution. A three-bit counter that counts from 0 through 7 can be used to address the multiplexer and demultiplexer in sequence as shown in Fig. 5.24. The two counters are synchronized with a separate synchronizing transmission line. When the counter at the transmitter is at count 0, the first teleprinter (TPR) is connected to the transmission line. At the receiving end the counter is also at its zeroth count. Hence while the clock is HIGH, the AND gate at the input of the demultiplexer is enabled and the transmission line is connected to the first receiving teleprinter. Usually one or more characters are sent from each device before switching to the next channel. When the counters count to 1, the second teleprinters at each end are connected together. A digital-to-analog converter (DAC) is used in receiver channel 7 to convert the digital temperature signal into an analog signal, which drives a temperature recorder. An analog-to-digital converter (ADC) is used in transmitter channel 7 to convert the analog temperature signal into a digital signal for the multiplexer.

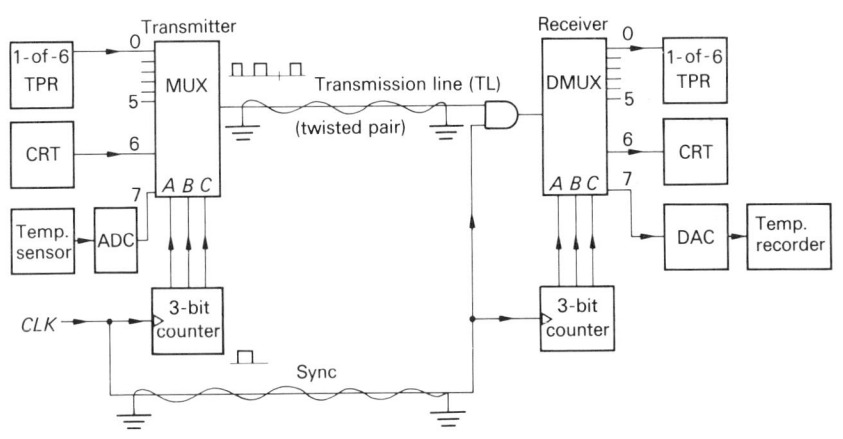

Fig. 5.24 Eight-channel digital communication system.

The transmission lines are twisted in order to minimize the amount of noise entering the system through these lines. If the transmission line is longer than a few feet, special line drivers and receivers are usually used at each end to help minimize noise, especially for high-speed ECL circuits. These devices are called **line drivers** and **line receivers** and they are available in IC form. In more sophisticated systems, the synchronizing pulse is sent along the data link with the data pulses. Parity bits can also be added to each channel. ∎

5.9 ENCODERS

An **encoder** is a device that converts from complex symbols such as A, $>$, 4, to a group of simpler symbols (each of which contains less information) such as 1001, 1110, and 1011. Thus encoding is the opposite of decoding.

Figure 5.25(a) shows an encoder, made from four OR gates, that codes decimal inputs into 8421 BCD coded outputs. For example, when the input corresponding to the decimal number 3 is HIGH, all other inputs being kept LOW, the 1 and the 2 outputs of the encoder become HIGH, that is, $3_{10} = 0011_2$.

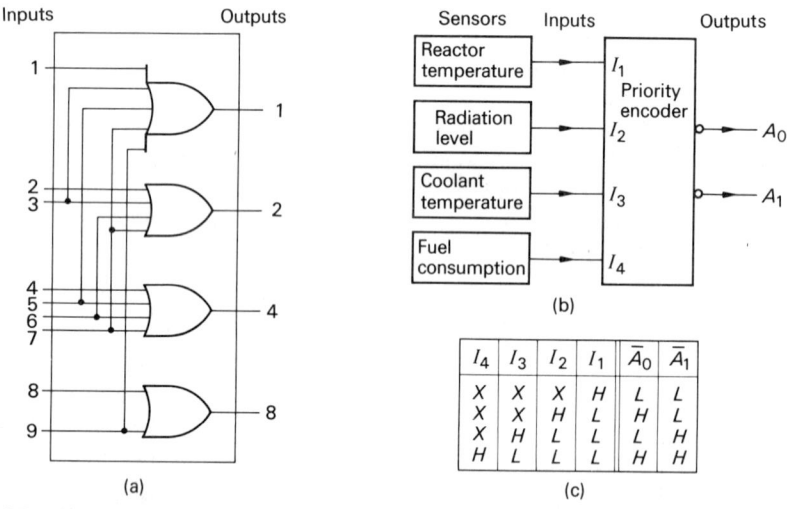

Fig. 5.25 Encoders: (a) decimal-to-8421 logic gate encoder; (b) block diagram of priority control system for Example 5.7; (c) priority encoder truth table.

I_4	I_3	I_2	I_1	$\bar{A}_0$	$\bar{A}_1$
X	X	X	H	L	L
X	X	H	L	H	L
X	H	L	L	L	H
H	L	L	L	H	H

The next example shows an application of a type of encoder called a priority encoder.

Example 5.7 *Nuclear Submarine Priority Control System.* A nuclear submarine is operating normally when suddenly a detection system indicates simultaneously that (1) the reactor temperature is excessive, (2) the radiation level in the cabin is above normal, (3) the temperature of the reactor coolant has exceeded normal limits, and (4) the fuel consumption rate is above normal. Devise a system that provides control signals that will remedy the malfunctions in the priority given above.

Solution. It is likely that if the reactor temperature is first reduced, the other malfunctions will also be remedied. In any case, if the reactor temperature is not reduced, the submarine will explode, so the other parameters will not matter. Therefore, first priority is assigned to reducing the reactor temperature. A four-input or 4-line-to-2-line **priority encoder** provides an output on two lines when any one of the four inputs becomes active (HIGH). If more than one input becomes HIGH simultaneously, it provides one output corresponding to the input with the highest priority.

Figure 5.25(b) shows the block diagram for the priority encoder warning system. The truth table for the encoder is given in Fig. 5.25(c). Note that if input 1 is HIGH, it doesn't matter whether inputs 2, 3, or 4 are HIGH or LOW; I_1 takes priority and the corresponding encoded outputs are $\overline{A}_0 = \overline{A}_1 = L$. ∎

The digital TV channel selector system (Example 5.4) uses a priority encoder for a digital channel number readout, as discussed in the problems.

MSI eight-input TTL, ECL, and CMOSL priority encoders are available. Priority encoders can be cascaded to provide more inputs as shown in the references [2, p. 4–6]. Encoders made from diode matrix arrays and read-only memories are also available [6]. An excellent discussion of keyboard encoders is given in the references [7].

SUMMARY

In this chapter we introduced several MSI devices and gave examples illustrating their use.

A digital comparator compares the magnitude of two N-bit numbers on one of three output lines. Comparators are useful for applications such as control circuits and arithmetic operations.

Information can be represented by digital waveforms. A binary code is used to specify a unique sequence of 0's and 1's for each character to be coded. One way of detecting errors in a digital transmission system is by adding a parity bit to each character. A parity generator is used to generate odd or even parity. A parity checker is used at the receiving end of a transmission system to detect parity errors.

Multiplexers are used to route messages from several sources along a data link. A demultiplexer is used to separate multiplexed messages into separate output channels.

Binary decoders convert binary inputs into more complex symbols, such as decimal numbers or alphabetic characters. Encoders convert complex symbols, such as alphameric characters, into simple symbols, such as binary numbers.

PROBLEMS

Section 5.1

1. *Binary Down-Counter.* (a) Using a TTL data book, draw the logic diagram for a two-bit binary down-counter using 74S112 flip-flops. (b) With the aid of a timing diagram, explain how this counter works.

2. *Four-Bit Up-Counter.* (a) Draw the logic diagram for a four-bit up-counter using JK flip-flops. (b) Explain how this counter works, using a timing diagram.

150   MSI DEVICES AND APPLICATIONS

Section 5.2

3. What are the three output logic levels of a four-bit digital comparator whose inputs are $A = 1101$ and $B = 1110$?

Section 5.3

4. Draw the pulse voltage waveform for the ASCII coded letter "B" (see Table 5.1). Let $+10$ V represent a logic-1 and -2.0 V represent a logic-0.

Section 5.4

5. Add a parity bit to the high-order position in the following bits to give them even parity.
   a) 0111001   b) 1101110   c) 1000111
6. (a) Draw the logic diagram for an eight-bit even-parity checker using EX-OR gates. (b) Show the logic levels (0 or 1) at the inputs and output of each gate in the parity checker for a 10011100 input.

Section 5.5

7. (a) Draw the block diagram for an eight-input data selector using four-input data selectors. (b) Explain in detail how it works. Use a truth table to aid in your explanation.
8. (a) Referring to the specifications for the 74LS253 multiplexer (Fig. 5.14), explain the meaning of the time parameters t_{ZH} and t_{LZ}. (b) Give their typical numerical values.
9. How many standard series TTL gates could the 74LS253 multiplexer drive in the HIGH state?
10. Draw the logic symbol of an 8-line-to-1-line data selector that will generate the pulse waveform 00111010.
11. Generate the pulse waveform 111111010101000000000000. Use periodic pulse waveform inputs, as needed, into a four-input multiplexer.
12. Generate the Boolean function $G = AB\bar{C} + \bar{A}BC + A\bar{B}\bar{C}$ using an eight-input multiplexer.
13. (a) What is the main advantage of multiplexing a multichannel digital data communication system as compared with an unmultiplexed system? (b) What is the main disadvantage?
14. *Digital Computer.* Referring to Appendix G, explain the purpose of each of the five main units in a micro-programmable computer.
15. *Microcomputer.* Refer to sheet 2 of the schematic diagram in Appendix G. (a) How is a printed circuit board edge connector pin shown on the schematic? (b) How are the signal sources to input multiplexer 5H pin 10 located on the schematic diagrams?
16. *Microcomputer.* Referring to Appendix G, study the description of the input multiplexer, data buffer, and address latch circuits. Briefly describe the purpose of these three circuits.
17. *Microcomputer.* (a) Referring to Problem 16 and to sheet 2 of the schematic diagrams in Appendix G, draw a logic diagram of (1) input multiplexer 5H, (2) data buffer 4H, and (3) address latch 5F. (b) Specifically, where do the inputs to the input multiplexer come from?

18. *Microcomputer.* Refer to Problem 17. Suppose the switch inputs to multiplexer 5H are SW00 = L, SW01 = H, SW02 = H, and SW03 = L. Give the outputs of (1) multiplexer 5H, (2) data buffer 4H, and (3) address latch 5F.

Section 5.6

19. Write the following decimal numbers in natural binary and in the 8421 *BCD* code.
    a) 7   b) 29   c) 256   d) 392
20. Write the following decimal numbers in the excess-3 *BCD* and 1's complement code.
    a) 3   b) 26   c) 123   d) 493
21. Find the 2's complement of the following.
    a) 2   b) 6   c) 29

Section 5.7

22. (a) Draw the logic symbol for a 1-of-8 decoder with active HIGH address and enable inputs and active LOW outputs. (b) Complete a truth table for this decoder. (c) With the aid of your truth table, explain how the decoder works.
23. *TV Digital Channel Selector.* (a) Referring to the schematic in Fig. P5.1, what is the purpose of the two *RS* latches made from IC201? (b) Draw a timing diagram for the inputs and output of the UP latch when the UP switch is off and when it is on (grounded).
24. *TV Digital Channel Selector.* Referring to Fig. P5.1, explain in detail how transistors Q201 to Q216 and diodes D249 to D264 operate as switches. Assume the decoder input is $ABCD = LLHL$.

Section 5.8

25. Draw the switch representation for a 1-of-4 demultiplexer.
26. *Digital Data Communication System.* Draw the block diagram for a four-channel serial TDM system that has two teleprinters and one nine-track digital magnetic tape unit at each end of the system. There should be an analog temperature sensor at one end and an analog temperature strip chart recorder at the other end of the system. Include a counter at each end of the system with a separate synchronization line. Also include a block symbol for a parity generator at one end, and a block symbol for a parity checker at the other end of the system. Between the parity generator (checker) and the multiplexer (demultiplexer), show a block symbol that converts the magnetic type data from parallel to serial form (serial to parallel). An ADC and a DAC should be used where appropriate. [*Hint*: Refer to Fig. 5.24.]

Section 5.9

27. Draw the logic diagram for a logic gate encoder that will encode decimal digits 0 through 5 into excess-3 coded outputs.
28. *Space Vehicle Control System.* A space vehicle is making a reentry into the earth's atmosphere. The reentry angle is the most important of four parameters that must be controlled. If the reentry angle is too small, the vehicle does not get captured by the earth's gravitational field. If the reentry angle is too large, atmospheric friction burns up the

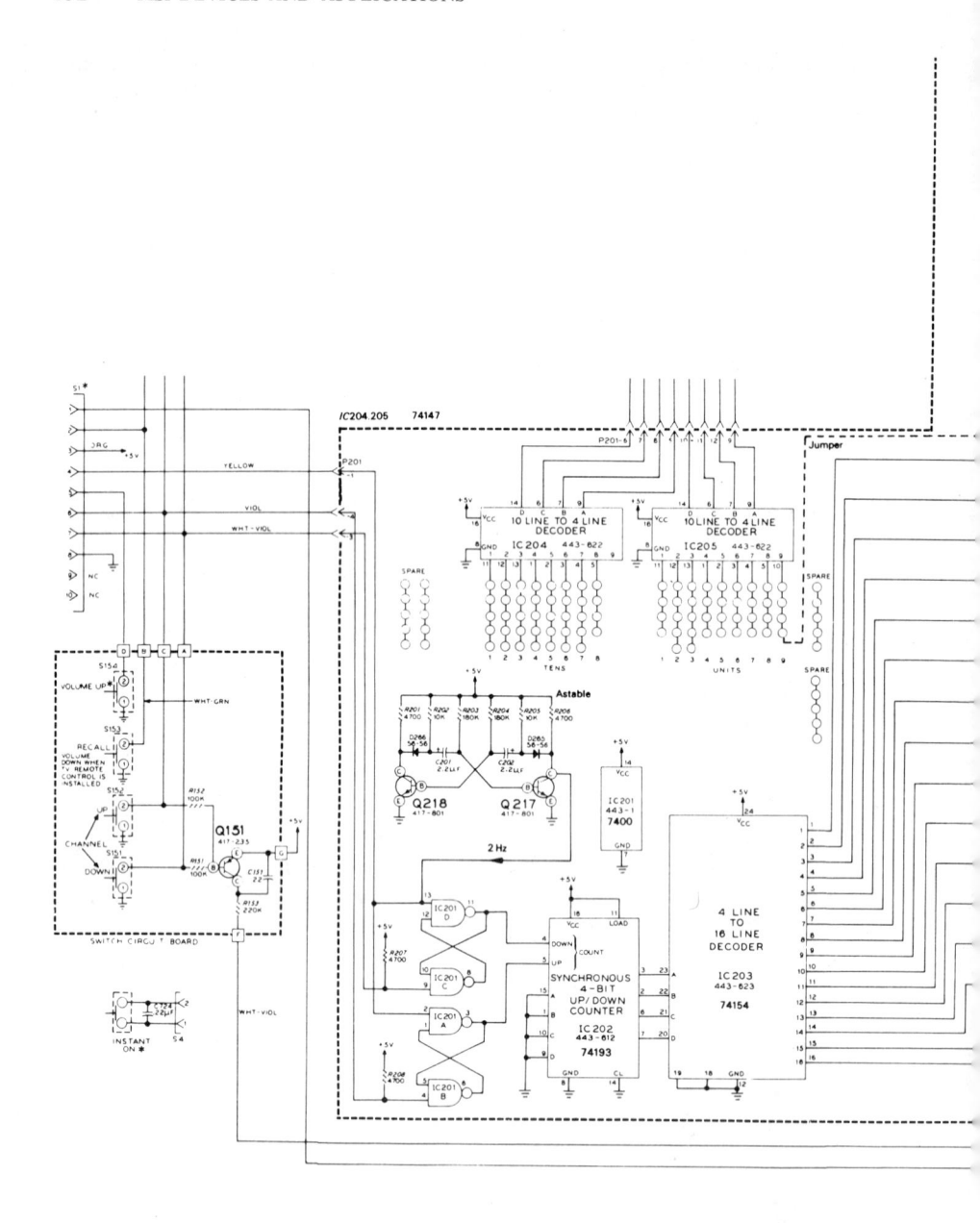

Fig. P5.1 TV digital channel selector. (Courtesy of Heath Company.)

PROBLEMS 153

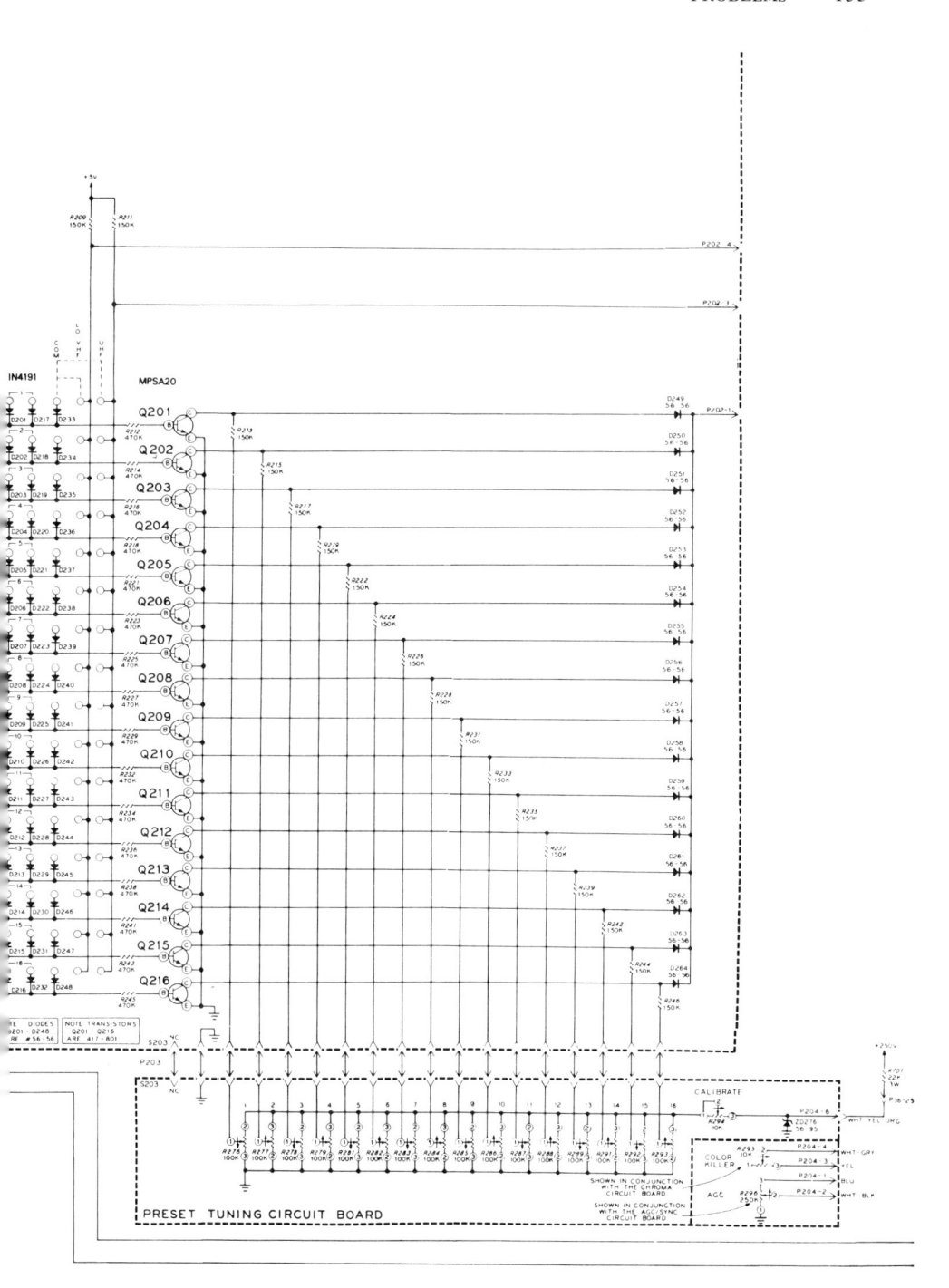

vehicle. In descending order of importance are: reentry angle, rotational orientation of the vehicle, fuel consumption for reentry angle motors, and cabin temperature. (a) Draw a block diagram for a control system that will issue the proper control signals. (b) What will the encoder outputs be if simultaneously the encoder receives signals indicating improper rotational orientation and excessive cabin temperature (use the truth table in Fig. 5.25c)?

29. *TV Digital Channel Selector* (a) Using a TTL data book, draw the connection diagram for a 74147 priority encoder. (b) Referring to the schematic in Fig. P5.1, explain specifically how encoders IC204 and IC205 can be used to provide encoded signals for displaying the selected channel number on the TV screen. [*Note*: Up to 16 channels can be preselected on the preset tuning board by setting the potentiometers. Jumper connections are made between diodes D201–D216 and the units inputs on IC205 or the tens inputs on IC204. For example, if R276 in the Preset Tuning Circuit Board (Fig. P5.1) is set for channel 9 then the jumper shown will encode channel 9 in BCD at the output of IC205.]

30. *Minimization Techniques and Computer Program* (a) Using a library reference, study the *Karnaugh map* method of minimizing (i.e., simplifying) Boolean functions. (b) Write the Boolean equation for A_0 and A_1 in terms of the inputs I_1, I_2, I_3, and I_4 for the priority encoder shown in Fig. 5.25(c). Minimize these two functions using the Karnaugh map technique. (c) Using a library reference, study the *Quine-McClusky* method of minimizing Boolean functions. (d) Write a computer program to simplify the functions A_0 and A_1 in part (b) above. (e) Compare the advantages of the Karnaugh map method with those of the Quine-McClusky method, and compare both with the Boolean algebra method explained in Chapter 4.

REFERENCES

1. William P. Davenport. *Modern Data Communication*. New York: Hayden Book Co., 1971.
2. Peter Alfke and Ib Larsen. *The TTL Applications Handbook*. Fairchild Semiconductor, August 1973.
3. *Delux Solid-State Color Television, Model GR2000* (Assembly Manual). Heath Co., 1973.
4. Larry Steckler (ed.). "STAR—New Kind of TV Remote Control." *Radio-Electronics*, p. 44, December 1974.
5. Eric Breeze. "TV Tuning and Channel Selection System." *Progress* (Fairchild Semiconductor), January 1974.
6. Jacob Millman and Christos Halkias. *Integrated Electronics: Analog and Digital Circuits and Systems*. New York: McGraw-Hill, 1972.
7. Peter Alfke. "Keyboard Encoders." *Progress* (Fairchild Semiconductor), August 1973.

6: SMALL-SCALE SOLID STATE DISPLAY SYSTEMS

6.1 INTRODUCTION

In many digital and analog systems it is desirable, and often necessary, to have a visual output. For example, the output of a digital multimeter is a set of decimal digits, signs, and a decimal point as shown in Fig. 6.1. The output of a small electronic calculator is also a set of decimal digits. The output of a temperature sensor for an automobile engine is a continuous voltage that may be converted to a digital signal (analog to digital conversion) and read out as decimal numbers or perhaps as a simple message such as "HIGH," "LOW," or "DANGER." These are examples of small-scale displays involving only a few characters, which is the subject of this chapter.

When a large amount of information is to be displayed visually, large-scale devices, such as CRTs (cathode ray tubes), are used.

In this chapter we shall consider small-scale light-emitting diode (LED) displays and methods for decoding and driving their inputs. Other types of displays, such as incandescent, neon (plasma), fluorescent, and liquid crystal displays (LCD) [12, 13], use the same principles of decoding and driving as LEDs and are treated in the references [1, Ch. 3; 2; 13; 14; 15]. Electrochromic displays are also discussed in the references [14, 16].

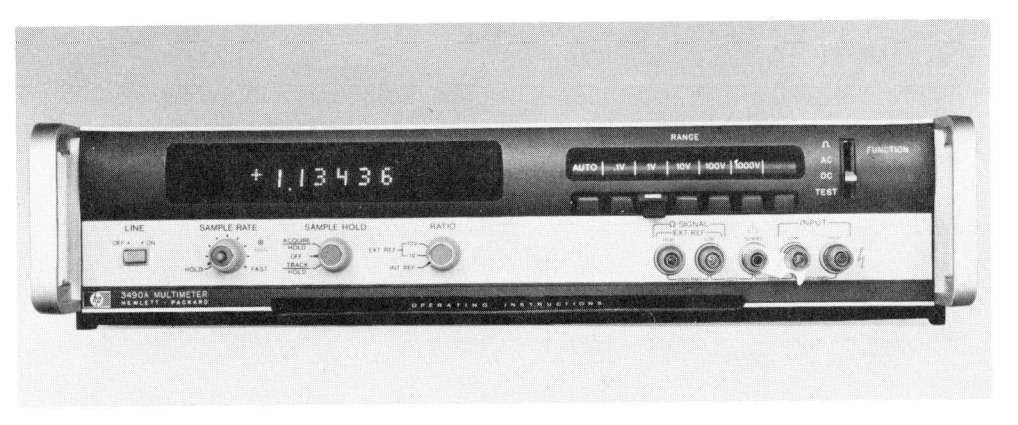

Fig. 6.1 Digital multimeter. (Courtesy of Hewlett-Packard Company.)

6.2 DISCRETE LEDs

6.2.1 Basic Principles

An **LED** is a semiconductor diode that emits light when it is sufficiently forward-biased. Figure 6.2(a) shows a typical LED. LEDs are available in several colors in the visible range such as red, yellow, and green. Infrared (IR) LEDs, which emit light invisible to the human eye, are also available. As with ordinary diodes, the forward current through an LED must be limited at the current source, or by a current-limiting resistor. The next example illustrates the use of an LED.

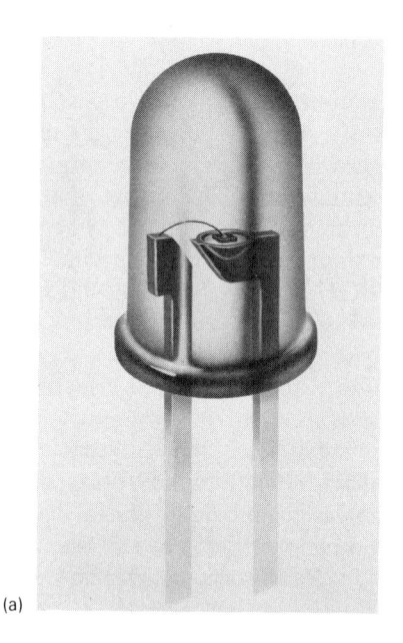

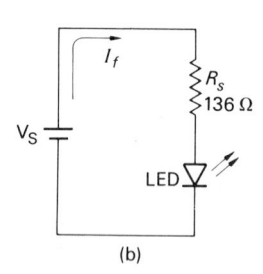

Fig. 6.2 (a) Light-emitting diode (courtesy of Hewlett-Packard Company); (b) circuit schematic for Example 6.1.

Example 6.1 *Calculation of LED Current-Limiting Resistance.* A red LED is used as the light source in the assembly line pill-monitoring system of Example 5.1, as shown in Fig. 6.2(b). Suppose the maximum dc forward current I_f through the LED used is 25 mA and the voltage drop V_f across the forward-biased LED is 1.6 V (at $I_f = 25$ mA). Calculate the value of resistor needed to limit the current to 25 mA.

Solution. Let's choose a convenient supply voltage V_s that is greater than V_f. Let $V_s = 5$ V. Then the resistance is given by

$$R_s = \frac{V_s - V_f}{I_f} = \frac{(5 - 1.6) \text{ V}}{25 \text{ mA}} = 136 \text{ }\Omega. \blacksquare$$

Next let's discuss some of the properties of LEDs.

ABSOLUTE MAXIMUM RATINGS

Forward DC Current	50 mA
Peak Forward Current (1 μsec, 300 pps)	1 A
Reverse Voltage	3 Volts
Power Dissipation—Derate 1.6 mW/°C above 25°C	120 mW
Storage Temperature	−55°C to 100°C
Operating Temperature	−55°C to 100°C
Relative Humidity @ 65°C	98%
Solder Temperature for 5 seconds 0.1″ from Case	260°C

ELECTRICAL CHARACTERISTICS (25°C)

SYMBOL	PARAMETERS	MIN.	TYP.	MAX.	UNITS
V_F	Forward Voltage @ 20 mA	—	1.7	2.0	V
BV_R	Reverse Breakdown Voltage @ 10 μA	3.0	8.0	—	V

OPTO-ELECTRONIC CHARACTERISTICS @ I_F = 20 mA (25°C)

SYMBOL	PARAMETERS	FLV 100 MIN.	FLV 100 TYP.	FLV 101 MIN.	FLV 101 TYP.	FLV 102 MIN.	FLV 102 TYP.	FLV 108 MIN.	FLV 108 TYP.	UNITS
I	Luminous Intensity	0.15	0.5	0.10	0.45	0.3	1.0	0.3	1.0	mcd (Note 1)
L	Luminance	—	800	—	2	—	5	—	5	mcd/cm² (Note 2)
A	Average Emitting Area	—	0.6 × 10⁻³	—	0.2	—	0.2	—	0.2	cm²
λpk	Wavelength @ Peak	—	665	—	665	—	665	—	665	nm
Δλ	Spectral Line Width	—	20	—	20	—	20	—	20	nm
t_r and t_f	Rise and Fall Time	—	10	—	10	—	10	—	10	ns (Note 3)

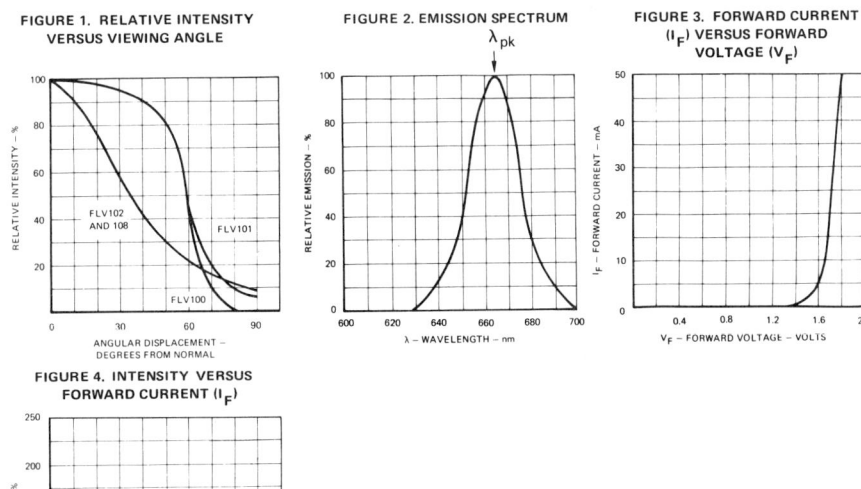

FIGURE 1. RELATIVE INTENSITY VERSUS VIEWING ANGLE

FIGURE 2. EMISSION SPECTRUM

FIGURE 3. FORWARD CURRENT (I_F) VERSUS FORWARD VOLTAGE (V_F)

FIGURE 4. INTENSITY VERSUS FORWARD CURRENT (I_F)

NOTES:
1. Measured on mechanical axis of package.
2. 1 cd/cm² = 2.92 × 10³ ft. lamberts.
3. Time for a 10%–90% change in light intensity with a step change in current.
4. Luminous intensity curve coincides with radiant intensity curve for pulse excitation (for average currents of 20 mA or less.)

Fig. 6.3 Discrete red LED specifications. (Courtesy of Fairchild Camera and Instrument Corporation.)

The brightness of light produced by an LED is called the **luminous intensity** I and is measured in units of **candela** (cd) (see Appendix E). A typical LED dc forward current of 10 mA can result in a luminous intensity of 1 millicandela (mcd). As the current through an LED increases, the luminous intensity increases until the LED saturates. After saturation no further increase in luminous intensity occurs. The curve in Fig. 6.3 shows the relative luminous intensity versus dc forward current I_f for an FLV100 LED. For example, at 10 mA the relative brightness is 50%, so the typical luminous intensity is 0.50×0.5 mcd $= 0.05$ mcd. The forward-biased voltage V_f across the diode depends on I_f; for example, at 20 mA, $V_f = 1.7$ V.

The brightness of an LED decreases as the viewing angle increases away from the LED axis as shown in Fig. 6.3. The luminous intensity is maximum at one wavelength, called the **peak wavelength** λ_{pk}, and decreases for all wavelengths greater than and smaller than λ_{pk}. Figure 6.3 shows a typical curve of wavelength versus luminous intensity, called a **spectral distribution curve**, for a red LED. For example, at about 652 nm the output is only one-half the output at a peak wavelength of 665 nm. More details on photometry [3] and LEDs [4, 5] are found in the references and in Appendix E.

The supply current for an LED may also be pulsed, in which case the peak current, for a given pulse width, must not be exceeded. Peak forward currents are considerably greater than dc forward currents. For example, the peak pulsed current for the FLV100 is 1A for 1 μs pulses whose frequency is 300 pulses per second, as given in Fig. 6.3.

6.2.2 Applications of LEDs

In this section we give several practical examples of LEDs and briefly discuss phototransistors and photodiodes.

Example 6.2 *Logic Gate LED Driver.* A hospital has a control panel that uses LEDs to detect whether either of two patients in a room has pushed the service button. Devise a warning system using TTL logic gates and an LED with a luminous intensity of 1 mcd. Use an FLV102 red LED (Fig. 6.3). Assume that a relative intensity of 100% results in the typical luminous intensity given in Fig. 6.3.

Solution. An OR gate will perform the desired logic. An OR gate and LED system are shown in Fig. 6.4(a). From Fig. 6.3, the typical luminous intensity for the FLV102 is 1 mcd. For a relative intensity of 100%, the forward current required is 20 mA. Since totem pole or open collector TTL gates source (I_{OH}) less than 1 mA of current (even for driver gates), the configuration shown in Fig. 6.4(a) will not produce the desired brightness. Most TTL gates will sink (I_{OL}) about 16 mA, which is almost enough current. If an AND gate with active LOW inputs was used, when either or both patients pressed their button (LOW) the AND output would go LOW, allowing up to 16 mA of current to flow into the output.

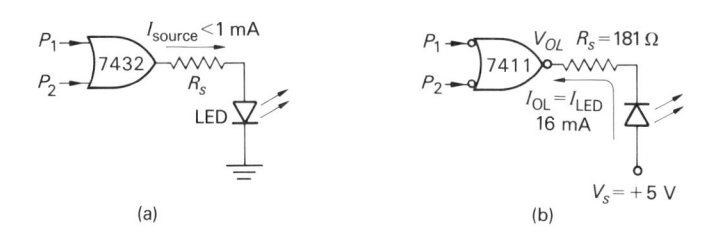

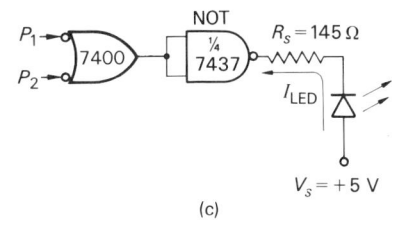

Fig. 6.4 Driving LEDs with TTL logic gates: (a) HIGH source current; (b) NOR gate solution of Example 6.2; (c) driver gate solution of Example 6.2.

Consider two solutions to the problem: (1) use a 7411 AND gate with 16 mA current and settle for a lower LED brightness (about 80% relative intensity) or (2) use a driver gate. The AND solution is shown in Fig. 6.4(b). For 16 mA of LED current, the curve of I_f versus V_f in Fig. 6.3 shows the voltage drop across the diode is about 1.7 V. The current-limiting resistor is calculated as follows:

$$R_s = \frac{(V_s - V_{LED}) - V_{OL(max)}}{I_{LED}} = \frac{5\text{ V} - 1.7\text{ V} - 0.4\text{ V}}{16\text{ mA}} = 181\text{ }\Omega,$$

where V_s is chosen greater than $V_{LED} + V_{OL}$ and less than the absolute maximum supply voltage V_{CC}.

Totem pole NAND buffer (driver) gates are available that will sink the required 20 mA of current. Let's illustrate the solution with a 7437 quad two-input NAND buffer. A NOR gate can be made from NAND gates as shown in Fig. 6.4(c). The 7437 can sink 48 mA of current. The current-limiting resistor is calculated from

$$R_s = \frac{(V_s - V_{LED}) - V_{OL(max)}}{I_{LED}} = \frac{(5 - 1.7 - 0.4)\text{ V}}{20\text{ mA}} = 145\text{ }\Omega. \blacksquare$$

CMOSL gates will generally source or sink only a few milliamperes of current using a 5-V supply. Therefore, to drive LEDs with CMOSL gates one can either use a higher supply voltage (10 to 15 V) or use a buffer gate, such as a 4010, which will sink 10 mA with a 10-V supply [6].

Next let's develop in detail the optoelectronics for the card sorter problem given in Chapter 4 (Example 4.2). This requires a very brief discussion of phototransistors.

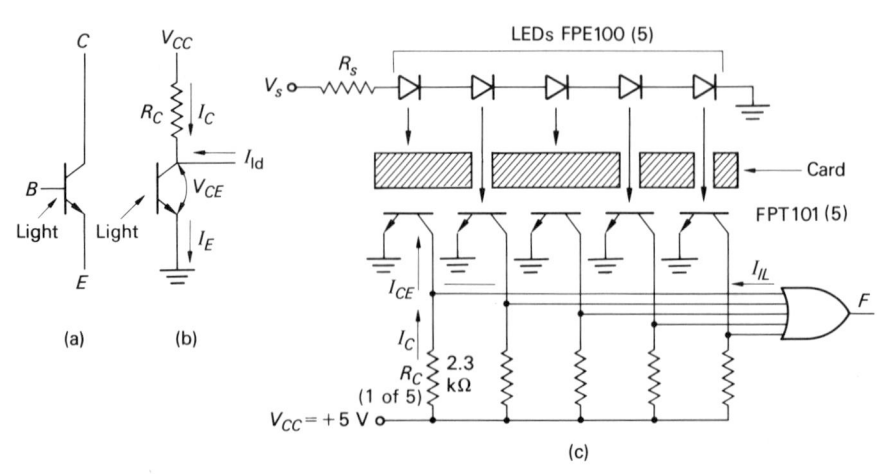

Fig. 6.5 Phototransistors: (a) schematic symbol; (b) currents and voltages; (c) use in a punched card reader.

A **phototransistor** is a transistor whose collector current depends on the intensity of the light of a given wavelength incident on the transistor [4]. Figure 6.5(a) shows the schematic symbol for a phototransistor. The base lead is normally left open and frequently is not shown on schematics. A phototransistor can be used as a **photodiode** by using the collector and base leads and leaving the emitter open. A photodiode has a faster switching speed than a phototransistor.

A phototransistor may be thought of as a switch (Fig. 6.5b). The phototransistor acts essentially like an open switch (cut off) when there is no light incident on it. The cutoff collector current, called the **dark current** I_{CEO} (where I_{CEO} stands for the collector (C)–emitter (E) current with the base open (O)), is on the order of a few nanoamperes. When sufficient light falls on the phototransistor it acts like a closed switch and the collector current is on the order of milliamperes. When the incident light on a phototransistor increases sufficiently the phototransistor becomes saturated and the collector–emitter voltage $V_{CE(sat)}$ is on the order of 0.25 V.

Now let's complete the punched card sorter example.

Example 6.3 *Punched Card Reader.* Draw the logic diagram for the punched card reader of Example 4.2 using TTL. Use FPE100 IR light sources with 40 mA of forward current. Use FPT101 phototransistors with $V_{CC} = +5$ V.

Solution. First let's design the light sources. Five LEDs can be connected in series as shown in Fig. 6.5(c). Referring to an FPE100 data sheet $V_f = 1.2$ V at $I_f = 40$ mA. Since each LED will drop 1.2 V, the total voltage drop across the five

LEDs is 5 × 1.2 = 6 V, so a reasonable power supply voltage V_s would be 10 V. Hence R_s is

$$R_s = \frac{V_s - 5 \times V_{LED}}{I_{LED}} = \frac{(10 - 5 \times 1.2)\ \text{V}}{40\ \text{mA}} = 100\ \Omega.$$

Consider the phototransistors next. If the phototransistors are connected in the common emitter configuration, they will act as inverters (just like ordinary transistors), so the input inverters used in Example 4.2 will not be needed. Let's use a five-input OR gate as shown in Fig. 6.5(c). When there is no hole in the card (missed lecture), no light falls on the phototransistor so it is cut off. In cutoff, the phototransistor output voltage is about $V_{CC} = +5$ V, i.e., the input to the OR gate is HIGH. The phototransistor cutoff collector current I_{CEO} is no more than 100 nA. When there is a hole in the card (attended lecture), sufficient light falls on the phototransistor to saturate it, so the input to the OR gate is LOW. In saturation, the specification sheet for the FPT101 gives $V_{CE(\text{sat})} = 0.4$ V and $I_{CE(\text{sat})} = 3.0$ mA. Therefore the FPT101 can easily sink the maximum TTL OR gate input source current $I_{IL} = 1.6$ mA.

Next let's calculate the value of the phototransistor collector resistor R_C. The total current flowing into the collector of a saturated phototransistor (Fig. 6.5b) is the sum of the current in R_C and the logic gate LOW current I_{OL}, that is, $I_{CE(\text{sat})} = I_C + I_{IL}$. For $V_{CC} = +5$ V, $V_{CE(\text{sat})} = 0.4$ V, $I_{IL(\text{typ})} = 1.0$ mA, and $I_{CE(\text{sat})} = 3$ mA, the collector resistance is

$$R_C = \frac{V_{CC} - V_{CE(\text{sat})}}{I_C} = \frac{(5 - 0.4)\ \text{V}}{(3 - 1)\ \text{mA}} = 2.3\ \text{k}\Omega.$$

This is also sufficient resistance for current limiting for the OR gate when the phototransistor is off, if a V_{CC} between 5.0 V and 7 V is used.

The complete schematic for the punched card sorter is shown in Fig. 6.5(c). If the available phototransistor cannot sink enough current for the logic family used, a discrete transistor switch will be required [4, p. 6–28]. Other phototransistor-to-logic-gate interfaces are shown in the references [4, p. 7–21]. ∎

LED light-sources and phototransistor sensors are available as arrays with various spacings between elements in the source and in the sensor [4].

LEDs have very fast response times (rise and fall times), typically 10 ns. Phototransistors are much slower, having rise and fall times on the order of 1 μs. Photodiodes have response times on the order of 1 ns. The response time of a phototransistor decreases with increasing collector current and decreasing load resistance.

Although LEDs are not very bright by incandescent lamp standards, their detectability can be enhanced by recessing the LED in its mounting panel and by painting the panel background black.

162   SMALL-SCALE SOLID STATE DISPLAY SYSTEMS   6.2

Example 6.4 *Troubleshooting Comparator.* A troubleshooting comparator is an instrument used to detect faults in digital circuits. The comparator compares each pin in a known good IC with the corresponding pin in the device under test (DUT). A light indicates a fault, as shown in the photograph of Fig. 6.6(a). Figure 6.6(b) shows a logic diagram of a reference IC, the DUT, and the circuit used to test one pin of the DUT. Identical circuits are used to test the remaining DUT IC pins. Explain how the comparator works.

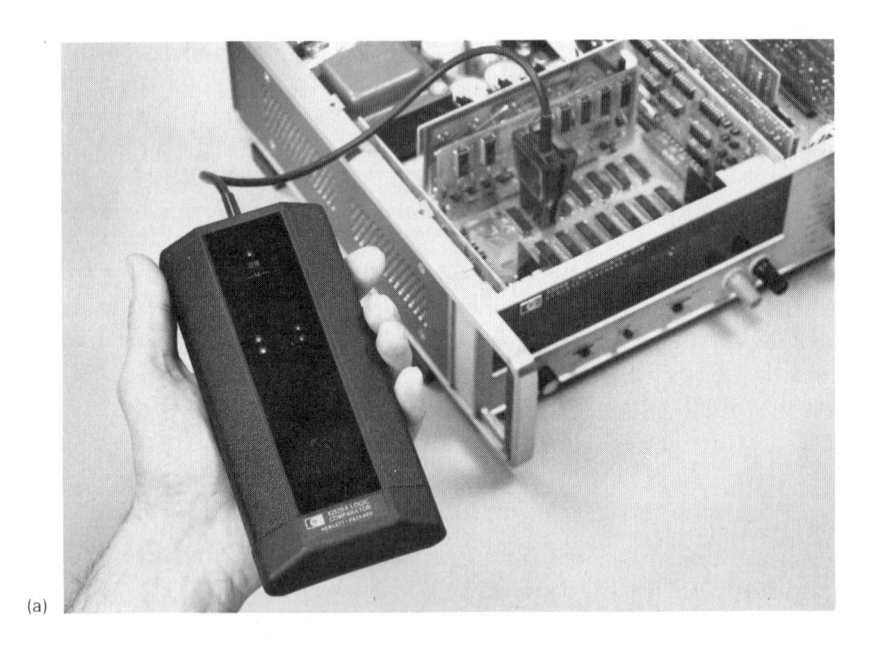

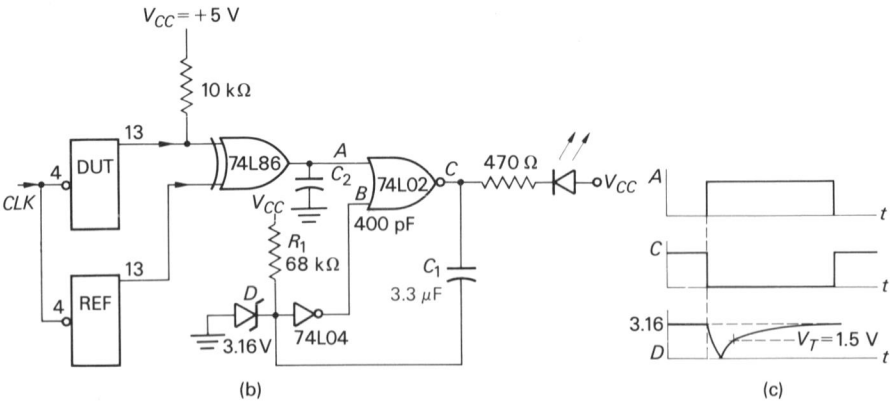

Fig. 6.6 Troubleshooting comparator: (a) photograph (courtesy of Hewlett-Packard Company); (b) logic diagram; (c) timing diagram.

Solution. The pulse stretcher circuit consisting of R_1, C_1, the NOT gate, and the 3.16-V Zener diode will be discussed in the problems and therefore will be ignored here. If the DUT has a different output than the reference device, the output of the EX-OR gate will go HIGH. The NOR gate output will go LOW, independent of the B input, and the LED will turn on, indicating a fault in the DUT. Capacitor C_2 is used to slow down the response time of the EX-OR gate so that it will ignore very short errors due to differences in the propagation delay times in the DUT and the reference device. Figure C.3, Appendix C, illustrates the use of a comparator for trouble shooting. ■

The next section shows how LEDs can be used to give a numerical readout.

6.3 SINGLE-DIGIT SEVEN-SEGMENT LED DISPLAYS

In a seven-segment LED display, seven bar-shaped LEDs are arranged in a pattern, as illustrated in Fig. 6.7(a). The decimal number 1 is generated by illuminating segments e and f while the remaining segments are kept off. The decimal 4 is generated by illuminating segments f, g, b, and c as illustrated in Fig. 6.7(b). Because of the limited number of segments in a seven-segment display, only decimal digits (0 through 9), a few special symbols (such as $-$,]), and a few alphabetic letters (such as A, C, P) can be displayed. Therefore, most seven-segment displays are used only for numerical readouts. Alphabetic character displays will be discussed in Section 6.7.

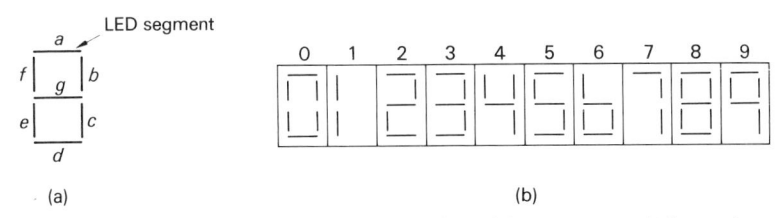

Fig. 6.7 Seven-segment LED display: (a) identification of the segments; (b) formation of digits 0 through 9.

A single seven-segment display is capable of generating one decimal digit, so a five-digit number would require five seven-segment displays. Multidigit displays will be discussed in Section 6.5. A decimal point (DP) can be illuminated along with a digit by illuminating a separate DP diode on an LED display. Thus an LED display that includes a decimal point actually contains eight LEDs, although it is usually called a seven-segment display.

A single-digit, seven-segment numeric display is shown in Fig. 6.8(a). It has eight red LEDs and is mounted in a DIP package. Each character is 0.3 inches in height and can be viewed from a distance of up to 10 feet. The display has a right-hand decimal point. Displays with characters 0.6 inch high are also available. Seven-segment displays are made in a variety of colors, such as red, green, and yellow.

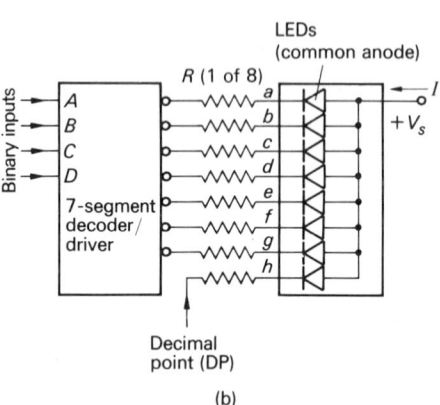

Fig. 6.8 LED displays: (a) DIP mounted display (courtesy of Hewlett-Packard Company); (b) block diagram of a seven-segment display with a decoder/driver; (c) decoder truth table.

Decimal	Inputs D C B A	a	b	c	d	e	f	g
0	L L L L	L	L	L	L	L	L	H
1	L L L H	H	H	H	L	L	L	H
2	L L H L	L	L	H	L	L	H	L
3	L L H H	L	L	L	L	H	H	L
4	L H L L	H	L	L	H	H	L	L
5	L H L H	L	H	L	L	H	L	L
6	L H H L	H	H	L	L	L	L	L
7	L H H H	L	L	L	H	H	H	H
8	H L L L	L	L	L	L	L	L	L
9	H L L H	L	L	L	H	H	L	L

(c)

The output of most digital systems is in binary; however, it is usually desirable to display the output in decimal form. Therefore a binary-to-decimal decoder is required. The outputs of the decoders described in Chapter 5 are able to source (or sink) only a few milliamperes of current. However, each segment of an LED display may require 10 mA or more for a total of 8×10 mA = 80 mA. Thus a driver must be supplied with the decoder in order to provide the required drive current. The decoder and driver are usually on one IC called a **display decoder/driver**. Display decoder/drivers will be discussed fully in the next section.

Figure 6.8(b) shows the block diagram for a typical seven-segment display along with a decoder/driver. The seven-segment decoder/driver takes four-bit 8421 coded data, converts it to seven-segment form, and provides enough current to drive each of the seven LEDs. When a decoder output goes LOW for segment b for example, current flows from V_s though the segment b LED and the current-limiting resistor R to ground (LOW).

6.3 SINGLE-DIGIT SEVEN-SEGMENT LED DISPLAYS

The truth table for an active LOW output seven-segment display decoder/driver is given in Fig. 6.8(c). For example, if the decoder inputs are $A = H, B = C = D = L$, the e and f outputs of the decoder go LOW, causing segments e and f to be illuminated.

The last example in this section shows how to make an optical tachometer.

Example 6.5 *Optical Tachometer.* Develop an optical system that will read out the rotational speed of an engine shaft in decimal form.

Solution. A strip on the rotating shaft is painted with a reflective paint as shown in Fig. 6.9(a). An LED light source is directed onto this reflective surface and a phototransistor is used to detect the reflected light. Each time the shaft rotates one

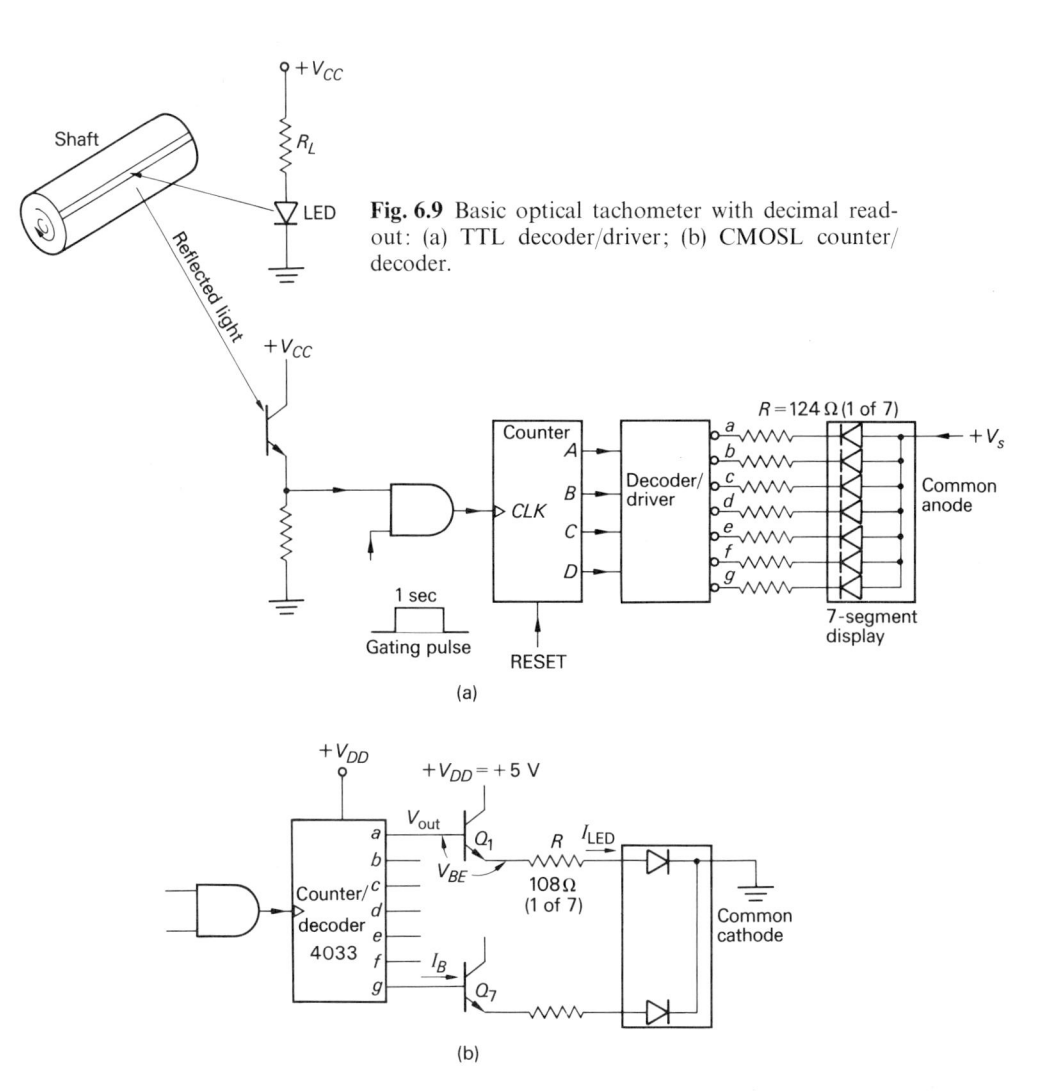

Fig. 6.9 Basic optical tachometer with decimal readout: (a) TTL decoder/driver; (b) CMOSL counter/decoder.

revolution, the photodetector will produce one output pulse. The photodetector pulses are then fed into the counter, which counts the total number of pulses (i.e., revolutions). An AND gate is used to enable the counter input for, say, 1 sec. The result is that the counter counts the total number of shaft revolutions per second. For example, suppose the shaft is rotating at six revolutions per second. The counter is initially reset to 0, and then the 1-sec gating pulse is applied. The counter counts to 6; the decoder inputs will be (Fig. 6.8c) $A = D = L$ and $B = C = H$. These inputs to the decoder will cause decoder outputs c, d, e, f, and g to go LOW, resulting in a 6 being displayed. ■

Current-limiting resistors are required with most LED displays. Note that there is one resistor for each LED (Figs. 6.8 and 6.9) rather than just one resistor in the power supply line V_s. Individual resistors provide approximately the same current through each diode regardless of the number of diodes that are ON or the variations in their voltage drops. Next let's calculate the value of the current-limiting resistors R for the optical tachometer in Fig. 6.9(a).

Example 6.6 *Calculation of Seven-Segment Display Resistor Values.* Determine the value of the LED current-limiting resistors R in Fig. 6.9(a).

Solution. The value of R is determined by the voltage across R divided by the current through R. The current through R is determined by the desired brightness of the LEDs and should be no more than $I_{LED, \, max}$ or the maximum current capability of the driver $I_{driver, \, max}$. Thus

$$R = \frac{(V_s - V_{LED}) - V_{out, \, driver}}{I_{LED}}.$$

As a specific example, for a TTL 7447 decoder/driver, $V_{out, \, driver} \equiv V_{on} = 0.3$ V typically. For an HP 5082–7750 seven-segment display, $I_{LED, \, max} = 25$ mA and $V_{LED} = 1.6$ V at 25 mA. Therefore, for $V_s = 5$ V,

$$R = \frac{(5 - 1.6) \text{ V} - 0.3 \text{ V}}{25 \text{ mA}} = 124 \, \Omega.$$

The tachometer could equally well be made from CMOSL devices, the main advantage over TTL being less power dissipation. The RCA CMOS 4033 is a counter and decoder on one IC chip, but requires a driver if more than about 5 mA of current are to be supplied to the LEDs [6, p. 477]. The logic for a CMOSL optical tachometer using an IC driver transistor array (such as an RCA CD 3082) is shown in Fig. 6.9(b).

The current-limiting emitter resistor values for the driver transistors are calculated from

$$R = \frac{(V_{out} - V_{BE}) - V_{LED}}{I_{LED}}.$$

For $V_{out} \approx V_{DD} = +5$ V, $V_{BE} = 0.7$ V, $V_{LED} = 1.6$ V, and $I_{LED} = 25$ mA; then $R = (5 - 0.7 - 1.6)$ V $/25$ mA $= 108 \, \Omega$. ■

6.4   SEVEN-SEGMENT DISPLAY DECODER/DRIVERS   167

Note in Fig. 6.9(b) that the 4033 decoder has active HIGH outputs, so all the LED cathodes are connected to ground. This is called a **common cathode** LED display because all the LED cathodes are internally connected together. The LEDs in Fig. 6.8(a) are called a **common anode** LED display because all the LED anodes are internally connected together. More details on LED displays are found in the references [4, 6, 7]. Unless inverters are used between the decoder and display ICs, decoder/drivers with active LOW outputs use common anode displays, whereas decoder/drivers with active HIGH outputs use common cathode displays.

Let's now look at the display decoder/driver in more detail.

6.4 SEVEN-SEGMENT DISPLAY DECODER/DRIVERS

A display decoder/driver takes a binary input, decodes it, and provides sufficient current output to drive a display device such as an LED display.

Figure 6.10(a) shows the logic symbol for a typical seven-segment display decoder/driver. It has four BCD data inputs (A_0, A_1, A_2, and A_3), and seven outputs labeled a through g that correspond to the seven segments of Fig. 6.7(a). The truth table for the decoder is shown in Fig. 6.10(b). Note that the display segments ($a, b, \ldots, g$) are active LOW (sink); that is, to produce a specified digit, the appropriate driver outputs go LOW. This decoder/driver is therefore used with common anode displays.

The inputs corresponding to decimal numbers 10 through 15 produce all HIGH outputs and hence no LED segments light up. These inputs are called **invalid inputs**. For some decoders the inputs corresponding to 10 through 15 will display special output symbols such as [,], or ⊔.

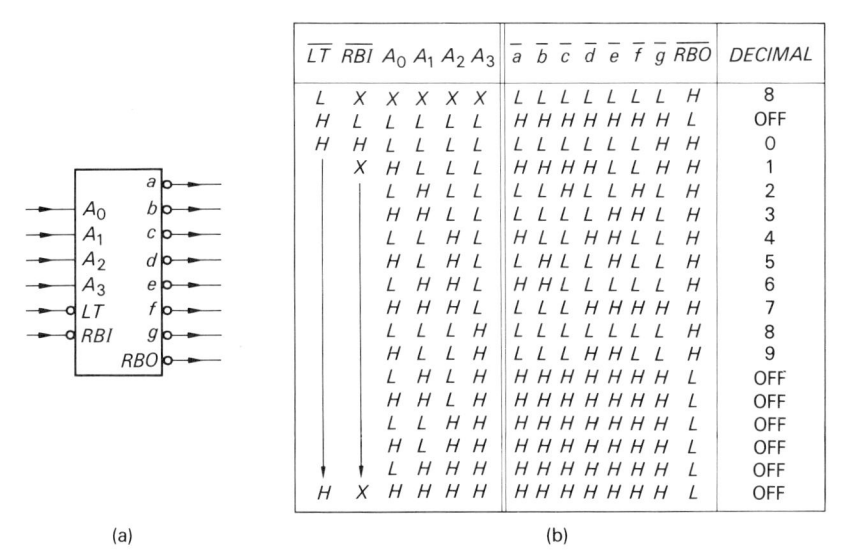

$\overline{LT}$	$\overline{RBI}$	A_0	A_1	A_2	A_3	$\bar{a}$	$\bar{b}$	$\bar{c}$	$\bar{d}$	$\bar{e}$	$\bar{f}$	$\bar{g}$	$\overline{RBO}$	DECIMAL
L	X	X	X	X	X	L	L	L	L	L	L	L	H	8
H	L	L	L	L	L	H	H	H	H	H	H	H	L	OFF
H	H	L	L	L	L	L	L	L	L	L	L	H	H	0
	X	H	L	L	L	H	H	H	L	L	H	H	H	1
		L	H	L	L	L	L	H	L	L	H	L	H	2
		H	H	L	L	L	L	L	L	H	H	L	H	3
		L	L	H	L	H	L	L	H	H	L	L	H	4
		H	L	H	L	L	H	L	L	H	L	L	H	5
		L	H	H	L	H	H	L	L	L	L	L	H	6
		H	H	H	L	L	L	L	H	H	H	H	H	7
		L	L	L	H	L	L	L	L	L	L	L	H	8
		H	L	L	H	L	L	L	H	H	L	L	H	9
		L	H	L	H	H	H	H	H	H	H	H	L	OFF
		H	H	L	H	H	H	H	H	H	H	H	L	OFF
		L	L	H	H	H	H	H	H	H	H	H	L	OFF
		H	L	H	H	H	H	H	H	H	H	H	L	OFF
		L	H	H	H	H	H	H	H	H	H	H	L	OFF
H	X	H	H	H	H	H	H	H	H	H	H	H	L	OFF

(a) (b)

Fig. 6.10 Display decoder/driver: (a) logic symbol; (b) truth table.

The lamp test (LT) input is used to check each of the seven segments for possible malfunction by using only a single input. When LT is LOW (L) all seven segments should light; if some segments do not light, either the decoder or the display is malfunctioning.

The two remaining terminals shown in Fig. 6.10, RBI and RBO, are used in multidigit displays, which we shall be discussing in detail in the next section. An eight-digit display will have eight seven-segment displays. If a number such as 5.1 were to be displayed in an eight-digit display, it might appear as 00005.100. The four leading zeros and the two trailing zeros make the display difficult to read. It is therefore often desirable to blank out these leading and trailing zeros automatically, so the number is easily read and appears as it would normally be written as 5.1. Blanking of undesired zeros also saves power. The **ripple blanking input** (RBI) and the **ripple blanking output** (RBO) are used to suppress undesired zeros automatically.

Suppose the address inputs to the decoder/driver are all LOW; then a 0 would normally be displayed. However, if $\overline{RBI} = L$, none of the diodes are turned on, as seen in the truth table of Fig. 6.10(b). Thus a LOW $\overline{RBI}$ input suppresses the 0 output. Further, note that the $\overline{RBO}$ output is LOW when the $\overline{RBI}$ input is LOW.

The automatic suppression of leading-edge and trailing-edge zeros is done as follows. To suppress the zero of the MSD (leading edge suppression) the $\overline{RBI}$ input is connected to a LOW. The next 0 is suppressed by connecting the $\overline{RBO}$ of the MSD to the $\overline{RBI}$ of the next digit. Figure 6.11 shows the suppression of the three leading-edge zeros and the two trailing-edge zeros. No other zeros are suppressed.

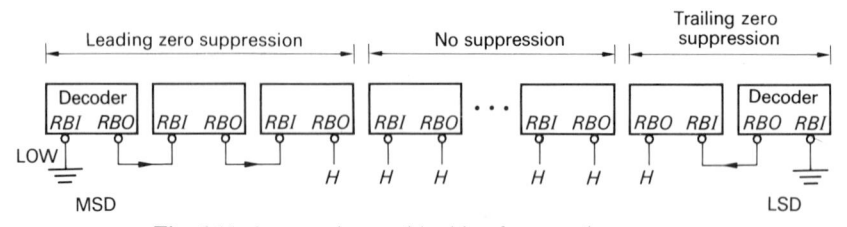

Fig. 6.11 Automatic zero blanking for a readout system.

Some TTL decoders have open collector outputs, instead of drivers, so that external drivers can be used. This may be desirable for driving high current devices. Some seven-segment decoder/drivers provide high voltage outputs (30 V) for driving devices such as relays. Some displays have a built-in decoder and driver.

6.5 MULTIDIGIT SEVEN-SEGMENT DISPLAYS

Let us now develop the display unit for a three-digit digital voltmeter.

Example 6.7 *Ramp Type Digital Voltmeter.* Develop a three-digit display for a digital voltmeter.

6.5 MULTIDIGIT SEVEN-SEGMENT DISPLAYS 169

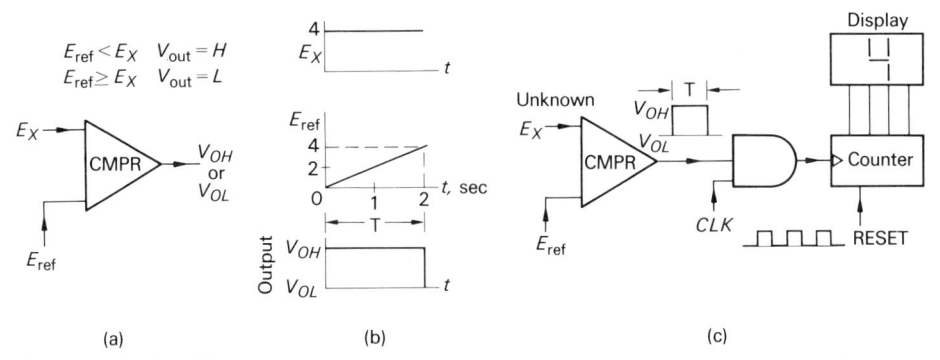

Fig. 6.12 Basic DVM concepts: (a) analog comparator; (b) comparator input and output waveforms; (c) simplified DVM block diagram.

Solution. A digital voltmeter (DVM) converts analog input voltage signals into digital signals, and displays the results in decimal form on a digital display. Consider first a dc analog voltage input to the DVM. One type of DVM converts the analog input into a pulse whose width T is proportional to the dc input voltage E_X. This type of DVM, called a **ramp type DVM**, uses an analog comparator [8].

An analog comparator has two analog inputs and one digital output, as shown in Fig. 6.12(a). If input E_{ref} is less than input E_X, the output of the comparator is HIGH. When the two inputs are equal the comparator output goes LOW. If the reference voltage E_{ref} is a ramp voltage, as shown in Fig. 6.12(b), then after a time T the ramp voltage will equal the unknown voltage E_X and the comparator's output voltage will go LOW. For example, if the reference voltage has a slope of 2 V/sec and if the $E_X = 4$ V, then T will be

$$T = \frac{E_X}{E_{ref}} = \frac{4 \text{ V}}{2 \text{ V/sec}} = 2 \text{ sec.}$$

The output of the comparator is connected to one input of an AND gate as shown in Fig. 6.12(c). The other input of the AND gate comes from a clock pulse generator. A counter counts the total number of clock pulses N_{CP} of frequency f_{CLK} during time T. The unknown voltage is $E_X = TE_{ref} = N_{CP}E_{ref}/f_{CLK}$. The counter is reset to zero at the end of each display period. Measurement of ac voltages is accomplished by first converting the ac voltage to a dc voltage and then using the above technique.

The readout shown in Fig. 6.12(c) will display only one decimal digit. Let's develop a three-digit display. Three counters can be connected together (cascaded) so that they will count from 0 through 999. Each counter (called a **decade counter**) counts from 0 through 9 and has four 8421 BCD outputs as shown in Fig. 6.13. Suppose that $T = 20$ sec, corresponding to an input voltage of $E_X = 40$ V; then if the clock produces two pulses per second the counters will read 40 at the end of 20 sec. Since the first (MSD) zero is blanked out the display will read 40, instead of 040. The decimal point (DP) location is selected by a range selector switch. The

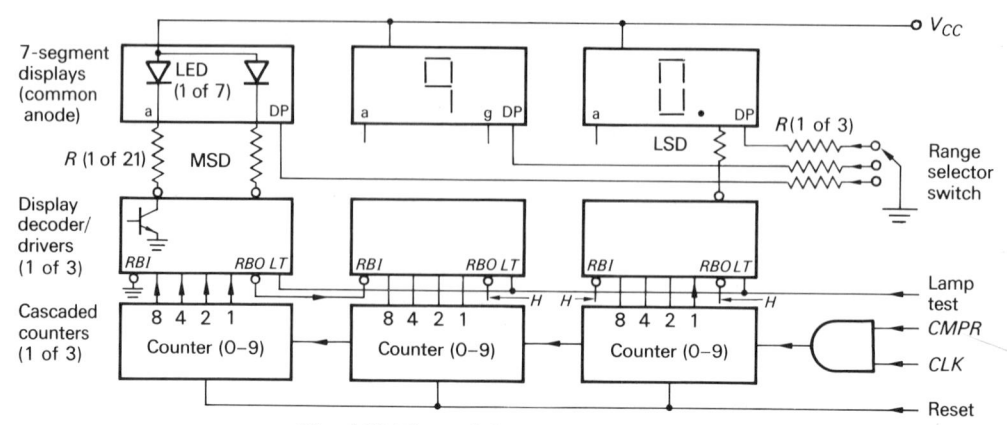

Fig. 6.13 Three-digit DVM readout system.

resistor values are calculated as shown in Example 6.6. A common lamp test input is used to check all three displays simultaneously. A common reset input resets all three counters to zero before the next comparison. The output can be displayed as long as desired by using latches between the counters and decoders to store the counter outputs, or by not clocking the counters during the display period. ∎

The last example in this section discusses a complete digital system that uses a digital readout.

Example 6.8 *Portable Heart-Rate Monitor* [9]. Figure 6.14 shows the logic diagram for a system that displays the heart rate of a patient by measuring the pulse rate from three electrodes attached to the patient's skin. This monitor uses an operational amplifier to convert the analog pulse rate information into digital signals. All the digital ICs use Motorola CMOSL devices. A three-digit display helps the patient to visually observe his heart rate and keep it below 100 beats per minute while he is doing exercises to increase his circulation. Using a Motorola CMOSL data book, describe the function of the following ICs in the display circuit: 14518 (frequency), 14511(2), and 14013(2).

Solution. The 14518 is a dual decade (10) up-counter with active HIGH reset inputs R_1 and R_2. The first counter is connected so that its outputs (Q_1, Q_2, Q_3, and Q_4) change on the $L \rightarrow H$ transition of clock input C_1. The second counter is clocked by input E_2 on the $H \rightarrow L$ transition of the Q_4 output of the first counter.

The outputs of the first counter (Q_1, Q_2, Q_3, and Q_4) are the address inputs (*A*, *B*, *C*, and *D*) for the 14511 latch/decoder/driver. This decoder/driver has a latch on each output so that when the latch enable (*LE*) input is HIGH the outputs of the decoder are latched, that is, its inputs (*A*, *B*, *C*, and *D*) do not have any effect on its outputs. Each decoder output drives one common cathode display digit. The hundreds digit is driven by a *D* flip-flop (14013) and discrete transistor MPS6515.

The detailed operation of this system will be discussed in the problems. ∎

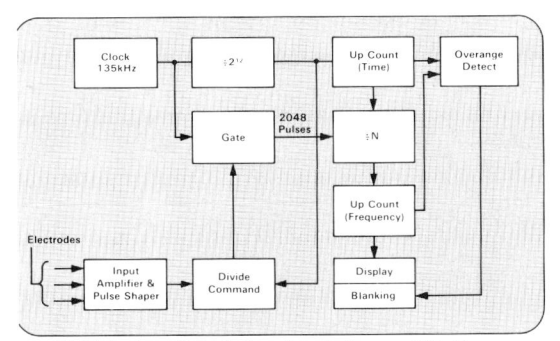

FIGURE 1 — A portable heart-rate monitor using low-power CMOS. The complete system package including batteries is the size of a bar of soap. There is no turn-on switch since the monitor operates only when electrodes are attached to the patient.

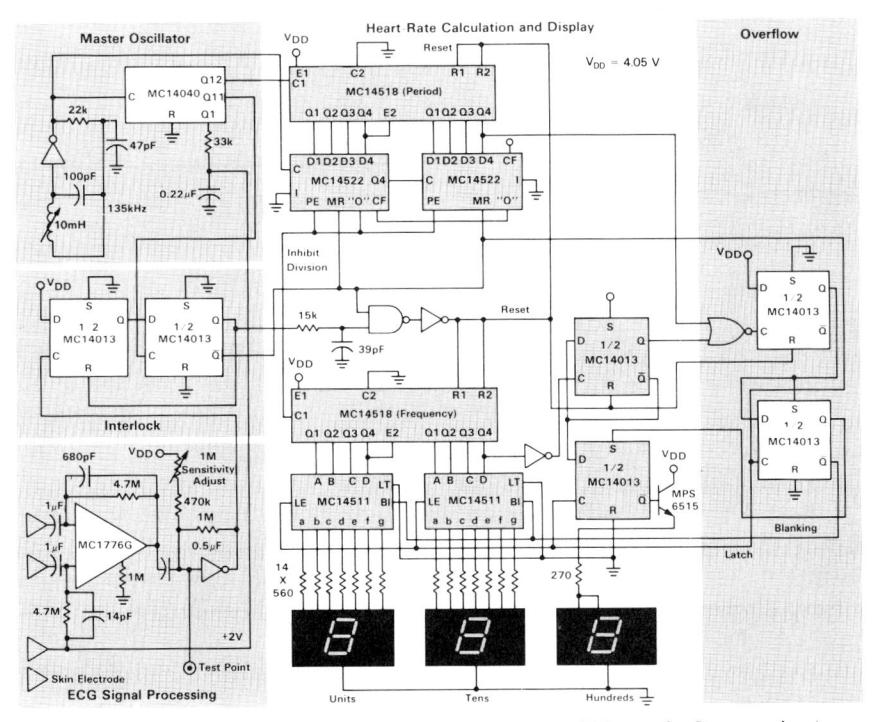

Fig. 6.14 Portable heart-rate monitor. (Courtesy of Motorola Corporation.)

This example of a three-digit display illustrates one method for displaying more than one decimal digit. It is called the **continuous method** because each digit is driven by a separate display decoder/driver, which continuously supplies current to that digit.

One disadvantage of the continuous method for multidigit displays is the relatively large dc power requirement. Another is the need for a separate decoder/

driver for each digit. Finally, the continuous method requires many interconnecting wires. Each decoder/driver can be multiplexed using the continuous method, with a saving in the number of interconnecting lines and the ability to use low multiplexing rates [1, p. 3–9]. The standard multiplexed method is discussed in the next section.

*6.6 MULTIPLEXED MULTIDIGIT SEVEN-SEGMENT DISPLAYS

6.6.1 Basic Principles

The standard multiplex method is based on the following principle. The human eye detects partially the peak and partially the average light output of an LED. If the digits in a display are illuminated for a relatively short period of time by relatively high current pulses, then the display has the same appearance as that produced by the continuous method, which uses lower, continuous currents.

Figure 6.15(a) shows the principle of a multiplexed seven-segment display. The seven-segment decoder/driver furnishes the necessary current through the current-limiting resistors R_S to the selected LEDs. Each column of LEDs represents one seven-segment LED display. The external scanner switches select digits 1 through N in succession. For example, the second digit of Fig. 6.15(a) will display the number 3 when the decoder switches a, b, c, d, and g are closed and the scanner switch for the second digit is closed. Note that each display is the common cathode type. Common anode multidigit displays are also available. The common segment lines in multidigit displays are usually internally connected, in which case the display cannot be used in the continuous method.

Figure 6.15(b) shows a five-digit display mounted in one DIP package. Multidigit displays 0.5" high are available; however, they cannot be mounted in a DIP due to their large size.

For multiplexed displays, as the number of digits increases, the length of time each digit receives current decreases. Therefore, the greater the number of digits, the higher the peak current that must be supplied to each LED in order to maintain display brightness. For example, eight digits require more current than a single digit would require for the same brightness.

The LED's relative efficiency in converting dc current into light (luminous intensity/applied current) increases for some LEDs with the peak applied current [10]. Therefore, using higher peak currents with a lower duty cycle, to give the same average current, can result in a higher light output from an LED. Multiplexed displays may thus require less dc power because there is only one decoder/driver, and because the LEDs can have a greater efficiency when pulsed.

One cannot state precisely when multiplexing is more economical than the continuous method. The following are among the factors to be considered when deciding on the method for connecting multidigit displays: (1) number of digits to be displayed, (2) brightness desired, (3) cost of decoder/drivers versus multiplexer

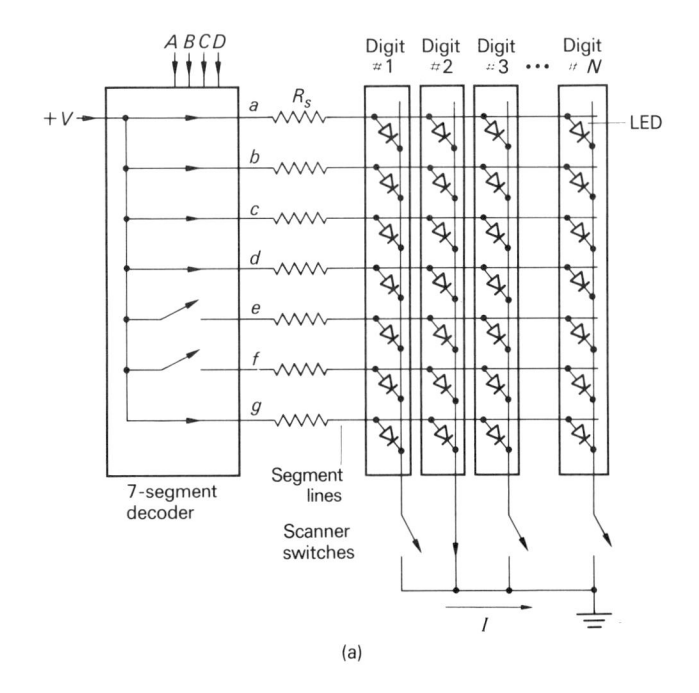

6.6   MULTIPLEXED MULTIDIGIT SEVEN-SEGMENT DISPLAYS   173

Fig. 6.15 Multiplexed seven-segment displays: (a) circuit illustrating basic principles; (b) five-digit LED displays (courtesy of Hewlett-Packard Company).

circuitry, (4) availability of data in serial or parallel form, (5) peak currents for multiplexing versus dc continuous currents, and (6) interface costs [4, p. 8–7].

The next section discusses a multiplexed display system.

6.6.2 Applications

Figure 6.16 shows the logic diagram for a common cathode three-digit multiplexed display. Note that a single active HIGH output display decoder/driver is used. The same segment in each digit is connected by a common bus to the output of the decoder/driver. For example, all the *a* segments are connected together by the *a* bus. A clock runs the counter, which counts in decimal from 0 through 2 repeatedly. When the clock output is 0, the LSD inputs of the four multiplexers (MUX) are connected to the four inputs (A_0, A_1, A_2, and A_3) of the decoder/driver. The binary inputs generally come from data stored in latches. The decoder/driver selects the same segments on *each* of the three display units. At the same time the counter selects the 0 output of the 1-of-4 decoder, which turns scanner switch transistor Q_0 on. This connects the cathode of the LSD display unit to ground. Thus only the

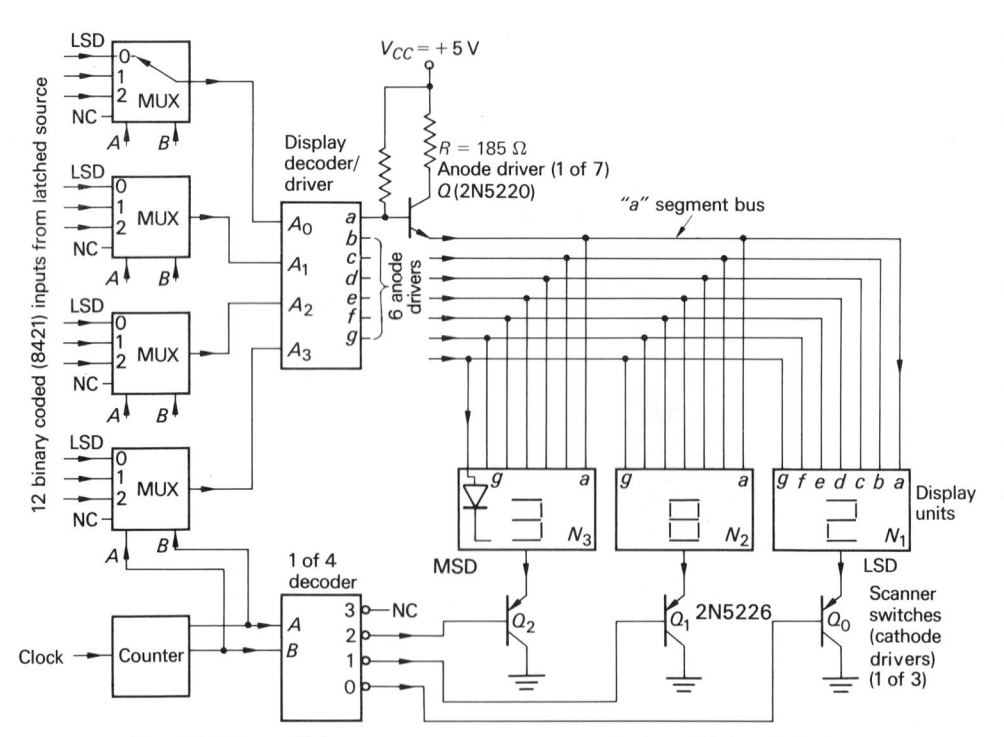

Fig. 6.16 Three-digit, seven-segment common cathode multiplexed display.

LSD is displayed. The cathode driver transistors are also called **digit drivers** since they determine which display is to be illuminated.

The counter next counts to 1. The 1 inputs to each of the four multiplexers are connected to the decoder/driver, which activates the appropriate segments on each display unit. Also, the 1-of-4 decoder connects the cathode of the next MSD display unit (N_2) to ground so that the N_2 digit is displayed. This sequential selection of display units continues for one more count and is then repeated, at the clock rate. A flicker-free display of the three digits results if the clock frequency is 100 Hz or greater. Ripple blanking, decimal points, and lamp test inputs are not shown. Anode driver transistors are used to supply sufficient current to the display units. A total of seven n-p-n anode driver transistors, such as 2N5220, are required. The 1-of-4 decoder has active LOW outputs, hence p-n-p cathode driver transistors are required so that they will turn on for a LOW decoder output. A total of three cathode driver transistors are required.

Transistors can be obtained separately or with several transistors in one IC chip, such as the SN75492 hex LED driver. Other specific suggestions for anode and cathode drivers, for both common anode and common cathode displays, are given in the references [1, pp. 3–22 and 3–23; 4, Chs. 8 and 9].

The collector resistor R in Fig. 6.16 is calculated as follows [7]:

$$R = \frac{V_{CC} - V_{\text{sat AD}} - V_{\text{LED}} - V_{\text{sat CD}}}{NI_{\text{avg}}}.$$

Here V_{CC} = the supply voltage, $V_{\text{sat AD}}$ = the collector–emitter voltage for the anode driver transistor when saturated at a current of $I_{pk} = NI_{\text{avg}}$, N is the number of digits in the display, V_{LED} = forward voltage drop across one LED segment at a current of I_{pk}, $V_{\text{sat CD}}$ = the collector–emitter voltage for the cathode driver transistor when saturated at MI_{pk} where M = number of segments (LEDs) per digit, and I_{avg} = average operating current per segment for the desired luminous intensity. Let's calculate the value of the collector resistors.

Example 6.9 Calculate the value of the collector resistor R in Fig. 6.16 using 2N5220 anode drivers, 2N5226 cathode drivers, and an HP 5082–7402 three-digit display. Assume a 10% duty cycle and a 100-μcd luminous intensity per segment.

Solution. The data books give $V_{\text{sat AD}} = 0.5$ V, $V_{\text{sat CD}} = 0.8$ V, $V_{\text{LED}} = 1.6$ V, and $I_{\text{avg}} = 3.8$ mA. Substituting in the above formula, using $V_{CC} = 5$ V, yields

$$R = \frac{(5 - 0.5 - 1.6 - 0.8) \text{ V}}{3 \times 3.8 \text{ mA}} = 185\ \Omega. \blacksquare$$

An eight-digit display similar to the three-digit display shown in Fig. 6.16, can be obtained by using four eight-input multiplexers, eight LED displays, a three-bit counter that counts to 8, a 1-of-8 decoder, eight cathode and eight anode drivers, and one display decoder/driver.

In comparing the continuous three-digit system of Fig. 6.13 with the multiplexed three-digit system of Fig. 6.16, we can see that this multiplexed system requires more ICs. However, as the number of digits increases the multiplexed system requires relatively fewer ICs than the continuous system. The multiplexed system can be more economical for four or more digits, depending on the system that generates the signals to be displayed.

Some advantages of multiplexing are: fewer ICs and hence lower cost, easier PC board layout, fewer interconnecting wires, and lower overall power consumption. Some disadvantages are: multiplexing rates must be fast enough to avoid flicker, more ICs are required for displays with only a few digits, if the clock or counter fails excessive current may damage the display, higher display operating current is required and a clock is required.

A discussion of errors produced by **display failures** is found in the references [1, p. 3–6]. Other display systems such as those using shift registers [7], clocks [3; 6, p. 8–12], an interface [4, p. 8–11], a digital voltmeter [4, p. 8–13], calculators [7; 4, p. 9–11], and decimal points [4, p. 9–12] are also discussed in the references.

*6.7 LED DOT MATRIX DISPLAYS

In this section alphameric displays are discussed. Suppose an automatic parking garage allows users with a special card to park in any of several controlled areas. Each card is similar to a plastic credit card except that a ferrite layer in the middle of the card contains the user's code in binary. The user inserts his card into a card reader, which determines whether the card is valid. If it is not valid, the reader captures the card and calls the attendant. If the card is valid, a multicharacter display unit displays one of several simple messages such as "ENTER," "LOT FULL GO TO NO. 7," etc. One display system that will display alphameric characters is the dot matrix display.

In order to make all the alphameric characters clearly distinguishable, displays must have more than seven segments. A **dot matrix** is an array of individual LEDs arranged in columns and rows called a matrix. A dot matrix usually contains from four to seven rows and from seven to nine columns. Thus a 4 (rows) × 7 (columns) dot matrix would contain 28 separate LEDs.

Dot matrix displays are available in single-character displays and in multicharacter displays. First let's consider single-character displays.

6.7.1 Single-Character Displays

Figure 6.17(a) shows a 4 × 7 dot matrix that is displaying the alphabetic character A, which is formed by illuminating the LEDs in the first and fourth columns and in the first and fourth rows of the matrix.

Figure 6.17(b) shows a typical single-character 4 × 7 dot matrix display that can form the alphameric characters 0 through 9 and A through F. The matrix

6.7 LED DOT MATRIX DISPLAYS 177

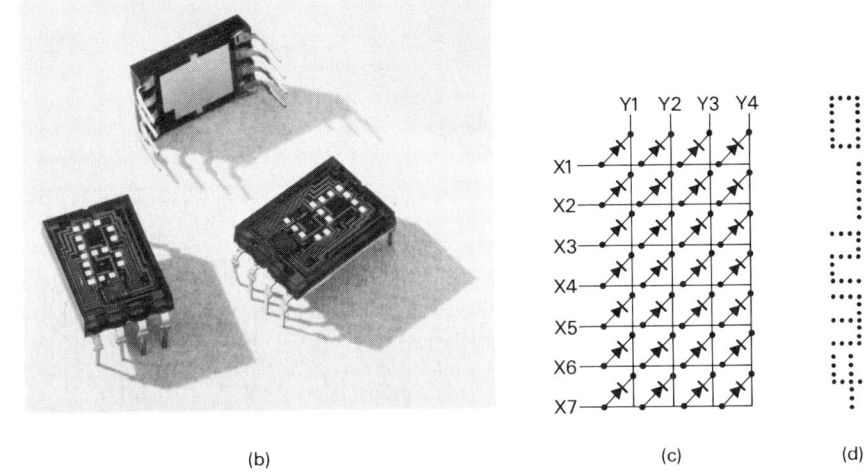

Fig. 6.17 Alphameric dot matrix displays: (a) character "A" displayed; (b) HP 7340 dot matrix displays (courtesy of Hewlett-Packard Company); (c) X-Y addressable 4 × 7 LED array layout; (d) 4 × 7 character font for numbers 0 through 4.

arrangement for the 28 diodes in a 4 × 7 display is illustrated in Fig. 6.17(c). The size and style, called the **font**, of the first five digits of a 4 × 7 display is shown in Fig. 6.17(d). To illuminate the number 1, for example, rows X1 through X7 (Fig. 6.17c) are connected to +5 V, and column Y4 is connected to ground.

The diodes are addressed using a system called X-Y addressing. In **X–Y addressing** each diode in the matrix has an address according to the row (X) and column (Y) in which it is located, just as in geometry a point in the X-Y plane is located by giving its X and Y coordinates. To turn on the LED in the second row and the third column, a positive voltage is applied to the X2 row, and column Y3 is grounded. To turn on the first row a positive voltage is applied to X1, and Y1 through Y4 are grounded. The advantage of X-Y addressing, as compared with using a decoder, is that it is simple and has the fewest interconnecting pins for the number of diodes addressed [7, p. 55].

The display shown in Fig. 6.17(b) (HP 5082–7340) has its own decoder/driver and a memory built into the base of the display so that no external decoder/driver is needed. Some typical characteristics of the 7340 are: character height 0.3 inch, supply voltage 5 V, power dissipation 470 mW, supply current 170 mA, luminous intensity 70 μcd, and peak wavelength 655 nm (red).

Displaying more than 16 characters with a 4 × 7 matrix results in poor readability. Hence 5 × 7 or larger matrices are generally used for displaying more than 16 characters.

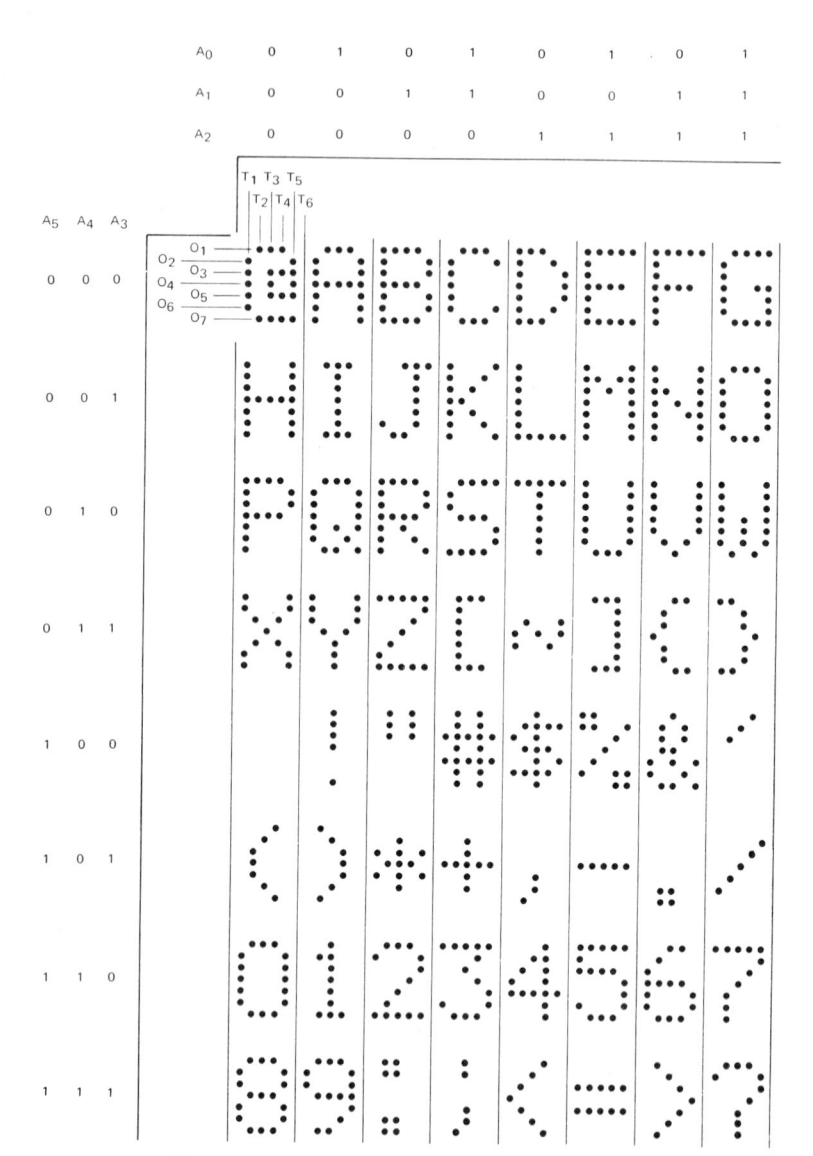

Fig. 6.18 5 × 7 dot matrix, standard ASCII font.

Many different fonts are available [11]. Figure 6.18 shows the ASCII character font for a 5 × 7 dot matrix. Each character in Fig. 6.18 has a binary address that is used when reading out that character. For example, the letter Z has the address 011011.

Now let's examine multicharacter dot matrix displays, which could be used in our parking lot example.

6.7.2 Multicharacter Displays

Figure 6.19(a) shows a five-character 5 × 7 dot matrix display in a DIP package. Several alphameric characters such as these can be displayed using a multiplexing scheme similar to that used for seven-segment multidigit displays.

Referring to Fig. 6.19(b), let us see how the characters *HI* could be formed using a multidigit dot matrix display. The logic circuitry for the display scans the rows in the array as follows. First, row X1 is addressed, i.e., switch X1 is closed, which applies a positive voltage to all the anodes of the LEDs in the first row. Then columns 1A, 1E, 2A, 2B, 2C, 2D, and 2E are connected to ground through the column switches. This addressing system illuminates the first row of LEDs in the characters *HI*. Next, the second row is scanned with switch X2 closed and

(a)

Fig. 6.19 5 × 7 dot matrix display: (a) HP 7102 five-character display (courtesy of Hewlett-Packard Co.); (b) formation of the characters "HI" by vertical strobing.

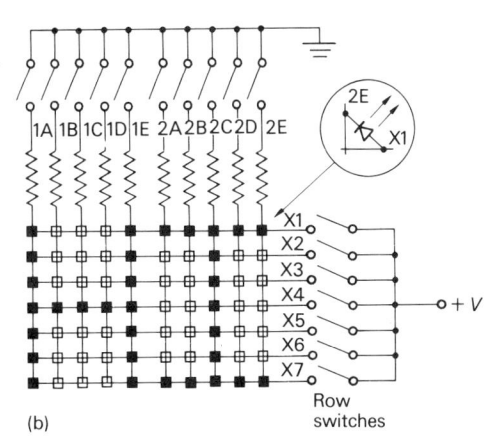

(b)

column switches 1A, 1E, and 2C closed. The remaining rows are scanned in sequence until the last row is scanned. The scanning is then repeated beginning with the first row, for the same characters or for a new set of characters. If the rows are scanned at least 100 times per second, a flicker-free display results.

The dot matrix display shown in Fig. 6.19(a) has the following characteristics: 0.27 inch character height, five characters per IC, 10 mA maximum average current per LED, 700 mW maximum power dissipation per character (all diodes lit), 10 ns typical rise and fall times, typical luminous intensity of 2.2 mcd, peak wavelength of 655 nm (red), and spectral line half-width of 30 nm.

There are two systems for addressing dot matrix displays. When the rows are addressed from top to bottom in sequence this is called **vertical scanning (strobing)**. When the columns are addressed from left to right, this is called **horizontal scanning (strobing)**. Vertical scanning is frequently more efficient, and is generally used for more than four characters [7, p. 75].

Using logic gates for decoding the address inputs of a dot matrix display would be very complicated, hence a read-only memory (ROM, see Chapter 10) is used to generate each character.

SUMMARY

In this chapter we first discussed LEDs and their uses; then we focused attention on two small-scale display systems, the seven-segment LED display and the dot matrix display. A seven-segment display consists of seven bar-shaped LEDs arranged in the pattern of an 8. Seven-segment displays are suitable for displaying numbers and a few letters or special characters. A dot matrix display consists of rows and columns of LEDs. A 7×9 dot matrix display is capable of displaying a complete range of alphameric characters.

Multicharacter displays use one of two methods for displaying characters. In the continuous method, each character is continuously illuminated by a separate decoder/driver. In the multiplexed method, each character is pulsed for a short period of time and one decoder/driver is time-shared among the display units. The multiplexed method is often cheaper for displays with four or more characters because fewer ICs, interconnecting lines, and connectors may be required.

PROBLEMS

Section 6.1

1. Name three examples of systems that use visible displays.

Section 6.2

2. (a) Using a TTL data book draw the logic diagram for an LED connected to the output of a TTL 7437 driver gate. (b) Calculate the value of the current-limiting resistor for an FLV108 LED (Fig. 6.3) with a luminous intensity of 1.5 mcd.

3. In Problem 2, what will the luminous intensity be at an angular displacement of 60° from normal?

4. (a) Draw the output circuit of a totem pole TTL gate that is connected to an LED through a current-limiting resistor to $V_s = V_{CC}$. (b) Explain why the LED does not light when the gate output is HIGH.

Section 6.2.2

5. *Intrusion Alarm.* A valuable piece of jewelry is located in a display case. At night it is protected by a warning system that consists of four infrared (IR) emitting diodes and four phototransistor detectors, which are connected so that four invisible light beams surround the jewelry case. When one or more beams are interrupted an alarm sounds. Develop the logic diagram for this intrusion alarm system. Use Fairchild FPE104 IR emitters and FPT130 phototransistors or equivalent devices. Use a 10-V supply voltage throughout and CMOSL gates.

6. *Troubleshooting Comparator.* (a) Explain with the aid of a timing diagram how the pulse stretcher circuit in the troubleshooting comparator in Example 6.4 works for a 500-ns and for a 1-sec error pulse at point *A*. [*Hint*: The time required for the voltage at point *D* to rise to the TTL threshold voltage $V_T = 1.4$ V is given by

$$t_{VT} = R_T C_1 \ln\left[\frac{V_{CC}}{V_{CC} - V_T}\right],$$

where R_T is the sum of R_1 and the input resistance of the NOT gate in the HIGH state. The Zener diode limits the voltage of point *D* to 3.16 V.] (b) Using a digital computer, calculate the value of t_{VT}.

7. *Beat Frequency Indicator.* Figure P6.1 shows the logic diagram of a device that indicates the relationship between the two input frequencies f_1 and f_2. The LEDs are arranged in a circular pattern. Only one LED is on at any one time. The LEDs light in clockwise or counterclockwise succession, depending on whether f_1 is greater or less than f_2. The rate of rotation of the light depends on the magnitude of the difference between f_1 and f_2. Explain in detail how the circuit works.

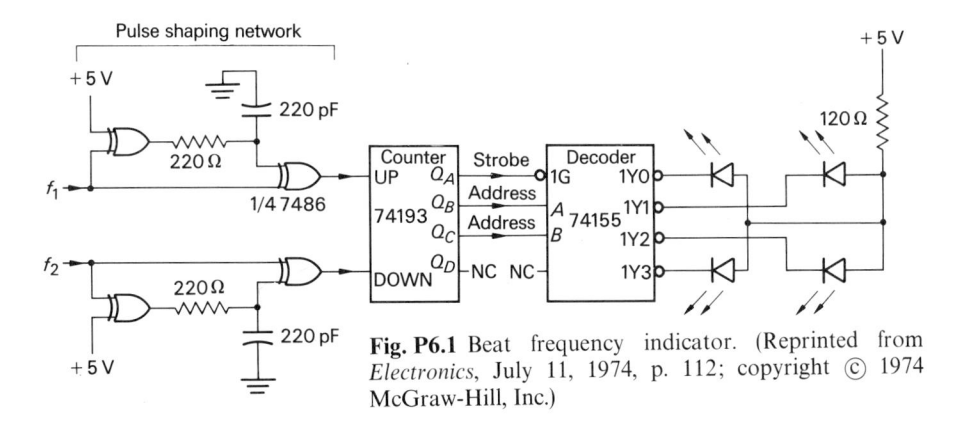

Fig. P6.1 Beat frequency indicator. (Reprinted from *Electronics*, July 11, 1974, p. 112; copyright © 1974 McGraw-Hill, Inc.)

182 SMALL-SCALE SOLID STATE DISPLAY SYSTEMS

Section 6.3

8. Draw the block logic diagram (similar to Fig. 6.8) for a one-digit seven-segment display system. Show the display decoder/driver, common cathode display, and current-limiting resistors.

9. (a) Using a Motorola CMOSL data book, calculate the value of the current-limiting resistors for a 14511 decoder/driver using an HP 5082–7740 seven-segment display with an average luminous intensity of 100 μcd per segment. (b) Use a digital computer to calculate the value of the current-limiting resistors. Have the computer program round off the calculated value to the nearest standard 10% resistor value.

10. From the decoder/driver truth table in Fig. 6.8(c), what are the decoder inputs and outputs for displaying a decimal 6?

Section 6.4

11. What are invalid inputs?

12. Draw the block diagram (similar to Fig. 6.11) for the decoder/drivers of a five-digit continuous display system that automatically blanks the two leading-edge zeros and one trailing-edge zero.

Section 6.5

13. The slope of the reference voltage input to the analog comparator of a DVM is 10 V/sec. What is the unknown dc voltage if the clock frequency is 2 Hz and the counter reads 4?

14. Name three disadvantages of the continuous display method as compared with the multiplexed method.

15. *Heart-Rate Monitor* (Fig. 6.14). Using a Motorola CMOSL data book, explain the operation of the interlock circuit (device number 14013).

16. *Heart-Rate Monitor* (Fig. 6.14). Using a Motorola CMOSL data book, explain the operation of the heart-rate calculation circuit (device numbers 14518, 14522(2), NAND and NOT gate).

Section 6.6

17. Explain how a two-digit multiplexed display can display the number 26. Draw a figure similar to Fig. 6.15(a) to aid in your explanation.

18. *DVM Readout.* Using a TTL data book, draw the connection diagram for a ramp type DVM with a four-digit TTL multiplexed readout. Use an HP 5082–7730 displays with a 100-μcd output. Use a 10% duty cycle. Use TTL device numbers 7447, 74153, 74490, and discrete components as needed.

Section 6.7.1

19. Explain how the number 4 can be generated with a 4 × 7 dot matrix display. Draw a circuit similar to Fig. 6.17(c) to help in your explanation.

Section 6.7.2

20. Explain how the message "TO" is generated by vertical-scanning a 5 × 7 array. Use the standard ASCII character font (Fig. 6.18) and a diagram similar to Fig. 6.19(b).

21. *Computer Program.* Write a computer program that simulates a 5 × 7 dot matrix display by printing out ASCII characters, in the font shown in Fig. 6.18, for the address inputs for characters *A, B, C, D,* and *E*.

REFERENCES

1. Peter Alfke and Ib Larsen (eds.). *The TTL Applications Handbook.* Fairchild Semiconductor, August 1973.
2. Owen Doyle. "The Right Numeric Readout: A Critical Choice to Designers." *Electronics,* pp. 65–72, May 24, 1971.
3. George Smith. "LEDs and Photometry." AN-1, Litronix Inc., March 1971.
4. *Opto Electronics Handbook.* Fairchild Semiconductor, February 1973.
5. Howard C. Borden and Gerald P. Pighini. "Solid State Displays." *Hewlett-Packard Journal,* February 1969.
6. *COS/MOS Digital Data Book and Application Notes.* AN ICAN 6733, RCA Corp., 1974.
7. *Solid State Display and Optoelectronics Designer's Catalog.* Hewlett Packard Co., July 1973.
8. *Hewlett-Packard Journal,* January 1971 and March 1969. An excellent discussion of the principles of digital voltmeters.
9. Milt Laflen. "Challenge, Contribution, Commitment." *Motorola Monitor,* pp. 5–10, October 1974.
10. David M. Barton. "Why Multiplex LEDs?" *IEEE Spectrum,* pp. 30–32, November 1972.
11. *American and European Fonts in Standard Character Generators.* AN-57, National Semiconductor Corp., January 1971.
12. *Liquid Crystal Applications Manual.* Hamlin Inc., Lake and Grove Streets, Lake Mills, Wisconsin 53551.
13. Michael J. Riezenman. "The New Displays Complement the Old." *Electronics,* pp. 91–99, April 12, 1973.
14. Howard Wolff. "Electrochromics Glow on the Horizon." *Electronics,* pp. 87–90, December 6, 1973.
15. Louis N. Heynick, *et al.* (guest editors). "Special Issue on Display Devices." *IEEE Transactions on Electron Devices,* vol. ED-20, No. 11, pp. 917–919, November 1973.
16. "Electrochromic Displays Shine in Bright or Dim Surroundings." *Digital Design,* pp. 10–18, August 1975.

7: Schmitt Triggers, Monostables, and Clocks

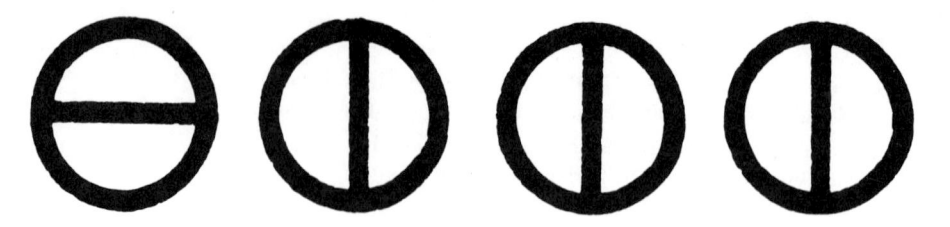

7.1 INTRODUCTION

In this chapter we shall discuss Schmitt triggers, monostables, astables, and clocks. The Schmitt trigger is useful for reshaping pulses. The monostable multivibrator can be used for producing time delays. Clocks and astable multivibrators are oscillators whose outputs are rectangular waves; their pulse outputs are used to control the timing in many digital systems. A knowledge of each of these devices will be helpful for understanding the digital systems described in Chapters 8, 9, and 10.

*7.2 THE SCHMITT TRIGGER

The **Schmitt trigger** is a device whose output goes HIGH when a positive-going input voltage exceeds the **positive-going threshold voltage** V_{T+}. The output will remain HIGH until the negative-going voltage decreases below the **negative-going threshold voltage** V_{T-}, at which time the output goes LOW. The threshold voltages for IC Schmitt triggers are generally fixed by the manufacturer and cannot be adjusted by the user. Figure 7.1(a) shows the logic symbol for a Schmitt trigger.

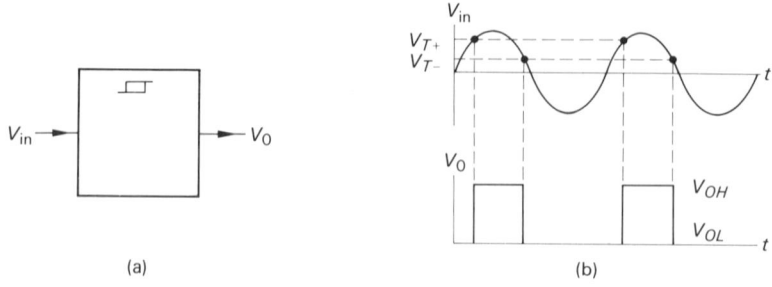

Fig. 7.1 The Schmitt trigger: (a) logic symbol; (b) input and output waveforms.

Figure 7.1(b) shows a sinusoidal input to a Schmitt trigger. At time $t = 0$ the output voltage V_0 is LOW. When the positive-going input voltage exceeds V_{T+} the output becomes HIGH. The output stays HIGH until the negative-going input voltage decreases below V_{T-}, at which time the output goes LOW. The output stays LOW until the input voltage again increases through V_{T+}.

The main application of the Schmitt trigger is to convert slowly varying input voltages into pulses that have very short rise and fall times, as illustrated in the following example.

Example 7.1 *Generation of a 60-Hz Rectangular Waveform.* A digital household clock uses the 60-Hz line frequency as its stable time base. Show how a Schmitt trigger can be used to convert 110-V line voltage into a rectangular waveform suitable for driving CMOSL devices ($V_{DD} = 5$ V).

Solution. The input voltage must first be reduced from 110 V rms to approximately 5 V peak. This can be done with a step-down transformer. The negative voltage swing can be eliminated by a diode as shown in Fig. 7.2. The output of the diode will be a half-wave rectified sine wave whose amplitude is approximately $+5$ V. The Schmitt trigger then converts this waveform into a 60-Hz rectangular waveform. The Schmitt trigger as used in this example is also a **zero crossing detector** since it produces a pulse each time the input voltage crosses the zero voltage axis. ∎

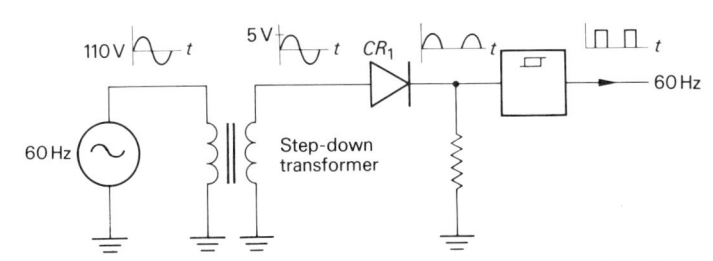

Fig. 7.2 Generation of a 60-Hz rectangular waveform using a Schmitt trigger.

Schmitt triggers are available in digital IC form in the TTL and CMOSL fami-families and can be made from ECL line drivers using external resistors and capacitors [1, p. 4–84]. Figure 7.3 shows the specifications for a CMOSL Schmitt trigger.

The **hysteresis voltage** V_H is the difference between the positive-going and the negative-going threshold voltage; that is, $V_H = V_{T+} - V_{T-}$. For the CMOSL 4093 the hysteresis voltage for a 5-V supply is $V_H = (2.6 - 2.0)$ V $= 600$ mV.

STATIC ELECTRICAL CHARACTERISTICS AT $T_A = 25°C$ (All Inputs . . . . $V_{SS} \leq V_I \leq V_D$
(Recommended DC Supply Voltage ($V_{DD}-V_{SS}$) . . . 3 to 15

CHARACTERISTIC	SYMBOL	TEST CONDITIONS V_O Volts	TEST CONDITIONS V_{DD} Volts	TYPICAL VALUES	UNITS	CURVE FIG. No.
Quiescent Device Current	I_L		5	0.001	μA	
			10	0.001		
Quiescent Device Dissipation/Package	P_D		5	0.005	μW	
			10	0.01		
Output Voltage: Low-Level	V_{OL}		5	0	V	
			10	0		
High-Level	V_{OH}		5	5	V	-
			10	10		
Noise Immunity (All Inputs)	V_{NL}	5	5	2.6	V	3
		10	10	5.2		
	V_{NH}	0	5	3		
		0	10	6.5		
Output Drive Current: n-Channel (Sink)	I_{DN}	0.4	5	0.8	mA	
		0.5	10	1.8		
p-Channel (Source)	I_{DP}	4.6	5	−0.8	mA	
		2.5	5	−1.8		
		9.5	10	−1.8		
Positive Threshold Voltage	V_P		5	2.6	V	2
			10	5.2		
Negative Threshold Voltage	V_N		5	2	V	2
			10	3.5		
Hysteresis Voltage (V_P-V_N)	V_H		5	0.6	V	2
			10	1.7		

DYNAMIC ELECTRICAL CHARACTERISTICS AT $T_A = 25°C$

CHARACTERISTIC	SYMBOL	TEST CONDITIONS V_{DD} Volts	TYPICAL VALUES $C_L = 15\,pF$	TYPICAL VALUES $C_L = 50\,pF$	UNITS
Propagation Delay Time	t_{PHL}, t_{PLH}	5	170	190	ns
		10	90	100	
Transition Time	t_{TLH}, t_{THL}	5	60	100	ns
		10	30	50	
Average Input Capacitance	C_I	Any Input	5		pF

Fig. 7.3 Specifications for CMOSL 4093 quad Schmitt triggers. (Courtesy of Solid State Division, RCA Corporation.)

The input logic to most IC Schmitt triggers is a built-in NAND gate, as illustrated in Fig. 7.4(a). This NAND gate allows the Schmitt trigger to be gated on or off by a control signal. A curve of the output of a Schmitt trigger versus its input is shown in Fig. 7.4(b). This curve is called the **transfer characteristic** or

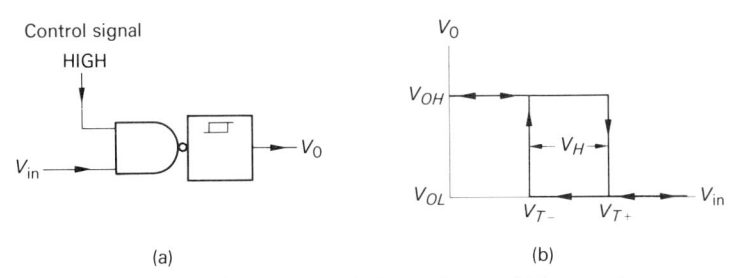

Fig. 7.4 Schmitt trigger: (a) NAND gate input; (b) hysteresis curve.

the **hysteresis curve** of the Schmitt trigger. When the input voltage is less than V_{T-} the output is HIGH. When the input voltage exceeds V_{T+} the output goes LOW. Note that the NAND gate inverts the logic of this Schmitt trigger.

The TTL dual 7413 Schmitt trigger has four inputs for each NAND gate, typical threshold voltages of 1.7 V and 0.8 V (900 mV hysteresis), and an average typical delay time of 16.5 ns. The 7413 also has an input clamp diode that allows a negative input voltage of up to -1.5 V. There are no restrictions on the input rise and fall times.

The threshold voltages of the IC logic Schmitt triggers are generally not adjustable. **Voltage comparators**, like the Schmitt trigger, are devices that compare input voltage levels; however, they have adjustable threshold voltages. They are available in IC form and have voltage outputs that are compatible with the TTL and CMOSL families. The next example illustrates an application of the voltage comparator.

Example 7.2 *Pulse Height Discriminator.* A physics experiment involves the measurement of neutrons in a specific range of energy levels. A detector converts the energy of neutrons into voltages that are proportional to their energy. Develop a logic system that will provide an output only when neutrons in a specific energy (voltage) range are detected.

Solution. Two voltage comparators can be used. Their inputs are both connected together as shown in Fig. 7.5. The positive-going threshold voltage of one

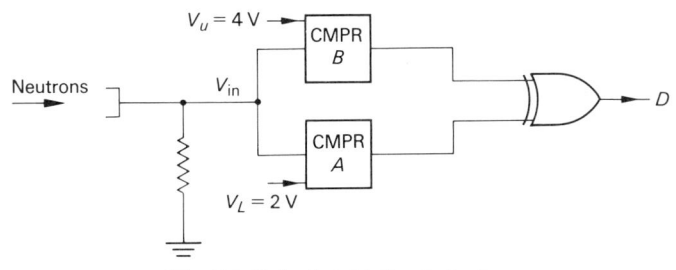

Fig. 7.5 Pulse height discriminator.

comparator is adjusted to the upper voltage (energy) level V_U and the positive-going threshold voltage V_{T+} of the second comparator is adjusted to the lower voltage level V_L. Suppose the threshold voltages are set at $V_U = 4$ V and $V_L = 2$ V. Then if the input voltage V_{in} is 3 V, the output of comparator A will be HIGH and the output of comparator B will be LOW; hence the output of the EX-OR gate will be HIGH. If $V_{in} = 1$ V, the output of both comparators will be LOW, so D is LOW. If $V_{in} = 6$ V, then the output of both comparators will be HIGH, so D is LOW. Thus this device will give a HIGH output only for pulses whose amplitude is between $V_L = 2$ V and $V_U = 4$ V. All other pulse heights will be discriminated against. A **pulse width discriminator** is discussed in the problems at the end of this chapter. ∎

The next section discusses some characteristics of pulses and illustrates the use of Schmitt triggers for pulse shaping.

7.3 PULSE CHARACTERISTICS [2, 3]

Recall that a nonideal pulse is one that has a nonzero rise time and nonzero fall time, as illustrated in Fig. 7.6(a). The **reference level** for a pulse is a specified fixed voltage, usually near the switching threshold voltage, or a voltage expressed as a percentage of the pulse amplitude E_p. The rise time t_r and fall time t_f of a pulse are usually measured between the 10% and 90% E_p points, as illustrated in Fig. 7.6(a), or between the 20% and 80% points.

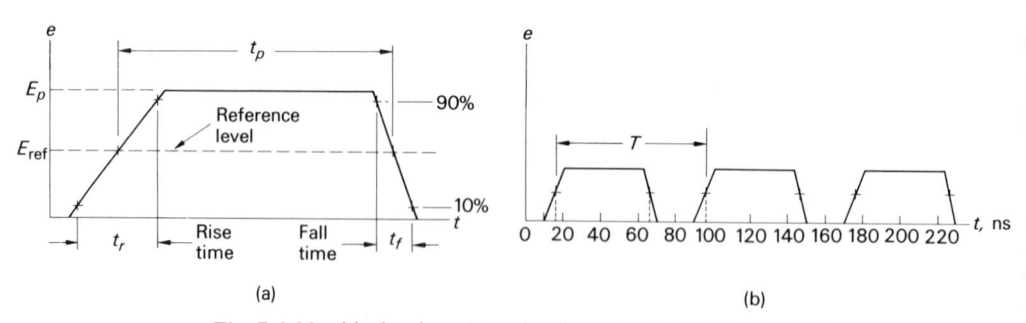

Fig. 7.6 Nonideal pulses: (a) pulse characteristics; (b) pulse train.

If pulses are repeated at regular intervals, the time interval from the leading (or trailing) edge of one pulse to the leading (or trailing) edge of the next pulse is called the **period** T or the **pulse repetition time** (prt). Figure 7.6(b) shows a pulse train (i.e., a series or train of pulses) and the prt. The **pulse repetition frequency** (prf) is

$$\text{prf} = \frac{1}{\text{prt}}.$$

7.3 PULSE CHARACTERISTICS [2, 3]

For example, if the pulse repetition time is 10 μs, the pulse repetition frequency is prf = 1/prt = 1/10 μs = $1/10^{-5}$ sec = 100 kHz. The **duty cycle** DC of a pulse train is the ratio of the pulse width to the prt,

$$DC = \frac{t_p}{prt}.$$

The duty cycle indicates the fraction of one time period that a pulse is present. For example, if the duty cycle is 0.1, then the pulse is present during only one-tenth of each cycle. A **square wave** is a pulse train whose duty cycle is 0.50, or 50%. If the duty cycle is greater than 10%, a pulse train is called a **rectangular wave**.

A typical pulse train in a TTL system might have the following characteristics: pulse width 20 ns, rise time 10 ns, fall time 5 ns, prf 10 MHz, and pulse amplitude 4 V. The following example illustrates some of these properties of pulses.

Example 7.3 *Pulse Characteristics.* For the pulse train shown in Fig. 7.6(b) determine the following: pulse width t_p, pulse repetition time prt, pulse repetition frequency prf, and duty cycle DC.

Solution. The pulse width is

$$t_p = (65 - 15) \text{ ns} = 50 \text{ ns}.$$

The pulse repetition time is

$$T = \text{prt} = (100 - 15) \text{ ns} = 85 \text{ ns};$$

therefore

$$\text{prf} = \frac{1}{\text{prt}} = 12 \text{ MHz}$$

and

$$DC = \frac{t_p}{\text{prt}} = \frac{50 \text{ ns}}{85 \text{ ns}} = 0.59 = 59\%. \blacksquare$$

A **glitch**, or a **spike**, is a narrow pulse in a logic circuit. Glitches are produced from sources such as current spiking and ringing (oscillations) in lines that connect high-speed devices.

As pulses travel down a transmission line they can become distorted; for example, their rise times can increase considerably. If pulses with long rise times are fed into logic devices two problems can occur. First, the logic devices may oscillate because the long rise time can cause the logic device to be biased in its transition region (between the HIGH and the LOW state) too long. Second, propagation delay times become unpredictable [4].

Let's illustrate the problem of feeding pulses having long rise and fall times into a TTL NOT gate. Figure 7.7(a) shows a nonideal pulse, having a rise and fall time of 10 μs, fed into the input of a NOT gate. When the input voltage to the NOT gate is between $V_{IL(max)}$ and $V_{IH(min)}$, the NOT gate is in the transition region of its transfer curve (Chapter 3). If the input to the NOT gate remains

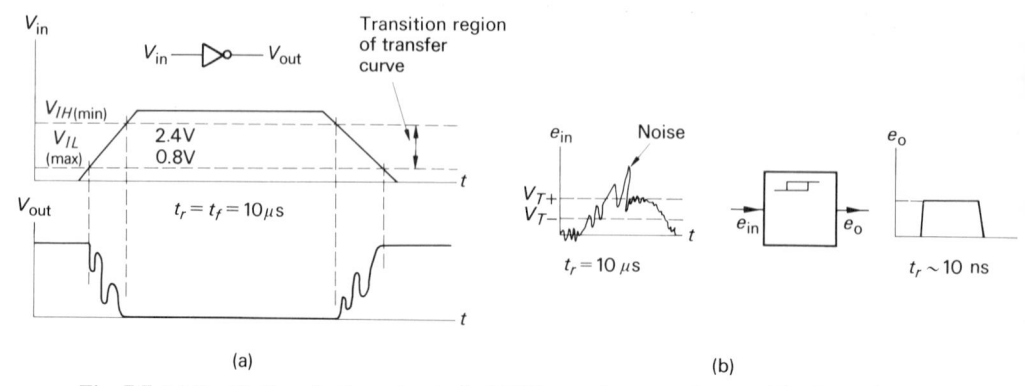

Fig. 7.7 (a) Oscillations in the output of a NOT gate due to an input with a long rise and fall time; (b) Schmitt trigger used to convert a long rise time to a short rise time.

in the transition region for longer than the gate propagation delay time, the gate can become unstable, i.e., its output can oscillate as shown in Fig. 7.7(a). To prevent standard series TTL devices from oscillating it is generally recommended that input rise and fall times be kept less than about 50 ns. Slower devices, such as those in the CMOSL family, can tolerate longer rise times. Faster devices, such as ECL devices, require relatively shorter rise times.

The next example illustrates one method for converting signals with long rise and fall times into ones having short rise and fall times.

Example 7.4 *Pulse Shaper.* Suppose the transmission line in the digital communication system in Example 5.6 (Section 5.8) is 2000 meters long. During transmission the pulses become noisy and their rise times increase from 50 ns to 10 μs, which in turn causes the input AND gate in the receiver to oscillate. Show a method for reducing the rise time at the receiver to acceptable TTL standards ($\sim$ 50 ns).

Solution. A Schmitt trigger can be used to convert the pulses with relatively long rise times to ones having short rise times. Figure 7.7(b) shows the logic diagram and the input and output waveforms. The Schmitt trigger also eliminates the noise and hence reshapes the input pulse to a relatively square, noise-free pulse. ∎

7.4 THE MONOSTABLE

A **monostable (multivibrator)** is a device whose output becomes HIGH when triggered by an input pulse. The output stays HIGH for a time determined by a resistor, which is usually external to the monostable, and an external capacitor. Output pulse widths ranging from a few nanoseconds to almost infinity can be

7.4 THE MONOSTABLE

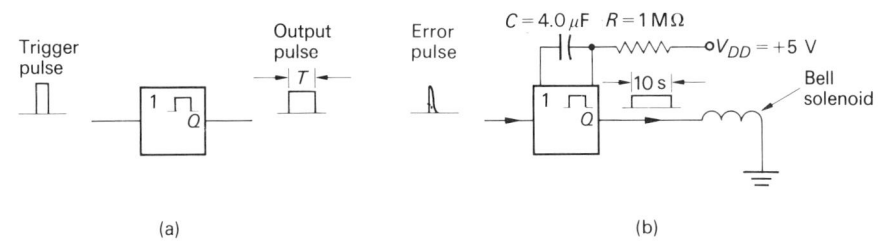

Fig. 7.8 Monostable multivibrator: (a) logic symbol; (b) intermittent error detector for Example 7.5.

generated with monostables. The logic symbol for a monostable is shown in Fig. 7.8(a).

Monostables, also called **one-shots**, are available in the TTL and CMOSL families in IC form. They are useful for applications such as delay elements, producing long pulses, and pulse generators.

The next example illustrates one application of a one-shot.

Example 7.5 *Computer Intermittent Error Detector.* In a computer occasional errors may occur that are only momentary and are automatically corrected by the computer. In order to alert the computer operator to these errors and thus to possible impending serious problems, we wish to devise a bell warning system that will indicate intermittent errors even if they last only a few microseconds.

Solution. The output of the error-detecting device, such as a parity checker, can be connected to the input of a monostable. The output of the monostable is connected to the solenoid of a bell as shown in Fig. 7.8(b). Suppose the bell is to ring for a time $T = 10$ sec. The values of the external timing components, R and C, are calculated (for a CMOSL 4047 monostable) from the timing formula $T = 2.5\,RC$, where T is the time the one-shot output is HIGH. Let's choose $R = 1$ MΩ; then $C = T/(2.5)\,R = 4.0\,\mu$F. The completed circuit is given in Fig. 7.8(b). ∎

Because of variations between ICs of a given type, timing formulas have inaccuracies that may range from ±1% to as much as ±10%. In addition, the tolerances in the external resistor and capacitor may cause further inaccuracies. Thus, in general, a monostable is not a very accurate timing device. An external trimmer (variable) resistor can be used to adjust the output pulse width to the desired width. But once the output pulse width is trimmed to the desired width, the stability (width) of the pulses will still be affected by IC temperature variations and power supply voltage variations. The output pulse width may also vary with the duty cycle of the input pulses. Therefore when accurate, stable pulses are required, other devices, such as counters or crystal oscillators, should be used.

However, many applications, such as the one in Example 7.5, do not require accurate timing, and the monostable is a very satisfactory inexpensive timer.

Another problem with monostables can occur during troubleshooting of a digital system. Frequently, troubleshooting a digital system is done by changing the inputs one step at a time and then checking for correct outputs. A monostable cannot be stepped with a clock; it is either on or off. Also a monostable may operate correctly when the trigger inputs occur at a low rate, but may not work correctly at high speeds. This may happen because several trigger pulses occur while the monostable output is HIGH, for example.

A monostable is partly an analog circuit, due to the R and C delay components, and partly a digital circuit. The analog part of the circuit is more susceptible to noise than an all-digital circuit. To minimize the introduction of noise into monostables the following precautions should be observed: (1) keep the R and C (and trimmer) lead lengths as short as possible to minimize stray capacitance; (2) bypass the power supply to ground with a high-frequency 0.1-μF ceramic disk capacitor placed within about 4 cm from the monostable; (3) keep the monostable output (Q) away from the capacitor (C) lead; (4) keep the monostable away from noise sources and conductors that contain noisy signals.

Some monostables have a Schmitt trigger built into their trigger input so that they can be triggered by trigger pulses with relatively long rise times. This prevents such monostables from being triggered by noise voltages, which usually have shorter rise times.

The maximum output pulse width for a monostable is determined by the maximum values of R and C that can be used with a given monostable. Electrolytic capacitors are usually used to obtain large capacitance. Large leakage currents in electrolytic capacitors usually limit C to about 1000 μF. Also, electrolytic capacitors have large tolerances, which may add to the inaccuracy of the calculated output pulse width [5, Ch. 13]. For CMOSL one-shots, R normally falls in the 10-kΩ to 1-MΩ range, while for TTL the range is on the order of 5–50 kΩ. Thus CMOSL one-shots can have longer output pulses than TTL one-shots for a given value of C.

Most monostables are not turned on by a continuous HIGH or LOW signal to the trigger input. A **leading-edge triggered** monostable is triggered (turned on) only by a trigger pulse that goes from a LOW to a HIGH ($L \rightarrow H$); a HIGH to LOW transition will not trigger such a monostable. A **trailing-edge triggered** monostable is triggered only by a trigger pulse that goes from a HIGH to a LOW ($H \rightarrow L$) level. A small circle at the trigger input on the logic symbol of a monostable indicates that it is trailing-edge triggered. Trailing-edge triggered monostables can be made from leading-edge monostables, and vice versa, by using an external inverter at the trigger input. The disadvantage of this is that an additional delay time is introduced by the NOT gate.

There are two basic types of monostables, (1) nonretriggerable and (2) retriggerable. Let us first discuss the nonretriggerable type.

7.4.1 The Nonretriggerable Monostable

In a monostable there is a delay time between input and output, just as there is in a logic gate. A timing diagram for a leading-edge triggered nonretriggerable monostable using ideal pulses is shown in Fig. 7.9(a). The output of the one-shot is initially LOW. One delay time after the leading edge of trigger pulse A, the output of the monostable goes HIGH. It stays HIGH for a total time T as determined by external timing components R and C. Input pulses B and C are ignored by the monostable while its output is HIGH, i.e., they do not lengthen the output pulse or change its state. After time T the monostable output goes LOW and stays LOW until one delay time after the leading edge of trigger pulse D. Trigger pulse E is ignored by the monostable, since its output is HIGH. Pulse E does not trigger the monostable once it becomes LOW because the trigger input did not undergo an $L \to H$ transition while the monostable output was LOW.

The 74221 is an example of a very stable ($\pm 0.5\%$) dual nonretriggerable TTL monostable.

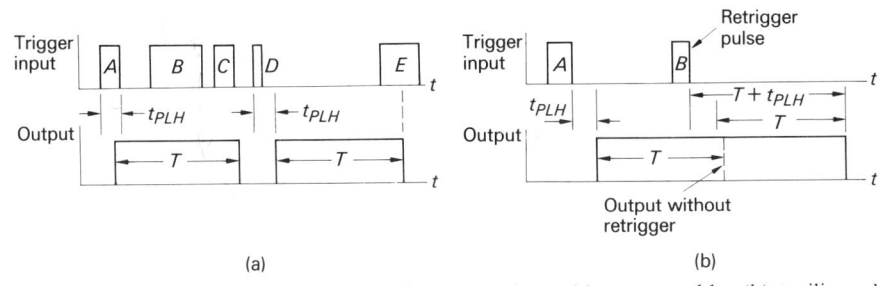

Fig. 7.9 Timing diagrams: (a) leading-edge nonretriggerable monostable; (b) trailing-edge retriggerable monostable.

7.4.2 The Retriggerable Monostable

A **retriggerable monostable** is a monostable that can be retriggered while the monostable output is HIGH. Thus, if a trigger pulse occurs during the time the output is HIGH, the monostable does not ignore the input but will stay HIGH for an additional length of time. The advantage of the retriggerable monostable is that long output pulses can be generated, and it is more versatile than the nonretriggerable one-shot.

Figure 7.9(b) shows the timing diagram for a trailing-edge retriggerable monostable. The output goes HIGH one delay time t_{PLH} after the trailing edge of trigger pulse A. In the absence of another trigger pulse the output would go LOW after a time T. However, a retrigger pulse B occurs while the monostable output is HIGH. This causes the output to stay HIGH for a time $T + t_{PLH}$ after the trailing edge of retrigger pulse B.

194 SCHMITT TRIGGERS, MONOSTABLES, AND CLOCKS 7.4

If a series of retriggering pulses are applied to the trigger input, timed so that they occur before the monostable output goes LOW, then the output will stay HIGH indefinitely.

Retriggerable monostables can be operated as nonretriggerable monostables by connecting the output to the trigger input. Most monostables have a complementary output $\bar{Q}$ that can be used in the conversion as shown in Fig. 7.10(a). Alternatively the Q output can be connected to an active HIGH input. Some monostables have AND and OR gates built into their inputs to provide the choice of leading-edge or trailing-edge triggering without an additional delay time incurred from external gates.

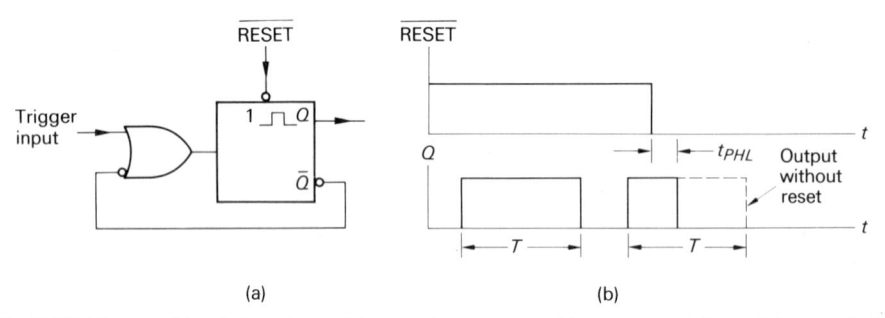

Fig. 7.10 Monostables: (a) retriggerable one-shot converted into a nonretriggerable one-shot; (b) timing diagram for the reset input.

A retriggerable monostable cannot be retriggered while the external timing capacitor is discharging. The discharge time t_d for the TTL 74122, for example, is $t_d(\text{ns}) \approx 0.22C$ (pF). So if $C = 4000$ pF then $t_d = 880$ ns; therefore a retriggering pulse should not be applied until at least 880 ns after the previous trigger pulse.

Most monostables have a reset input that allows the output to be reset LOW regardless of the trigger input at that time. Figure 7.10(b) shows a timing diagram for an active LOW reset pulse. The second output pulse would have normally stayed HIGH for a time T. However, when the reset input goes LOW, the Q output goes LOW one reset-to-output delay time later. Some monostables start a new timing cycle when the reset input becomes inactive.

Example 7.6 *Delayed Pulse Generator.* The timing in a digital system requires that a 1-μs pulse be generated from a 40-kHz square-wave signal. Also, the 1-μs pulse must be delayed 10 μs after the leading edge of the 40-kHz pulses. Draw the logic diagram for a system that will generate these delayed pulses. Use TTL one-shots.

Solution. The period of a 40-kHz square wave is 1/40 kHz = 25 μs. Figure 7.11(a) shows the timing diagram for this problem. Two monostables will be required,

7.4 THE MONOSTABLE

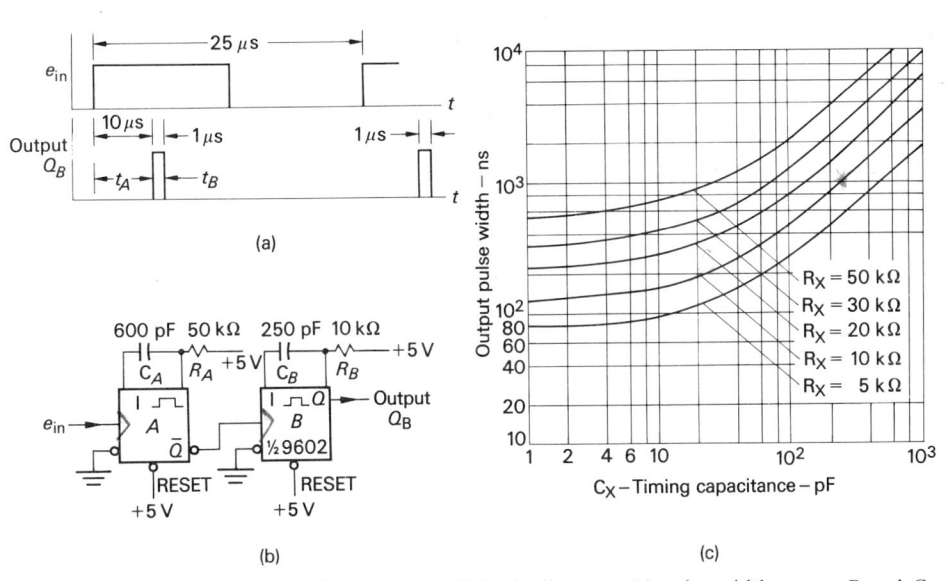

Fig. 7.11 Example 7.6: (a) timing diagram; (b) logic diagram; (c) pulse width versus R and C for a TTL 9602 monostable (courtesy of Fairchild Camera and Instrument Corporation).

one to produce the 10-μs delay, another to produce the 1-μs output pulse. The logic diagram using one dual 9602 monostable is shown in Fig. 7.11(b). Note that the 9602 has a choice of leading- or trailing-edge trigger inputs. When the leading edge of the input pulse to monostable A arrives, the $\bar{Q}_A$ output of monostable A goes LOW. The $\bar{Q}_A$ output goes HIGH after 10 μs, at which time it triggers monostable B to turn on for 1 μs.

The R and C timing components are determined from the curves shown for the 9602 in Fig. 7.11(c). For monostable A, if $R_A = 50$ kΩ is chosen then the curves give $C_A = 600$ pF for $t_A = 10$ μs. If 10 kΩ is chosen for R_B then Fig. 7.11(c) gives $C_B = 250$ pF for $t_B = 1$ μs. ∎

Figure 7.12 shows the electrical characteristics of the TTL 9602 dual retriggerable resettable monostable. Note that the timing resistor R_x is restricted to the range 5–50 kΩ. This means that for wide output pulses t, large-value capacitors are required. The consequences of large capacitances have already been discussed. The output pulse width t can be calculated from the formula $t = 0.31 R_x C_x (1 + 1/R_x)$, for $C_x \geqslant 10^3$ pF. For example, if $C_x = 1$ μF and $R_x = 10$ kΩ then $t = 3.1$ ms.

196 SCHMITT TRIGGERS, MONOSTABLES, AND CLOCKS 7.4

SYMBOL	PARAMETER	0°C MIN.	0°C MAX.	+25°C MIN.	+25°C TYP.	+25°C MAX.	+75°C MIN.	+75°C MAX.	UNITS	CONDITIONS (Note 1)
V_{OH}	Output HIGH Voltage	2.4		2.4	3.4		2.4		Volts	$V_{CC} = 4.75$ V, $I_{OH} = -0.96$ mA (Note 2)
V_{OL}	Output LOW Voltage		0.45		0.2	0.45		0.45	Volts	$V_{CC} = 4.75$ V, $I_{OL} = 11.3$ mA (Note 2) $V_{CC} = 5.25$ V, $I_{OL} = 12.8$ mA
V_{IH}	Input HIGH Voltage	1.9		1.8			1.65		Volts	Guaranteed Input HIGH Threshold Voltage
V_{IL}	Input LOW Voltage		0.85			0.85		0.85	Volts	Guaranteed Input LOW Threshold Voltage
I_{IL}	Input LOW Current		-1.6 -1.41		-1.0	-1.6 -1.41		-1.6 -1.41	mA mA	$V_{CC} = 5.25$ V, $V_{IN} = 0.45$ V $V_{CC} = 4.75$ V, $V_{IN} = 0.45$ V
I_{IH}	Input HIGH Current				10	60		60	µA	$V_{CC} = 5.25$ V, $V_{IN} = 4.5$ V
I_{SC}	Short Circuit Current					-35			mA	$V_{CC} = 5.25$ V, $V_{OUT} = 1.0$ V (Note 2)
I_{PD}	Quiescent Power Supply Drain		52		39	50		52	mA	$V_{CC} = 5.0$ V, Ground Pins 1 and 2
t_{PLH}	Negative Trigger Input to True Output				25	40			ns	$V_{CC} = 5.0$ V $R_X = 5.0$ kΩ $C_X = 0$, $C_L = 15$ pF
t_{PHL}	Negative Trigger Input to Complement Output				29	48			ns	$V_{CC} = 5.0$ V $R_X = 5.0$ kΩ $C_X = 0$, $C_L = 15$ pF
$t_{(min)}$	Minimum True Output Pulse Width				72	100			ns	$V_{CC} = 5.0$ V $R_X = 5.0$ kΩ
	Minimum Complement Output Pulse Width				78	110			ns	$C_X = 0$, $C_L = 15$ pF
t	Pulse Width	3.08		5.0	3.42	3.76	5.0		µs	$V_{CC} = 5.0$ V, $R_X = 10$ kΩ, $C_X = 1000$ pF
C_{STRAY}	Maximum Allowable Wiring Cap. (Pins 2 and 14)		50			50		50	pF	Pins 2 and 14 to Ground
R_X	Timing Resistor	5.0	50	5.0		50	5.0	50	kΩ	

Fig. 7.12 Electrical characteristics of the TTL 9602 dual monostables. (Courtesy of Fairchild Camera and Instrument Corporation.)

7.4 THE MONOSTABLE

The 9602 will tolerate a maximum of 50 pF of stray capacitance C_{stray} at the *RC* terminal with respect to ground without affecting its specifications. More stray capacitance than this can be tolerated, but the timing equations and timing graphs will be less accurate. The minimum typical output pulse width is 72 ns for the true (uncomplemented) output and 78 ns from the complemented output. The output pulse width t for $C_x = 1000$ pF and $R_x = 10$ kΩ has been measured and is guaranteed to be between 3.08 and 3.76 μs. The minimum trigger input pulse width that will trigger the 9602 (not shown in Fig. 7.12) is 40 ns.

The 9602 dissipates about 125 mW of power. The low-power 96L02 dissipates only 25 mW, but is slower than the 9602. By comparison the CMOSL 4047 monostable dissipates only 145 nW.

One disadvantage of CMOSL one-shots is that their minimum output pulse widths are larger than those of TTL monostables. The 555 IC timer (Fig. 7.13) can be connected as a resettable nonretriggerable monostable with a typical accuracy of 0.5%, drift with temperature of only 50 ppm/°C, and drift with supply voltage of 0.005%/V. It can produce output pulse widths ranging from microseconds to hours and has TTL-compatible supply and output voltages.

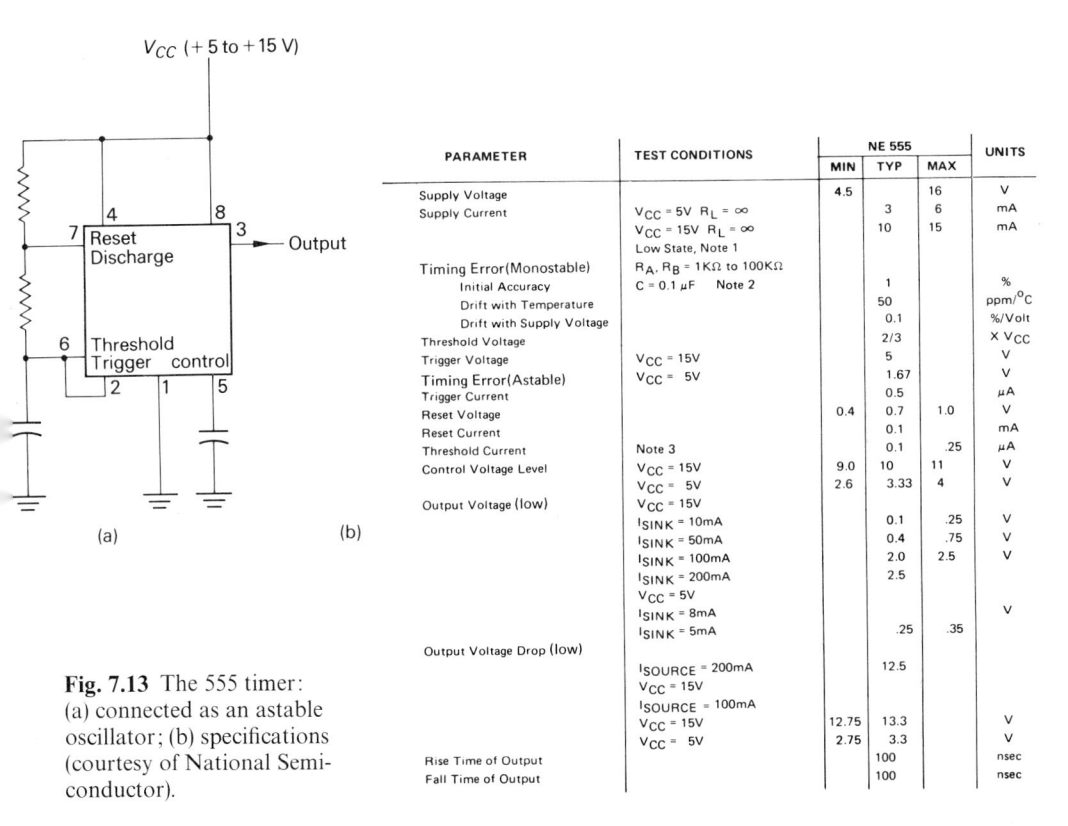

PARAMETER	TEST CONDITIONS	NE 555 MIN	NE 555 TYP	NE 555 MAX	UNITS
Supply Voltage		4.5		16	V
Supply Current	$V_{CC} = 5V$ $R_L = \infty$		3	6	mA
	$V_{CC} = 15V$ $R_L = \infty$		10	15	mA
Timing Error (Monostable)	Low State, Note 1 $R_A, R_B = 1K\Omega$ to $100K\Omega$ $C = 0.1 \mu F$ Note 2				
Initial Accuracy			1		%
Drift with Temperature			50		ppm/°C
Drift with Supply Voltage			0.1		%/Volt
Threshold Voltage			2/3		× V_{CC}
Trigger Voltage	$V_{CC} = 15V$		5		V
	$V_{CC} = 5V$		1.67		V
Timing Error (Astable)					
Trigger Current			0.5		μA
Reset Voltage		0.4	0.7	1.0	V
Reset Current			0.1		mA
Threshold Current	Note 3		0.1	.25	μA
Control Voltage Level	$V_{CC} = 15V$	9.0	10	11	V
	$V_{CC} = 5V$	2.6	3.33	4	V
Output Voltage (low)	$V_{CC} = 15V$				
	$I_{SINK} = 10mA$		0.1	.25	V
	$I_{SINK} = 50mA$		0.4	.75	V
	$I_{SINK} = 100mA$		2.0	2.5	V
	$I_{SINK} = 200mA$		2.5		
	$V_{CC} = 5V$				
	$I_{SINK} = 8mA$				V
	$I_{SINK} = 5mA$		.25	.35	
Output Voltage Drop (low)	$I_{SOURCE} = 200mA$ $V_{CC} = 15V$		12.5		
	$I_{SOURCE} = 100mA$				
	$V_{CC} = 15V$	12.75	13.3		V
	$V_{CC} = 5V$	2.75	3.3		V
Rise Time of Output			100		nsec
Fall Time of Output			100		nsec

Fig. 7.13 The 555 timer: (a) connected as an astable oscillator; (b) specifications (courtesy of National Semiconductor).

7.5 THE ASTABLE MULTIVIBRATOR

An **astable multivibrator** is a device that has two output states, neither of which is stable. The result is that the output oscillates back and forth between these two states, yielding an approximately square wave output. An astable is also called a **free-running multivibrator**.

IC astables are available in the ECL and CMOSL families. The 555 timer can be used as a TTL-compatible astable. It has an adjustable duty cycle and operates at frequencies from 0.1 Hz to 100 kHz. Figure 7.13(a) shows the 555 connected as an astable multivibrator (pins 2 and 6 connected together). An external capacitor C and two external resistors R_1 and R_2 determine the frequency $f = 1.44/(R_1 + 2R_2)C$. The duty cycle DC is determined by DC $= R_2/(R_1 + R_2)$. The capacitor C charges and discharges between $\frac{1}{3}V_{CC}$ and $\frac{2}{3}V_{CC}$. The output frequency is independent of the supply voltage. The specifications for the 555 timer are shown in Fig. 7.13(b). The problems illustrate an application of the 555 timer.

The CMOSL 4047 monostable/astable multivibrator, operating in the astable mode, has an approximate frequency range of 0.1 Hz to 1 MHz, and a 50% non-adjustable duty cycle.

7.6 CLOCKS

A digital **clock** is a pulse generator (oscillator) that produces a rectangular pulse train. Clocks are used for controlling the times at which device outputs can change. In this section we shall discuss synchronous and asynchronous digital systems and several methods for generating clock pulses.

7.6.1 Introduction

Digital systems may be either synchronous or asynchronous. In **synchronous systems** clock pulses control the exact time at which any output in the system can change. Figure 7.14(a) shows an AND gate with two logic inputs A and B,

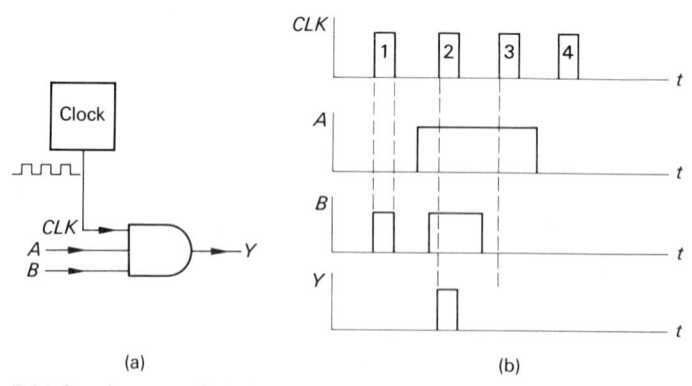

Fig. 7.14 Synchronous digital system: (a) logic diagram; (b) timing diagram.

and the third input (*CLK*) connected to a clock. The timing diagram in Fig. 7.14(b) shows the clock input, inputs *A* and *B*, and the output *Y*. Ideal pulses are used and delay times are not shown. Note that the AND gate output does not go HIGH during clock pulse 1 because *A* is LOW then. The output goes HIGH only at a time when *A* and *B* are HIGH and *CLK* is also HIGH, such as during clock pulse 2. Thus the clock controls the timing of the output of this simple system.

In an **asynchronous system**, the output of logic devices can change state whenever there is an appropriate change in the inputs. For example, in Fig. 7.14(a), if no clock pulse input is used the output of the AND gate can change whenever inputs *A* and *B* become HIGH.

Many digital systems are synchronous. A synchronous system is easier to troubleshoot (see Appendix C) since the system can be halted at any clock pulse and the value of the logic signals determined. Asynchronous systems are used in some data transmission systems, in some digital computer systems, and in applications where timing is not important. Some systems use a combination of synchronous and asynchronous devices.

Clock pulses can be produced by a variety of devices such as astables (Section 7.5), sine wave oscillators combined with Schmitt triggers (Example 7.1), logic gates, IC clocks, and monostables. Logic gate clocks and IC clocks will be discussed in the following sections. Monostable clocks are discussed in the problems.

7.6.2 Logic Gate Clocks

The basic principle of an oscillator is as follows. An amplifier amplifies an input signal as shown in Fig. 7.15(a). A portion of this amplified signal is used as the output of the oscillator. The remaining portion of the amplified signal is fed back into the input through a feedback network. If the amplifier produces a 180° phase shift from input to output, then the feedback network must produce an additional 180° phase shift so that the output of the feedback network is in phase

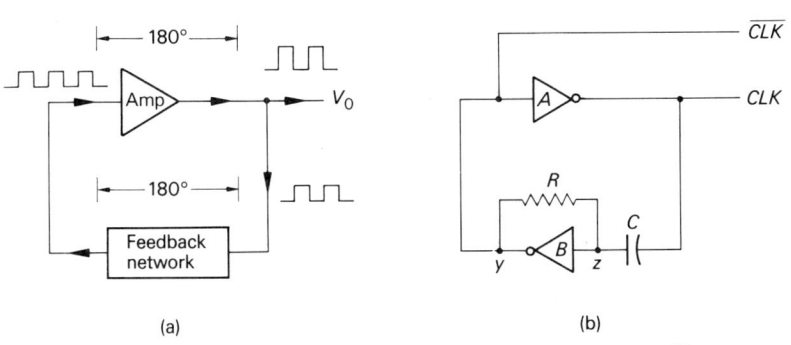

Fig. 7.15 Oscillators: (a) feedback principle; (b) logic gate oscillator.

(360°) with the amplifier input, thereby reinforcing it. An output signal V_0 will be continuously produced if the gain of the amplifier is sufficient to overcome losses in the feedback network.

The feedback network can be made from passive components (R, L, C) or it can be an active network such as another amplifier.

Figure 7.15(b) shows a clock made from two inverters. The feedback network consists of a capacitor C, a resistor R, and an inverting amplifier B. Resistor R biases inverter B away from its saturation and cutoff states so that it acts like an ordinary (linear) inverting amplifier. Inverter A acts like a switch, so the output waveform is approximately a square wave.

The oscillator works as follows. Assume that C is discharged and point z is LOW. Then the oscillator output CLK is LOW and point y is HIGH, due to inverting amplifier B. Therefore C charges from point y (V_{OH}) through R toward the potential of the output (CLK), which is LOW (V_{OL}). The voltage across C increases until point z becomes HIGH, at which time point y goes LOW and CLK goes HIGH. Now C discharges from point z through R toward the potential of point y (V_{OL}). When the voltage at point z becomes LOW, the output is LOW and the cycle repeats itself.

The values of R and C determine the frequency of the clock. The values of R and C for a given frequency are listed in the references for TTL [5, p. 10–5; 6], ECL [7], and CMOSL [8]. For TTL, R can vary over a small range from about 200 to 400 Ω, and $f \approx 1/(3RC)$. For example, if R = 330 Ω and C = 1 nF then $f = 1/(3 \times 330 \, \Omega \times 10^{-9} \, F) \approx 1$ MHz.

Notice in Fig. 7.15(b) the availability of a second clock pulse waveform $\overline{CLK}$, which is 180° out of phase with the CLK output.

The main advantage of logic gate clocks is that they are inexpensive, self-starting, and easy to make. Their main disadvantage is that the outputs do not generally have exactly a 50% duty cycle, and their duty cycle cannot be adjusted.

The advantage of CMOSL gate oscillators is that they require very little power, while ECL gate clocks are capable of much higher frequencies than TTL or CMOSL clocks.

The references show how to make a capacitance meter from a CMOSL logic gate clock, and a monostable from CMOSL gates [9]. The next example shows how to make a flashing warning light system.

Example 7.7 *Oscillator Warning System.* The landing gear of a jet aircraft is lowered by a switch in the cockpit. Devise a system in which a flashing light will indicate that the gear is being lowered. The light should go off when the gear is fully extended and latched. Use CMOSL logic.

Solution. An oscillator can be made from two NOR gates connected as inverters. A 4000 NOR gate has two three-input NOR gates in it. Two of the inputs on one NOR gate can be used to gate the oscillator on and off as shown in Fig. 7.16(a). The landing gear switch in the cockpit is normally open (N.O.) and closes when

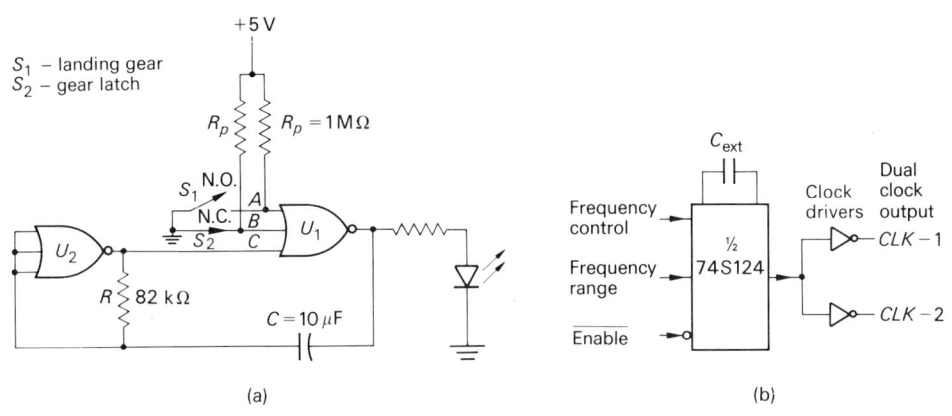

Fig. 7.16 Clock generators: (a) logic gate clock with two enable inputs; (b) TTL dual 74S124 voltage controlled oscillator.

the landing gear switch is activated. The gear latching switch, which indicates when the gear is down and latched, is normally closed (N.C.). The oscillator is disabled by connecting the two enable inputs (A, B) of NOR gate U_1 to a HIGH through pull-up resistors R_p. When the gear is up, the landing gear switch S_1 is open, thereby making the A input to gate U_1 HIGH. With A HIGH, gate U_1 is disabled and hence no oscillations occur because the output of gate U_1 stays LOW independent of input C. When the landing gear switch is closed, enable inputs A and B to gate U_1, become LOW (0 V), so the oscillator oscillates and the light flashes. When the gear is down and latched, the gear latching switch opens, making input B to U_1 HIGH which disables the oscillator.

The frequency f of this oscillator is approximately $f = 1/1.4\ RC$. Let's choose $f = 1$ Hz and $C = 10\ \mu F$; then $R = 1/1.4\ fC = 71$ kΩ. A standard value 82-kΩ resistor and 1-MΩ pull-up resistors are used. ∎

7.6.3 IC Clocks

Several IC clocks are available in each logic family. The main disadvantage of most IC oscillators is that they have a 50% duty cycle (square wave), whereas many systems require a rectangular waveform. IC clocks are also generally more expensive than logic gate oscillators. However, IC clocks are easy to use and offer many features not found in logic gate clocks. Astable oscillators, which we have already discussed, are also often used for clocks.

Figure 7.16(b) shows the logic diagram for one-half of a dual TTL 74S124 square-wave oscillator, which has a frequency range of 0.12 Hz to 85 MHz. It requires only an external timing capacitor C_{ext}. The center frequency f of the oscillator is approximately f (MHz) $= 500/C_{ext}$ (pF). The frequency can be controlled over the range from 1 Hz to 60 MHz with an external dc voltage; this is therefore called a **voltage controlled oscillator** (VCO).

When the active LOW enable input is LOW the oscillator is enabled, and when it is HIGH the output of the oscillator is HIGH (disabled). A dc frequency range voltage of 3 to 4.5 V determines the range of frequencies that can be obtained by applying dc voltages of 1.5 to 5 V to the frequency control input.

recommended operating conditions

	SN54S124 MIN	SN54S124 NOM	SN54S124 MAX	SN74S124 MIN	SN74S124 NOM	SN74S124 MAX	UNI
Supply voltage, V_{CC}	4.5	5	5.5	4.75	5	5.25	V
Input voltage at frequency control or range input, $V_{I(freq)}$ or $V_{I(rng)}$	1		5	1		5	V
High-level output current, I_{OH}			−1			−1	mA
Low-level output current, I_{OL}			20			20	mA
Output frequency (enabled), f_o	1			1			Hz
			60			60	MHz
Operating free-air temperature, T_A	−55		125	0		70	°C

electrical characteristics over recommended operating free-air temperature range (unless otherwise noted)

PARAMETER		TEST CONDITIONS[†]		MIN	TYP[‡]	MAX	UNI
V_{IH}	High-level input voltage at enable			2			V
V_{IL}	Low-level input voltage at enable					0.8	V
V_I	Input clamp voltage at enable	V_{CC} = MIN, I_I = −18 mA				−1.2	V
V_{OH}	High-level output voltage	V_{CC} = MIN, V_{IH} = 2 V, I_{OH} = −1 mA	SN54S'	2.5	3.4		V
			SN74S'	2.7	3.4		
V_{OL}	Low-level output voltage	V_{CC} = MIN, V_{IL} = 0.8 V, I_{OL} = 20 mA				0.5	V
I_I	Input current	Freq control or range, V_{CC} = MAX	V_I = 5 V		10	50	μA
			V_I = 1 V		1	15	
I_I	Input current at maximum input voltage	Enable, V_{CC} = MAX, V_I = 5.5 V				1	mA
I_{IH}	High-level input current	Enable, V_{CC} = MAX, V_I = 2.7 V				50	μA
I_{IL}	Low-level input current	Enable, V_{CC} = MAX, V_I = 0.5 V				−2	mA
I_{OS}	Short-circuit output current[§]	V_{CC} = MAX		−40		−100	mA
I_{CC}	Supply current, total into pins 15 and 16	V_{CC} = MAX, See Note 3			105	150	mA
		V_{CC} = MAX, T_A = 125°C, See Note 3	W package only			110	

[†] For conditions shown as MIN or MAX, use the appropriate value specified under recommended operating conditions.
[‡] All typical values are at V_{CC} = 5 V, T_A = 25°C.
[§] Not more than one output should be shorted at a time and duration of the short circuit should not exceed one second.
NOTE 3: I_{CC} is measured with the outputs disabled and open.

switching characteristics, V_{CC} = 5 V, R_L = 280 Ω, C_L = 15 pF

PARAMETER		TEST CONDITIONS		MIN	TYP	MAX	UNIT
f_o	Output frequency	C_{ext} = 2 pF	$V_{I(freq)}$ = 4 V, $V_{I(rng)}$ = 1 V	60	85		MHz
			$V_{I(freq)}$ = 1 V, $V_{I(rng)}$ = 5 V	25	40		
	Output duty cycle	C_{ext} = 8.3 pF to 500 μF			50%		
t_{PHL}	Propagation delay time, high-to-low-level output from enable	f_o = 1 Hz to 20 MHz			$\dfrac{1.4}{f_o(Hz)}$		s
		f_o > 20 MHz			70		ns

Fig. 7.17 Specifications for the TTL dual 74S124 voltage controlled oscillator. (Courtesy of Texas Instruments Incorporated.)

The 74S124 has a frequency stability of 0.4% for supply voltage variations ranging from 4.5 V to 5.5 V and temperature variations ranging from $-30°C$ to $100°C$.

Figure 7.17 gives the specifications for the 74S124. Note that the output currents are $I_{OH} = -1$ mA and $I_{OL} = 20$ mA. For large digital systems, the clock may have to drive more circuits than it is capable of driving by itself. In this case two or more drivers can be used, as shown in Fig. 7.16(b), to increase the fan-out of an oscillator. The disadvantage of using several drivers is that any difference in delay time between the drivers will result in a phase difference between the clock outputs, which could cause errors in the system. This time difference can be minimized by using high-speed gates or selecting two gates with nearly equal delay times or using specially designed clock drivers.

Some digital systems require two or more clocks with a certain phase relation between the clock signals. These are called **multiphase clocks**. A two-phase (2ϕ) clock, with an approximately 180° phase difference, can be obtained using an inverter at the oscillator output as shown in Fig. 7.18(a). The two clock signals will not be exactly 180° out of phase with respect to each other, due to the inverter delay time. The quad 74265 complementary-output elements have a typical time difference (skew) of only 0.5 ns. Other phase relations can be obtained using counters or shift registers (Chapter 9) or monostables (Chapter 7, Problems).

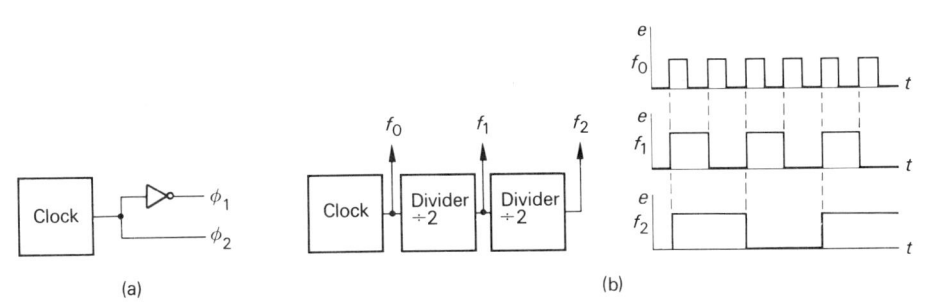

Fig. 7.18 Clock systems: (a) two-phase clock; (b) block diagram of a clock divider chain.

Among the multiphase IC clocks that are available is the National MH8803 2ϕ MOS oscillator. This is a VCO with a frequency range of 100–500 kHz, adjustable pulse widths of 260 ns to 1.4 μs, and duty cycle up to 50%.

Many digital systems require several clock frequencies for controlling different devices. A **frequency divider chain**, made from flip-flops, is usually used for this purpose. Figure 7.18(b) shows the logic diagram of a divider chain and the waveform out of the clock at frequency f_0 and the output waveforms f_1 and f_2 from two divide-by-2 ($\div 2$) circuits. The frequency f_2 is one-quarter of f_0. So if $f_0 = 1$ kHz, then $f_1 = 500$ Hz and $f_2 = 250$ Hz. Frequency division can continue down to less than 1 Hz by using a sufficient number of frequency dividers.

The Motorola 14451 is a combined oscillator and frequency divider chain that can divide by 2^{11} to 2^{19} and has a variable pulse width control.

*7.6.4 Crystal Controlled Clocks

Some digital systems require very precise, stable clock pulses, which can be provided by crystal controlled oscillators. A **crystal controlled oscillator** is an oscillator in which the frequency is controlled by a crystal rather than by an *RC* timing network or an *LC* resonant circuit. A crystal is made from quartz and is mounted in a sealed package that has two (or three) leads for connecting it into the circuit. An excellent discussion of crystals is given in the references [8].

A **stable oscillator** is one whose output frequency changes very little when its temperature changes or its supply voltage changes. Frequency stabilities of 0.02% (200 parts per million, ppm) are readily achievable for temperature and voltage variations of $\pm 10\%$ with ECL [10] and CMOSL [8, p. 147; 11, 12] logic gate crystal oscillators. Figure 7.19(a) shows the logic diagram for a CMOSL crystal oscillator that will operate in the frequency range from 500 kHz to 10 MHz [11]. This oscillator has a frequency change of less than 0.2% for $\pm 50\%$ supply voltage variations over a $-50°C$ to $+70°C$ temperature range. There are now on the market some inexpensive ($5) miniature crystals with good frequency

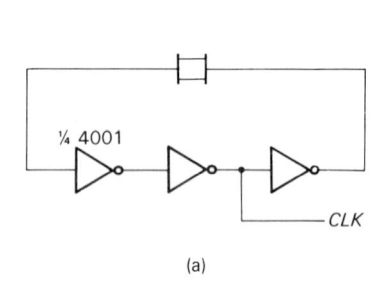

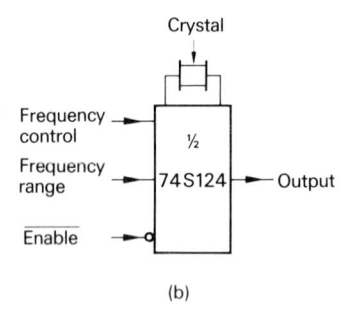

Fig. 7.19 Crystal oscillators: (a) CMOSL gate oscillator; (b) TTL dual 74S124; (c) DIP crystal oscillator (courtesy of Erie Frequency Control).

accuracy (0.01%) and low frequency drift (0.05%) within the temperature range of $-40°C$ to $80°C$ [13]. These make crystal controlled oscillators an attractive alternative to other oscillators.

The TTL 74S124 oscillator can be crystal controlled by replacing the capacitor by a crystal as shown in Fig. 7.19(b).

Several manufacturers offer crystal controlled oscillators with various logic level outputs covering a wide variety of frequencies [14]. Crystal oscillators are also available in DIP packages. For example, the Erie 20A01111 has a TTL-compatible output, 4 MHz to 20 MHz ranges, 0.01% stability, and is mounted in a 14-pin DIP package as shown in Fig. 7.19(c). It requires a 5-V dc supply voltage at 15 mA.

SUMMARY

In this chapter we have discussed several useful special-purpose devices. The Schmitt trigger is a device whose output is HIGH for input voltages exceeding the positive-going threshold voltage. It is very useful for converting voltages with long rise times into pulses with short rise times. A voltage comparator is similar to a Schmitt trigger, but has the advantage of adjustable threshold voltages.

Nonideal pulses are characterized by their amplitude, pulse width, rise time, and fall time. The pulse repetition frequency of a pulse train is the inverse of the pulse repetition time. The duty cycle is the ratio of the pulse width to the pulse repetition time.

Monostable multivibrators have a LOW output until turned on for a predetermined time by a trigger pulse. They are useful for time delay elements.

An astable multivibrator is an oscillator that is useful for generating pulse waveforms.

Clocks are oscillators that have rectangular pulse outputs. Clocks are used to control the timing of events in synchronous systems. Systems that require more than one clock are called multiphase clock systems.

PROBLEMS

Section 7.2

1. Draw a logic diagram and timing diagram that show how a Schmitt trigger can be used to square up pulses from an integrator.
2. (a) Why is the propagation delay time in the CMOSL 4093 Schmitt trigger greater for $V_{DD} = 5$ V than for $V_{DD} = 10$ V? (b) Why is it greater for a load capacitance $C_L = 50$ pF than for $C_L = 15$ pF?

Section 7.3

3. Draw and label a single nonideal pulse that has a pulse width of 100 ns, rise time of 10 ns, fall time of 5 ns, and amplitude of $+10$ V.

206   SCHMITT TRIGGERS, MONOSTABLES, AND CLOCKS

4. What are the prf and DC of a pulse train whose pulse width is 2 μs and whose prt is 10 μs?

5. Explain how a Schmitt trigger can be used to eliminate oscillations in a TTL gate due to pulses that have a long rise time. Use a diagram to aid in your explanation.

Section 7.4

6. List three problems that can occur with monostables in a digital system.
7. List four precautions that should be observed when using monostables.

Section 7.4.1

8. Draw a timing diagram for a nonretriggerable monostable that uses leading-edge triggering. Illustrate how it ignores trigger inputs when its output is HIGH.

Section 7.4.2

9. Show the logic diagram for converting a retriggerable monostable into a nonretriggerable monostable using the Q output.
10. Draw a timing diagram that illustrates how a resettable monostable works.
11. *Double Pulse Detector.* With the aid of a timing diagram explain how the circuit shown in Fig. P7.1 detects the presence of extra pulses in a sequence of regularly spaced pulses.

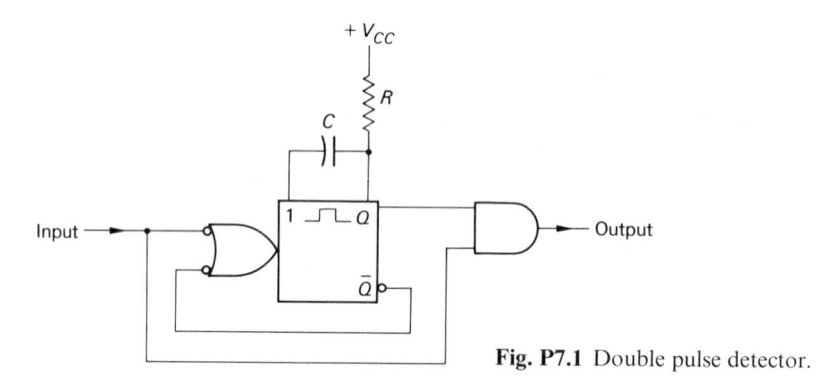

Fig. P7.1 Double pulse detector.

12. (a) It is desired to produce an accurate 10-μs timing pulse using a TTL 74LS122 monostable. With the aid of a TTL data book, calculate the value of C, using the maximum allowed value for R. (b) Draw a logic diagram. Include a trimmer resistor for adjusting the pulse width. (c) Using a digital computer, calculate a set of five values for R and C which give a 10-μs pulse.

13. *Pulse Width Discriminator.* It is sometimes desirable to know whether a pulse represents a likely event or not. The pulse width discriminator in Fig. P7.2 gives an output for input pulses that are less than or greater than the width T determined by the monostable. Draw an input waveform with different pulse widths. With the aid of a timing diagram explain how the circuit works.

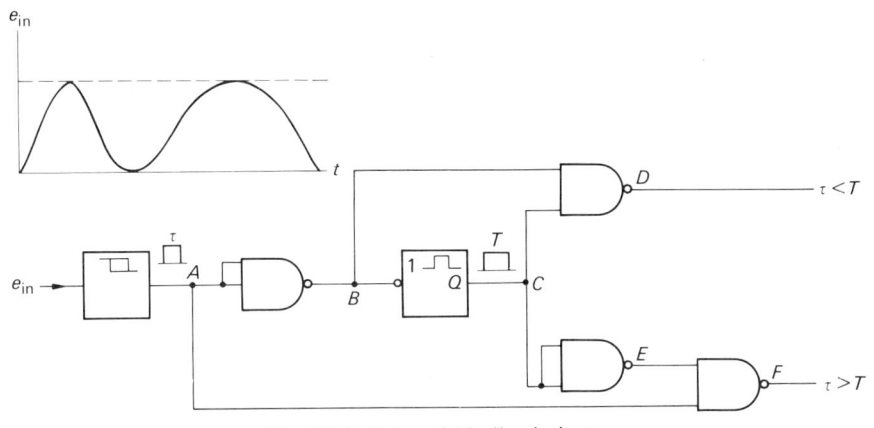

Fig. P7.2 Pulse width discriminator.

Section 7.5

14. (a) Using a linear IC data book, draw the connection diagram for a 10-kHz square wave astable using a 555 timer. (b) Summarize the important electrical properties of the 555.

15. *Intensity Control by Pulse Width Modulation* (PWM). It is desirable to automatically adjust the intensity of a multidigit display with respect to the ambient light conditions in applications such as a bedroom clock. Excessive brightness or insufficient display brightness causes eyestrain and fatigue. Figure P7.3 shows the logic diagram of a display system in which the duty cycle of the supply voltage V_{CC} is proportional to the ambient light. When the ambient light decreases, the resistance of the photoresistor R_x increases, which in turn increases the duty cycle of the 555 timer. (a) Using a data book, show the pin numbers on each device in Fig. P7.3. (b) Explain in detail how the circuit works. Use a timing diagram in your explanation.

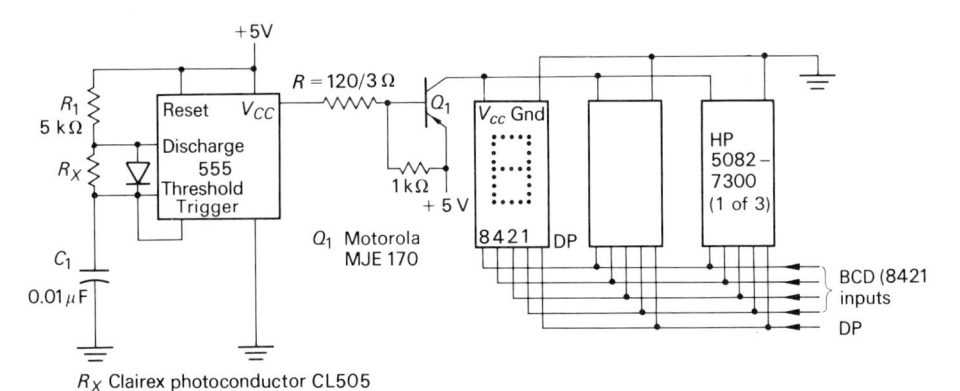

Fig. P7.3 Intensity control by PWM. (Courtesy of Hewlett-Packard Company.)

Section 7.6.1

16. Draw a logic diagram and timing diagram that illustrate a synchronous digital system.

Section 7.6.2

17. Explain the basic principle of a feedback oscillator.
18. Using a TTL data book, draw the connection diagram for a 100-kHz oscillator made from 7400 NAND gates. Provide for an active LOW enable input.
19. Draw the block diagram of a clock divider chain that produces frequencies of 50 MHz, 25 MHz, and 12.5 MHz from a 100-MHz clock.

Section 7.6.4

20. What is a stable oscillator?
21. *Single-Phase Monostable Clock.* (a) Explain with the aid of a timing diagram how the monostable clock in Fig. P7.4 works. (b) Calculate the values of R_1, R_2, C_1, and C_2 using a TTL 9602 that will generate 1-MHz pulse train with pulses 350 ns wide. Use trimmer resistors.
22. *Computer Program.* Write a computer program that calculates the capacitance C_x for a 9602 (Section 9.4) monostable for given resistor values R_x and "on" time t. Calculate C_x for $R_x = 5, 10, 20, 40,$ and 50 kΩ and $t = 100$ ms.

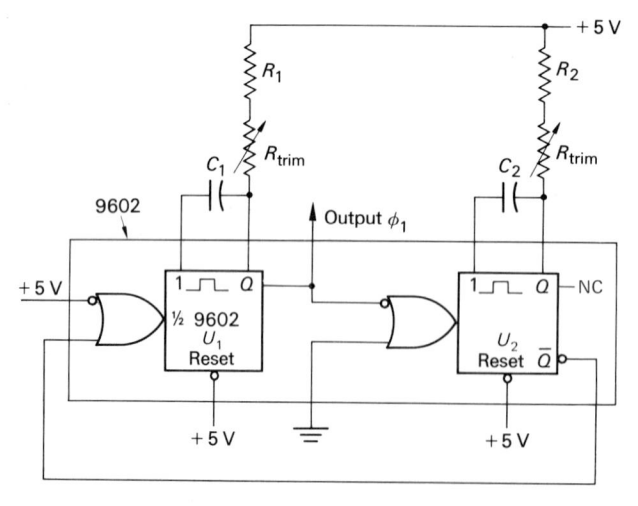

Fig. P7.4 Monostable clock.

REFERENCES

1. *MECL Integrated Circuits Data Book.* Motorola Inc., 1974.
2. Leonard Strauss. *Wave Generation and Shaping*, 2nd edition. New York: McGraw-Hill, 1970.
3. Brinton B. Mitchell. *Semiconductor Pulse Circuits.* New York: Holt, Rinehart and Winston, 1970.

4. Robert L. Morris and John R. Miller (eds.). *Designing with TTL Integrated Circuits.* New York: McGraw-Hill, 1971.
5. Peter Alfke and Ib Larsen (eds.). *The TTL Applications Handbook.* Fairchild Semiconductor, August 1973.
6. C. J. Krall. "Oscillator Clock Circuits." *Digital Design*, pp. 34–35, August 1973.
7. *MECL System Design Handbook.* Motorola Inc., 1972.
8. *COS/MOS Integrated Circuit Manual.* CMS-271, RCA Corp., June 1972.
9. Harry Garland and Roger Melen. "Single-IC Capacitance Meter." *Popular Electronics*, pp. 44–45, February 1974.
10. Chuck Byers and Bill Blood. *IC Crystal Controlled Oscillators.* AN 417A, Motorola Inc., 1972.
11. Sukhendu Das. "CMOS Holds Down Parts Count for Digital Clocks." *Electronics*, pp. 118–119, June 7, 1973.
12. S. S. Chuang. "C-MOS Minimizes the Size of Crystal Oscillators." *Electronics*, pp. 117–118, June 7, 1973.
13. Statek Corporation, 1200 Alvarez Ave., Orange Ca. 92668.
14. Erie Technological Products Incorporated, 453 Lincoln Street, Carlisle Pa. 17013.

8: FLIP-FLOPS

8.1 INTRODUCTION

One common device used in most digital systems is the logic gate, which we have treated in the previous chapters. In this chapter we discuss in further detail the second major device used in most digital systems, the flip-flop. A **flip-flop** is a device which has two stable output states, HIGH and LOW. The output remains in a given state until inputs cause the output to change. An ordinary light switch is an example of a mechanical flip-flop. It has two stable states, ON and OFF, and an input that can change the output to either state. Unlike the logic gate, the flip-flop is an example of a **sequential device**, i.e., it has a feedback circuit so that its output depends on previous inputs as well as on the present inputs.

Flip-flops are used primarily (1) to store information (memories), (2) to count (counters), (3) to alter the form of data, such as by serial-to-parallel conversion (shift registers), and (4) to control other devices. Flip-flops are available in IC form as SSI devices in the TTL, CMOSL, and ECL families. However, in this chapter we shall describe flip-flops as if they were made from discrete logic gates, in order to gain a firm understanding of their behavior. We have already discussed topics of a basic nature, such as loading rules, logic levels, and propagation delay, and will not dwell further on these here. We begin with a review of the RS latch, which is a special case of a flip-flop. We then go on to consider other latches and flip-flops, and conclude with a discussion of triggering methods and timing problems. Numerous practical examples are given along the way.

8.2 THE RS LATCH

An **RS latch** is a sequential device that consists of two cross-coupled gates. It has two inputs, reset (R) and set (S), and two outputs, Q and $\bar{Q}$. Figure 8.1(a) shows an RS latch made from two cross-coupled NAND gates. Latches are used for applications such as the temporary storage of data between a data source and an indicator, and for debouncing switches (see Example 8.1).

Let's develop the truth table (Fig. 8.1b) for this latch. Suppose $R = H, S = H$, and assume $Q = L$. The $Q = L$ output is one input to NAND gate B, so the output of B is HIGH, that is, $\bar{Q} = H$. The inputs to NAND gate A are $S = H$ and $\bar{Q} = H$, so its output is LOW as assumed initially. If $R = H$, $S = H$, and Q is assumed

8.2 THE *RS* LATCH 211

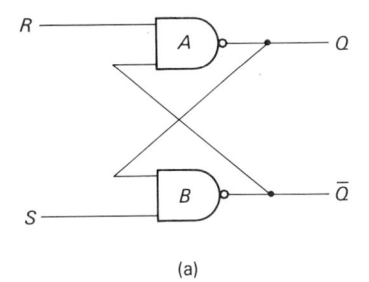

Fig. 8.1 The *RS* NAND latch: (a) logic diagram; (b) truth table.

Inputs		Outputs		
R	S	Q	$\bar{Q}$	Name
L	L	H	H	Indet.
L	H	H	L	Set
H	L	L	H	Reset
H	H	No change		Remember

HIGH (i.e., $Q = H$), then both inputs to NAND gate *B* are HIGH, so $\bar{Q}$ is LOW and *Q* is HIGH as assumed initially. Thus, when $R = H$ and $S = H$ the outputs of an *RS* latch do not change state, i.e., they remember their last state, as shown in the last line of the truth table in Fig. 8.1(b).

Suppose $R = H$ and $S = L$. This causes the output of gate *B* to go HIGH, i.e., $\bar{Q} = H$. The inputs to gate *A* are then $R = H$ and $\bar{Q} = H$, so its output is LOW, i.e., $Q = L$, as shown in line 3 of Fig. 8.1(b).

When $R = L$ and $S = H$, the $R = L$ input to NAND gate *A* causes its output to go HIGH, i.e., $Q = H$ and $\bar{Q} = L$ as shown in line 2 of Fig. 8.1(b). Recall that when the *Q* output of a device, such as a latch, is HIGH the device is said to be **set**. When the *Q* output is LOW, the device is said to be **reset**.

Finally, if *S* and *R* are both LOW, then outputs *Q* and $\bar{Q}$ both become HIGH. This is the only case in which the outputs are not the complements of each other. Now let's consider what happens if *Q* and $\bar{Q}$ are both HIGH and then the *S* and *R* inputs change. If both outputs are HIGH, and then both the *R* and *S* inputs become HIGH simultaneously, there is no way of predicting whether the *Q* or $\bar{Q}$ outputs will go HIGH or LOW, hence, this is called the **indeterminate state**. If the *Q* and $\bar{Q}$ outputs are HIGH and then the *S* and *R* inputs do not both go HIGH at the same time, then the last input to go HIGH will determine the output. For example, if *R* goes HIGH last, the *Q* output will become HIGH. The indeterminate state is avoided, because of its unpredictable behavior, by gating or by mechanical means.

The order in which the inputs and outputs of a latch are labeled is purely arbitrary (see the problems).

We have already discussed an *RS* latch made from two cross-coupled NOR gates (Section 2.7); its logic diagram and truth table are reproduced in Fig. 8.2. Note that the indeterminate and remember states for the NOR latch are reversed from the NAND latch. The NOR latch is useful in applications where the inputs to the latch are normally LOW.

The following example shows how an *RS* latch can be used to eliminate contact bounce in a mechanical switch.

212 FLIP-FLOPS

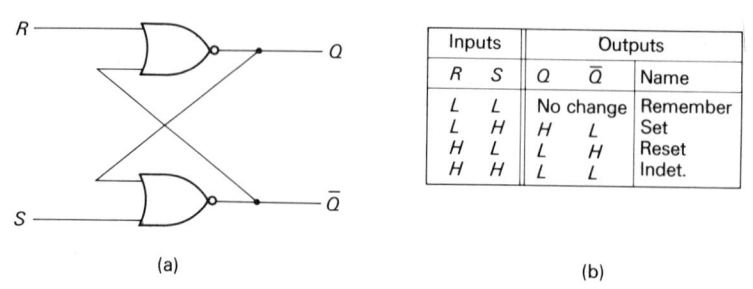

Fig. 8.2 The RS NOR latch: (a) logic diagram; (b) truth table.

Inputs		Outputs		
R	S	Q	$\bar{Q}$	Name
L	L	No change		Remember
L	H	H	L	Set
H	L	L	H	Reset
H	H	L	L	Indet.

Example 8.1 *Switch Contact Bounce Eliminator.* A single-pole double-throw (SPDT) mechanical switch is used to reset the internal states of the central processing unit (CPU) of a computer. Draw the logic diagram for a TTL interface latch that will eliminate the contact bounce of this reset switch.

Solution. When a mechanical or electromechanical switch is closed, the metal contacts alternately collide and bounce apart a number of times before settling to the closed position a few milliseconds later. This phenomenon is called **contact bounce**. A typical voltage waveform produced across a bouncing switch is shown in Fig. 8.3(a), where ideal pulses are assumed. The length of time the switch stays open decreases with each bounce until, after three bounces, the switch settles to its closed state. This bouncing switch will produce a 0101010 input to the CPU reset input, instead of the required 01 input, resulting in incorrect resetting of the CPU.

Figure 8.3(b) shows how a TTL RS NAND latch can be used to eliminate contact bounce. The 4-kΩ pull-up resistors ensure that the inputs are not floating open. When the switch is first switched to the reset position the inputs to the latch are $R = L$, $S = H$, so the Q output goes HIGH. When the switch bounces

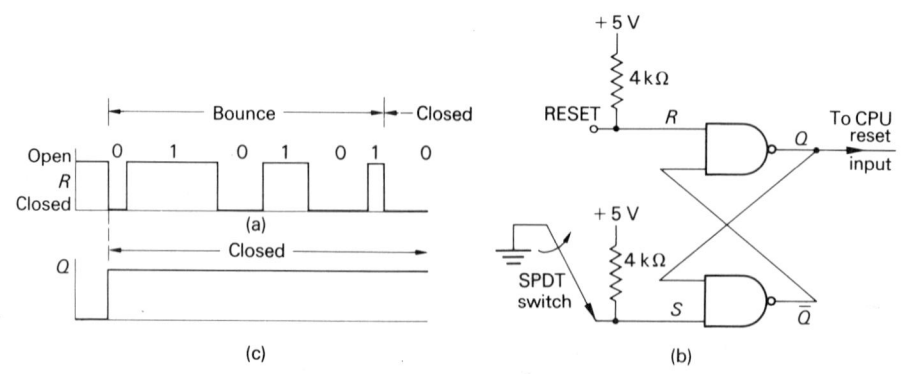

Fig. 8.3 Contact bounce eliminator: (a) voltage waveform across the switch; (b) bounce eliminator circuit; (c) output waveform from RS flip-flop.

open, the latch inputs are $R = H$ and $S = H$, so the Q output does not change, i.e., Q stays HIGH. When the switch closes the second and third time the inputs are $R = L$, $S = H$, so the latch remains set, as shown in the output waveform of Fig. 8.3(c). The switch must stay closed long enough on the first closure to allow the latch to set. ∎

Now let's develop a timing diagram for an *RS* NAND latch (Fig. 8.4). Assume that the input pulses are ideal and that both gates have equal delay times t_p.

Initially assume that $Q = H$, $\bar{Q} = L$, and both inputs are HIGH. Then the Q output stays HIGH and $\bar{Q}$ stays LOW (interval 1 in Fig. 8.4). When $S = L$ is applied to the set input for a time $4t_p$ (interval 2) then $\bar{Q}$ goes HIGH after a time t_p measured from the falling edge of S. When $\bar{Q}$ goes HIGH it will then force Q to go LOW one additional delay time t_p after the falling edge of S.

Next suppose S returns HIGH while R is HIGH (interval 3); then there is no change in the outputs because the $Q = L$ output of gate A will hold the output of gate B HIGH. Now when R goes LOW (interval 4) for a time $2t_p$, Q becomes HIGH one delay time after the falling edge of R. When Q goes HIGH this forces $\bar{Q}$ LOW $2t_p$ after the falling edge of R, since both inputs to gate A are HIGH. $\bar{Q} = L$ will then hold $Q = H$ even after R returns HIGH. Note that the S input pulse width (interval 4) is just barely wide enough to hold the outputs unchanged if R returns HIGH after two delay times. If input pulse widths are less than $2t_p$ the output of the latch can oscillate. Thus with ideal pulses $2t_p$ is the minimum set (and reset) pulse width that will allow predictable operation of the latch. In actual circuits, where the rise and fall times are on the order of t_p, sustained oscillations cannot occur.

Finally, suppose both R and S go LOW simultaneously (interval 5), so that both outputs become HIGH one delay time later. If both inputs now go HIGH simultaneously, the outputs can go to either state.

The outputs of a latch should not be connected back into their inputs since this may cause the latch to oscillate.

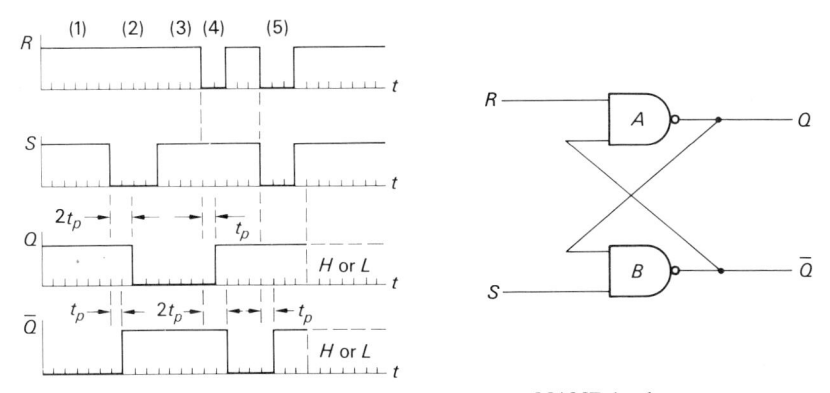

Fig. 8.4 Timing diagram for an *RS* NAND latch.

214 FLIP-FLOPS 8.2

When a latch changes state, the output rise and fall times are limited only by the propagation delay times of the gates, and not by the rise and fall times of the inputs.

The R and S inputs to a latch are sometimes called **direct, or level, inputs** because the latch reacts only to the dc level of the voltage present at these inputs.

RS latches are available as IC devices. Some typical properties of the TTL 74279 quad RS latch are: delay times, S to output 10.5 ns and R to output 15 ns; power dissipation 90 mW. The CMOSL 4043 quad RS latch for $V_{DD} = 5$ V typically has delay times of 175 ns for either input, a 50-nW power dissipation, and a three-state output.

The next example illustrates the use of an RS latch in a warning system.

Example 8.2 *Pressure Warning System.* A manned satellite uses a warning system to indicate when the air pressure inside the satellite falls below a safe limit. Devise a warning system that consists of a green safe light, which indicates safe pressure, and a red warning light and buzzer, which indicate low pressure. A technician who monitors the control panel can turn off the buzzer while the pressure is unsafe with a pushbutton switch. However, the red light must stay on until the trouble is corrected. A pressure sensor (transducer) is used to convert air pressure into a dc voltage.

Solution. Assume that the pressure sensor provides a +5-V output when the pressure is unsafe and 0 V when it is safe. A latch can be used to remember that the buzzer is on, until it is turned off by the technician. Figure 8.5(a) shows the

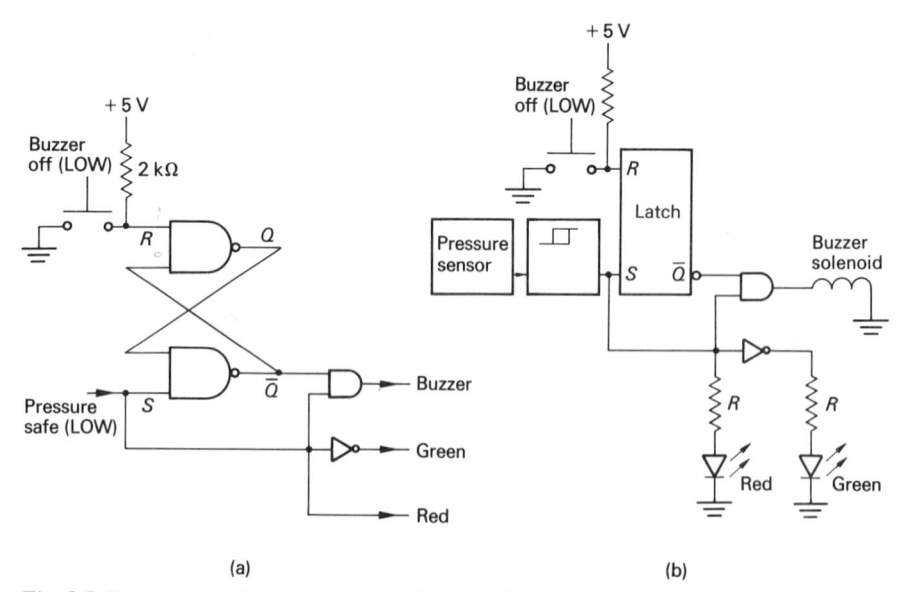

Fig. 8.5 Pressure warning system: (a) basic logic diagram; (b) more detailed logic diagram.

warning system. The R input to the latch is normally HIGH, and when the pressure is safe $S = L$, so the latch is reset; that is, $\bar{Q} = H$, the green light is on, and the red light and buzzer are off. When the pressure falls below the safe limit the S input to the latch becomes HIGH, so the buzzer and the red light turn on and the green light turns off. The buzzer can be turned off during an unsafe condition (i.e., when $S = H$) by momentarily making the reset input LOW. Figure 8.5(b) shows a more detailed drawing of the pressure sensor and warning lights. The standard logic symbol for a latch with active HIGH inputs is shown. The Schmitt trigger may be used to convert the slowly varying analog signal from the pressure sensor to a digital signal with a fast rise time. This example is pursued further in the problems. ∎

8.3 THE STROBED LATCH

A **strobed latch** is a latch with a strobe input, also called an enable input, that determines when the latch inputs become effective. The advantage of the strobed latch over the nonstrobed latch is that the outputs can be held (latched) for as long as desired even if the inputs change. The disadvantage is a longer delay time due to additional gates A and B. The outputs of a nonstrobed latch change almost immediately in response to the inputs. There are two types of strobed latches, the RS and the D types.

8.3.1 The Strobed RS Latch

A strobed RS latch is an RS latch with two additional gates and a strobe (or enable) input as shown in Fig. 8.6(a).

When the enable input E is LOW the outputs of NAND gates A and B are both HIGH, regardless of the S and R inputs, and therefore the output of the latch cannot change from its previous state.

In the truth table for the strobed latch, Fig. 8.6(b), the notation Q_n designates the output state of the latch before the nth enable input. The notation Q_{n+1} designates the Q output after the nth enable input. Thus when $E = L$, the notation $Q_{n+1} = Q_n$ means the output is the same after the nth enable input as it was before the enable input, i.e., there is no change in the output.

When E is HIGH the strobed latch acts just like the RS NAND latch dis-

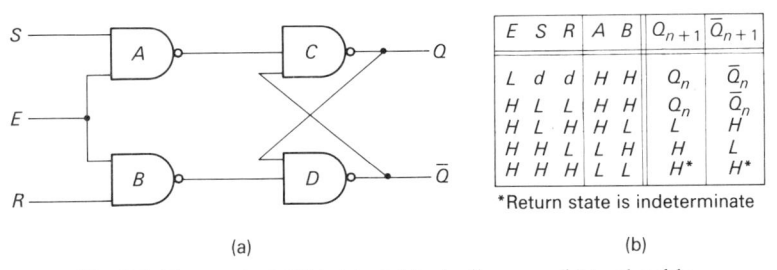

E	S	R	A	B	Q_{n+1}	$\bar{Q}_{n+1}$
L	d	d	H	H	Q_n	$\bar{Q}_n$
H	L	L	H	H	Q_n	$\bar{Q}_n$
H	L	H	H	L	L	H
H	H	L	L	H	H	L
H	H	H	L	L	H*	H*

*Return state is indeterminate

(a)   (b)

Fig. 8.6 The strobed RS latch: (a) logic diagram; (b) truth table.

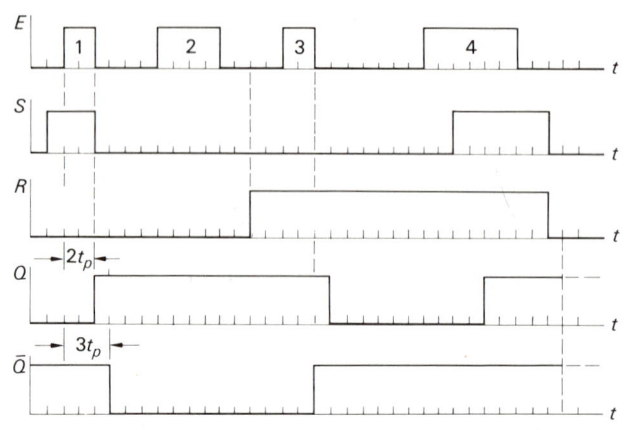

Fig. 8.7 Timing diagram for a strobed *RS* latch.

cussed in the previous section, but note that here the upper input, corresponding to the Q output, has been labeled S instead of R because NAND gates A and B invert the inputs. With this change in labeling the truth tables for the nonstrobed and strobed latch are the same (when $E = H$) for the set and reset states.

Figure 8.7 shows the timing diagram for a strobed *RS* latch with ideal input pulses and a propagation delay time t_p. The enable input can be a periodic waveform, or it can be nonperiodic as shown in Fig. 8.7. Initially the outputs are $Q = L$ and $\bar{Q} = H$. For enable pulse 1, $S = H$ and $R = L$, so Q goes HIGH a time $2t_p$ after the leading edge of enable pulse 1, and $\bar{Q}$ goes LOW $3t_p$ after the leading edge of enable pulse 1, as shown by line 4 in the truth table (Fig. 8.6b). For enable pulse 2, $S = L$ and $R = L$, so the output does not change, i.e., the outputs before enable pulse 2 are the same as after enable pulse 2. This is shown in the truth table as $Q_{n+1} = Q_n$ and $\bar{Q}_{n+1} = \bar{Q}_n$. The rest of the timing diagram in Fig. 8.7 is discussed in the problems.

Strobed *RS* latches are not generally available in IC form, but can be readily made from logic gates or by gating IC *RS* latches.

The second type of strobed latch is the strobed *D* latch, which is discussed next.

8.3.2 The Strobed *D* Latch

The **strobed *D* latch** has one input D, called the **data** input, an enable input E, and two outputs Q and $\bar{Q}$. A strobed *D* latch can be made from a strobed *RS* latch by connecting the S input through an inverter to the R input as shown in Fig. 8.8(a). The logic symbol for a strobed *D* latch is shown in Fig. 8.8(b).

Since the R input is always the complement of the S input, the simplified truth table of Fig. 8.8(c) results. When the enable input E is LOW the output cannot change (latched), i.e., $Q_{n+1} = Q_n$. When the enable input is HIGH and the D input is LOW, that is, $S = L$ and $R = H$, then $Q = L$. When $E = H$ and $D = H$, then

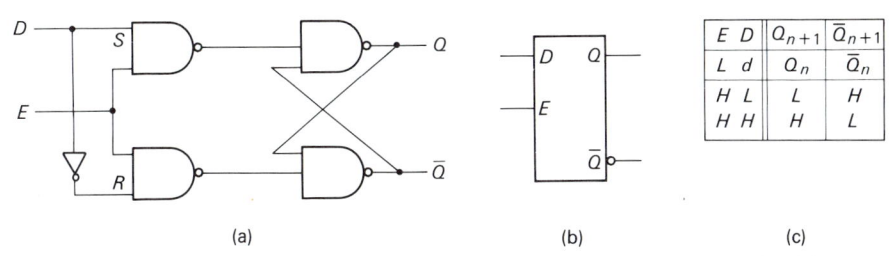

Fig. 8.8 The strobed D latch: (a) logic diagram; (b) logic symbol; (c) truth table.

$Q = H$. Thus when the latch is enabled, the output follows the input. When the latch is disabled, that is, when $E = L$, the latch retains the last state that it had before being disabled. Note that the indeterminate state of the strobed RS latch cannot exist in the strobed D latch.

A D latch can also be made from a nonstrobed RS NAND or NOR latch by connecting the R input to the S input through a NOT gate. The D latch is often used as a memory for storing one bit of information. Arrays of latches are used to store more than one bit. Strobed D latches are available in IC form in the ECL, CMOSL, and TTL families. Addressable multibit latches are discussed in the problems.

The next example describes one use of an IC strobed D latch.

Example 8.3 *Latched Multidigit Display.* Suppose data entering a multiplexed display, such as the one described in Section 6.6.2, changes too rapidly to be displayed clearly. Develop a system that will allow data to be displayed as long as the observer desires.

Solution. Strobed quad D latches at the inputs to each display decoder/driver can be used as shown in Fig. 8.9. When the correct data is present at each latch

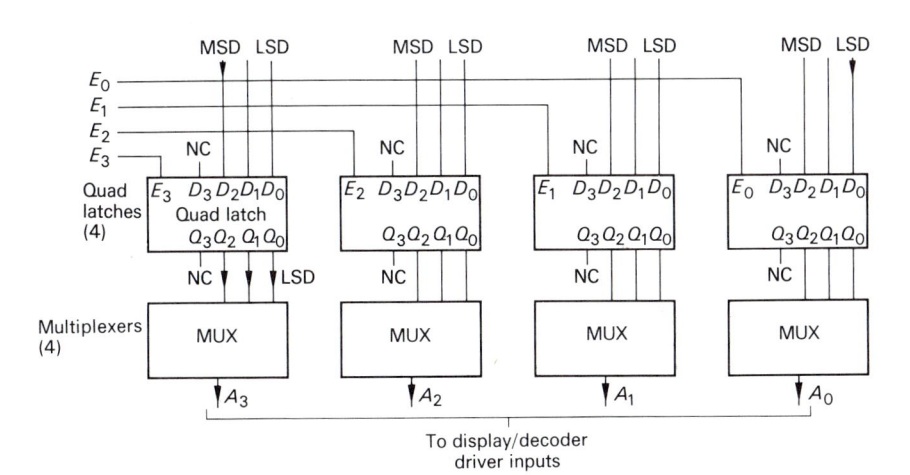

Fig. 8.9 Latched multidigit display system.

DC CHARACTERISTICS: $V_{EE} = -5.2$ V, $V_{CC} = $ GND

SYMBOL	CHARACTERISTIC	LIMITS B	TYP	LIMITS A	UNITS	T_A	CONDITIONS	
V_{OH}	Output Voltage HIGH	-1000 -960 -900		-840 -810 -720	mV	0°C 25°C 75°C	$V_{IN} = V_{IHA}$ or V_{ILB} per Truth Table	
V_{OL}	Output Voltage LOW	-1870 -1850 -1830		-1665 -1650 -1625	mV	0°C 25°C 75°C		Loading is 50 Ω to -2.0 V
V_{OHC}	Output Voltage HIGH	-1020 -980 -920			mV	0°C 25°C 75°C	$V_{IN} = V_{IHB}$ or V_{ILA} per Truth Table	
V_{OLC}	Output Voltage LOW			-1645 -1630 -1605	mV	0°C 25°C 75°C		
V_{IH}	Input Voltage HIGH	-1145 -1105 -1045		-840 -810 -720	mV	0°C 25°C 75°C	Guaranteed Input Voltage HIGH for All Inputs	
V_{IL}	Input Voltage LOW	-1870 -1850 -1830		-1490 -1475 -1450	mV	0°C 25°C 75°C	Guaranteed Input Voltage LOW for All Inputs	
I_{IH}	Input Current HIGH Data Common Enable Gate and Enable			245 265 350	µA	25°C	$V_{IN} = V_{IHA}$	
I_{IL}	Input Current LOW	0.5			µA	25°C	$V_{IN} = V_{ILB}$	
I_{EE}	Power Supply Current	75			mA	25°C	Inputs and Outputs Open	

SWITCHING CHARACTERISTICS: $V_{EE} = -5.2$ V, $T_A = 25$°C

SYMBOL	CHARACTERISTIC	LIMITS B	TYP	A	UNITS
t_{PLH}	Propagation Delay Enable to Output	1.0	4.0	5.4	ns
t_{PHL}	Propagation Delay Data to Output	1.0	4.0	5.4	ns
t_{PLH}	Propagation Delay Gate to Output	1.0	2.0	3.1	ns
t_{TLH}	Output Transition Time LOW to HIGH (20% to 80%)	1.5	2.0	3.5	ns
t_{THL}	Output Transition Time HIGH to LOW (80% to 20%)	1.5	2.0	3.5	ns
t_s	Set-Up Time Data		0.7	2.5	ns
t_h	Hold Time Data		0.7	1.5	ns

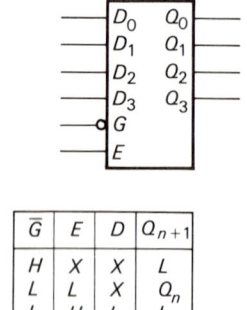

$\overline{G}$	E	D	Q_{n+1}
H	X	X	L
L	L	X	Q_n
L	H	L	L
L	H	H	H

Fig. 8.10 Specifications for the ECL quad 10133 strobed D latch. (Courtesy of Fairchild Camera and Instrument Corporation.)

input, that latch can be enabled momentarily and then disabled, thereby storing the data for as long as desired. If the data to all display decoder/drivers arrives simultaneously, all four latches can be enabled and disabled simultaneously. ∎

The specifications for an ECL quad 10133 strobed D latch are shown in Fig. 8.10. Note that the limits of each characteristic are denoted by A and B, instead of as maximum (the value closest to positive infinity) and minimum (the value closest to negative infinity), respectively. The minimum power dissipation for the 10133 with a supply voltage $V_{EE} = 5.2$ V is $I_{EE}V_{EE} = 75$ mA $\times$ 5.2 V = 390 mW. The typical enable or data input-to-output propagation delay time is 4 ns. The gate input is an additional control input that forces the outputs LOW when it is HIGH independent of the enable input and has no effect when it is LOW. The gate input-to-output delay time is 2 ns.

On the output waveforms for the 10133 latch, the transition times t_{TLH} and t_{THL} (rise and fall times) are measured between the 20% and the 80% voltage levels as shown in Fig. 8.11(a). The typical transition times for the 10133 are $t_{THL} = t_{TLH} = 2$ ns.

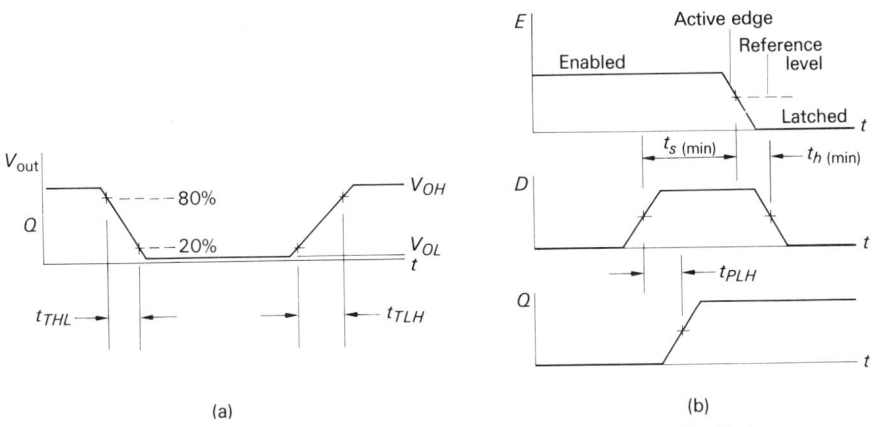

Fig. 8.11 Timing diagrams: (a) transition times; (b) setup and hold times.

The setup and hold times are important parameters for many logic devices. The **setup time** t_s is the time interval immediately preceding the active transition of a timing or control pulse during which data must be maintained at the proper level in order to ensure its recognition by the device. The active transition for a latching input is the falling edge of the enable E input. In order for a HIGH to be recognized by a D latch, the D input must stay HIGH for a minimum (also called maximum by some manufacturers) time t_s prior to the falling edge of the E input, as illustrated in Fig. 8.11(b). The minimum setup time for the D input of the 10133 latch is $t_s = 2.5$ ns.

220   FLIP-FLOPS   8.4

The **hold time** t_h is the time interval immediately following the active transition of a timing or control pulse during which data must be maintained at the proper level in order to ensure its recognition by the device. The hold time is illustrated in Fig. 8.11(b). For the 10133, $t_h = 1.5$ ns. Many logic devices have a zero hold time, which means that the instant the active edge of the timing or control pulse passes through its reference level, the inputs may change level. A **negative hold time**, as used in this text, means that the logic signal may be changed prior to the active edge of the enable input. Like all device parameters, the setup and hold times have a certain spread in them due to variations in temperature, supply voltage, and manufacturing tolerances.

The remainder of this chapter is devoted to flip-flops.

8.4 THE *D* FLIP-FLOP

The ***D* flip-flop** is a flip-flop with a single data input *D*, a clock input *CLK*, and complementary outputs Q and $\bar{Q}$ as shown in Fig. 8.12(a). The clock controls *when* the outputs can change, while the data inputs control *what* the outputs will change to. The truth table for the *D* flip-flop, which is the same as for the strobed *D* latch with $E = H$, is given in Fig. 8.12(b). If the *D* input is LOW the *Q* output will go LOW after the *CLK* input goes from HIGH to LOW ($H \rightarrow L$). If the *D* input is HIGH, the *Q* output will go HIGH after *CLK* goes from HIGH to LOW. Recall (Section 4.8.1) that the small circle on any dynamic input, such as on the *CLK* input in Fig. 8.12(a), means the outputs can change on the $H \rightarrow L$ transition of that input. The operational difference between the *D* flip-flop and the strobed *D* latch is that the output of the flip-flop can change *only* after the clock input is in transition, say from HIGH to LOW, whereas the output of the latch can change *anytime* while it is enabled.

Many *D* flip-flops have one or two additional inputs called the **preset** S_D and **clear** C_D inputs, as shown in Fig. 8.12(a). The preset and clear inputs act like set and reset inputs, respectively, and are useful for controlling the outputs of the flip-flop independently from the *D* input. The preset and clear inputs may be synchronous or asynchronous. An **asynchronous** or **direct input** is an input that

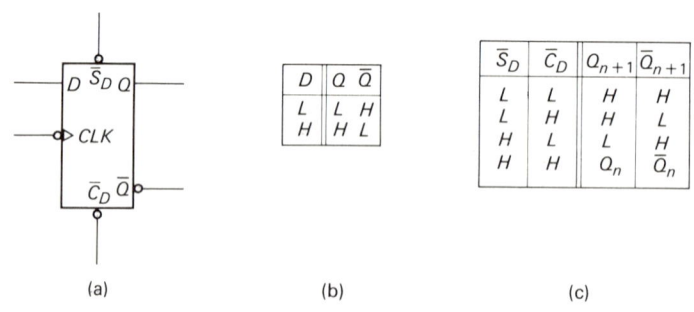

D	Q	$\bar{Q}$
L	L	H
H	H	L

$\bar{S}_D$	$\bar{C}_D$	Q_{n+1}	$\bar{Q}_{n+1}$
L	L	H	H
L	H	H	L
H	L	L	H
H	H	Q_n	$\bar{Q}_n$

(a)   (b)   (c)

Fig. 8.12 The *D* flip-flop: (a) logic symbol; (b) data input truth table; (c) preset and clear truth table.

8.4 THE D FLIP-FLOP

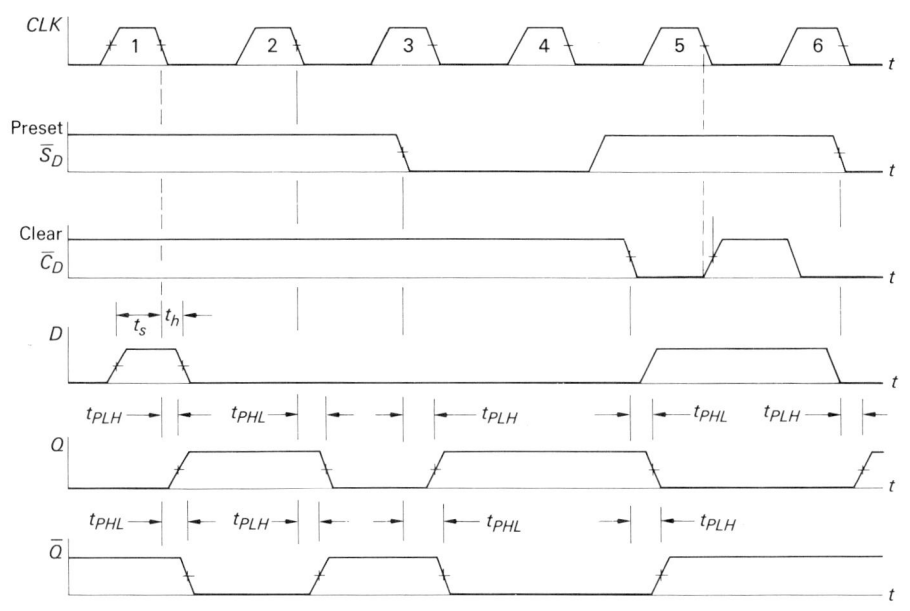

Fig. 8.13 Timing diagram for a trailing-edge D flip-flop.

carries out its function regardless of whether the clock input is HIGH or LOW. A **synchronous input** is an input that depends on the clock. The D input is a synchronous input because the output can change only when the clock input changes; i.e., the D input is in synchronism with the clock. The truth table for the asynchronous active LOW preset and clear inputs, for the D flip-flop shown in Fig. 8.12(a), is given in Fig. 8.12(c). If $\bar{S}_D = L$ and $\bar{C}_D = L$, then both outputs go HIGH, that is, $Q = H$ and $\bar{Q} = H$. If $\bar{S}_D = L$ and $\bar{C}_D = H$ the Q output goes HIGH. If $\bar{S}_D = H$ and $\bar{C}_D = L$ the Q output goes LOW. When $\bar{S}_D = H$ and $\bar{C}_D = H$, the preset and clear inputs have no effect on the outputs Q and $\bar{Q}$.

Figure 8.13 shows a timing diagram for a D flip-flop whose output changes on the $H \to L$ transition of the clock. Assume the output is initially LOW, $\bar{S}_D = H$, and $\bar{C}_D = H$. If the D input goes HIGH one setup time t_s prior to the active (i.e., falling edge) of clock pulse 1 and is held there for one hold time t_h, the Q output goes HIGH one delay time t_{PLH} later, measured relative to the reference level of the falling edge of clock pulse 1. The Q output goes LOW one delay time t_{PHL} after the $H \to L$ transition of clock pulse 2 as shown in Fig. 8.13.

When the preset input goes LOW, the output goes HIGH one preset-to-output delay time later, measured relative to the falling edge of the preset pulse. The output stays HIGH as long as $\bar{S}_D = L$ and $\bar{C}_D = H$, independent of whether the clock or D inputs are HIGH or LOW. When $\bar{S}_D$ returns HIGH with $\bar{C}_D = H$, then the output once again changes after the falling edge of the clock pulse as determined by the D input.

222   FLIP-FLOPS   8.4

The D flip-flop is useful for applications that require only one input, such as in shift registers (Chapter 9) and control circuits (Example 8.4, which follows). D flip-flops are available in IC form in the TTL, ECL, and CMOSL families.

Figure 8.14 shows the specifications for the TTL 74LS175 quadruple rising-

recommended operating conditions

				SN54LS174 SN54LS175			SN74LS174 SN74LS175			UN
				MIN	NOM	MAX	MIN	NOM	MAX	
Supply voltage, V_{CC}				4.5	5	5.5	4.75	5	5.25	V
High-level output current, I_{OH}						−400			−400	µ
Low-level output current, I_{OL}						4			8	m
Clock frequency, f_{clock}				0		30	0		30	MH
Width of clock or clear pulse, t_w				20			20			n
Setup time, t_{setup}			Data input	20			20			n
			Clear inactive-state	25			25			n
Data hold time, t_{hold}				5			5			n
Operating free-air temperature, T_A				−55		125	0		70	°C

electrical characteristics over recommended operating free-air temperature range (unless otherwise noted)

PARAMETER		TEST CONDITIONS†			SN54LS174 SN54LS175			SN74LS174 SN74LS175			UN
					MIN	TYP‡	MAX	MIN	TYP‡	MAX	
V_{IH}	High-level input voltage				2			2			V
V_{IL}	Low-level input voltage						0.7			0.8	V
V_I	Input clamp voltage	V_{CC} = MIN, I_I = −18 mA					−1.5			−1.5	V
V_{OH}	High-level output voltage	V_{CC} = MIN, V_{IH} = 2 V, V_{IL} = V_{IL}max, I_{OH} = −400 µA			2.5	3.5		2.7	3.5		V
V_{OL}	Low-level output voltage	V_{CC} = MIN, V_{IH} = 2 V, V_{IL} = V_{IL} max	I_{OL} = 4 mA			0.25	0.4				V
			I_{OL} = 8 mA						0.35	0.5	
I_I	Input current at maximum input voltage	V_{CC} = MAX, V_I = 5.5 V					0.1			0.1	m
I_{IH}	High-level input current	V_{CC} = MAX, V_I = 2.7 V					20			20	µ
I_{IL}	Low-level input current	Clock input	V_{CC} = MAX, V_I = 0.4 V				−0.4			−0.4	m
		Other inputs					−0.36			−0.36	
I_{OS}	Short-circuit output current§	V_{CC} = MAX			−6		−40	−5		−42	m
I_{CC}	Supply current	V_{CC} = MAX, See Note 2	'LS174			13	22		13	22	m
			'LS175			9	15		9	15	

† For conditions shown as MIN or MAX, use the appropriate value specified under recommended operating conditions.
‡ All typical values are at V_{CC} = 5 V, T_A = 25°C.
§ Not more than one output should be shorted at a time.
NOTE 2: With all outputs open and 4.5 V applied to all data and clear inputs, I_{CC} is measured after a momentary ground, then 4.5 V applied to clock.

switching characteristics, V_{CC} = 5 V, T_A = 25°C

PARAMETER		TEST CONDITIONS	MIN	TYP	MAX	UN
f_{max}	Maximum clock frequency		30	40		MH
t_{PLH}	Propagation delay time, low-to-high-level output from clear (SN54LS175, SN74LS175 only)	C_L = 15 pF, R_L = 2 kΩ,		16	25	ns
t_{PHL}	Propagation delay time, high-to-low-level output from clear			23	35	ns
t_{PLH}	Propagation delay time, low-to-high-level output from clock			20	30	ns
t_{PHL}	Propagation delay time, high-to-low-level output from clock			21	30	ns

Fig. 8.14 Specifications for the TTL 74LS175 quadruple rising-edge triggered D flip-flop. (Courtesy of Texas Instruments Incorporated.)

8.4   THE D FLIP-FLOP   223

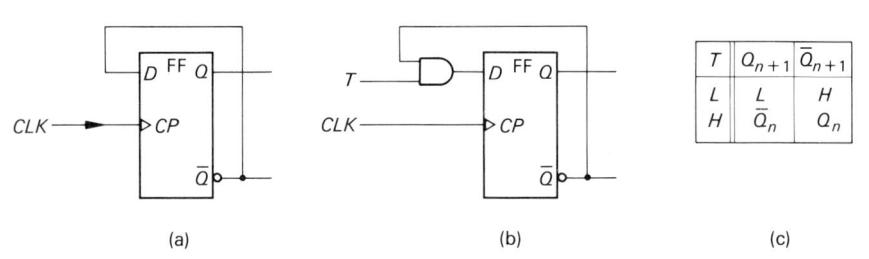

Fig. 8.15 Toggle flip-flop made from a D flip-flop: (a) logic diagram—with control gate; (b) logic diagram—with input AND gate; (c) truth table.

edge ($L \rightarrow H$) triggered D flip-flop. It has a clear input, but no preset input. Note that the data and clear input setup times and the hold times are listed as minimums, whereas some manufacturers would show maximum values. Also note that the data and clear setup times, 20 ns and 25 ns, respectively, do not have the same values. There is a positive data hold time of +5 ns. The **maximum clock frequency** f_{max} is the highest frequency that can be applied to the clock input of a device with a guarantee that the device will operate normally. The 74LS175 has an f_{max} of at least (minimum) 30 MHz, and typically it will operate at 40 MHz.

The CMOSL 14013 dual D flip-flop has the following typical properties for $V_{DD} = +5$ V: quiescent power dissipation 25 nW, setup and hold times 40 ns, and maximum clock frequency 4 MHz.

The D flip-flop can be made to **toggle** (i.e., to repeatedly alternate between HIGH and LOW) by connecting the $\bar{Q}$ output to the D input as shown in Fig. 8.15(a). If an AND gate is used in the D input, as shown in Fig. 8.15(b), the output will toggle when the T input is HIGH and will be forced to the reset state when $T = L$ (Fig. 8.15c).

Example 8.4 *Microcomputer Interrupt Handler.* A computer must have a provision for being stopped (interrupted) while it is doing one of its internal processes, such as addition, to respond to an external request, such as entering new data from a temperature sensor through an analog-to-digital converter (ADC). Referring to Appendix G, study sheet 1 of the IMP-16C schematic diagram (Fig. G.5) and the circuit description (Section G.3.6), and then explain how the interrupt handler circuit works.

Solution. Let's first redraw the logic diagram of the interrupt handler as shown in Fig. 8.16. The circuit consists of two AND gates, an OR gate, and a 74H74 (high-speed TTL series) rising-edge D flip-flop. As we see from the nomenclature table in Appendix G (Table G.3), the signal lines have the following meanings: INTRA—interrupt request signal coming from a source external to the computer; STF—stack full signal line, indicates that the 16-bit memory (called a stack) is full; C23—clock, indicates that signal line is HIGH during clock periods T_2 and T_3 (as explained in Section G.3.1); INTEN—interrupt enable flag signal, indicates when no other interrupt is currently being serviced; INT-Q1—delayed interrupt signal.

The interrupt circuit works as follows. Assume initially the output of gates

Fig. 8.16 Microcomputer interrupt handler (Appendix G).

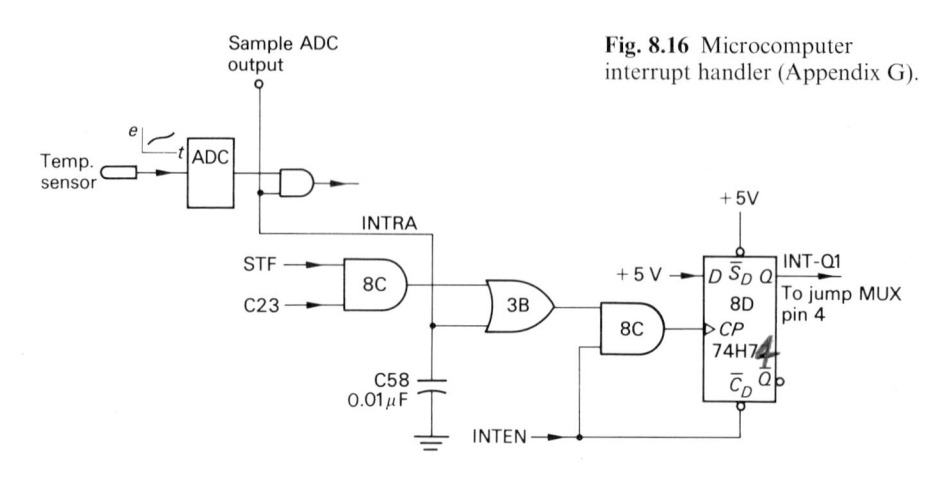

8C and 3B are LOW, and the flip-flop output is LOW. If no other interrupt is currently being serviced, INTEN is HIGH, which enables the clock input to the D flip-flop and inactivates the clear input $\bar{C}_D$. Now when the output of the ADC is sampled the INTRA input to OR gate 3B goes HIGH. Capacitor C58 provides a delay and smooths out any transients in the INTRA signal line. The output of AND gate 8C goes from a LOW to a HIGH, which clocks the D flip-flop so its output INT-Q1 goes HIGH. The microprogram (a set of instructions stored in a memory), which processes interrupts, resets the INTEN flag LOW, which prevents any further interrupts from being serviced. The interrupt enable flag INTEN then returns HIGH after the interrupt request has been serviced. The flip-flop is also set when the stack is full during clock periods T_2 and T_3, that is, when STF and C23 go HIGH. ∎

8.5 THE *JK* FLIP-FLOP

The problem of the indeterminate state of an *RS* latch is eliminated in the *JK* flip-flop. A **JK flip-flop** is a flip-flop that has two synchronous data inputs *J* and *K*, a clock input (*CLK*), and two outputs *Q* and $\bar{Q}$. The labels *J* and *K* have no symbolic meaning.

The logic symbol for a *JK* flip-flop with active HIGH data inputs is shown in Fig. 8.17(a). The small circle at the clock input indicates that the outputs can change one delay time after the $H \to L$ transition of the clock.

The truth table for the data inputs is the same as for the strobed *RS* latch, with *J* acting as the set input and *K* as the reset input, except when $J = H$ and $K = H$. In this case, the *JK* flip-flop toggles, i.e., changes state with each clock pulse, indicated in Fig. 8.17(b) by $Q_{n+1} = \bar{Q}_n$.

The asynchronous active LOW preset ($\bar{S}_D$) and clear ($\bar{C}_D$) inputs obey the same truth table as for the *D* flip-flop discussed in the previous section. For some flip-flops, such as the 7470, the preset and clear inputs are active only when the clock is LOW.

8.5   THE *JK* FLIP-FLOP   225

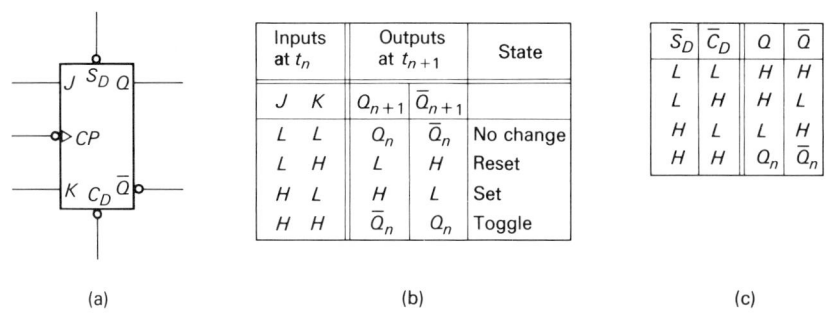

Fig. 8.17 The *JK* flip-flop: (a) logic symbol; (b) truth table for the data inputs; (c) truth table for active LOW preset and clear inputs.

(b)

Inputs at t_n		Outputs at t_{n+1}		State
J	K	Q_{n+1}	$\bar{Q}_{n+1}$	
L	L	Q_n	$\bar{Q}_n$	No change
L	H	L	H	Reset
H	L	H	L	Set
H	H	$\bar{Q}_n$	Q_n	Toggle

(c)

$\bar{S}_D$	$\bar{C}_D$	Q	$\bar{Q}$
L	L	H	H
L	H	H	L
H	L	L	H
H	H	Q_n	$\bar{Q}_n$

The *JK* flip-flop is a widely used, very versatile flip-flop and is available in the TTL, ECL, and CMOSL families in IC form.

The following example illustrates the use of a *JK* flip-flop to control other devices, in this case by extending the period of a clock pulse in a computer.

Example 8.5 *Microcomputer Clock Stretcher.* Taking information from memories is called reading from memory. Reading from memory is usually done in synchronism with a clock and requires several clock periods to complete. It may require two extra clock periods to read out information stored in slow memories. Referring to the schematic diagram (Fig. G.5) and description (Section G.3.5) in Appendix G, describe how the clock phase-4 stretcher works.

Solution. Begin by redrawing the clock stretcher as shown in Fig. 8.18. The clock

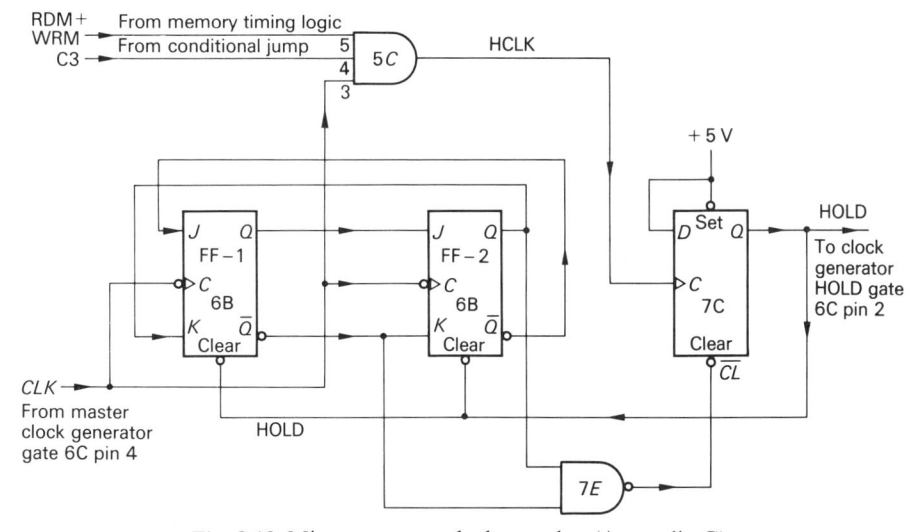

Fig. 8.18 Microcomputer clock stretcher (Appendix G).

stretcher consists of one AND gate, one NAND gate, two JK falling-edge flip-flops, and one rising-edge D flip-flop.

Let's first consider the circuit consisting of the two JK flip-flops. Note that the Q output of FF-2 is connected to the K input of FF-1, and the $\bar{Q}$ output of FF-2 is connected to the J input of FF-1. Assume the clear input HOLD is LOW and then goes HIGH prior to the trailing edge of clock pulse 1 as shown in Fig. 8.19(a). Thus Q_1 and Q_2 are initially LOW. At the falling edge of clock pulse 1, the inputs to FF-1 are $J_1 = \bar{Q}_2 = H$ and $K_1 = Q_2 = L$, so FF-1 sets, that is, $Q_1 = H$ and $\bar{Q}_1 = L$. These outputs are the J and K inputs to FF-2, but FF-2 cannot change in response to these inputs until the falling edge of the next clock pulse, due to the setup time required. At the falling edge of clock pulse 2, the inputs to FF-2 are $J_2 = H$ and $K_2 = L$, so FF-2 sets. At the trailing edge of clock pulse 3, $J_1 = \bar{Q}_2 = L$ and $K_1 = Q_2 = H$, so FF-1 resets. At clock pulse 4, Q_2 goes LOW and $\bar{Q}_2$ goes HIGH. This counting cycle repeats itself as long as the clear input is HIGH. Note that this two-bit counter counts in the sequence 00, 01, 11, 10 and not in the natural binary sequence. It is called a shift or Johnson counter (see Chapter 9).

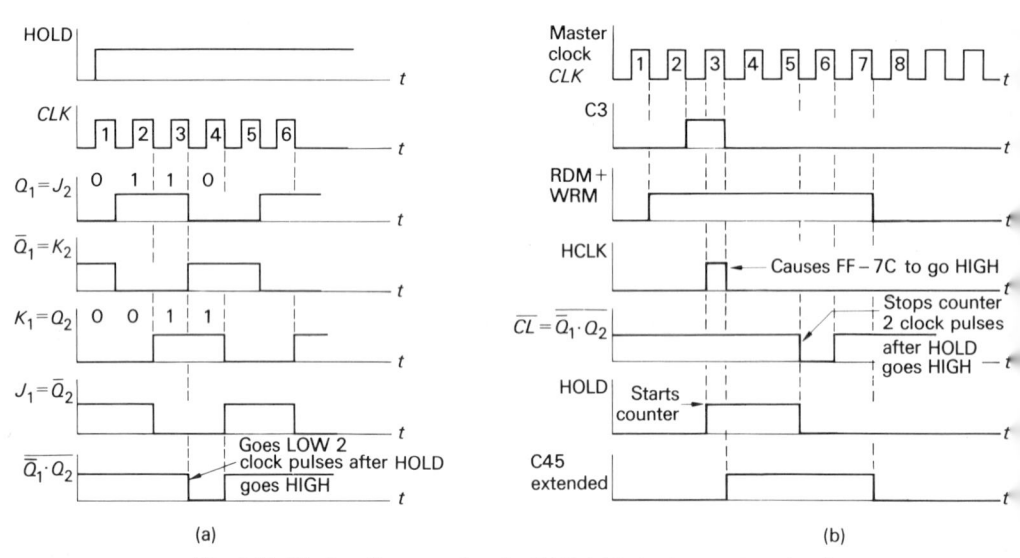

Fig. 8.19 Timing diagrams for the IMP-16C clock stretcher circuit.

By NANDing $\bar{Q}_1$ and Q_2, with NAND gate 7E, a LOW pulse is generated at the falling edge of the second clock pulse (clock pulse 3) after the falling edge of the clock pulse for which the HOLD input goes HIGH.

Figure 8.19(b) shows the timing diagram for the entire clock stretcher circuit. The D flip-flop has a permanent HIGH, at its D input which is transferred to its Q output on the rising edge of its clock signal $HCLK$. As shown in Fig. 8.18,

8.5   THE JK FLIP-FLOP   227

$HCLK$ is the output of AND gate 5C, i.e.,

$$HCLK = (CLK) \cdot (C3) \cdot (RDM + WRM).$$

The waveform for CLK, C_3, and memory timing signal $RDM + WRM$ are shown in Fig. 8.19(b). When these signals are ANDed, it is seen that $HCLK$ goes HIGH when clock pulse 3 is HIGH. This clocks the D flip-flop, so its output, called $HOLD$, goes HIGH and stays HIGH until its $\overline{CL}$ input goes LOW two clock pulses later.

When HOLD goes HIGH, this causes the two-bit shift counter to start counting. Two clock pulses later, the output of NAND gate 7E goes LOW, which clears FF-7C, and HOLD goes LOW, which stops the counter. While HOLD is HIGH, the output of the four-phase clock generator circuit is held HIGH, resulting in an extended pulse C_{45} during the T_4 computer timing interval as shown in Fig. 8.19(b). ∎

This example has illustrated the use of flip-flops in controlling other devices. When a JK flip-flop is operated with its J and K inputs connected together, as shown in Fig. 8.20(a), it becomes a toggle flip-flop. Often the J and K inputs are relabeled simply as T, as shown in Fig. 8.20(b), where it is assumed that J and K are connected together, i.e., $J = K = T$.

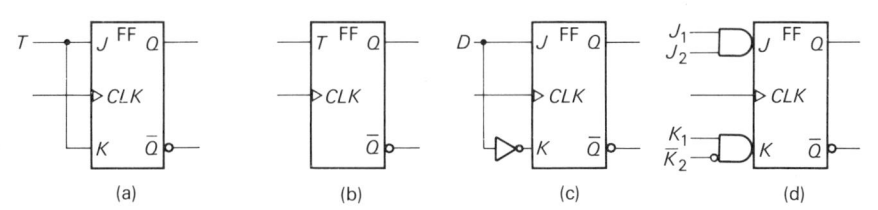

Fig. 8.20 The JK flip-flop: (a) converted to a toggle flip-flop; (b) simplified JK toggle symbol; (c) converted to a D flip-flop; (d) with gated inputs.

The JK flip-flop can also be converted into a D flip-flop by connecting the J input to the K input through an inverter, as illustrated in Fig. 8.20(c).

In many applications logic gates are used in conjunction with flip-flops for controlling the J and K inputs. In order to avoid using external logic gates, with the attendant gate delay times in such applications, many flip-flops have logic gates built into their inputs. Figure 8.20(d) shows a JK flip-flop with two built-in AND gates. When the logic gate symbols touch the logic symbol to which they are connected, as in Fig. 8.20(d), this means the gates are internal to the device, and not external gates supplied by the user. The J input will be HIGH when $J_1 = H$ and $J_2 = H$, that is, when $J = J_1 \cdot J_2$. There is an inverter built into the K_2 input, so $K = K_1 \cdot \bar{K}_2$. A D flip-flop can easily be made by connecting J_1 and J_2 to K_2 with K_1 HIGH. Unused J and K inputs in TTL and CMOSL should not be left open, but should be terminated as with logic gates (see Chapter 3).

*8.6 METHODS OF CLOCKING FLIP-FLOPS

RS, *D*, and *JK* flip-flops can be clocked by several different methods. Here we shall discuss three methods: (1) master/slave, (2) edge-triggering, and (3) capacitive coupling. For each method we shall discuss a simplified functional logic diagram. The preset and clear inputs will be discussed only for the master/slave clocking method.

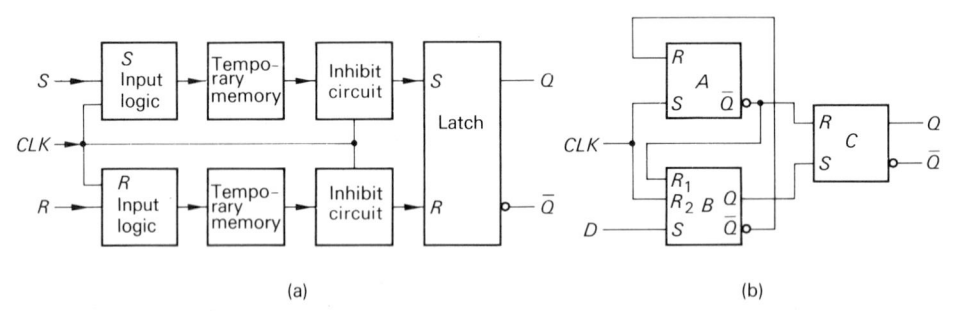

Fig. 8.21 Generalized flip-flop block diagram: (a) *RS*, *D*, or *JK* type; (b) alternate form of some *D* types.

The generalized block diagram for an *RS*, *D*, or *JK* flip-flop is shown in Fig. 8.21(a) [1]. Such a diagram is useful for explaining clocking methods and clarifying the details of flip-flop logic diagrams and circuit schematics. Note that it consists of input logic, temporary memory, inhibit circuit, and an *RS* latch. Data enters the temporary (short-term) memory through the input logic, which is turned on when the clock becomes active. The clock enables the inhibit circuit, which disconnects the memory from latch during the time data is stored in the temporary memory. At the end of the active clock pulse the input logic is disabled. After a short delay, the memory is connected to the latch and the data in the memory is transferred to the latch.

The *R* and *S* input logic consists of logic gates. The temporary memory can be made from (1) a latch, (2) charge stored on a capacitor (discrete capacitor, gate capacitance of a MOS transistor, or depletion region of a *p-n* junction), or (3) base charge stored in a conducting (on) transistor or charge stored in a conducting diode. The inhibit circuit can be made from (1) logic gates, (2) switches made from diodes, MOS or bipolar transistors.

The generalized block diagram for some types of *D* flip-flops is shown in Fig. 8.21(b). It consists of three *RS* latches. Latch *A* is used to store a HIGH input, and latch *B* to store a LOW input. Latch *C* serves as the output stage latch, which isolates the input switching circuitry from the load on the output.

8.6 METHODS OF CLOCKING FLIP-FLOPS

8.6.1 The Simplified Master/Slave Flip-Flop

The **master/slave** (or **pulse triggered**) **flip-flop** consists of two internally interconnected latches, called the **master** and the **slave**, as shown in Fig. 8.22. Gates 1 and 2 in Fig. 8.22 are the R and S input logic gates of Fig. 8.21; gates 3 and 4 are the temporary memory (master latch), battery D_1 and gates 5 and 6 are the inhibit circuit, and gates 7 and 8 are the RS (slave) latch.

In practice a diode D_1 is used in place of the battery so that the master latch and slave latch may be activated at different instants of time.

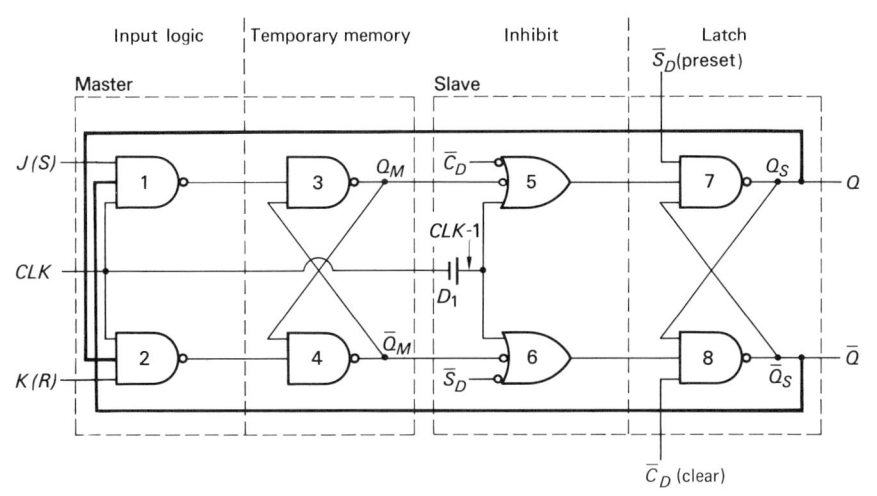

Fig. 8.22 Simplified logic diagram of a master/slave flip-flop (timing diagram shown in Fig. 8.23).

For a JK master/slave flip-flop (Fig. 8.22) the Q output of the slave (gate 7) is internally connected to the K input (gate 2), and the $\bar{Q}$ output (gate 8) is internally connected to the J input (gate 1), i.e., the slave outputs are cross-coupled and fed back to the master inputs. This coupling eliminates the indeterminate state of the RS flip-flop and allows a controlled predictable oscillation (toggle) when $J = H$ and $K = H$. The JK master/slave flip-flop obeys the JK truth table (Fig. 8.17).

In an RS master/slave flip-flop the outputs are not cross-coupled; in other words, the heavy lines in Fig. 8.22 are deleted. The RS master/slave flip-flop obeys the RS latch truth table, as shown in Table 8.1 at the end of this chapter. However, its outputs, unlike RS latch inputs, can be connected back to its inputs, unless R and S are high simultaneously, without danger of unpredictable outputs.

Let's see how the JK master/slave flip-flop shown in Fig. 8.22 works. Assume TTL gates are used with a threshold voltage of 1.5 V, a LOW of 0 V and a HIGH of 3.5 V, and $\bar{S}_D = H$ and $\bar{C}_D = H$. The two clock waveforms CLK and CLK-1

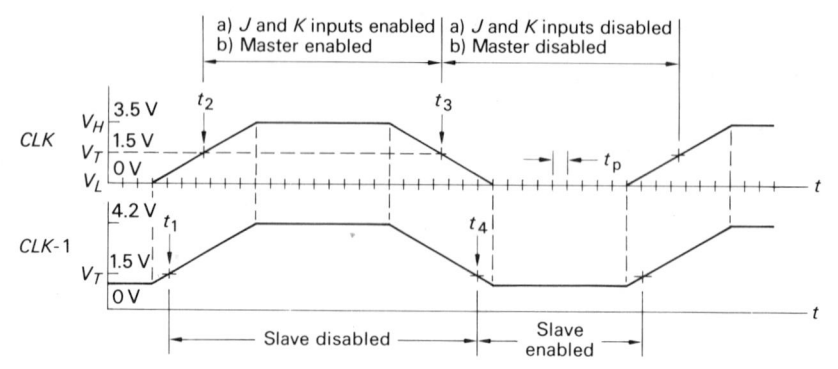

Fig. 8.23 Timing diagram for the clock inputs for a trailing-edge master/slave flip-flop.

used in this flip-flop are shown in Fig. 8.23. Note that when the *CLK* voltage goes from a LOW to a HIGH, the offset voltage provided by D_1 causes *CLK*-1 to reach the threshold voltage before (at time t_1) *CLK* reaches the threshold voltage (at time t_2). On the rising edge of the *CLK* waveform, *CLK*-1 reached the threshold voltage at time t_1. This forces the outputs of gates 5 and 6 HIGH, which prevents the Q and $\bar{Q}$ outputs of the slave latch from changing. In other words, the slave is disabled.

At time t_2 *CLK* reaches the threshold voltage so gates 1 and 2 are enabled. Therefore the Q_M and $\bar{Q}_M$ outputs of the master latch go to a state which is determined by the *J* and *K* inputs, and the previous state of the slave outputs Q and $\bar{Q}$ (due to cross coupling of Q and $\bar{Q}$ into the inputs of gates 1 and 2).

At time t_3 *CLK* goes LOW so gates 1 and 2 are disabled and the *J* and *K* inputs can have no further effect on the slave, i.e., the master is disconnected from the slave. At time t_4 the *CLK*-1 inputs to gates 5 and 6 go LOW so the slave is enabled and data at the Q_M and $\bar{Q}_M$ outputs of the master are passed on to the slave latch. From the user's point of view, the flip-flop in Fig. 8.22 appears to be sensitive to the falling (trailing) edge of the clock input, since it is during the $H \rightarrow L$ clock transition that the output can change.

The preset and clear inputs for the *JK* flip-flop shown in Fig. 8.22 work as follows. When the active LOW preset input $\bar{S}_D$ is LOW and the active LOW clear input $\bar{C}_D$ is HIGH, the output of gate 7 is forced HIGH, i.e., $Q = H$, and the output of gate 6 goes HIGH. This causes all inputs to gate 8 to be HIGH so $\bar{Q}$ becomes LOW. Note that these slave outputs occur independent of the clock, or the *J* and *K* inputs, i.e., they are asynchronous inputs. When $\bar{S}_D = H$ and $\bar{C}_D = L$ the flip-flop outputs are $Q = L$ and $\bar{Q} = H$. Thus the preset input sets the Q output and the clear input resets the Q output. When $\bar{S}_D = H$ and $\bar{C}_D = H$ the preset and clear inputs have no effect on the output. When $\bar{S}_D = L$ and

8.6   METHODS OF CLOCKING FLIP-FLOPS   231

$\bar{C}_D = L$ both outputs are forced HIGH, i.e., $Q = H$ and $\bar{Q} = H$. The truth table in Fig. 8.17(c) summarizes these results. Setup times for inactivating the clear and preset inputs are specified for some flip-flops (e.g. the 74LS175 shown in Fig. 8.14).

There are two clocking systems for a master/slave flip-flop, the falling-edge system and the rising-edge system. A master/slave flip-flop with a falling-edge clock input, as just described, accepts data while the clock input is HIGH and changes outputs after the falling edge of the clock input. A master/slave flip-flop with a rising-edge clock input accepts input data when the clock input is LOW and changes its output after the rising edge of the clock input. Since most clocked MSI devices use a rising-edge clock input, an external inverter is usually required to convert the clock input of a falling-edge flip-flop into a rising-edge type, unless an offset triggering is desired.

Now, let's discuss the timing diagram for a master/slave flip-flop. Figure 8.24 shows a timing diagram for a falling-edge master/slave JK flip-flop. Before clock pulse 1, $Q = H$ and $\bar{Q} = L$. One minimum setup time $t_{s(min)}$ before the falling edge of clock pulse 1, $J = L$, and K goes HIGH and stays HIGH for the minimum hold time $t_{h(min)}$ after the falling edge of clock pulse 1. The Q and $\bar{Q}$ outputs change one delay time after the falling edge of clock pulse 1. The minimum setup time for most master/slave flip-flops is equal to the width of the clock pulse $t_{p(clock)}$.

When $J = H$ and $K = L$ at clock pulse 2, then Q goes HIGH one delay time t_{PLH} after the falling edge of clock pulse 2. When $J = H$ and $K = H$ (clock pulses 3 and 4) the flip-flop toggles. A more detailed timing diagram will be developed in the problems.

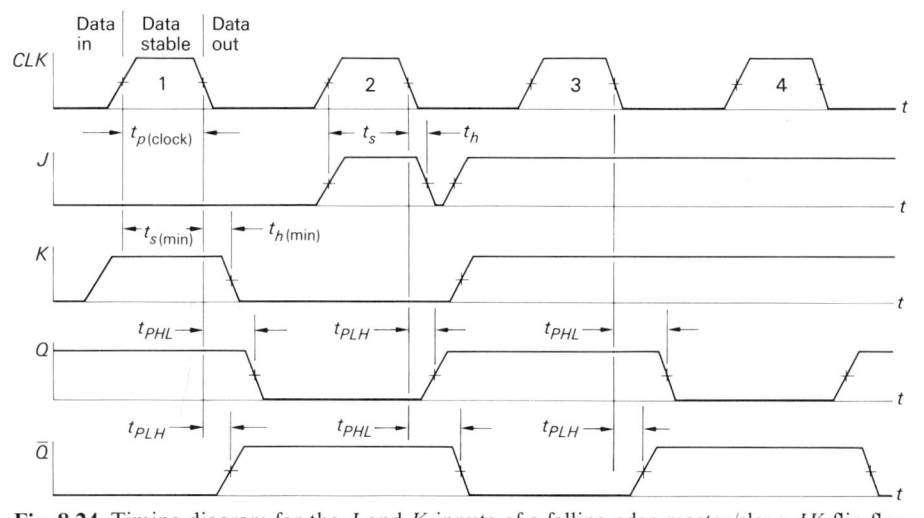

Fig. 8.24 Timing diagram for the J and K inputs of a falling-edge master/slave JK flip-flop.

232   FLIP-FLOPS   8.6

Suppose that while the clock is active, the J and K inputs to a master/slave flip-flop change the output of the master to a state opposite to that of the slave. Then if the J or K inputs change again while the clock is active the master will not be affected until the beginning of the next active clock pulse. The phenomenon occurs in many master/slave flip-flops and is called **1's catching**. In effect, if data

RECOMMENDED OPERATING CONDITIONS

PARAMETER	9N76XM/5476XM MIN.	TYP.	MAX.	9N76XC/7476XC MIN.	TYP.	MAX.	UN
Supply Voltage V_{CC}	4.5	5.0	5.5	4.75	5.0	5.25	Vo
Operating Free-Air Temperature Range	−55	25	125	0	25	70	°C
Normalized Fan Out from Each Output, N			10			10	U.L
Width of Clock Pulse, $t_{p(clock)}$ (See Fig. E)	20			20			ns
Width of Preset Pulse, $t_{p(preset)}$ (See Fig. F)	25			25			ns
Width of Clear Pulse, $t_{p(clear)}$ (See Fig. F)	25			25			ns
Input Setup Time, t_{setup} (See Fig. E)	≥ $t_{p(clock)}$			≥ $t_{p(clock)}$			
Input Hold Time, t_{hold}	0			0			

X = package type; F for Flatpak, D for Ceramic Dip, P for Plastic Dip. See Packaging Information Section for packages available on this produ

ELECTRICAL CHARACTERISTICS OVER OPERATING TEMPERATURE RANGE (Unless Otherwise Noted)

SYMBOL	PARAMETER	MIN.	TYP. (Note 2)	MAX.	UNITS	TEST CONDITIONS (Note 1)	TES FIGU
V_{IH}	Input HIGH Voltage	2.0			Volts	Guaranteed Input HIGH	46 &
V_{IL}	Input LOW Voltage			0.8	Volts	Guaranteed Input LOW	46 &
V_{OH}	Output HIGH Voltage	2.4	3.5		Volts	V_{CC} = MIN., I_{OH} = −0.4 mA	46
V_{OL}	Output LOW Voltage		0.22	0.4	Volts	V_{CC} = MIN., I_{OL} = 16 mA	47
I_{IH}	Input HIGH Current at J or K			40	µA	V_{CC} = MAX., V_{IN} = 2.4 V	49
				1.0	mA	V_{CC} = MAX., V_{IN} = 5.5 V	
	Input HIGH Current at Clear, Preset or Clock			80	µA	V_{CC} = MAX., V_{IN} = 2.4 V	49
				1.0	mA	V_{CC} = MAX., V_{IN} = 5.5 V	
I_{IL}	Input LOW Current at J or K			−1.6	mA	V_{CC} = MAX., V_{IN} = 0.4 V	48
	Input LOW Current at Clear, Preset, or Clock			−3.2	mA	V_{CC} = MAX., V_{IN} = 0.4 V	48
I_{OS}	Output Short Circuit Current (Note 3)	−20		−57	mA	9N76/5476 V_{CC} = MAX.	51
		−18		−57	mA	9N76/7476 V_{IN} = 0 V	
I_{CC}	Supply Current		20	40	mA	V_{CC} = MAX.	49

SWITCHING CHARACTERISTICS (T_A = 25°C)

SYMBOL	PARAMETER	MIN.	TYP.	MAX.	UNITS	TEST CONDITIONS	TES FIGU
f_{max}	Maximum Clock Frequency	15	20		MHz		E
t_{PLH}	Turn Off Delay Clear or Preset to Output		16	25	ns	V_{CC} = 5.0 V	F
t_{PHL}	Turn On Delay Clear or Preset to Output		25	40	ns	C_L = 15 pF	F
t_{PLH}	Turn Off Delay Clock to Output	10	16	25	ns	R_L = 400Ω	E
t_{PHL}	Turn On Delay Clock to Output	10	25	40	ns		E

Fig. 8.25 Specifications for the TTL 7476 dual JK master/slave flip-flop. (Courtesy of Fairchild Camera and Instrument Corporation.)

inputs to a 1's catching master/slave flip-flop change while its clock is active, the output will not obey the *JK* truth table given in Fig. 8.17(b). The output can be predicted using another form of a truth table, called an excitation table; however, the simplest procedure is to avoid changing data inputs while the clock is active. The 1's catching phenomenon is a disadvantage of this type of master/slave flip-flop because data inputs may change inadvertently, due to noise for example, while the clock is active, resulting in an unpredictable output. For 1's catching flip-flops, the clock pulses should be as short as possible, to minimize the chance that data inputs will change during an active clock.

Some master/slave flip-flops are not 1's catching. For example, the TTL 74110 has a **data lockout** feature that eliminates the need to hold the data inputs stable while the clock is active. If the setup and hold times are observed, the data inputs can change while the clock is active without any effect on the outputs.

The timing diagram for the asynchronous preset and clear inputs for a *JK* flip-flop is the same as for the *D* flip-flop discussed in Section 8.4.

Some devices that use flip-flops, such as shift registers (Chapter 9) always have their flip-flop inputs in the opposite state, in other words $R = H$ and $S = H$ never occur, so *RS* master/slave flip-flops can be used. The advantage of the *RS* master/slave flip-flop over the *JK* master/slave flip-flop is that it is simpler (no cross-coupled outputs), it has a lower power dissipation, and the data inputs can change while the clock is active (as long as the setup and hold times are observed).

The advantage of the *D* flip-flop, due to its simpler circuit (as compared with making a *D* flip-flop from a *JK* flip-flop), is its shorter delay times (higher speed); it may also have a lower power dissipation. The disadvantage of the *D* flip-flop is that it is less flexible, i.e., it cannot be converted into a *JK* flip-flop.

Figure 8.25 gives the specifications for the 1's catching TTL 7476 falling-edge dual *JK* master/slave flip-flop. The minimum clock pulse width is 20 ns, so the minimum setup time is also 20 ns since $t_s \geq t_{clock}$. The minimum preset and clear pulse widths are 25 ns and the hold time is 0 ns. The 7476 has a maximum clock frequency f_{max} of at least (minimum) 15 MHz. The clock, clear, and preset-to-output turn-on (t_{PHL}) and turn-off (t_{PLH}) delay times are 25 ns and 16 ns, respectively.

The ECL 10135 dual *JK* master/slave flip-flop typically has a maximum clock (toggle) frequency of 140 MHz, delay times of 3 ns, setup time of 1 ns, hold time of 1 ns, and power dissipation of 300 mW.

8.6.2 The Simplified Edge Triggered Flip-Flop

An **edge triggered flip-flop** is one in which data enters the data inputs just prior to the active edge of a clock pulse and appears at the output one delay time after the *same* edge of that clock pulse. This is in contrast to the master/slave flip-flop, in which data enters the flip-flop on one edge of the clock pulse but is not available at the output until after the next edge of that clock pulse.

234   FLIP-FLOPS   8.6

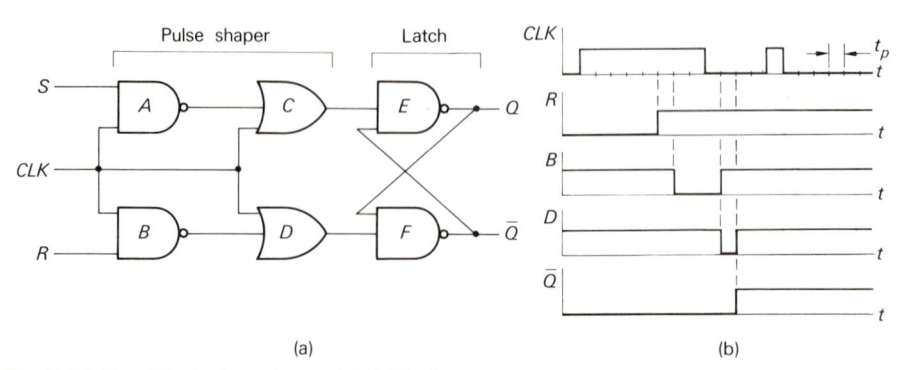

Fig. 8.26 Simplified edge triggered RS flip-flop: (a) logic diagram; (b) pulse-shaping timing diagram.

Figure 8.26(a) shows the logic diagram of a simplified edge triggered RS flip-flop. Gates A and B in Fig. 8.26 are the R and S input logic gates of the generalized flip-flop block diagram (Fig. 8.21a). Gates C and D are the inhibit gates. The temporary memory is essentially the pulse-shaping circuit, which consists of gates A, B, C, and D. In effect, this shaping circuit "remembers" the falling edge of the clock input.

Let's see how the pulse-shaping circuit works. Assume initially that $Q = H$, $CLK = H$, $S = L$, and all gate delay times have the same value t_p. Now apply a reset pulse $R = H$ as shown in Fig. 8.26(b). The output of gate B stays HIGH until the HIGH reset pulse R coincides with the HIGH clock pulse, then one delay time later B goes LOW. Gate B stays LOW until one delay time after either CLK or R goes LOW. Gate D goes LOW one delay time after the falling edge of the clock pulse. The $\bar{Q}$ output goes HIGH two delay times after the falling edge of the clock pulse. Actual edge triggered circuits differ somewhat from the one shown, but this simplified circuit shows the basic principles, using ideal pulses.

Like the master/slave flip-flop, the edge triggered flip-flop has two clocking systems, the falling-edge type and rising-edge type. In an edge triggered flip-flop with a falling-edge clock input, illustrated in Fig. 8.27(a), data must be present at the data inputs the minimum setup time before the falling edge of the clock input and must be held the minimum hold time after the same falling clock edge. For a rising-edge triggered flip-flop (Fig. 8.27b), data must be present at the data inputs the minimum setup time before, and held the minimum hold time after, the rising edge of the clock input. Note that the HIGH and LOW setup times t_{sHL} and t_{sLH}, respectively, are not necessarily equal.

Short clock transition times are especially desirable for edge triggered flip-

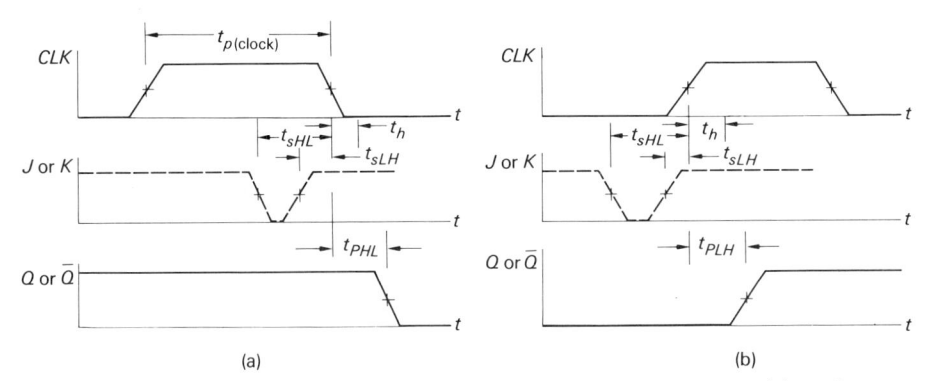

Fig. 8.27 Edge triggered flip-flop timing diagrams: (a) falling-edge type; (b) rising-edge type.

flops in order to improve their noise immunity. With very long clock transition times, the clock voltage may be nearly equal to the flip-flop threshold voltage for a period that is longer than the flip-flop delay time. This can result in multiple clocking from noise on the clock line or from cross talk on the IC chip. For standard series TTL flip-flops, for example, it is recommended that clock rise and fall times be less than 50 ns.

Now let's compare the master/slave and the edge triggered flip-flop. The edge triggered flip-flop does not require that the data be set up the entire time the clock is active, as the 1's catching master/slave does. This means greater freedom in the application of data inputs (D, RS, or JK) relative to the clock when using edge triggered flip-flops, and hence simpler system design. For example, the control of the clock width and the duty cycle is not critical. The edge triggered flip-flop can also have a higher maximum clock frequency because the output can change after the same clock edge on which data entered the flip-flop. In general, master/slave flip-flops are not as sensitive to clock transition times as edge triggered flip-flops are. The 1's catching phenomenon that occurs in some master/slave flip-flops is generally an undesirable characteristic, because this means one must ensure that the data is stable during the entire time the clock is active.

Figure 8.28 shows the specifications for the dual CMOSL 14027 rising-edge clock, edge triggered JK flip-flop. Notice the long delay times as compared with those for TTL and ECL flip-flops. The maximum clock frequency is typically only 3 MHz ($V_{DD} = +5$ V). The hold time is 0 ns, and the maximum setup time is 200 ns. Note that the clock input rise and fall times should not exceed a minimum of 15 μs.

The last type of triggering to be considered is the capacitively coupled flip-flop.

ELECTRICAL CHARACTERISTICS

Characteristic	Figure	Symbol	MC14027AL −55°C Min	Max	+25°C Min	Typ	Max	+125°C Min	Max	MC14027CL/CP −40°C Min	Max	+25°C Min	Typ	Max	+85°C Min	Max	Unit
Output Voltage																	Vdc
(V_{DD} = 5.0 Vdc) "0" Level		V_{OL}	—	0.01	—	0	0.01	—	0.05	—	0.01	—	0	0.01	—	0.05	
(V_{DD} = 10 Vdc)			—	0.01	—	0	0.01	—	0.05	—	0.01	—	0	0.01	—	0.05	
(V_{DD} = 5.0 Vdc) "1" Level		V_{OH}	4.99	—	4.99	5.0	—	4.95	—	4.99	—	4.99	5.0	—	4.95	—	Vdc
(V_{DD} = 10 Vdc)			9.99	—	9.99	10	—	9.95	—	9.99	—	9.99	10	—	9.95	—	
Noise Immunity																	Vdc
(V_{DD} = 5.0 Vdc)		V_{NL}	1.5	—	1.5	2.25	—	1.4	—	1.5	—	1.5	2.25	—	1.4	—	
(V_{DD} = 10 Vdc)			3.0	—	3.0	4.5	—	2.9	—	3.0	—	3.0	4.5	—	2.9	—	
(V_{DD} = 5.0 Vdc)		V_{NH}	1.4	—	1.5	2.25	—	1.5	—	1.4	—	1.5	2.25	—	1.5	—	Vdc
(V_{DD} = 10 Vdc)			2.9	—	3.0	4.5	—	3.0	—	2.9	—	3.0	4.5	—	3.0	—	
Output Drive Current																	mAdc
(V_{OH} = 2.5 Vdc, V_{DS} = 5.0 Vdc) Source	4	I_{OH}	−0.62	—	−0.5	−1.5	—	−0.35	—	−0.23	—	−0.2	−1.5	—	−0.16	—	
(V_{OH} = 9.5 Vdc, V_{DS} = 10 Vdc)			−0.62	—	−0.5	−1.0	—	−0.35	—	−0.23	—	−0.2	−1.0	—	−0.16	—	
(V_{OL} = 0.4 Vdc, V_{DS} = 5.0 Vdc) Sink	5	I_{OL}	0.5	—	0.4	0.8	—	0.28	—	0.23	—	0.2	0.8	—	0.16	—	mAdc
(V_{OL} = 0.5 Vdc, V_{DS} = 10 Vdc)			1.1	—	0.9	1.2	—	0.65	—	0.6	—	0.5	1.2	—	0.4	—	
Input Current	—	I_{in}	—	—	—	10	—	—	—	—	—	—	10	—	—	—	pAdc
Input Capacitance (V_{in} = 0)	—	C_{in}	—	—	—	5.0	—	—	—	—	—	—	5.0	—	—	—	pF
Quiescent Dissipation	2, 3	P_D															μW
(V_{DD} = 5.0 Vdc)			—	5.0	—	0.025	5.0	—	300	—	50	—	0.05	50	—	700	
(V_{DD} = 10 Vdc)			—	20	—	0.05	20	—	1200	—	200	—	0.2	200	—	2800	
Output Rise Time	6, 7	t_r															ns
(C_L = 15 pF, V_{DD} = 5.0 Vdc)			—	—	—	100	175	—	—	—	—	—	100	200	—	—	
(C_L = 15 pF, V_{DD} = 10 Vdc)			—	—	—	35	75	—	—	—	—	—	35	110	—	—	
Output Fall Time	6, 7	t_f															ns
(C_L = 15 pF, V_{DD} = 5.0 Vdc)			—	—	—	100	175	—	—	—	—	—	100	200	—	—	
(C_L = 15 pF, V_{DD} = 10 Vdc)			—	—	—	35	75	—	—	—	—	—	35	110	—	—	
Turn-On Delay Time	6, 8	t_{PHL}															ns
(C_L = 15 pF, V_{DD} = 5.0 Vdc)			—	—	—	150	300	—	—	—	—	—	150	400	—	—	
(C_L = 15 pF, V_{DD} = 10 Vdc)			—	—	—	75	110	—	—	—	—	—	75	150	—	—	
Turn-Off Delay Time	6, 8	t_{PLH}															ns
(C_L = 15 pF, V_{DD} = 5.0 Vdc)			—	—	—	150	300	—	—	—	—	—	150	400	—	—	
(C_L = 15 pF, V_{DD} = 10 Vdc)			—	—	—	75	110	—	—	—	—	—	75	150	—	—	
Min Clock Pulse Width	6	PW															ns
(C_L = 15 pF, V_{DD} = 5.0 Vdc)			—	—	—	125	300	—	—	—	—	—	125	400	—	—	
(C_L = 15 pF, V_{DD} = 10 Vdc)			—	—	—	50	100	—	—	—	—	—	50	150	—	—	
Max Clock Pulse Frequency	6	PRF															MHz
(C_L = 15 pF, V_{DD} = 5.0 Vdc)			—	—	1.5	3.0	—	—	—	—	—	1.0	3.0	—	—	—	
(C_L = 15 pF, V_{DD} = 10 Vdc)			—	—	4.5	8.0	—	—	—	—	—	3.0	8.0	—	—	—	
Clock Pulse Rise and Fall Times	6	t_r, t_f															μs
(C_L = 15 pF, V_{DD} = 5.0 Vdc)			—	—	—	15	—	—	—	—	—	—	15	—	—	—	
(C_L = 15 pF, V_{DD} = 10 Vdc)			—	—	—	5.0	—	—	—	—	—	—	5.0	—	—	—	
Set and Reset Propagation Delay Times	—	t_{PHL}, t_{PLH}															ns
(C_L = 15 pF, V_{DD} = 5.0 Vdc)			—	—	—	175	225	—	—	—	—	—	175	350	—	—	
(C_L = 15 pF, V_{DD} = 10 Vdc)			—	—	—	75	110	—	—	—	—	—	75	150	—	—	
Set and Reset Pulse Width	—	PW_H															ns
(C_L = 15 pF, V_{DD} = 5.0 Vdc)			—	—	—	125	200	—	—	—	—	—	125	300	—	—	
(C_L = 15 pF, V_{DD} = 10 Vdc)			—	—	—	50	80	—	—	—	—	—	50	120	—	—	
Setup Time	—	t_{setup}															ns
(C_L = 15 pF, V_{DD} = 5.0 Vdc)			—	—	—	70	150	—	—	—	—	—	70	200	—	—	
(C_L = 15 pF, V_{DD} = 10 Vdc)			—	—	—	25	50	—	—	—	—	—	25	75	—	—	

*DC Noise Margin (V_{NH}, V_{NL}) is defined as the maximum voltage change from an ideal "1" or "0" input level, that the circuit will withstand before producing an output state change.

Fig. 8.28 Electrical characteristics of the dual CMOSL 14027 rising-edge triggered JK flip-flop. (Courtesy of Motorola Corporation.)

8.6.3 The Capacitively Coupled Flip-Flop

Figure 8.29 shows a simplified logic diagram of a capacitively coupled RS flip-flop. With reference to Fig. 8.21, gates A and B in Fig. 8.29 form the R and S input logic. Capacitors C_1 and C_2 are the temporary memory. The inhibit circuit consists of gates C, D, and E, and the latch is formed by cross-coupled NAND gates F and G.

Charge stored in capacitors C_1 and C_2 cause the outputs of gates A and B to change more slowly, due to the RC time constant, than they would without the capacitors. Thus C_1 and C_2 act as a temporary memory.

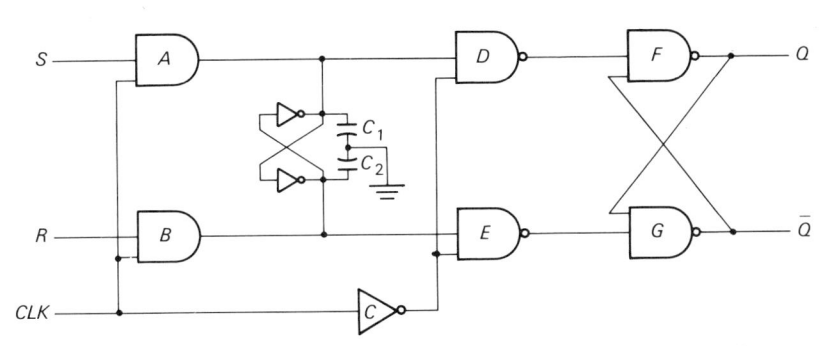

Fig. 8.29 Simplified logic diagram of a capacitively coupled *RS* flip-flop.

The main disadvantage of capacitively coupled flip-flops is that they are sensitive to clock rise and fall times.

To conclude this chapter let us examine some important timing problems that can occur in digital systems.

8.7 RACES AND CLOCK SKEW

A **race condition** occurs in a digital system when the stable output state of that system occurs only after a race between two variables that have been required to change state at the same time. Races can occur both in combinational and in sequential logic devices.

Figure 8.30 illustrates a race condition in a NAND gate whose output F is controlled by two variables A and B. Suppose the delay times for the inverting and noninverting gates U_1 and U_2 are not the same. Ideally the output F should always be HIGH because one input is always LOW. However, during a transition let us say that A goes from LOW to HIGH before B goes from HIGH to LOW (Fig. 8.30b), due to differences in the propagation delay time through gates U_1

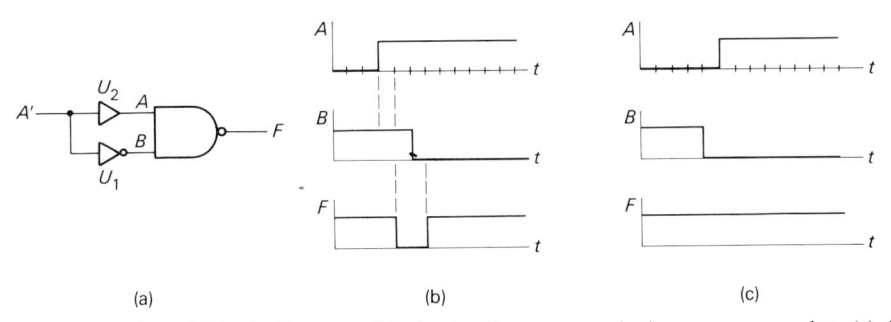

Fig. 8.30 Racing: (a) logic diagram; (b) A wins the race, producing a narrow pulse; (c) B wins the race, producing a constant output.

and U_2. Then at the output F a pulse appears whose width equals the offset in time between the A and B inputs. If input B wins the race to the NAND gate input, as shown in Fig. 8.30(c), the output never goes LOW. Of course, actual pulses have nonzero rise and fall times, but the same principles apply.

The *RS* latch and the strobed *RS* latch can race if their outputs are connected back into their inputs, or when both the *R* and *S* inputs change simultaneously while the latch is in the indeterminate state. The outputs of all flip-flops can, however, be connected back into their synchronous inputs without racing. This is due to the memory-inhibit-latch configuration (Fig. 8.21a) of all *RS*, *D*, and *JK* flip-flops, or the coupling used in some flip-flops (Fig. 8.21b). A race condition can occur if flip-flop outputs are connected into asynchronous inputs.

Race conditions can be avoided by designing the system so that no more than one variable is required to change at the same time. The design of race-free asynchronous systems can be quite tedious, so that synchronous systems, which are not as prone to racing, are generally preferred.

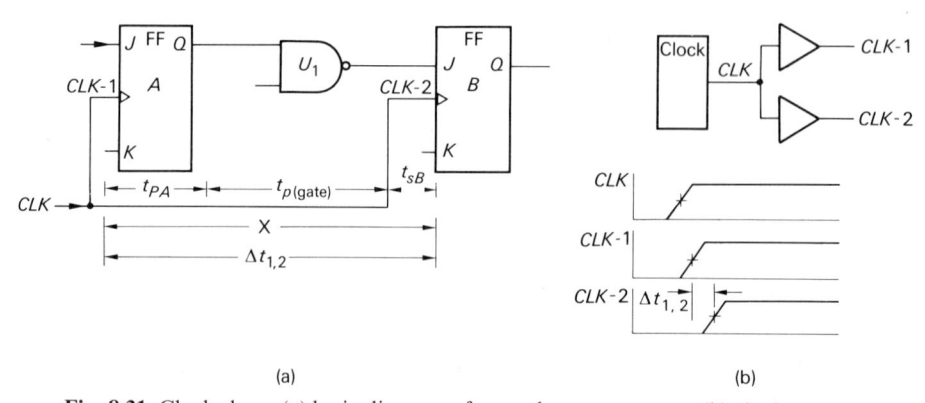

Fig. 8.31 Clock skew: (a) logic diagram of a synchronous system; (b) clock source.

Clock skew refers to the difference in the time of arrival of clock signals at the clock inputs of two or more devices. Consider the circuit shown in Fig. 8.31. Suppose that a time difference $\Delta t_{1,2}$ exists between the time a clock pulse arrives at FF-*A* and the time a clock pulse arrives at FF-*B*, due to delays. If the clock signal arrives at FF-*B* later than at FF-*A*, then the clock signal to FF-*B* can arrive no later than time t after the arrival of the clock pulse for FF-*A*, where t is given by

$$t \leqslant \Delta t_{1,2} = t_{pA} + t_{p(\text{gate})} + t_{sB}.$$

If it arrives later than this, FF-*B* may respond erroneously to new data from FF-*A*.

The time difference at $\Delta t_{1,2}$ could be due to the physical distance x between the two flip-flops, *RC* time delays, propagation delays, or setup times. If loading of the clock output is a problem, then two clock drivers might be used as shown in

Fig. 8.31(b). If the delay times of these two drivers are not the same, a time difference $\Delta t_{1,2}$ will then occur between the output pulses.

Clock skew can be prevented by several means, such as (1) using one clock source, (2) equalizing the delay between multiple clock outputs, or (3) arranging the circuit so the clock delays run in the direction opposite to that of data delays.

SUMMARY

In this chapter we have discussed two important sequential devices, latches and flip-flops. An RS latch is made from two cross-coupled gates; it has one indeterminate state, which must be avoided. Strobed latches are useful in applications that require the latch to be enabled independent of a clock or clock transitions. The strobed latch can be of the RS or D type, and is available in IC form.

Setup and hold times are important specifications for IC latches and flip-flops. They refer to the times during which data must be present prior to and after, respectively, the active edge of a latching or clock input.

An RS, D, or JK flip-flop is a device that generally consists of input logic, a temporary memory, an inhibit circuit, and an RS latch. Three methods for clocking flip-flops were (1) the master/slave, (2) edge triggering, and (3) capacitive coupling. In each method the clocks may be of the rising-edge or the falling-edge type. Table 8.1 summarizes the truth table of the flip-flops we have discussed.

Table 8.1 Truth Table for the D, T, RS, and JK Flip-Flops.

D	Q_{n+1}	T	Q_{n+1}	S	R	Q_{n+1}	J	K	Q_{n+1}
L	L	L	Q_n	L	L	Q_n	L	L	Q_n
H	H	H	$\bar{Q}_n$	L	H	L	L	H	L
				H	L	H	H	L	H
				H	H	*	H	H	$\bar{Q}_n$

* Indeterminate return state if R and S go LOW simultaneously.

A race occurs when two signals change simultaneously. Clock skew, a time difference between the arrival of clock signals at two or more devices, may result in erroneous operation of a synchronous system.

PROBLEMS

Section 8.2

1. Explain in detail how an RS NAND gate latch works. Draw four separate logic diagrams and show the logic levels on all inputs and outputs.
2. Complete a truth table for the latch shown in Fig. 8.1(a), with the input to gate A relabeled S and the input to gate B relabeled R.

240   FLIP-FLOPS

3. (a) Draw a timing diagram for a NOR gate latch. (b) Explain the significance of an R input that is less than two gate delay times wide. Use ideal pulses but include delay times.
4. Draw a timing diagram for an RS NAND latch. Use ideal pulses. Initially let $R = S = L$, $Q = L$, and $\bar{Q} = H$.
5. Discuss the effect of putting a delay element with delay time $2t_p$ in series with the feedback line between the Q output and its input to gate B in Fig. 8.1.
6. *Pressure Warning System.* (a) Draw a timing diagram for Fig. 8.5(b) that will illustrate typical waveforms at the output of each device. (b) Add a pulse height discriminator circuit to Fig. 8.5(b) that will turn on a red light when the pressure decreases below a safe limit, a yellow light when the pressure exceeds a certain limit, and a green light otherwise.
7. *Microcomputer Address Latches.* Referring to Appendix G, study Section G.3.4, which describes the address latches shown on sheet 2 of the schematic diagram (Fig. G.6). What is the purpose of the latches? To aid in your explanation, draw a simplified connection diagram that shows latches 5F and 6F and how they are connected to (1) input multiplexers 5H and 6H, (2) read/write memory H2B (Sheet 3 in Appendix G), and (3) read-only memory 3H (Sheet 3). Omit all connections to edge connectors.
8. (a) Complete the truth table and discuss the operation for the logic diagram shown in Fig. P8.1. (b) Discuss the operation for the RS latch circuit shown in Fig. P8.1(b).

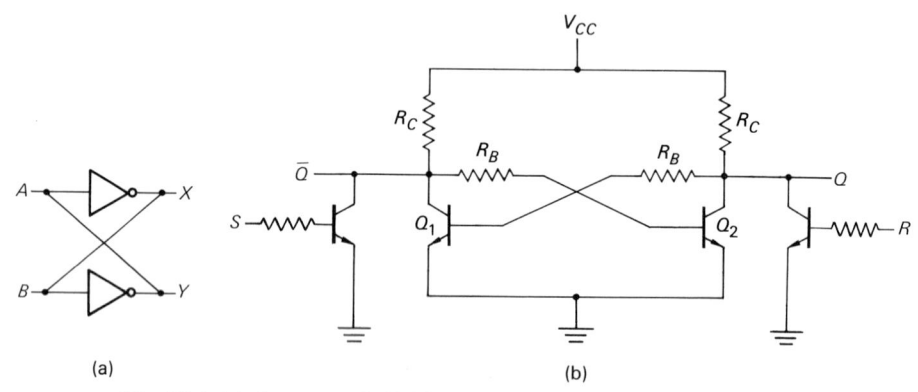

Fig. P8.1 (a) Cross-coupled NOT gates; (b) RS NOR latch schematic.

Section 8.3.1

9. Refer to the timing diagram for the strobed RS latch (Fig. 8.7). Explain in detail the output waveforms for enable pulses 3 and 4.
10. Explain why the circuit shown in Fig. P8.2 will not work as a strobed RS latch.
11. Suppose the outputs of a strobed RS latch are coupled back into the inputs through a delay element τ as shown in Fig. P8.3. The propagation delay time t_p of all gates is the same. Draw a timing diagram for inputs E, R, S, outputs of gates A and B, and the Q output. Use ideal pulses and a 33% duty cycle for the enable E input. The enable pulse

Fig. P8.2 Latch circuit.

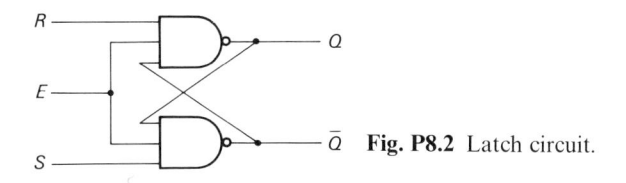

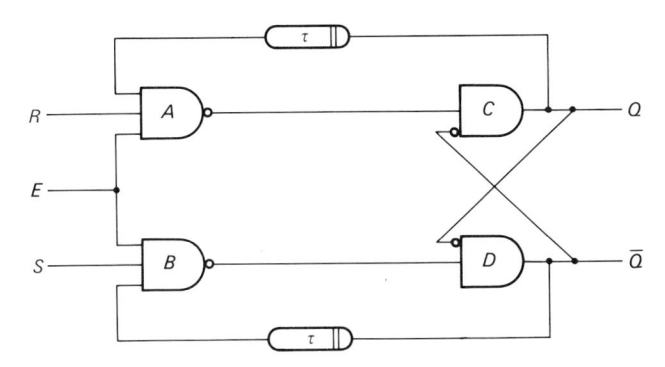

Fig. P8.3 Strobed *RS* latch with outputs coupled back into the inputs.

width is $3t_p$ and $\tau = t_p$. Show all four possible combinations of R and S on the timing diagram.

12. Referring to Fig. P8.3, show that this circuit will not operate properly if the enable pulse is shorter than $2t_p$ or greater than $6t_p$ for $\tau = t_p$. Use ideal pulses.

Section 8.3.2

13. *Addressable Latch.* (a) Referring to a TTL data book, draw the logic symbol for a 9334 (National, Fairchild) eight-bit addressable latch. (b) Explain how this latch works. (c) Explain how it can be used as a 1-of-8 decoder.

14. *Microcomputer Control Flag Circuit.* Referring to Appendix G, study Section G.3.3, which describes the control flag circuit shown on sheet 1 of the schematic drawing (Fig. G.5). (a) Draw the logic diagram for the addressable latches (Problem 13) and the memory timing logic circuit (Fig. G.5). Label all signal lines and give their meaning (see Table G.3). (b) What address inputs to the addressable latch are required for the WRP and WRM signals to go HIGH? (c) Under what input conditions does the output of gates 5C and 8C in the memory timing logic circuit go HIGH?

Section 8.4

15. Draw the timing diagram which will compare the operation of an active HIGH enable D latch and a falling-edge triggered D flip-flop. The timing diagram should contain: one enable (clock) input of 50% duty cycle, one D input, a Q_L output for the latch, and a Q_{FF}

output for the flip-flop. Use ideal pulses. For given D inputs demonstrate on the timing diagram how the Q_L and Q_{FF} can differ and how they can be the same. Include set up times, hold times and delay times on your diagram.

16. Referring to the specifications for the TTL 74LS175 D flip-flop (Fig. 8.14), what is the maximum number of standard 7400 NAND gates the Q output of the latch could drive in its LOW state?

Section 8.5

17. *Clock Stretcher Troubleshooting.* What could cause a stuck-HIGH fault (i.e., a permanent HIGH) at the output of gate 7E in Fig. 8.18?

Section 8.6

18. Explain the function of each block in the generalized flip-flop block diagram of Fig. 8.21(a).
19. Draw the block diagram for the generalized D flip-flop as shown in Fig. 8.21(b). On this diagram show the input and output levels for each flip-flop when $D = H$.

Section 8.6.1

20. Explain how a JK master/slave operates in the toggle mode. Refer to the simplified logic diagram in Fig. 8.22.
21. Explain why the outputs of an RS master/slave flip-flop cannot be connected back into its inputs in certain cases.

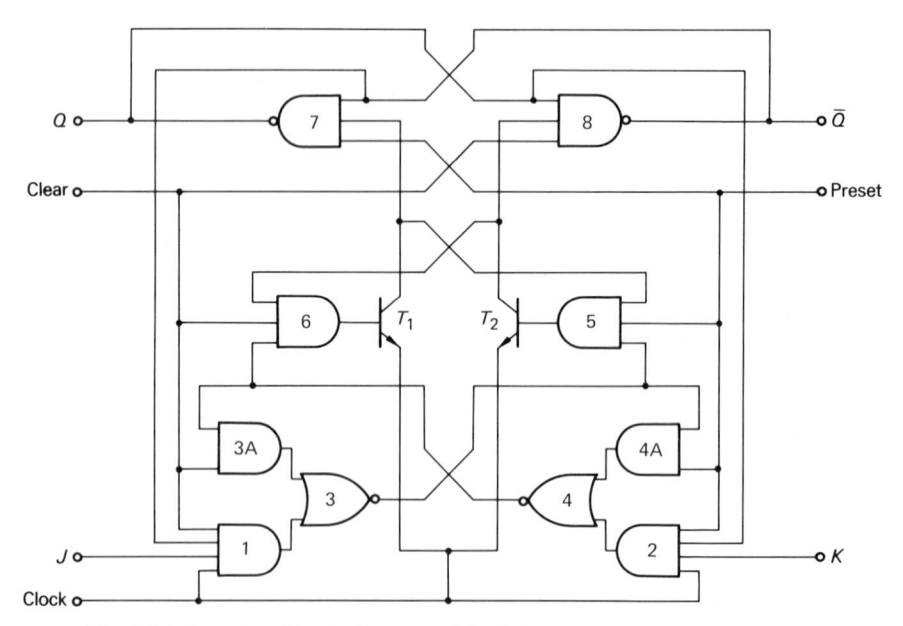

Fig. P8.4 Functional logic diagram of the TTL 7476 master/slave flip-flop.

22. Draw a detailed timing diagram for the simplified JK master/slave flip-flop shown in Fig. 8.22 for the case $J = H$, $K = L$, $Q_n = L$, $\bar{Q}_n = H$. Assume that all gates have the same delay time t_p. Show just one CLK and one $CLK\text{-}1$ pulse, as in Fig. 8.23, with relatively long rise and fall times. Show all other pulses as ideal pulses. Include a waveform for the output of each gate (gates 1 through 8 in Fig. 8.22). Show all delay times.

23. Figure P8.4 shows the functional logic diagram for the TTL 7476 master/slave flip-flop. (a) Redraw this logic diagram in a form similar to the simplified master/slave diagram (Fig. 8.22). Omit all preset and clear signal lines. Identify the following sections of your logic diagram: input logic, temporary memory, inhibit circuit, latch (refer to Fig. 8.21). (b) Make a table which shows the logic levels at the outputs of each gate in Fig. P8.4 after the falling edge of the clock for all possible combinations of the J and K inputs. Omit the preset and clear inputs.

24. Figure P8.5 shows the schematic diagram for the TTL 7476 master/slave flip-flop. Study this circuit and try to correlate the transistors that make up each gate in the logic diagram shown in Problem 23. Make a table that relates transistors T_4, T_9, T_{10}, T_{13}, T_{14}, T_{15}, and T_{16} to the gates labeled 1 through 8 in Fig. P8.4.

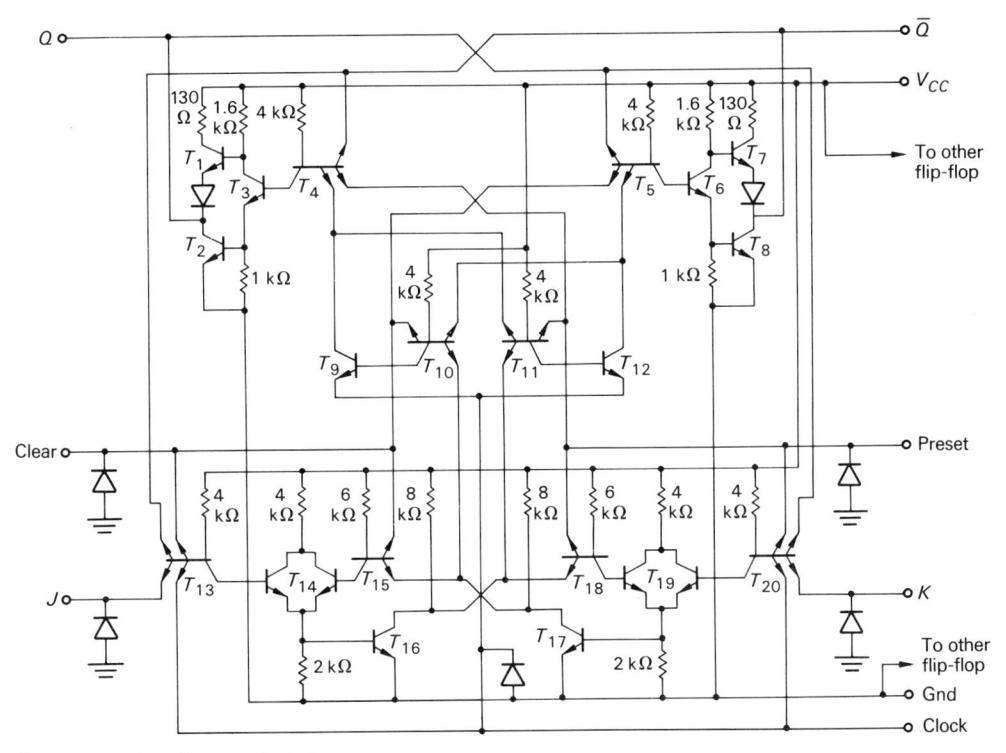

Component values shown are typical

Fig. P8.5 Schematic diagram of the TTL 7476 JK master/slave flip-flop. (Courtesy of Fairchild Camera and Instrument Corporation.)

244   FLIP-FLOPS

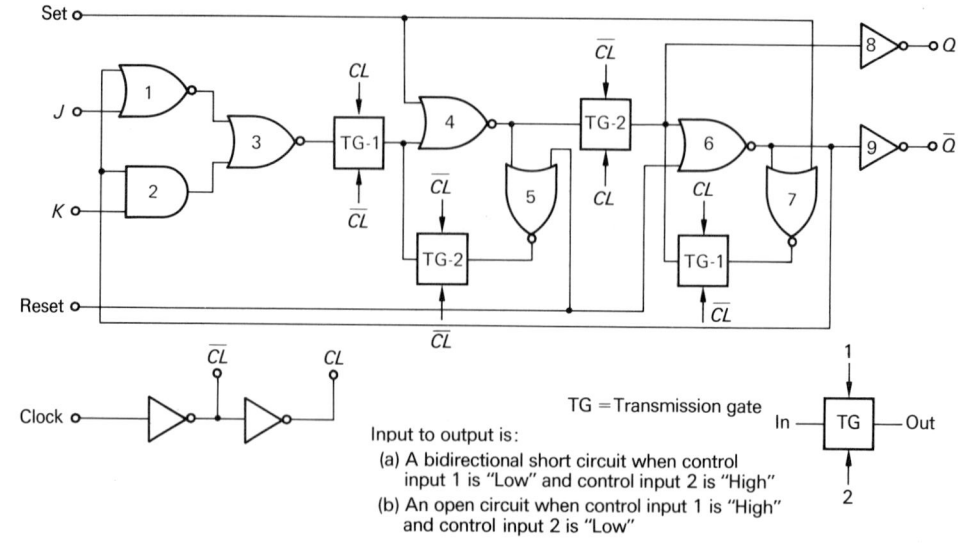

Fig. P8.6 CMOSL 14027 master/slave JK flip-flop. (Courtesy of Motorola Corporation.)

Input to output is:
(a) A bidirectional short circuit when control input 1 is "Low" and control input 2 is "High"
(b) An open circuit when control input 1 is "High" and control input 2 is "Low"

TG = Transmission gate

25. Explain how the CMOSL 14027 master/slave JK flip-flop shown in Fig. P8.6 works by completing a table that shows the outputs of each gate after the active rising edge of a clock pulse. The active HIGH set and reset inputs are kept LOW.

Section 8.6.2

26. Compare the advantages and disadvantages of master/slave and edge triggered JK flip-flops.

Section 8.6.3

27. *Computer Program.* (a) Write a computer program that will plot the following: (1) N ideal clock pulses with period T and 50% duty cycle, (2) arbitrary J and K input waveforms, and (3) the resultant Q and $\bar{Q}$ outputs. The plotting can be done with a line printer or x-y plotter. (b) Run your program for $N = 10$ and $T = 50$ ns, which shows the Q and $\bar{Q}$ outputs for all four combinations of the J and K inputs shown in the JK flip-flop truth table.

28. Draw a timing diagram for $S = H, R = L, Q_n = L$ for one clock pulse in the capacitively coupled RS flip-flop shown in Fig. 8.29.

Section 8.7

29. Show how a race condition can occur in a two-input NOR gate. Use nonideal input pulses.

30. Using a TTL data book calculate the worst-case time difference $\Delta t_{1,2}$ in Fig. 8.31 for two 7476 flip-flops and one 7400 NAND gate.

31. *Computer Program.* (a) Write a computer program that will plot NAND gate and flip-flop logic symbols at any specified location. (b) Have your program plot the logic diagram shown in Fig. 8.31.

REFERENCES

1. Leonard Strauss. *Wave Generation and Shaping.* New York: McGraw-Hill, 1970.
2. *COS/MOS Integrated Circuits Manual.* CMS-271, RCA Corporation, June 1972.
3. *Transmission and Multiplexing of Analog or Digital Signals Utilizing the CD4016A Quad Bilateral Switch.* AN-6601, COS/MOS Data Book, RCA Corporation, 1974.
4. Norman Doyle and Robert Hood. "Wheeling with Electronics." *Progress* (Fairchild), January 1974, Vol. 2, No. 1.
5. Gregory M. Cinque. "Simplified Bidirectional Filtering of Switch Transitions." *Computer Design*, p. 84, August 1973.

9: COUNTERS AND REGISTERS

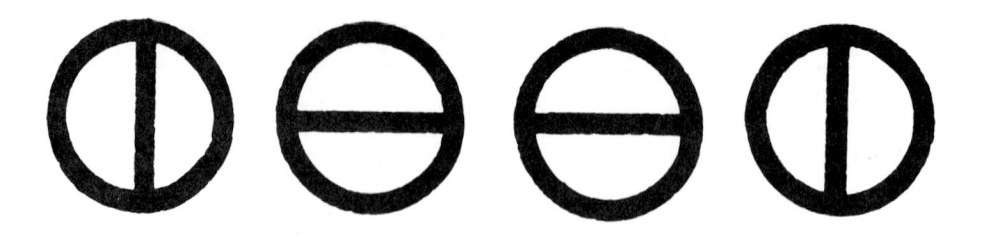

9.1 INTRODUCTION

Counters and registers are sequential circuits that are used in many digital systems. Counters perform tasks such as simple counting, controlling the sequence of several events, and frequency division. Registers are used as temporary storage (memory) devices, as delay devices, and for conversion between serial and parallel forms of data. The pulmonary function analyzer shown in Fig. 9.1 is a medical electronics system used to analyze a patient's breath. It uses counters for frequency division and in its digital readouts, and it uses a 256-bit register as a delay element.

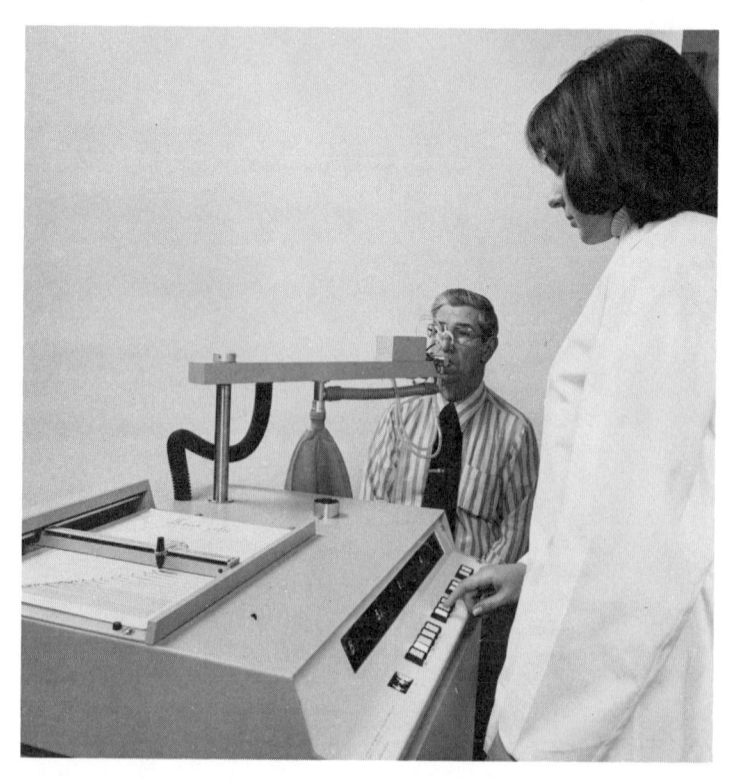

Fig. 9.1 Pulmonary function analyzer. (Courtesy of Hewlett-Packard Company.)

An n-bit counter or register is made from n interconnected flip-flops. A wide variety of MSI counters and registers are available in the TTL, CMOSL, and ECL families. This chapter covers the basics of counters and registers from a flip-flop point of view, rather than a logic symbol (black box) approach, in order to present a clear picture of how these IC devices work.

A **binary counter** is a device that counts the number of input pulses it receives. Its output is in *some* binary code, such as the excess-3 code or the 8421 code. In this book a binary counter specifically means one whose output is in the natural binary code. A counter can be classified according to whether all of its n flip-flops are clocked simultaneously (**synchronous counter**) or whether each flip-flop is clocked by the output of the previous flip-flop (**asynchronous counter**). Let's consider asynchronous counters first.

9.2 ASYNCHRONOUS BINARY COUNTERS

An n-bit counter can have 2^n different output states, since each flip-flop has two states. For example, a four-bit counter can have $2^4 = 16$ output states, called **count states**, that is, it can count up to 16 input pulses. A four-bit counter may, however, be designed so that it has only ten count states. The **modulus** M of a counter is the number of count states that a counter uses. Thus a modulo-10 (mod-10 for short) counter has 10 count states. A mod-16 counter, also called a **hexadecimal counter**, has 16 count states.

Sections 9.2, 9.3, and 9.4 discuss counters with count states that are integer powers of 2, in other words, those whose modulus is $M = 2^n$. All counters with $M = 2^n$ are binary counters, i.e., their output is in the natural binary code. Section 9.5 discusses counters for which $M \neq 2^n$.

9.2.1 Basic Principles

The following example illustrates an application of a four-bit asynchronous counter.

Example 9.1 *Pulmonary Function Analyzer.* The analyzer shown in Fig. 9.1 requires, in its time base generator, an asynchronous counter that can count up to 16 input pulses. Draw the logic diagram for this counter. Include a method for generating a single pulse when the fourth and eighth input pulses have been counted.

Solution. Since 16 pulses are to be counted, four flip-flops will be required to count $2^4 = 16$ pulses. Let's use four falling-edge master/slave JK flip-flops connected as shown in Fig. 9.2. The J and K inputs of each flip-flop are permanently HIGH, so the flip-flops toggle on the falling edge of their clock inputs. The clock input of FF-A is connected to the external clock (CLK). Each succeeding flip-flop is clocked by the Q output of the previous flip-flop. A common $\overline{CLEAR}$ input can be used simultaneously to reset the Q outputs of each flip-flop.

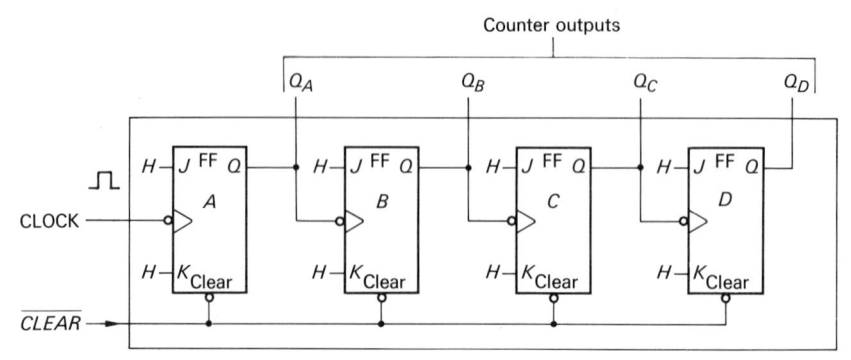

Fig. 9.2 Asynchronous four-bit binary counter (timing diagram shown in Fig. 9.3).

Now let's develop the timing diagram for this four-bit counter. Initially all the flip-flops are cleared, as shown in Fig. 9.3(a). The output of FF-A changes state one delay time after the falling edge of each clock pulse. The delay times are exaggerated to illustrate their effect. Since the output of FF-A is the clock input to FF-B, the output of FF-B toggles one delay time after the falling edge of Q_A. Similarly, FF-C toggles on the falling edge of Q_B, and FF-D toggles on the falling edge of Q_C.

After clock pulse 1 takes effect the outputs of flip-flops D, C, B, and A, respectively, are 0001, after clock pulse 2 they are 0010, etc., as summarized in the truth table in Fig. 9.3(b). Note that the Q outputs that represent a given clock pulse occur *after* the falling edge of that clock pulse. The decimal equivalent of each set of outputs is also shown in Fig. 9.3(a). Thus we see that this counter has 16 different output (count) states. At the end of the 15th count (16 input pulses) the counter begins counting over again from 0 through 15 without being externally reset. Figure 9.3(c) shows the logic symbol for the counter in this example.

The last part of this example is to generate a single-pulse output corresponding to the fourth and eighth pulses. The fourth pulse, like all other pulses, has a unique output; it is $Q_A = 0$, $Q_B = 0$, $Q_C = 1$, and $Q_D = 0$. The four-input AND gate shown in Fig. 9.3(d) will decode the fourth pulse and produce a pulse output as required for this example. Decoding for the eighth pulse is also shown. ∎

9.2.2 Characteristics of Asynchronous Counters

Note in Fig. 9.3(a) that the output of each stage in the counter occurs one delay time after the output of the previous stage changes. For example, for clock pulse 8, Q_A changes one delay time after the falling edge of clock pulse 8. Flip-flops B, C, and D change $2t_P$, $3t_P$, and $4t_P$, respectively, after clock pulse 8. Thus the outputs appear to ripple through the counter, and for this reason an asynchronous counter is also called a **ripple counter**.

9.2 ASYNCHRONOUS BINARY COUNTERS 249

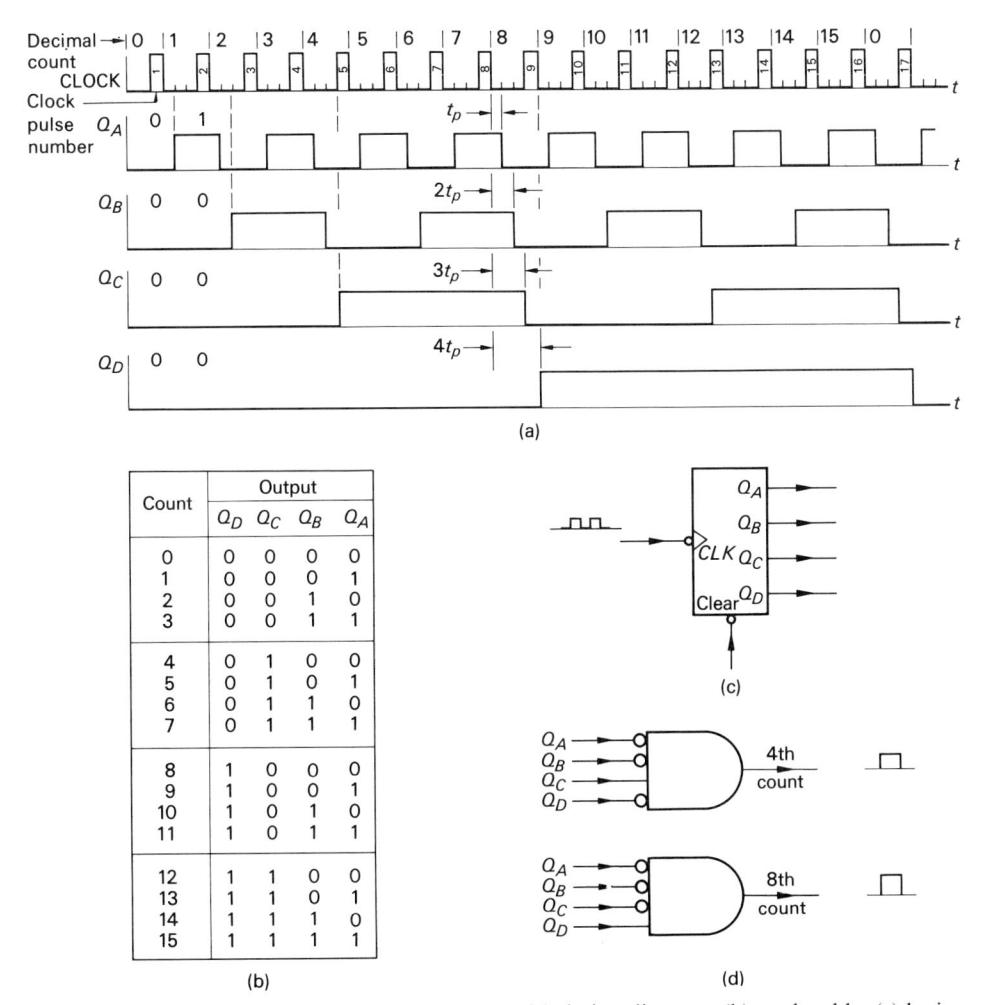

Fig. 9.3 Asynchronous four-bit binary counter: (a) timing diagram; (b) truth table; (c) logic symbol for Fig. 9.2; (d) decoding of the fourth and eighth count.

From Fig. 9.3(a) we see that the frequency of FF-A is one-half the clock frequency. The frequency of FF-B is in turn one-half the frequency of FF-A, or one-fourth the clock frequency. Similarly, flip-flops C and D produce frequencies $(\frac{1}{2})(\frac{1}{2})(\frac{1}{2}) = \frac{1}{8}$ and $(\frac{1}{2})(\frac{1}{2})(\frac{1}{2})(\frac{1}{2}) = \frac{1}{16}$, respectively, of the clock frequency. Hence, the square wave (50% duty cycle) outputs of four-bit ripple counter can also be used as a divide-by-2, 4, 8, and 16 frequency divider.

Usually when a counter is used as a frequency divider, only one output frequency is required, so only one counter output is used. Therefore, when we refer to a **frequency divider** or a **divide-by-N counter**, we mean a counter with only one primary output, which is some function of the input (clock) frequency. In all other cases a counter has multiple outputs, which are interpreted as an indication of the number of input pulses received by the counter.

The frequency at the output of the *n*th stage of a binary divide-by-*N* counter (asynchronous or synchronous) is

$$f_N = \frac{1}{2^n} f_{CLK}, \quad n = 0, 1, 2, 3, \ldots,$$

where *n* is the number of stages (flip-flops) and f_{CLK} is the clock frequency externally applied to the first stage of the counter. When $M = 2^n$, the duty cycle at all *n* outputs is 50%. When $M \neq 2^n$ the duty cycle is generally not 50%.

The two main disadvantages of ripple counters are their relatively low speed and the generation of decoding spikes if the outputs are decoded. Let's discuss speed first.

Speed Asynchronous counters are relatively slow because the *n*th output cannot change until *n* delay times after a clock pulse. Thus in Fig. 9.3(a) if the average delay time per flip-flop is 20 ns, then for clock pulse 8 the output Q_D will not change until 4×20 ns = 80 ns later. This is the worst case, because for many counts only the outputs of the first two or three flip-flops change state.

In many applications, high speed is of no consequence so the speed limitation of ripple counters is not important. For example, counting parts on a conveyor belt is a low-speed process.

Decoding The second disadvantage of ripple counters is the generation of false outputs, called **decoding spikes**, if the counter outputs are decoded. It is frequently desirable to decode the outputs of a counter so that each count state will have a unique output. Let's develop the decoding for the two-bit ripple counter shown in Fig. 9.4(a). The timing diagram for this counter is given in Fig. 9.4(b). The counter outputs for the 0 count are $Q_A = 0$ and $Q_B = 0$. Hence, if $\bar{Q}_A$ and $\bar{Q}_B$ are fed into AND gate G_1 in Fig. 9.4(a), its output will be HIGH for the 0 count. The remaining count states are decoded in a similar way. (Ignore the strobe input to the gates temporarily.)

The 0 output of the decoder becomes HIGH whenever $Q_A = Q_B = L$, which occurs not only after clock pulse 1 but also after clock pulse 2, thus producing a false output spike. Similarly, the 2 output of the decoder becomes HIGH when $A = L$ and $B = H$, which occurs after clock pulse 2 and also, erroneously, after clock pulse 4. These false outputs can result in incorrect operation of subsequent devices such as flip-flops, counters, and registers. Spikes in the outputs of the decoding gate can be avoided by using a strobe (enable) pulse on one input of

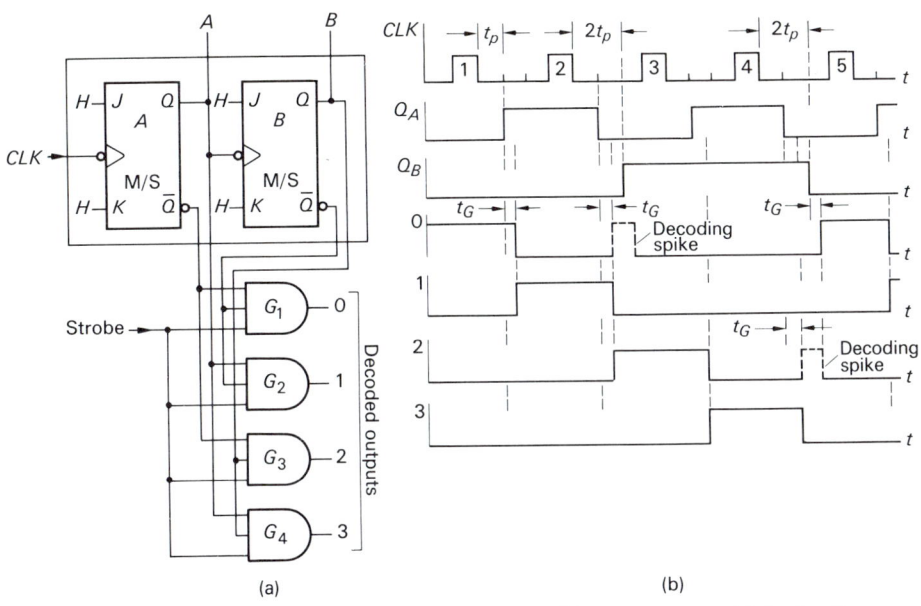

Fig. 9.4 Decoding spikes in a two-bit asynchronous binary counter: (a) logic diagram; (b) timing diagram.

each decoding gate, as shown in Fig. 9.4(a). The strobe pulse is applied after all the flip-flop outputs become stable.

In order to decode all outputs of an n-bit counter a total of 2^n decoding gates are required with n inputs for each gate (not including the strobe input, if one is used). For example, full decoding of a four-bit counter requires 16 four-input AND gates.

The next section introduces synchronous counters.

9.3 SYNCHRONOUS BINARY COUNTERS

In general, synchronous systems are easier to design, troubleshoot, and service, and they are more reliable than asynchronous systems. Therefore, in most of the remainder of our discussion of counters we shall stress synchronous counters.

Figure 9.5 shows a four-bit synchronous binary counter made from four rising-edge JK flip-flops which are converted to rising-edge flip-flops by inverter U_1. Note that the clock inputs of each flip-flop are connected to a single external clock input (CLK), so each flip-flop is clocked at the same time, (i.e., synchronously). In a synchronous counter, the J and K inputs are not all permanently HIGH but are controlled with logic gates, such as gates U_4 and U_7 in Fig. 9.5.

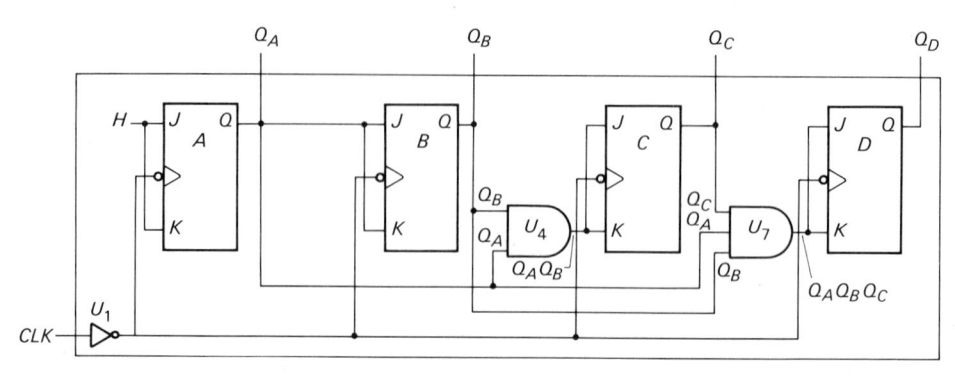

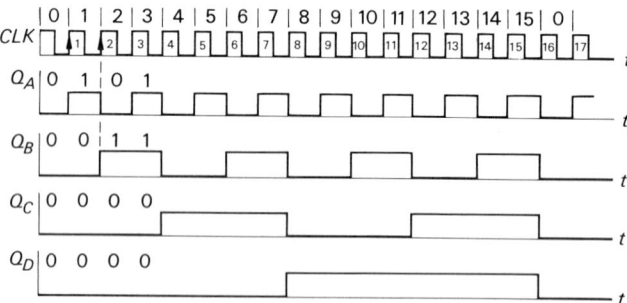

Fig. 9.5 Synchronous four-bit binary counter.

Let's see how the J and K inputs to each stage can be determined from the timing diagram in Fig. 9.5 (delay times have been omitted for simplicity). The timing diagram (Fig. 9.5) shows that FF-A toggles on the rising edge (due to inverter U_1) of each clock pulse. To make FF-A toggle the J and K inputs are connected to a permanent HIGH, that is, $J_A = H$ and $K_A = H$. FF-B toggles when Q_A is HIGH, that is, $J_B = K_B = Q_A$. FF-C toggles only when $Q_A = H$ and $Q_B = H$. These *two* conditions are needed because each flip-flop can toggle on the rising edge of any clock pulse. Hence $J_C = K_C = Q_A \cdot Q_B$. Finally, FF-$D$ toggles only when $Q_A = H$ and $Q_B = H$ and $Q_C = H$; in other words, $J_D = K_D = Q_A \cdot Q_B \cdot Q_C$.

Let's compare the synchronous counter with the asynchronous counter. The counting rate (clock frequency) in a synchronous counter is limited by the delay time for only *one* flip-flop, plus the delay time for the control gates (such as U_4 and U_7). Synchronous counters, when decoded, generally do not have decoding spikes. Decoding spikes may occur if the propagation delays vary between stages. However, the width of these spikes will not exceed the difference in delay times between the slowest and the fastest flip-flop. Thus a synchronous counter can be faster and easier to decode than an asynchronous counter. Finally, most synchronous counters are triggered by a rising-edge clock input, which makes them compatible with most MSI devices. Most ripple counters use a falling-edge clock

9.4 SYNCHRONOUS AND ASYNCHRONOUS UP/DOWN-COUNTERS

All the counters discussed so far were **up-counters**, i.e., they counted from a smaller number to a larger number. A **down-counter** counts from a larger number to a smaller number. For example, a three-bit down-counter would count 7, 6, 5, 4, 3, 2, 1, 0, 7, 6, etc. Down-counters are useful in applications such as counting the number of items remaining. Let's investigate the basic principles of down-counters.

9.4.1 Basic Principles

Asynchronous Down-Counters Figure 9.6 shows the logic diagram and timing diagram for a three-bit up/down ripple binary counter. If the count-up C_U input is made HIGH and the count-down input C_D is made LOW, then the AND

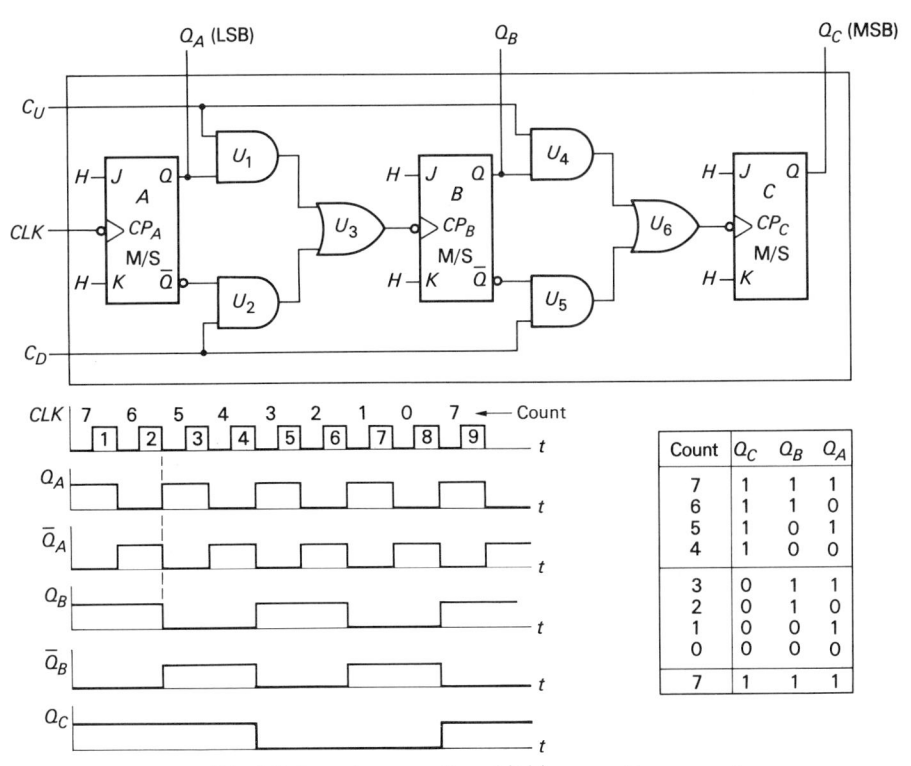

Fig. 9.6 Asynchronous three-bit binary up/down-counter.

Count	Q_C	Q_B	Q_A
7	1	1	1
6	1	1	0
5	1	0	1
4	1	0	0
3	0	1	1
2	0	1	0
1	0	0	1
0	0	0	0
7	1	1	1

gates U_1 and U_4 are enabled. Then the Q output of each flip-flop becomes the clock input of the next flip-flop, so the counter counts up from 000 through 111 repeatedly. The timing diagram and truth table for this up-counter are identical to the one shown in Fig. 9.3, for the first eight counts.

When $C_U = L$ and $C_D = H$, then gates U_1 and U_4 are disabled, and gates U_2 and U_5 are enabled, so the $\bar{Q}_A$ is connected to CP_B, and $\bar{Q}_B$ is connected to CP_C. With $C_D = H$, $C_U = L$, and the Q outputs set, the counter output is $Q_A = Q_B = Q_C = 1$ as shown in the timing diagram in Fig. 9.6 (delay times have been omitted for simplicity). After the falling edge of each clock pulse, FF-A toggles, so the $\bar{Q}_A$ waveform shown in Fig. 9.6 results. FF-B toggles on the falling edge of the $\bar{Q}_A$ waveform. The $\bar{Q}_B$ waveform is obtained by complementing the Q_B waveform as shown. The Q outputs of this down-counter produce the down-count 111, 110, 101, 110, 011, 010, 001, 000, 111, etc., as shown in the count table.

Output decoding for asynchronous down-counters is done in the same way as for up-counters, as shown in Fig. 9.7. Strobe inputs to the decoding gates are not shown.

Most IC up/down-counters are synchronous counters, as described in the next section.

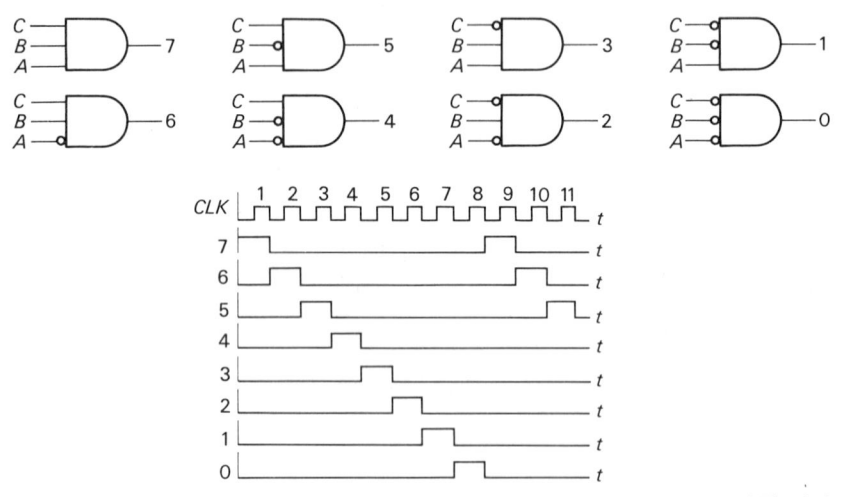

Fig. 9.7 Decoding gates and waveforms for the three-bit up/down-counter of Fig. 9.6.

Synchronous Down-Counters Figure 9.8 shows the logic diagram for a four-bit synchronous up/down binary counter. The letter T at the clock input of each flip-flop denotes that these are toggle flip-flops, that is, the J and K inputs are not shown, but are all connected to a permanent HIGH. Thus FF-A will toggle on the *rising* edge of the C_U (or C_D) clock input. Flip-flop B toggles on the rising edge of the C_U (or C_D) input when $Q_A = H$ (or $\bar{Q}_A = H$), etc.

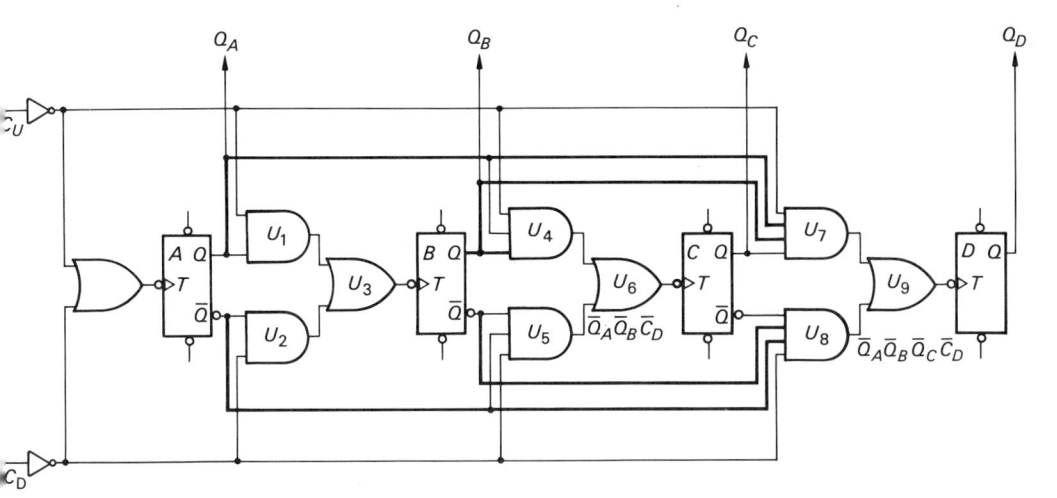

9.4 SYNCHRONOUS AND ASYNCHRONOUS UP/DOWN-COUNTERS   255

Fig. 9.8 Synchronous four-bit binary up/down-counter.

Figure 9.8 is similar to the up/down-counter in Fig. 9.6 in that it uses AOI gates between stages. However, the synchronous counter is clocked from the C_U or C_D inputs. To count up, the clock signal is applied to the C_U input with C_D held HIGH. To count down, the clock signal is applied to the C_D input with $C_U = H$. The heavy lines and gates U_4 and U_7 in Fig. 9.8 correspond to those in the synchronous counter in Fig. 9.5.

9.4.2 Applications

Our first practical application of a down-counter involves the use of a stepper motor (Fig. 9.9a), so before proceeding let's look at how this device operates.

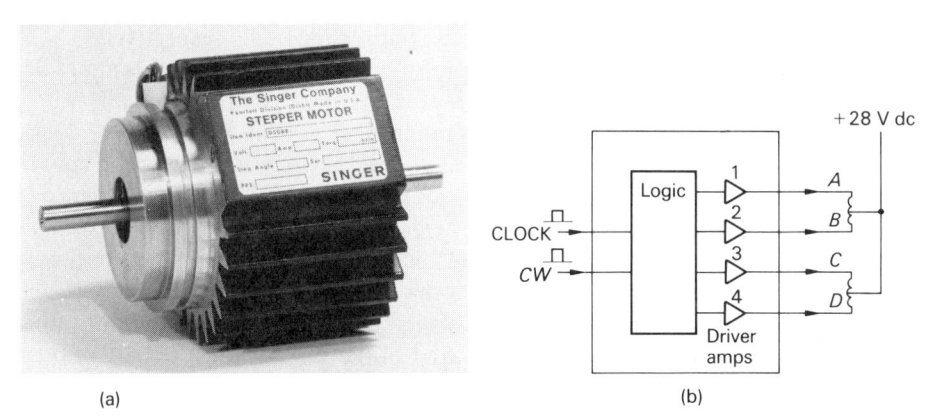

(a)   (b)

Fig. 9.9 Stepper motor: (a) photograph (courtesy of the Singer Company); (b) block diagram of logic and drivers.

An electric **stepper motor** is an electric motor whose shaft rotates a specified amount for each input pulse it receives [1, 2]. Stepper motors are useful for producing precise rotational or linear motion such as required for X-Y plotters and numerical control (NC) machines.

There are two inputs to the stepper motor logic (Fig. 9.9b). For each clock input pulse, the motor shaft rotates clockwise (CW) or counterclockwise (CCW) a specified fraction of a revolution called the **stepping angle** θ_s. For example, if $\theta_s = 15°$, then 10 pulses in the clock input, with the CW input HIGH, will cause the shaft to rotate $10 \times 15° = 150°$ clockwise. Similarly, each pulse in the clock input, with the CW input LOW, causes the shaft to rotate counterclockwise through an angle θ_s. The rotational velocity of the shaft is proportional to the frequency of the clock pulses.

The logic circuit shown in Fig. 9.9(b) converts each input pulse into four pulses that have the required phase relation between them. These four pulses are then amplified by the four driver amplifiers (1, 2, 3, and 4) and sent to the four motor windings A, B, C, and D. The windings are center tapped and connected to the motor drive voltage $+28$ V dc. When the output of amplifier 1 goes LOW, winding A is energized, causing rotation of the motor shaft.

Stepper motors are available with stepping angles of 0.1 to 120 degrees/step; the most common angles are $1.8°$, $15°$, $22\frac{1}{2}°$, $45°$, and $90°$. Maximum stepping rates range from 2 to 4000 steps/sec.

Now let's see how a stepper motor and a down-counter work together in a numerical control example.

Example 9.2 *Numerical Controlled Drill Press.* A small electric drill press is used to drill holes in a printed circuit board (PCB) with a precision of 0.001 in. Develop the control logic for a numerical control system that will control the movable table along one axis, using a stepper motor with a stepping angle of $15°$ and maximum stepping rate of 330 steps/sec.

Solution. Figure 9.10(a) shows the drill press and PCB mounted on a movable table. The table moves right or left along the x-axis depending on whether the x-axis stepper motor rotates clockwise or counterclockwise, respectively.

Figure 9.10(b) shows a block diagram for the NC system. For this simplified system, data for specifying the x position of the table is entered manually with a thumbwheel switch, via the manual data input block (Fig. 9.10b).

The manual control determines when the table will move. The cycle control block determines in what sequence the various operations will be performed in more complex systems, which contain x-, y-, and z-axis motors. The indexer contains a pulse generator, down-counter, and logic/driver for driving the x-axis motor.

The movable work table is geared so that one step ($15°$) produces a 0.001-in. movement in the table. Suppose the table is to be moved 0.01 in. to the right; then 10 pulses must be applied to the pulse input of the stepper logic circuit with the CW input HIGH. One way of supplying the correct number of pulses to the stepper motor is to gate a pulse generator off at the end of 10 pulses, using a down-counter preset to a count to 10. Figure 9.11 shows the logic diagram for the NC system.

9.4 SYNCHRONOUS AND ASYNCHRONOUS UP/DOWN-COUNTERS

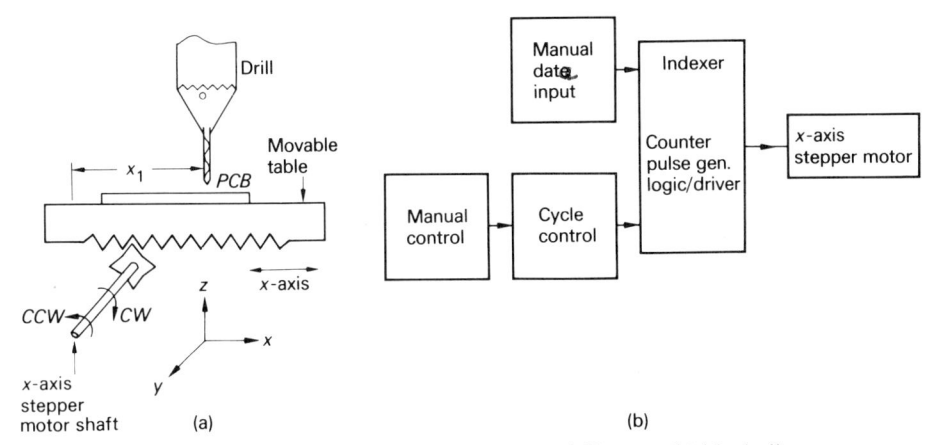

Fig. 9.10 Numerical control of a drill press: (a) drill press; (b) block diagram.

The system works as follows. The *CW* switch is positioned HIGH (open) and the thumbwheel switch is set to 10 (10 steps). When the load/count switch is momentarily depressed (LOW) the initial count of the counter is set to $10_{10} = 1010$. This is called loading the counter. When the count switch is released (HIGH), it causes the down-counter to start counting down from 10 to 0. Then output of OR gate U_1 is HIGH for all counter outputs except 0. This enables gate U_2, so pulses are fed to the clock input of the logic driver, causing the stepper motor to

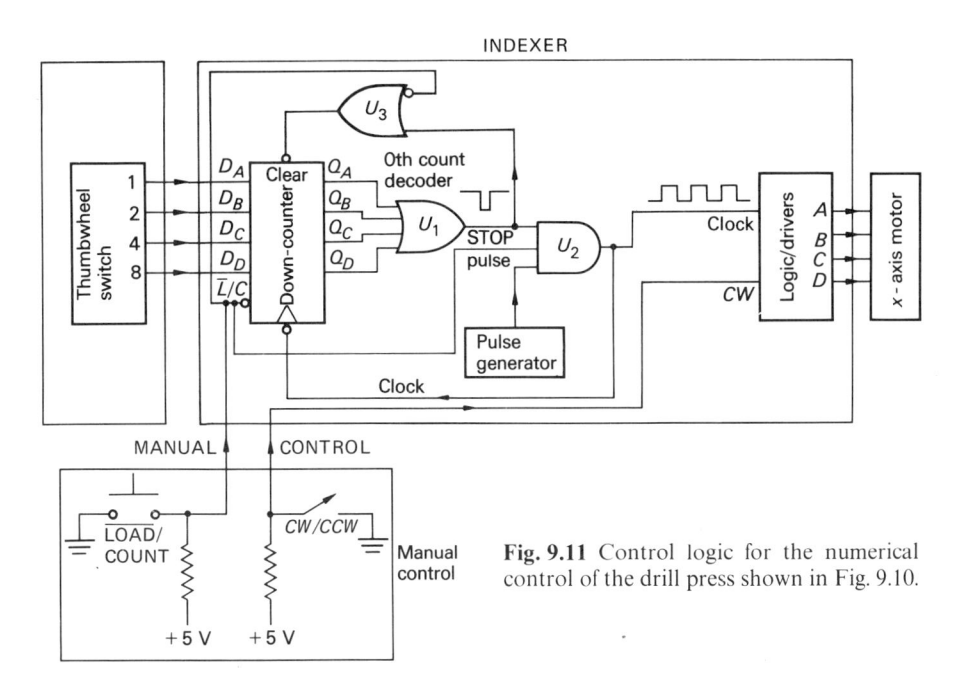

Fig. 9.11 Control logic for the numerical control of the drill press shown in Fig. 9.10.

rotate 15° clockwise for each pulse. At the zeroth count the output of OR gate U_1 goes low, so no more pulses are fed to the stepper motor and the counter is stopped from counting by the STOP pulse. The STOP pulse holds the counter at a count of zero until the LOAD switch is depressed again. The purpose of OR gate U_3 is to disable the clear input during loading, since the clear input to most counters overrides the preset inputs. This is an example of an **open loop** control system, since there is no position information fed back from the movable table to the controller. ∎

9.5 SYNCHRONOUS AND ASYNCHRONOUS MODULO COUNTERS

The counters we have discussed so far are able to count only to integer powers of 2, that is, $M = 2^1, 2^2, 2^3, 2^4$, etc. Frequently it is necessary to be able to count to numbers that are not powers of 2; that is, M must be 3, 7, 10, 13, etc. We are now ready to consider how this can be accomplished. A technique for making modulo counters is important because IC counters are available with only a few moduli such as 2, 4, 5, 6, 10, 12, and 16.

First, let's investigate the basic principles used in all synchronous and asynchronous modulo counters.

9.5.1 Basic Principles

Suppose a mod-7 counter is needed in a certain application. A mod-7 counter has a total of $M = 7$ count states. Three flip-flops will be required, because two flip-flops can provide only $2^2 = 4$ count states. However three flip-flops have $2^3 = 8$ count states, which is one too many. Therefore, one count state must be eliminated. The next example shows the possible ascending sequential counts that can be chosen for a mod-7 counter.

Example 9.3 List the possible ascending sequential counts that could be used in a mod-7 counter.

Solution. Since three flip-flops are required, there are eight possible count states, 0, 1, 2, 3, 4, 5, 6, and 7. Any one of these states can be omitted, as shown below.

Possible sequences	Skip count ...	... and use counts
1	0	1, 2, 3, 4, 5, 6, 7
2	1	0, 2, 3, 4, 5, 6, 7
3	2	0, 1, 3, 4, 5, 6, 7
4	3	0, 1, 2, 4, 5, 6, 7
5	4	0, 1, 2, 3, 5, 6, 7
6	5	0, 1, 2, 3, 4, 6, 7
7	6	0, 1, 2, 3, 4, 5, 7
8	7	0, 1, 2, 3, 4, 5, 6

9.5 SYNCHRONOUS AND ASYNCHRONOUS MODULO COUNTERS

Thus there are eight ways of using these sequences to implement a mod-7 counter using three flip-flops. The count sequence is generally chosen to minimize the number of devices used. The choice also depends on the output binary code and waveform desired. Note that only when count 7 is skipped does the natural binary code for the output result. For frequency division only one pulse is required for each seven clock pulses, so the output code is of no consequence unless a 50% duty cycle is required. ∎

A mod-6 counter made from three flip-flops has 28 possible ascending sequential counts. In the next section three methods will be discussed for making asynchronous or synchronous modulo counters: (1) feedback, (2) reset, (3) preset.

*9.5.2 Developing Modulo Counters

Feedback Method The feedback method eliminates the undesired count states by feeding some flip-flop outputs back into the *J*, *K*, or toggle inputs of other flip-flops in the counter. The feedback method for making a mod-*M* counter applies to asynchronous or synchronous counters designed from individual flip-flops. It does not apply to IC counters because their *J(S)* and *K(R)* flip-flop inputs are not available at the IC pins. The feedback method applies to binary counters as well as to those whose output is not in the natural binary code. Let's illustrate the feedback method by building a synchronous mod-5 counter from flip-flops.

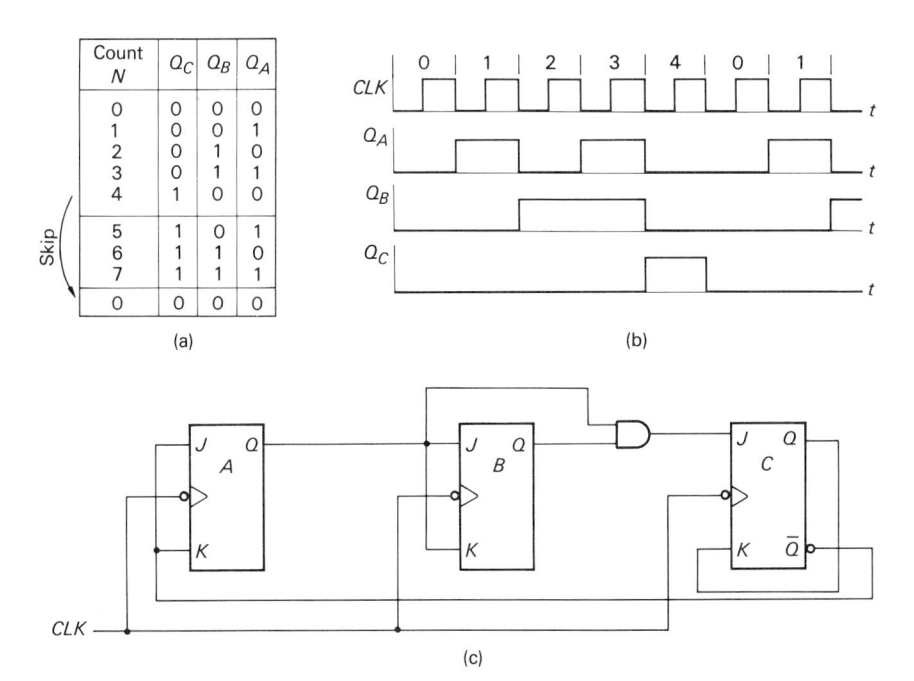

Fig. 9.12 Mod-5 synchronous counter: (a) count table; (b) timing diagram; (c) logic diagram.

260   COUNTERS AND REGISTERS   9.5

A synchronous mod-5 counter will require three flip-flops clocked by a common clock. The count table for a mod-5 binary counter (i.e., one with a natural binary count) is shown in Fig. 9.12(a). The timing diagram shown in Fig. 9.12(b) is drawn from the count table. Observe from the timing diagram the following facts. FF-*A* toggles with each clock pulse if $Q_C = 0$, so the inputs to FF-*A* are $J_A = K_A = \bar{Q}_C$. FF-*B* toggles with the output of FF-*A*, so $J_B = K_B = Q_A$. FF-*C* toggles only when Q_A and Q_B are HIGH, and it stays reset otherwise, so $J_C = Q_A \cdot Q_B$ and $K_C = \bar{Q}_C$, as shown in Fig. 9.12(c).

A widely used form of mod-5 binary counter is the one shown in Fig. 9.13. It counts in natural binary as shown by the count table. Note that flip-flops *A* and *C* are clocked synchronously, while flip-flop *B* is clocked asynchronously by Q_A. This is called a **semisynchronous counter**. It is frequently easier to implement than a pure synchronous or asynchronous counter.

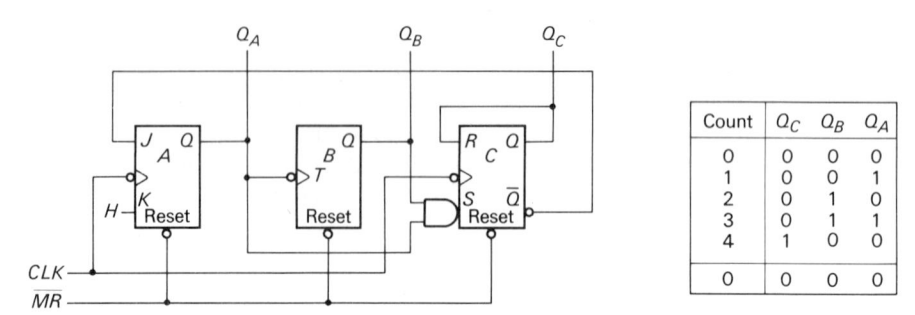

Fig. 9.13 Semisynchronous mod-5 binary counter and count table.

Count	Q_C	Q_B	Q_A
0	0	0	0
1	0	0	1
2	0	1	0
3	0	1	1
4	1	0	0
0	0	0	0

Figure 9.14(a) shows a mod-10 binary counter made from a mod-2 counter (FF-*A*) connected to a semisynchronous mod-5 counter (FF-*B*, *C*, and *D*). This is a mod-10 counter because FF-*A* has two count states for each mod-5 count state,

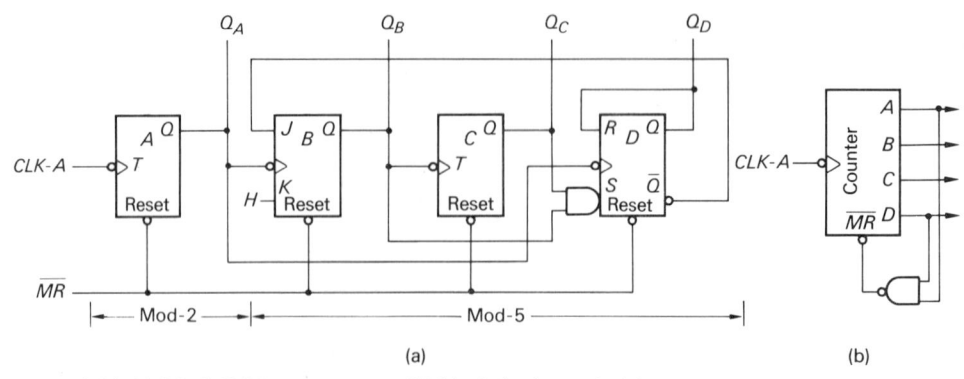

Fig. 9.14 (a) Mod-10 binary counter; (b) block logic symbol for a mod-9 counter using the reset method.

or ten in total. Thus a mod-2 counter connected before a mod-5 counter, denoted by mod-2 × mod-5, generates a mod-10 counter. In general,

$$\text{mod-}M_1 \times \text{mod-}M_2 = \text{mod-}(M_1 \times M_2)$$

and

$$\text{mod-}M_2 \times \text{mod-}M_1 = \text{mod-}(M_1 \times M_2).$$

Reset Method [3]   The reset method applies to asynchronous or synchronous counters, designed from flip-flops or IC counters, that have a master reset or separate reset (clear) inputs. To make a mod-M counter by the reset method, each flip-flop in the counter which has a 1 output on the $(M + 1)$st count (in the normal count sequence of the counter) is connected to an input of a NAND gate. The output of this NAND gate is fed into the active LOW master reset input of the counter. The initial count state, whether it be 0 or not, is included in determining $M + 1$. It is assumed that the reset count state (i.e., all zeros) is one of the count states in the normal sequence of counts for the counter. Let N represent the output of the counter on the $(M + 1)$st count.

First let's apply the reset method to a binary counter. Consider converting the mod-10 counter in Fig. 9.14(a) to a mod-9 counter. This counter has natural binary outputs 0, 1, 2, 3, 4, 5, 6, 7, 8, 9, 0, 1, etc. To convert it to a mod-9 counter it must be reset on the $M + 1 = $ 10th count (beginning with the initial count state of 0) when its outputs ($Q_D Q_C Q_B Q_A$) are $N = 9 = 1001$. Since Q_A and Q_D are HIGH for $N = 9$, a NAND gate with inputs Q_A and Q_D is connected to the active LOW master reset input $\overline{MR}$ (Fig. 9.14b). The outputs of this mod-9 counter will be in the 8421 BCD code.

Some counters have internal logic gates on their reset or preset inputs, thus eliminating or minimizing the need for external reset gates. Figure 9.15 gives the logic diagram for the TTL 7490, 7492, and 7493 asynchronous counters. The 7490 is a mod-2 and a mod-5 counter, similar to Fig. 9.14(a), except that it has built-in reset-to-0 ($R_{0(1)}$ and $R_{0(2)}$) and preset-to-9 ($R_{9(1)}$ and $R_{9(2)}$) asynchronous inputs. The 7492 is a mod-2 and a mod-6 counter. The 7493 is a mod-2 and a mod-8 counter. Connecting the Q_A output to input B of the 7490, 7492, or 7493 results in a mod-10 (2 × 5), mod-12 (2 × 6), or a mod-16 (2 × 8) counter, respectively. A mod-24 counter can be made by adding an external flip-flop to the 7492 (mod-2 × mod-2 × mod-6 = mod-24). Other moduli can also be obtained using the reset method with these counters.

When the number of count states in a counter is shortened by the reset method voltage spikes may occur on some outputs. For example, consider using the 7490 to make a mod-6 counter. The counter must be reset on the $M + 1 = $ 7th count when its outputs are $N = 6 = 0110$, as shown in Fig. 9.16(a). Hence Q_B and Q_C are connected to the reset inputs ($R_{0(1)}$ and $R_{0(2)}$) as shown in Fig. 9.16(b). If the counter was not being reset, the Q_B output would go HIGH after the falling edge of clock pulse 6 and Q_C would stay HIGH (Fig. 9.16a, c). However, as soon as Q_B and Q_C go HIGH, the output of the reset NAND gate (R_0) goes LOW. This

'90A, 'L90

'92A

'93A, 'L93

RESET/COUNT FUNCTION TABLE

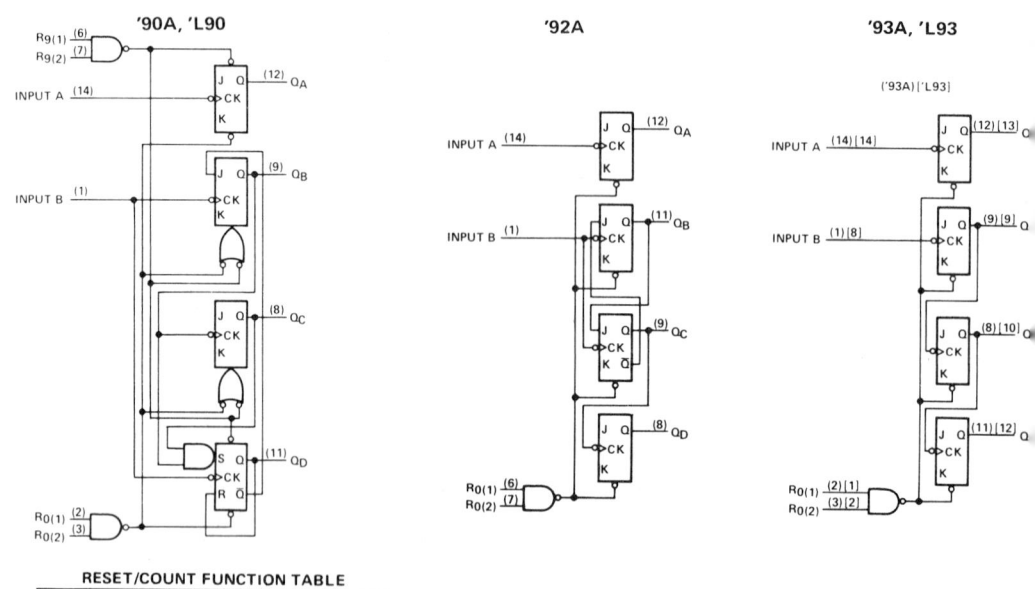

RESET INPUTS				OUTPUT			
$R_{0(1)}$	$R_{0(2)}$	$R_{9(1)}$	$R_{9(2)}$	Q_D	Q_C	Q_B	Q_A
H	H	L	X	L	L	L	L
H	H	X	L	L	L	L	L
X	X	H	H	H	L	L	H
X	L	X	L	COUNT			
L	X	L	X	COUNT			
L	X	X	L	COUNT			
X	L	L	X	COUNT			

'92A, '93A, 'L93 RESET/COUNT FUNCTION TABLE

RESET INPUTS		OUTPUT			
$R_{0(1)}$	$R_{0(2)}$	Q_D	Q_C	Q_B	Q_A
H	H	L	L	L	L
L	X	COUNT			
X	L	COUNT			

Fig. 9.15 Functional logic diagram and reset/count table for the TTL 7490, 7492, and 7493 counters. (Courtesy of Texas Instruments Incorporated.)

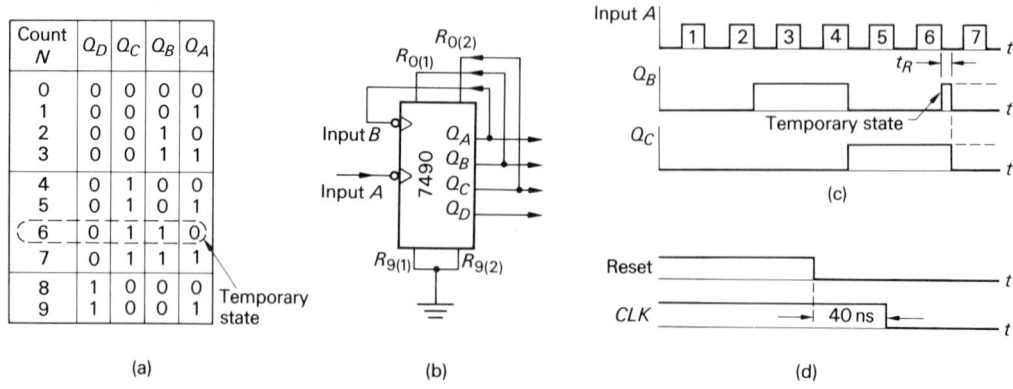

Count N	Q_D	Q_C	Q_B	Q_A
0	0	0	0	0
1	0	0	0	1
2	0	0	1	0
3	0	0	1	1
4	0	1	0	0
5	0	1	0	1
6	0	1	1	0
7	0	1	1	1
8	1	0	0	0
9	1	0	0	1

Temporary state

(a)   (b)   (c)   (d)

Fig. 9.16 Mod-6 counter using the reset method for a TTL 7490 mod-10 counter.

9.5 SYNCHRONOUS AND ASYNCHRONOUS MODULO COUNTERS

causes the Q_B and Q_C outputs to reset, one reset-to-output delay time t_R later, thus generating a spike (Fig. 9.16c). The maximum reset-to-output delay time for the 7490, 7492, and 7493 is 40 ns. Hence the next clock pulse after a reset should not go LOW less than 40 ns after the reset input goes LOW (Fig. 9.16d).

To use the 7490 for a mod-7 counter would require a three-input NAND gate since when $N = 7$, $Q_A = Q_B = Q_C = 1$. However a mod-7 counter can be made from a 7490 without using any external gates by connecting Q_B to $R_{9(1)}$ reset input and Q_C to the $R_{9(2)}$ reset input, as shown in the problems. When the 7490 is connected as a mod-2 × mod-5 counter it is a binary counter, but does not have a 50% duty cycle. Figure 9.17(a) shows the 7490 connected as a mod-5 × mod-2 counter with a 50% duty cycle, which is useful for frequency division by 10. The output code in this case is not the natural binary code, but a code called the **biquinary (2–5) code** shown in Fig. 9.17(a). Note that the MSB is at the Q_A output.

A table showing the connections for a 7490 for moduli 2 though 10 is developed in the problems.

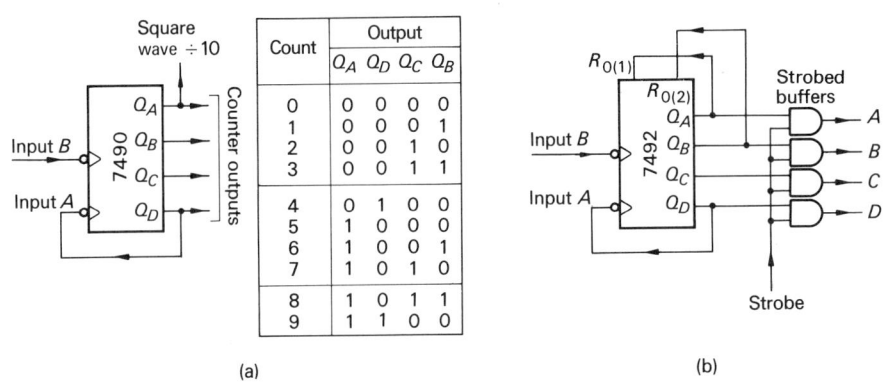

Fig. 9.17 Reset counters: (a) divide-by-10 with 50% duty cycle and biquinary coded outputs; (b) mod-7 with strobed output buffers.

The reset method also applies to counters whose output is not in the natural binary code. For example, when the 7492 is connected as a mod-6 × mod-2 counter its equivalent decimal output count states are 0, 1, 2, 4, 5, 6, 8, 9, 10, 12, 13, 14, 0, 1, etc. (see the problems). As a mod-6 × mod-2 counter the 7492 flip-flops are connected in the order Q_A, Q_D, Q_C, Q_B, where Q_B is the LSB and Q_A is the MSB. The output for the $M + 1 = $ 8th count is $N = 9 = 1001$ $(Q_A Q_D Q_C Q_B)$. Hence the Q_A and Q_B outputs are connected to the 7492 reset inputs as shown in Fig. 9.17(b).

The reset method is quite simple but can be unreliable. If the counter outputs are not equally loaded there can be a substantial difference between the reset times for the various flip-flops in the counter. The fastest flip-flop may reset so quickly that the reset pulse from the NAND gate isn't wide enough to reset the slower

264   COUNTERS AND REGISTERS   9.5

flip-flops. This problem can be cured with an *RS* latch placed in the reset line to hold the reset pulse, as discussed in the Problems. Buffering the outputs of the counter, as shown in the mod-7 counter in Fig. 9.17(b), helps to solve the loading problem.

The reset method also applies to synchronous counters. Most synchronous counters do not have internal gates with uncommitted inputs for resetting the counter. Therefore external gates must be used for resetting the counter which increases the package count and power consumption of the system. The reset technique for synchronous counters is identical to that for asynchronous and semi-synchronous counters described above. A preferable way to make a mod-*M* counter from a synchronous counter that has a count-down mode is described next under the preset method.

Preset Method [3]   The preset method can be used on asynchronous or synchronous counters, designed from flip-flops or IC counters, that have separate preset inputs for each flip-flop. To make a mod-*M* counter using the preset method, each flip-flop that has a 1 output on the *M*th count (in the normal count sequence of the counter) is fed into the inputs of a NAND gate. The counter's clock pulse is also fed into one input of the NAND gate. The output of this NAND gate is fed into the active LOW preset inputs of all flip-flops that have 0 outputs on the *M*th count. It is assumed in this method that the last count in the natural sequence of count states of the counter is all 1's. Let *N* represent the output of the counter on the *M*th count.

To illustrate the preset method for an asynchronous counter, let's convert the four-bit ripple counter shown in Fig. 9.18(a) to a mod-9 counter. On the $M = $ 9th

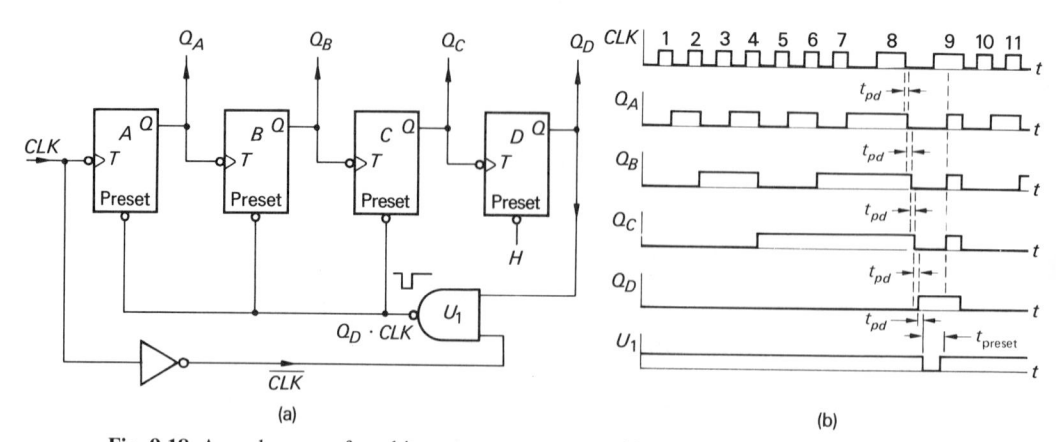

Fig. 9.18 Asynchronous four-bit up-counter converted into a mod-9 counter using the preset method: (a) logic diagram; (b) timing diagram.

count the outputs of the counter are $N = 8 = 1000$. So FF-D and the clock are fed to a NAND gate as shown in Fig. 9.18(a). For the first eight clock pulses, the counter counts in binary from 0 through 7 and the output of gate U_1 is HIGH. After the falling edge of clock pulse 8 the inputs to U_1 are $Q_D = H$ and $\overline{CLK} = H$, so its output goes LOW one gate delay time after Q_D goes HIGH. This presets flip-flops A, B, and C to a 1 (one preset-to-output delay time later), so $Q_A = Q_B = Q_C = Q_D = 1$. The output of U_1 goes HIGH one delay time after the leading edge of clock pulse 9, thereby deactivating the preset inputs. Just prior to the falling edge of clock pulse 9 (if it doesn't arrive too soon) the counter is at its largest count 1111, and the next count in its sequence is 0000. This sequence then repeats itself.

Note that during one-half of the ninth output state, while clock pulse 9 is HIGH, the counter outputs do not represent binary 9. However, it is a unique state (all 1's) and can be decoded with a four-input AND gate.

The preset method also works well with synchronous counters. If a synchronous counter has a down-count mode, and internally generates a pulse at the minimum count, then no external gates are required to make a modulo counter using the preset method. Consider the TTL 74LS193 synchronous four-bit up/down-counter shown in Fig. 9.19(a). It has an asynchronous active HIGH clear input, which forces all outputs LOW when a HIGH level is applied. The external clock signal is applied to the down-count or up-count input depending on whether a down- or an up-counter is desired. The unused count input must stay HIGH. When the load input is LOW, the output of each flip-flop may be made HIGH or LOW by applying a HIGH or LOW, respectively, to the data inputs. The flip-flop outputs will change to agree with the data inputs independent of the clock input. Counters that have data inputs for setting each flip-flop are called **programmable counters** because they can be programmed to a given count state. For example, if the load input in Fig. 9.19(a) is LOW and the data inputs are $D_A = 1$, $D_B = 0$, $D_C = 1$, $D_D = 1$, the counter outputs will be $1101 = 13$. In the count-up mode, when the load input becomes HIGH, the counter will count 13, 14, 15, 0, 1, 2, ...

The borrow output goes LOW on the falling edge of the count-down pulse which corresponds to the 0 state (0000), and returns HIGH on the rising edge of the next count-down clock pulse, as shown in Fig. 9.19(b). The borrow output signal is generated by the upper NAND gate (Fig. 9.19a), whose Boolean function is

$$\text{borrow output} = \overline{\overline{Q_A} \cdot \overline{Q_B} \cdot \overline{Q_C} \cdot \overline{Q_D} \cdot \text{count-down}}.$$

The carry output goes LOW on the falling edge of the count-up pulse which represents the maximum count (1111), and returns HIGH on the rising edge of the next count-up clock pulse. These outputs are used for cascading counters, which we shall discuss in Section 9.7.

266   COUNTERS AND REGISTERS

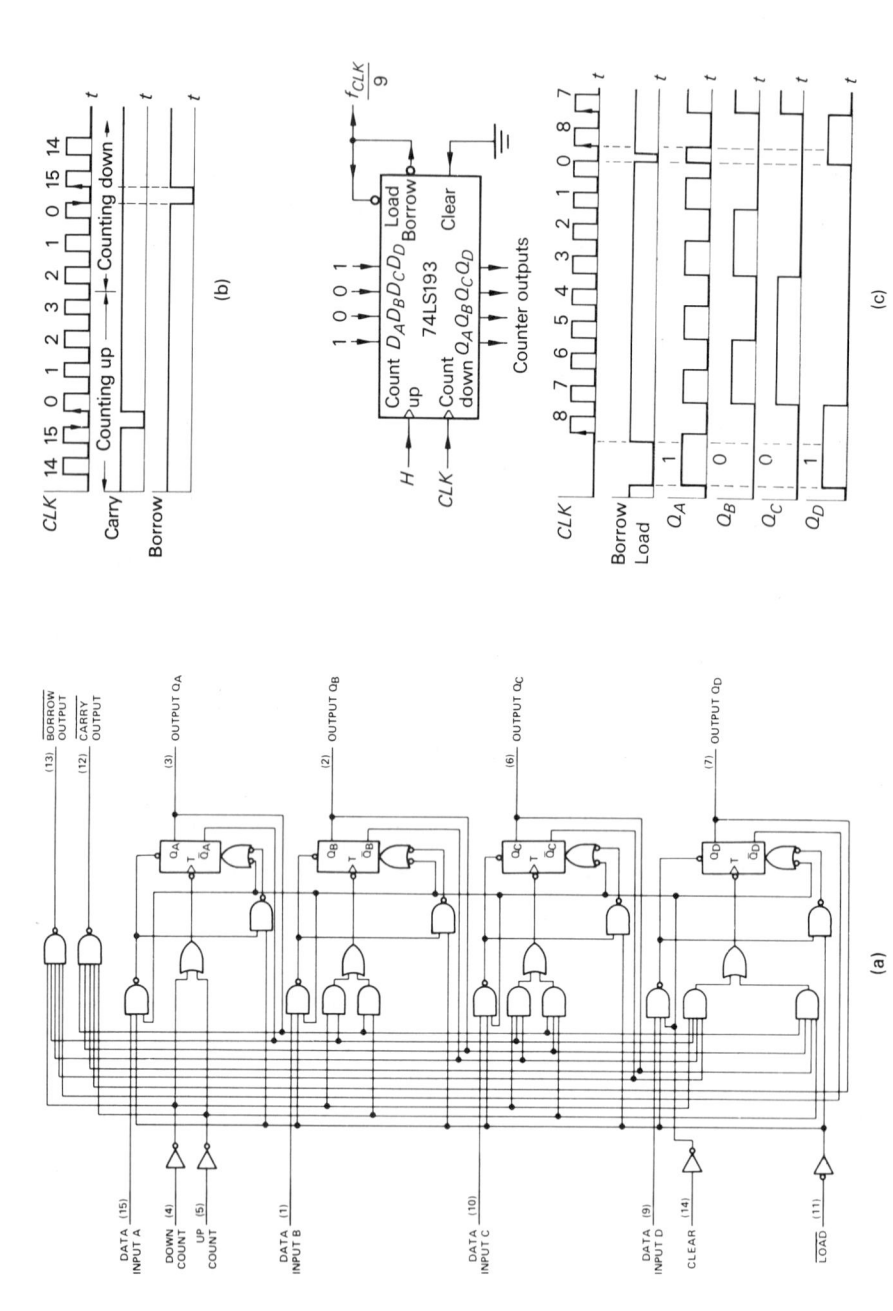

Fig. 9.19 TTL 74LS193 binary synchronous up/down-counter (courtesy of Texas Instruments Incorporated); (b) timing diagram for the borrow and carry outputs; (c) mod-9 down-counter.

Now let's consider how to make a mod-9 counter from a programmable synchronous down counter.

Example 9.4 Draw the block diagram for making a mod-9 counter using a 74LS193 counter.

Solution. The counter is programmed with the maximum count desired, i.e., $DCBA = 1001 (= M = 9)$. The borrow output is connected to the load input as shown in Fig. 9.19(c). The clock signal is applied to the count-down input. Initially we clear the counter and then load it with 1001, with the load input LOW. As soon as the LOAD input goes HIGH, the counter begins counting down (8, 7, 6, 5, 4, 3, 2, 1, 0). On the falling edge of clock pulse 0 the borrow output goes LOW. This loads a 1001 into the counter. On the rising edge of the next count-down clock pulse, the counter begins counting down again (8, 7, 6, ..., 0). It presets again when it reaches its minimum count of 0. An extra pulse occurs in the Q_A output between count 0 and count 8 whose width is approximately one-half the clock period.

The borrow output produces one pulse for every nine clock pulses, so this output can be used for a divide-by-9 output. This output pulse will be quite narrow, i.e., less than one-half of the clock period, because the borrow output goes HIGH as soon as any one Q output becomes preset (HIGH). However, the slower flip-flops will still preset because the preset signal has to propagate through the flip-flop *and* the borrow output gate before the preset inputs are cut off. Additional delay between the load input and borrow output can be generated, if needed, with logic gate delay elements. The borrow pulse can be widened if necessary by any of the pulse-stretching circuits discussed in earlier chapters. ■

Note that the count-up mode on a synchronous up/down-counter cannot be easily used to make a mod-M counter because no internally generated pulse is generated for resetting the counter to 0 when the maximum count of M is reached. An external NAND gate could be used; however, this would increase the package count of the system. Alternatively, one can load the number P = maximum count $- M$ into an up-counter and use the carry output for loading. For example, for a mod-9 counter using the 74LS193, $P = 16 - 9 = 7$ is loaded into the counter.

Some synchronous counters have synchronous data, load, reset, or enable inputs that may be 1's catching, if 1's catching master/slave flip-flops are used. For example, the 74163 is 1's catching on the enable, reset, and clear inputs.

9.6 ELECTRICAL CHARACTERISTICS OF IC COUNTERS

Let's consider some of the electrical characteristics of the TTL 74LS193 counter shown in Fig. 9.19(a). Data applied to the data inputs appears at the Q outputs one delay time after the falling edge of the load input. Data must be present at the data inputs one setup time prior to the rising edge of the load input, as shown in Fig. 9.20. Figure 9.21 shows that the minimum data setup time is 20 ns and the

recommended operating conditions

	SN54LS192 / SN54LS193 MIN	NOM	MAX	SN74LS192 / SN74LS193 MIN	NOM	MAX	UNIT
Supply voltage, V_{CC}	4.5	5	5.5	4.75	5	5.25	V
High-level output current, I_{OH}			−400			−400	μA
Low-level output current, I_{OL}			4			8	mA
Count frequency, f_{count}	0		25	0		25	MHz
Width of any input pulse, t_W	20			20			ns
Data setup time, t_{setup} (see Figure 1)	20			20			ns
Data hold time, t_{hold}	0			0			ns
Operating free-air temperature range, T_A	−55		125	0		70	°C

electrical characteristics over recommended operating free-air temperature range (unless otherwise noted)

PARAMETER	TEST CONDITIONS†	SN54LS192 / SN54LS193 MIN	TYP‡	MAX	SN74LS192 / SN74LS193 MIN	TYP‡	MAX	UNIT
V_{IH} High-level input voltage		2			2			V
V_{IL} Low-level input voltage				0.7			0.8	V
V_I Input clamp voltage	V_{CC} = MIN, I_I = −12 mA			−1.5			−1.5	V
V_{OH} High-level output voltage	V_{CC} = MIN, V_{IH} = 2 V, V_{IL} = V_{IL} max, I_{OH} = −400 μA	2.5	3.4		2.7	3.4		V
V_{OL} Low-level output voltage	V_{CC} = MIN, V_{IH} = 2 V, V_{IL} = V_{IL} max, I_{OL} = 4 mA		0.25	0.4				V
	I_{OL} = 8 mA					0.35	0.5	
I_I Input current at maximum input voltage	V_{CC} = MAX, V_I = 5.5 V			0.1			0.1	mA
I_{IH} High-level input current	V_{CC} = MAX, V_I = 2.7 V			20			20	μA
I_{IL} Low-level input current	V_{CC} = MAX, V_I = 0.4 V			−0.4			−0.4	mA
I_{OS} Short-circuit output current§	V_{CC} = MAX	−6		−40	−5		−42	mA
I_{CC} Supply current	V_{CC} = MAX, See Note 2		17	31		17	31	mA

†For conditions shown as MIN or MAX, use the appropriate value specified under recommended operating conditions for the applicable type
‡All typical values are at V_{CC} = 5 V, T_A = 25°C.
§Not more than one output should be shorted at a time.
NOTE 2: I_{CC} is measured with all outputs open, clear and load inputs grounded, and all other inputs at 4.5 V.

switching characteristics, V_{CC} = 5 V, T_A = 25°C

PARAMETER¶	FROM INPUT	TO OUTPUT	TEST CONDITIONS	MIN	TYP	MAX	UNIT
f_{max}				25	32		MHz
t_{PLH}	Count-up	Carry			17	26	ns
t_{PHL}					16	24	
t_{PLH}	Count-down	Borrow	C_L = 15 pF,		16	24	ns
t_{PHL}			R_L = 2 kΩ,		16	24	
t_{PLH}	Either Count	Q	See Figures 1 and 2		25	38	ns
t_{PHL}					31	47	
t_{PLH}	Load	Q			27	40	ns
t_{PHL}					29	40	
t_{PHL}	Clear	Q			22	35	ns

¶f_{max} ≡ maximum clock frequency
t_{PLH} ≡ propagation delay time, low-to-high-level output
t_{PHL} ≡ propagation delay time, high-to-low-level output

Fig. 9.20 Specifications for the TTL 74LS193 synchronous four-bit binary up/down-counter. (Courtesy of Texas Instruments Incorporated.)

9.6 ELECTRICAL CHARACTERISTICS OF IC COUNTERS

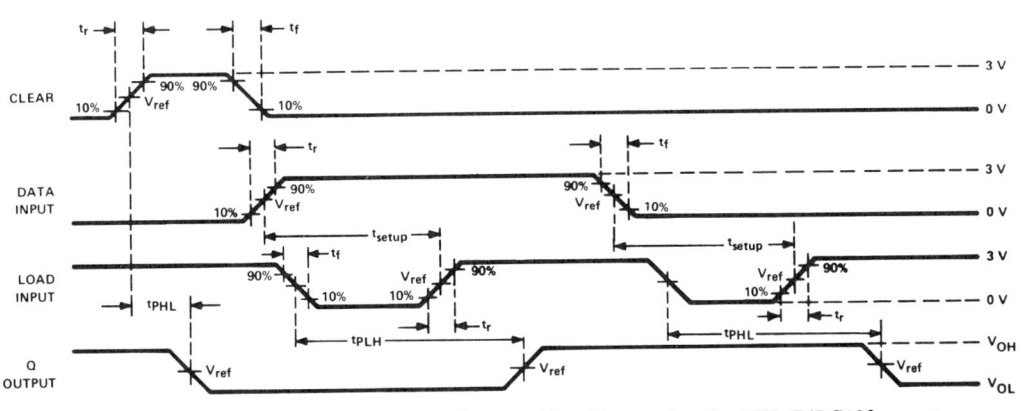

Fig. 9.21 Timing diagram for the clear, data, and load inputs for the TTL 74LS193 counter. (Courtesy of Texas Instruments Incorporated.)

minimum load input pulse width is 20 ns. The input current requirements are kept low by using buffer gates on the input lines.

MSI and LSI devices generally have very complex circuits. The 74LS193 has 55 equivalent gates. The user of the device need not attempt to understand all the circuit details. However, it is often necessary to know what the equivalent circuits for the input and output pins are, in order to interface these pins with other devices. For this purpose manufacturers frequently give the equivalent circuits for inputs and outputs only. Figure 9.22 shows the input and output equivalent circuits for the 74LS193.

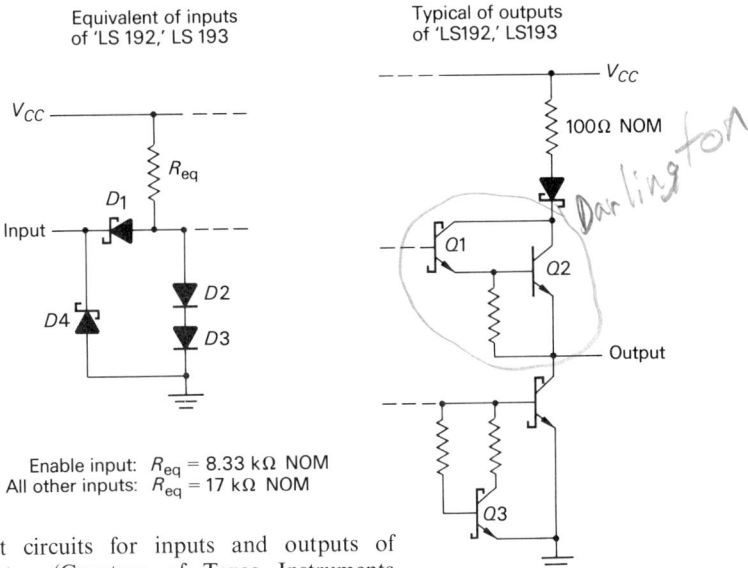

Enable input: $R_{eq} = 8.33$ kΩ NOM
All other inputs: $R_{eq} = 17$ kΩ NOM

Fig. 9.22 Equivalent circuits for inputs and outputs of the 74LS193 counter. (Courtesy of Texas Instruments Incorporated.)

For example, the equivalent circuit for any input, such as the load input, is a Schottky diode D_1 connected to the input transistor (diode-transistor logic) base–emitter junction represented by D_2. D_2 is connected to a second transistor, whose base–emitter junction is represented by diode D_3. Resistor R_{eq} is a bias resistor. Diode D_4 is the input protection diode. All output circuits have a totem pole structure with Darlington pull-ups (Q_1, Q_2) and an active pull-down (Q_3).

9.7 CASCADING COUNTERS

To perform large counts or frequency division by a large number, several counters can be connected together. The next example illustrates cascaded counters in a digital voltmeter (DVM).

Example 9.5 *Digital Voltmeter* [4]. Draw the logic diagram of a DVM that has three seven-segment LED readouts. Use the continuous method for multidigit displays and a step function reference voltage (refer to Section 6.5).

Solution. Figure 9.23(a) shows the principles of a DVM in which a digital-to-analog converter (DAC) produces the staircase reference voltage waveform shown in Fig. 9.23(b). An eight-bit DAC will produce $2^8 = 256$ voltage steps. If each step is 10 mV, then the maximum reference voltage is 2.55 V, with a ± 5-mV resolution. The output of the comparator is HIGH when the reference voltage is less than the unknown input voltage V_x, and LOW when $V_{ref} \geqslant V_x$. Initially both counters are

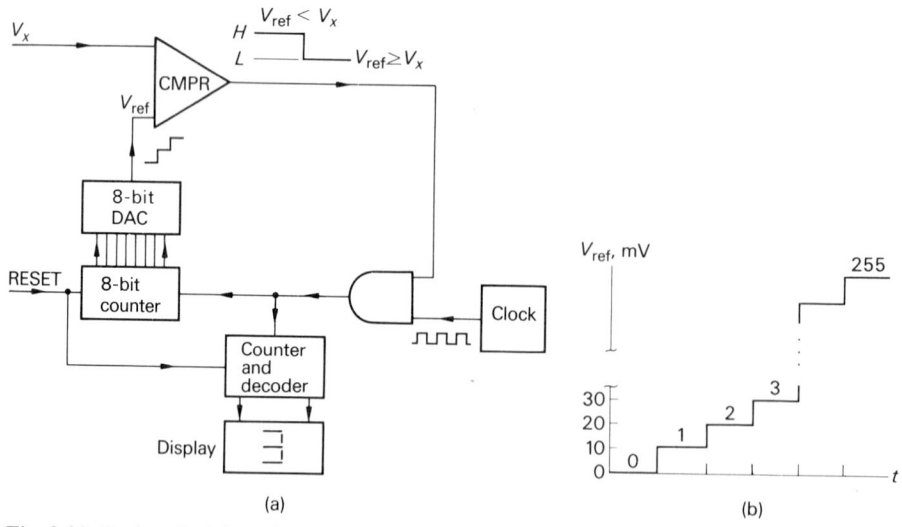

Fig. 9.23 Basic principles of a DVM: (a) block diagram; (b) staircase output waveform from the DAC.

9.7　　　　　　　　　　　　　　　　　　　　　　CASCADING COUNTERS　271

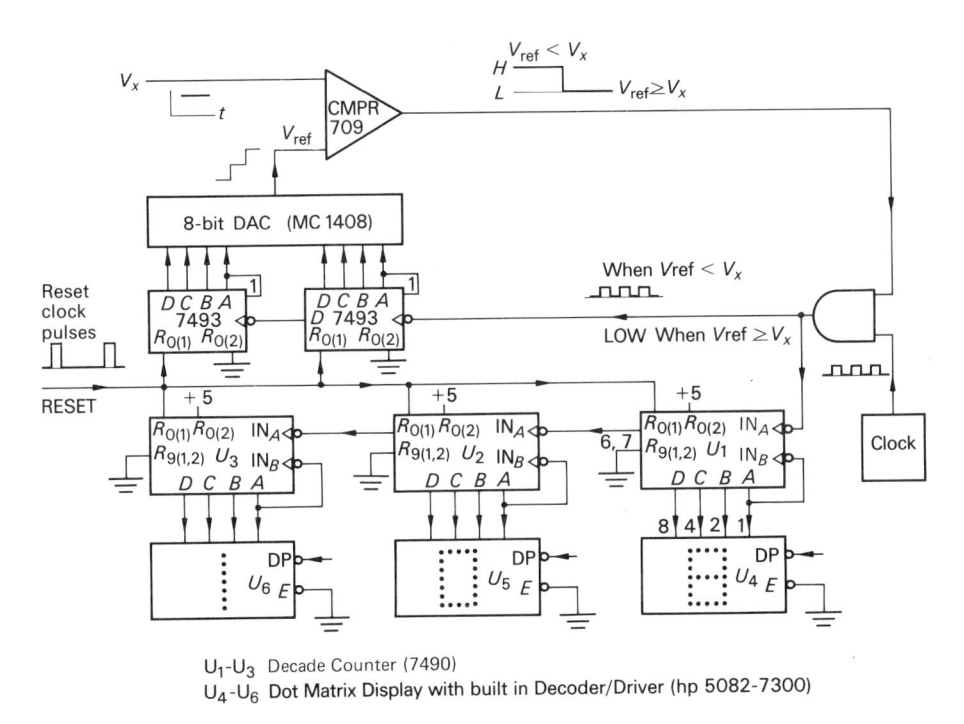

Fig. 9.24 DVM connection diagram.

reset to 0, so the output of the DAC is 0 V. Then suppose an input voltage, say 30 mV, is applied and the reset removed. Since $V_{ref} < V_x$ the output of the comparator is HIGH, so both counters begin counting. On the third count $V_{ref} = V_x$, the comparator goes LOW, and the counters stop counting. A decimal 3 will then be displayed on the display unit.

Figure 9.24 shows the details of this DVM using TTL. Three 7490 decade counters U_1, U_2, and U_3 are cascaded. When U_1 reaches its maximum count (9), on the next clock pulse U_1 goes to its next state (0) while FF-A in U_2 toggles and becomes a 1, thus displaying a decimal 10. The 7493s are four-bit ripple counters cascaded to form one eight-bit counter having 256 count states. The reset input determines how long an input is displayed. The HP 7300 is a 4 × 7 dot matrix display with a built-in latch and decoder/driver. This DVM will be discussed in further detail in the problems. ∎

Now let's consider an example of cascading synchronous counters.

Many synchronous counters, such as the 74LS193, have outputs for the maximum count (for up-counters) and for the minimum count (for down-counters). Figure 9.25 shows three 74LS193 counters cascaded to give a binary count from

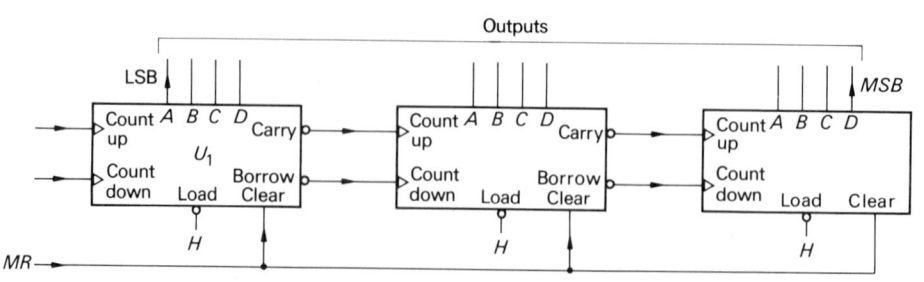

Fig. 9.25 Three-stage cascaded synchronous up/down-counter.

0 through 4096 (16 × 16 × 16). A divide-by-N (where N is between 1 and 4,096), frequency divider can be made using a similar configuration.

The TTL 74490 has two four-bit decade counter in one package capable of division by 100. The CMOSL 4020 is capable of division by 2^{14}.

*9.8 RATE MULTIPLIERS

A **rate multiplier** is a programmable counter whose output is decoded so that the output frequency f_0 is a predetermined fraction of its input frequency f_{in}. The output of a six-bit **binary-rate multiplier** (BRM) is

$$f_0 = \frac{M \cdot f_{in}}{64},$$

where the rate input $M = F \cdot 2^5 + E \cdot 2^4 + D \cdot 2^3 + C \cdot 2^2 + B \cdot 2^1 + A \cdot 2^0$. For example, if $F = 1$, $E = 0$, and $D = C = B = A = 1$, then $M = 47$ and the output is $\frac{47}{64} f_{in}$. Such fractional division rates are useful in performing arithmetic operations, in radar, in digital and analog conversion, and in other conversion operations [5; 6, p. 350]. Decade-rate multipliers are also available, with $f_0 = M \cdot f_{in}/10$. By cascading rate multipliers, it is possible to divide by any fraction, such as 0.0941.

Rate multipliers are available in the TTL and CMOSL families.

The advantages of a rate multiplier over a modulo counter for frequency division are: (1) the rate multiplier can be easily programmed through inputs $(A, B, C, D, E, \ldots)$ for the desired rate, and (2) its output is already decoded. The design of rate multipliers, including a specific example of their use in radar, is discussed in the references [7, p. 226].

9.9 REGISTERS

An n-bit or n-stage **register** is an MSI or LSI storage device tht consists of a set of n interconnected flip-flops. It is more sophisticated than a set of n flip-flops in its complexity and ability to perform a variety of functions. A **shift register** is a register

9.9   REGISTERS   273

in which the outputs of all flip-flops are clocked simultaneously. Outputs may be connected back into inputs without danger of racing, since flip-flops rather than latches are used. Shift registers have many applications, such as storage of data, serial and parallel conversion, and counting. They are available in capacities ranging from four bits in TTL, ECL, and CMOSL, to 2048 bits in the MOSL family. In this section, after describing the basic shift register we shall discuss general-purpose shift registers, cascaded shift registers, MOS shift registers, and circulating memories. We shall conclude the chapter with a discussion of shift register counter in Section 9.10.

9.9.1 Shift Register Principles

Shift registers can be classified as follows: (1) serial input to serial output, (2) serial input to parallel output, (3) parallel input to serial output, and (4) parallel input to parallel output (called **buffer registers**). A wide variety of IC shift registers are available. They are made from D or RS master/slave bipolar transistor flip-flops or from MOS transistors. Some shift registers have asynchronous parallel data inputs, and care must be exercised in feeding register outputs to these inputs in order to avoid racing.

Figure 9.26(a) shows the logic diagram of a basic shift register that uses four rising-edge RS flip-flops. The first flip-flop is connected as a D flip-flop; the remaining three flip-flops are RS flip-flops. An active LOW clear input ($\overline{CLR}$) resets all flip-flops to their LOW state independent of the clock.

Let's see how the shift register in Fig. 9.26 works. Each flip-flop is cleared by applying a LOW to the $\overline{CLR}$ input, and then $\overline{CLR}$ is returned HIGH. Clock pulses are applied to the CLK and data inputs. Suppose the four-bit word to be entered in the shift register is $LLHH$, where the righthand HIGH enters the D_S input first. Prior to the rising edge of the first clock pulse, the input to FF-A is

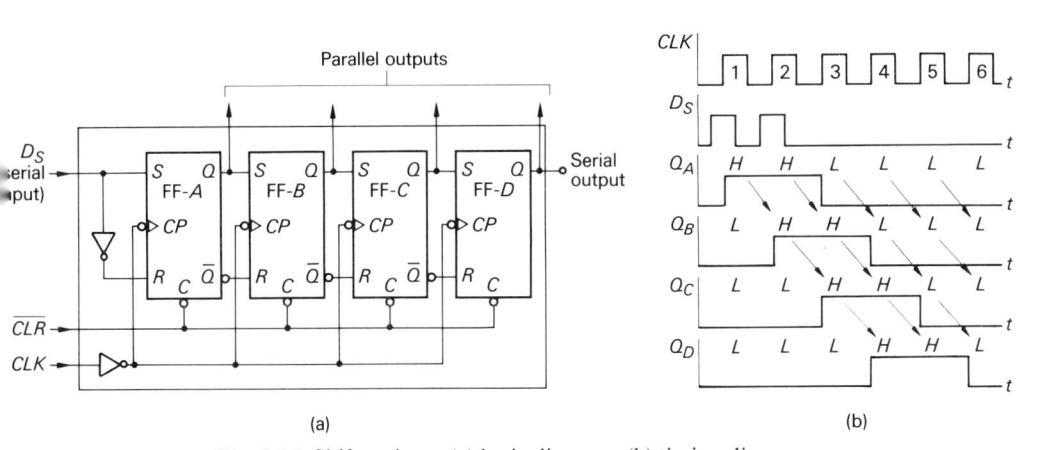

Fig. 9.26 Shift register: (a) logic diagram; (b) timing diagram.

HIGH, so after the rising edge of clock pulse 1, $Q_A = H$, as shown in Fig. 9.26(b). Since the outputs of FF-A, B, and C were all LOW just prior to clock pulse 1, the outputs of FF-B, C, and D remain LOW after clock pulse 1.

Prior to clock pulse 2, the inputs to FF-A and FF-B are both HIGH, so after clock pulse 2, $Q_A = Q_B = H$ while $Q_C = Q_D = L$. Prior to clock pulse 3, a LOW is present at the D_S input and a HIGH is present at the set inputs of FF-B and FF-C. Hence, after clock pulse 3 the outputs are $Q_A = L$, $Q_B = Q_C = H$, and $Q_D = L$. Finally, after clock pulse 4 the outputs are $Q_A = Q_B = L$ and $Q_C = Q_D = H$.

Notice in Fig. 9.26(b) how the input data ($LLHH$) has been shifted toward the right in the register until, after four clock pulses, the four data bits ($LLHH$) are stored in flip-flops A, B, C, and D, respectively. A **shift-right shift register** is one in which data entering the register is shifted toward the right in the register, i.e., away from the external clock input. An n-bit shift register consists of n flip-flops and requires n clock pulses to shift one bit to the rightmost flip-flop in the register. Note in Fig. 9.26(b) that after the fifth clock pulse, the outputs become $LLLH$, i.e., a HIGH bit has been shifted out of the register and lost permanently. After the sixth clock pulse the outputs will be LLLL.

The shift register in Fig. 9.26(a) is also a serial-to-parallel converter, because after four clock pulses the flip-flop outputs are $Q_A = L$, $Q_B = L$, $Q_C = H$, and $Q_D = H$. These outputs are available simultaneously, i.e., in parallel. Hence a serial input of $LLHH$ has been converted, in four clock pulses, to a parallel $LLHH$

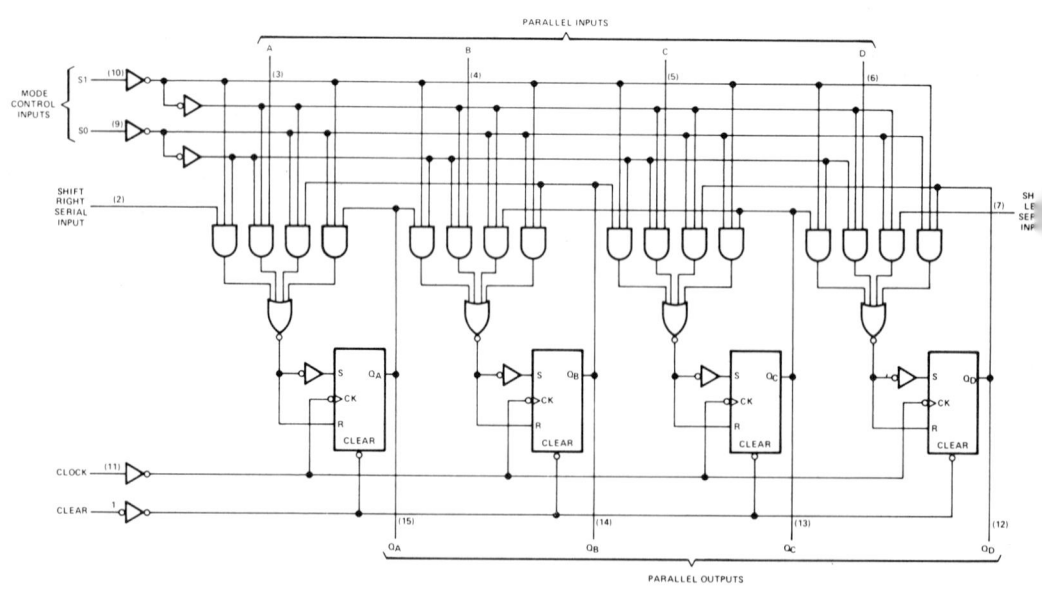

Fig. 9.27 TTL 74LS194 four-bit universal shift register. (Courtesy of Texas Instruments Incorporated.)

output. In a similar manner a 16-bit shift register could convert a 16-bit serial word into parallel in 16 clock pulses. Serial-to-parallel converters are very useful in the transmission of data, which is often sent serially and then converted into parallel form for computer processing or storage.

Figure 9.27 shows the logic diagram of a shift register that can perform the following functions: shift-right, shift-left, conversion from serial to parallel, conversion from parallel to serial, and parallel-in to parallel-out transfer of data.

When the mode control inputs are $S_0 = H$ and $S_1 = L$, data entering the shift-right serial input is shifted to the right (i.e., toward FF-D). When $S_0 = L$ and $S_1 = H$, data entering the shift-left serial input is shifted toward the left (i.e., toward FF-A). When $S_0 = H$ and $S_1 = H$, data entering the parallel inputs becomes available at the parallel outputs. By inhibiting the clocking with $S_0 = L$ and $S_1 = L$, data is stored in the register. In the parallel-in to parallel-out mode, a shift register is called a **buffer register**, and it can serve as a temporary storage device between two digital devices.

The following example shows how long it takes to convert data from serial form to parallel form.

Example 9.6 *Conversion Time.* Determine the total time required to convert a four-bit word from serial form to parallel form using a CMOSL 4035 and an ECL 94000 shift register. Their maximum typical clock input frequencies are 2.5 MHz ($V_{DD} = 5$ V) and 200 MHz, respectively.

Solution. It requires four clock pulses to put four bits into a four-bit shift register so that it can be read out in parallel. For the CMOSL 4035, one clock period takes $T = 1/f_{\text{clock}} = 1/2.5$ MHz $= 400$ ns. Hence four clock periods require

$$4T = 4(1/f) = 4 \times 400 \text{ ns} = 1600 \text{ ns}.$$

The conversion time for the ECL shift register is

$$4T = 4(1/f) = 4/200 \text{ MHz} = 20 \text{ ns}.$$

Thus the larger the shift frequency, the faster data can be converted from serial to parallel form. Short conversion times are necessary when a large quantity of data must be converted rapidly, as in a data communication system that can operate at mega-bit rates. ■

9.9.2 Shift Register Applications

In addition to its use as a converter, the shift register is useful for introducing a time delay into a digital system. For example, the pulmonary analyzer shown in Fig. 9.1 uses a 256-bit MOS shift register as a delay element.

If a train of pulses enters the serial input of an *n*-bit shift register, whose clock period is t_{clock}, the first pulse of that train does not leave the serial output until time $T = (n - 1)t_{\text{clock}}$ later. The following example illustrates the use of this principle.

Example 9.6 *Digital Delay Line.* Determine the time delay for a 256-bit shift register that has a shift frequency of 1 MHz.

Solution. Using the above formula, with $n = 256$ and $t_{clock} = 1/f_{clock} = 1$ μs, the delay time is $T = (256 - 1) \times 1.0$ μs $= 255$ μs. By cascading shift registers much longer delay times can be achieved. ∎

Another way of using shift registers is as frequency dividers [3]. They can also generate Boolean functions, and are then called **sequence generators**, as discussed in the references [8, p. 635; 9, p. 3–15]. A **pseudorandom sequence generator**(PRSG) is a pulse generator that generates a sequence of 0's and 1's that appear to be in random order [10; 11]. A PRSG is useful for testing digital systems [11], for generating timing pulses, and as a programmable frequency divider [12]. PRSGs can be made from shift registers and EX-OR/NOR gates, as shown in the problems.

The next section deals with cascaded shift registers, which can be used to increase the capacity of registers.

*9.9.3 Cascaded Shift Registers

Shift registers can be cascaded to increase their capacity, just as counters can be cascaded to increase their counting capacity. The following example illustrates the general method.

Example 9.7 *Cascaded 12-Bit Shift Register.* Draw the connection diagram for a 12-bit shift register made from three four-bit 74S194 shift registers (same logic diagram as the 74LS194).

Solution. A 12-bit shift register is made essentially by chaining 12 flip-flops together. Figure 9.28 shows three four-bit shift registers connected to form a 12-bit shift-right shift register. A 12-bit shift-left shift register could be made using similar connections. The mode inputs for each register are connected to common S_0 and S_1 mode inputs. The clock and clear lines in each register are also connected to

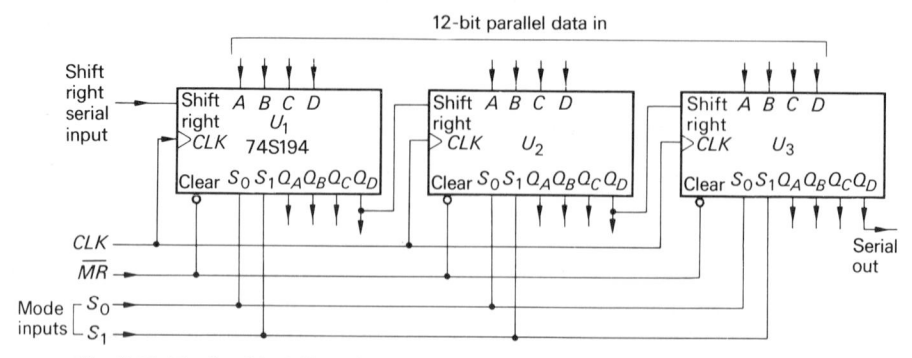

Fig. 9.28 Twelve-bit shift register made by cascading three four-bit shift registers.

common lines. Serial data enters the shift-right serial input of register U_1 and flows successively from Q_D in U_1 to U_2, from the Q_D output of U_2 to U_3, and finally to the Q_D output of U_3. Parallel-to-serial conversion can also be performed with this circuit using the parallel data inputs. ∎

A second way of obtaining large-capacity registers is by using high-density MOS shift registers, which we shall discuss next.

*9.9.4 MOS Shift Registers [8, p. 650; 13]

A **MOS** (metal-oxide semiconductor) **shift register** is an IC shift register made from MOS transistors. MOS shift registers usually contain a large number of bits. They therefore generally operate only in the serial-in to serial-out mode, because all the outputs cannot be brought out to external IC pins. They offer the following advantages over bipolar shift registers: (1) lower power dissipation, (2) smaller size, for the same number of bits, and (3) lower cost, for a large number of bits. MOS shift registers are commonly available with capacities ranging from 32 bits to more than 2048 bits. They are used in applications such as calculators, display systems, scratch-pad (temporary) memories, and data communication systems. Either PMOS or NMOS transistors can be used, with the NMOS transistors yielding higher shift frequencies.

Some MOS shift registers can be driven directly from TTL logic circuits and can drive TTL circuits directly. Others require an interface circuit between the TTL and MOS circuits.

There are two types of clocking for MOS shift registers, dynamic and static. The gate-to-source input of a MOS transistor is essentially a capacitor which can be used to store electric charge. Hence it can be used to store binary information (i.e., a bit). The **dynamic MOS shift register** stores a data bit in the gate capacitor of a MOS transistor on one clock pulse, having a phase ϕ_1. This stored data is transferred to the gate capacitor of the output transistor when a second clock pulse, having a different phase ϕ_2, is applied. Some MOS shift registers require two external clocks that have different phases. Others require only one external clock, and an internal generator produces the necessary clock signals. Dynamic shift registers require that the data be shifted at some minimum (nonzero) clock rate in order to avoid losing the data through gate capacitor leakage currents.

Static MOS shift registers use latches to store data indefinitely, as long as power is supplied. The disadvantage of static MOS shift registers are that they require more space on the IC chip to store each bit, and they consume more power.

*9.9.5 Shift Register Memories

Suppose the serial output of an n-bit shift register is connected to its serial input as shown in Fig. 9.29(a). Then the previous contents of the shift register will circulate around in the register as long as the clock pulses are applied. Since this type of shift register stores n bits of data for as long as desired, it is called a **circulating**

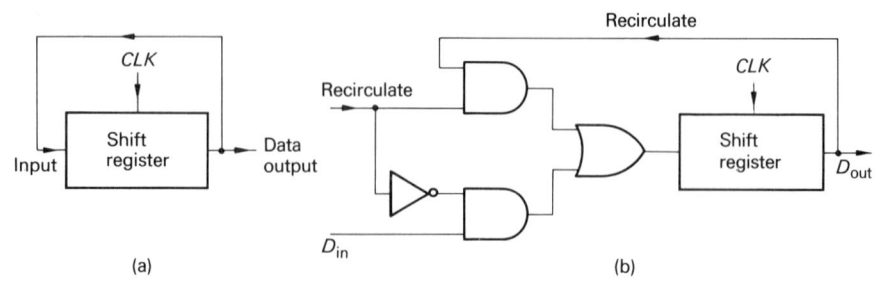

Fig. 9.29 Circulating memory: (a) block diagram; (b) logic diagram.

memory. Figure 9.29(b) shows the logic diagram for a circulating memory. It allows external serial data to be written into (i.e., enter) the memory. Data can also be read out of (i.e., leave) the memory, or be stored (recirculated).

Circulating memories, also called **recirculating shift registers**, are useful as temporary memories in applications such as the refreshing of CRT displays.

An instrument that is extremely useful in troubleshooting digital systems is shown in Fig. 9.30. The logic state analyzer can display up to 16 consecutive 12-bit

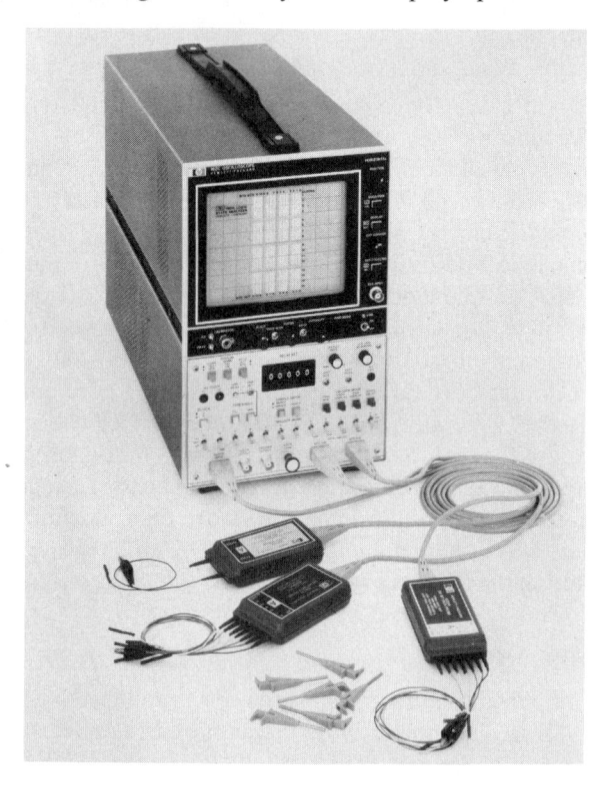

Fig. 9.30 Troubleshooting logic state analyzer. (Courtesy of Hewlett-Packard Company.)

*9.10 SHIFT REGISTER COUNTERS

An n-bit **Johnson counter** is a counter, made from an n-bit shift register, that can count up to $2n$. A Johnson counter, also called a **twisted ring, shift**, or **Moebius counter**, is formed by connecting the $\bar{Q}$ output of the last shift register stage to the serial input. Figure 9.31(a) shows a four-bit Johnson counter made from a four-bit shift register by connecting the complementary output of FF-D to the serial data input D_S.

The counter works as follows. The master reset MR is used first to clear all outputs LOW, as shown in the timing diagram of Fig. 9.31(b). Just before the leading edge of the first clock pulse the input D_S to FF-A is $D_S = \bar{Q}_D = 1$. So after the leading edge of clock pulse 1 the output of FF-A changes from 0 to 1 as shown in Fig. 9.31(b).

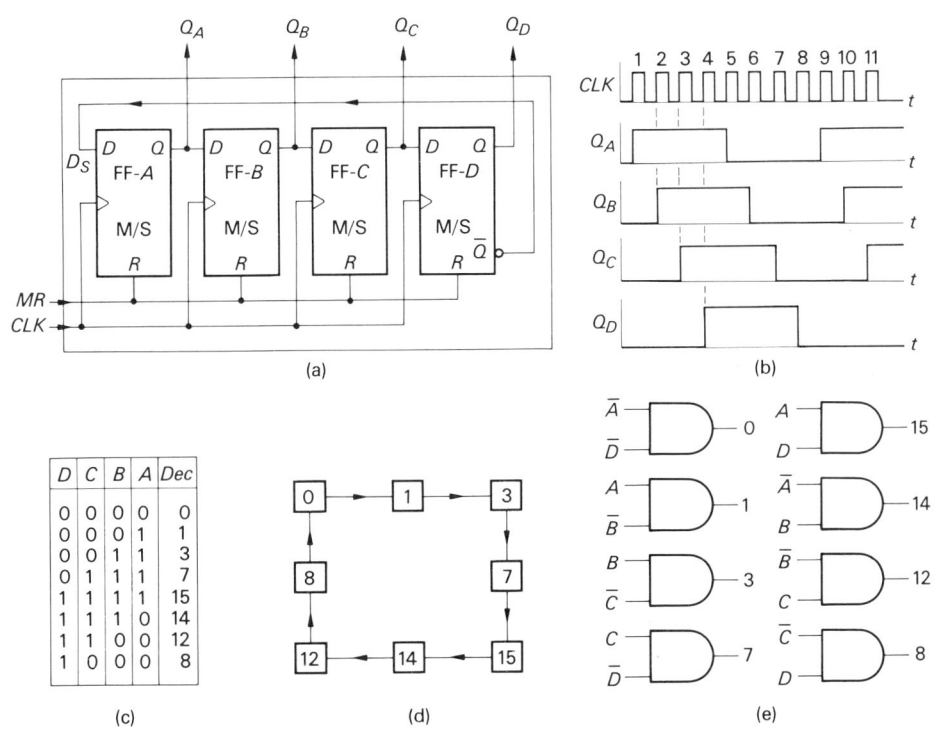

Fig. 9.31 Four-stage Johnson counter: (a) logic diagram; (b) timing diagram; (c) truth table; (d) state diagram; (e) decoding gates.

Before clock pulse 2 the input to FF-B is $Q_A = 1$, so after the leading edge of clock pulse 2 Q_B becomes a 1 while Q_A remains HIGH since $\bar{Q}_D = 1$. After clock pulse 3, Q_C becomes a 1, while $Q_A = Q_B = 1$. After clock pulse 4, $Q_D = 1$ as well as $Q_A = Q_B = Q_C = 1$.

Prior to clock pulse 5, $\bar{Q}_D = 0$, so after clock pulse 5, Q_A goes to 0. After clock pulse 6, $Q_B = 0$, etc. This cycle of events repeats itself after clock pulse 8. The binary outputs ($DCBA$) of the counter, and their decimal equivalents, are: 0000 (0), 0001 (1), 0011 (3), 0111 (7), 1111 (15), 1110 (14), 1100 (12), 1000 (8), and 0000 (0). Figure 9.31(c) shows the eight states of a four-bit Johnson counter along with their 8421 code decimal equivalents. Since there are eight distinct states for the device in Fig. 9.31(a), it is a counter that can count to 8. Figure 9.31(d) shows the state diagram for a four-bit Johnson counter. A **state diagram** is a diagram that shows the output states of a system. In this case, the only states are 0, 1, 3, 7, 15, 14, 12, and 8.

This four-stage counter is also a divide-by-8 frequency divider whose output waveforms are square waves (50% duty cycle).

Any n-stage Johnson counter can be decoded with $2n$ two-input logic gates. For example, the four-stage Johnson counter in Fig. 9.31(a) can be decoded using the eight two-input AND gates shown in Fig. 9.31(e). Only two-input AND gates are required because each of the eight states is uniquely determined by only two conditions. Thus the 1 state is the only state in which $Q_A = 1$ and $Q_B = 0$. The inputs to the AND gate that decodes the 3 state are $Q_B = 1$ and $Q_C = 0$, etc.

A four-bit Johnson counter can count only to $2n = 8$, even though it uses four flip-flops, which have a total of 16 states. Thus a four-bit Johnson counter has $16 - 8 = 8$ unused states, called **illegal states**. If a Johnson counter gets into an illegal state, due to noise, power failure, or not being preset to a legal state before starting to count, then it may remain in an invalid sequence of counts, or may not count at all. To prevent a Johnson counter from remaining in an illegal sequence, external feedback circuits, called **antilock circuits**, are required. The references show several feedback circuits, state diagrams, and modulo Johnson counters [3, p. 294; 14, pp. 9–35 and 10–8].

The advantages of an n-stage Johnson counter over a modulo counter are: (1) only two-input decoding gates are required, (2) there are no decoding spikes since only one flip-flop changes at any one time, and (3) the outputs are always square waves. The disadvantage of a Johnson counter is the inefficient use of flip-flops, since there are always $2^n - 2n$ illegal states. To count to 20 an ordinary counter requires five flip-flops, whereas a Johnson counter would require 10 flip-flops, a relative efficiency of $\frac{5}{10} = 50\%$.

An n-stage **ring counter** is made from an n-stage shift register by connecting the Q output of the nth stage to the J input of the first stage and the $\bar{Q}$ output of the nth stage to the K input of the first stage. The main disadvantage of a ring counter, as compared with a twisted ring counter, is that the ring counter will produce decoded outputs only if a single 1 is in the counter. No 1's cause it to lock up and more than one 1 produces nondecoded output waveforms. A ring counter does not have square wave outputs.

SUMMARY

Counters are devices that perform useful functions such as simple counting and frequency division. In synchronous counters the flip-flops are clocked from one external clock. In asynchronous counters, or ripple counters, each flip-flop is clocked by the output of the preceding flip-flop. Synchronous counters can be faster than asynchronous counters, and generally do not have decoding spikes.

Asynchronous counters, with modulus $M = 2^n$ count in natural binary and produce square-wave outputs. Modulo counters can count to any number such as 3, 10, 19, etc., and do not necessarily have square-wave outputs. A mod-10 counter is called a decade counter. An up/down-counter can count up or down. Rate multipliers are decoded counters that are useful for producing fractional counts.

Registers are $1 \times N$ memories that are useful for storing data bits, for converting between serial and parallel forms of data, as digital delay lines, and as counters. MOS shift registers generally have a large number of bits and are usually the serial-in to serial-out type. The Johnson counter is an easily decoded shift register counter that produces no decoding spikes and has a square-wave output.

PROBLEMS

Section 9.2

1. Draw the logic diagram (flip-flops) for a 3-bit binary ripple counter. With the aid of a timing diagram explain how it works. Use falling-edge *JK* flip-flops. Use ideal pulses and show delay times.

2. (a) Draw the logic diagram for logic gates that will decode all outputs of the counter in Problem 1. (b) Draw the timing diagram for the decoded outputs. Show false outputs where they occur.

Section 9.3

3. Compare the advantages and disadvantages of asynchronous and synchronous counters.

Section 9.4

4. (a) Draw the logic diagram (flip-flops) for a synchronous four-bit binary up/down-counter with a rising-edge clock input. Include decoding gates. (b) Draw the timing diagram for the Q outputs of each stage, and for the decoded outputs when operated as a down-counter.

5. *Numerical Control Drill Press.* Referring to Example 9.2, draw the logic diagram for controlling the *x*-axis and the *z*-axis motion of the drill press table. Include limit switches that will stop the table when the $\pm x$ or $\pm z$ axis limits are reached.

Section 9.5

6. *Illegal States.* Suppose the mod-5 counter shown in Fig. 9.12 has been preset to the illegal state 6. (a) Will the counter return to its normal sequence? (b) What states does it go through in returning to its normal sequence (0, 1, 2, 3, 4)?

282 COUNTERS AND REGISTERS

7. Draw the block diagram for a mod-20 synchronous counter made from a mod-5 counter and two flip-flops.
8. Draw the logic (block) symbol for a mod-7 counter using the reset method with a TTL 7493 counter.
9. Make a table which shows the connections on a block logic symbol of a TTL 7490 counter for moduli 2 through 10 using the reset method.
10. Explain how the R_9 inputs to a 7490 counter can be used to obtain a mod-7 counter without using any external gates.
11. (a) Draw the block diagram for a TTL 7492 counter connected as a mod-6 × mod-2 counter and as a mod-2 × mod-6 counter. (b) Give the count table for the mod-2 × mod-6 counter.
12. (a) Draw the logic (block) symbol for a mod-7 counter made from a TTL 74LS193 counter using the preset method. (b) Draw the timing diagram for all outputs.
13. Make a table that shows advantages and disadvantages of the feedback, reset, and preset methods for obtaining modulo counters. Consider factors for asynchronous and synchronous counters such as: availability of preset inputs or clear inputs, accessibility to the $R(K)$ and $S(J)$ inputs, whether applicable to IC counters, need for external gates, and whether an up- or a down-counter is required.

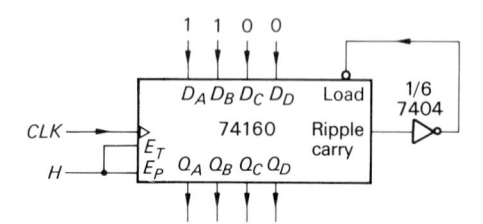

Fig. P9.1 Divide-by-7 counter.

14. (a) Using a TTL data book, explain how the decade up-counter in Fig. P9.1 operates as a divide-by-7 counter using the preset method. The programmed inputs are $D_A = D_B = 1$, and $D_C = D_D = 0$. (b) Compare this technique with using a down-counter. Note that the required data inputs are $10 - 7 = 3 = 0011$.

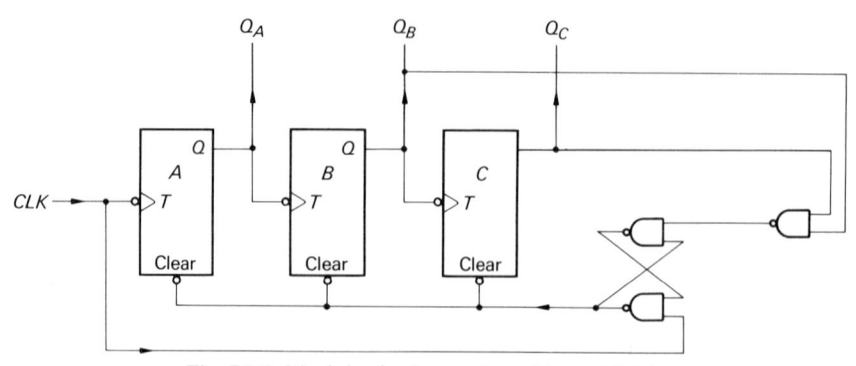

Fig. P9.2 Modulo ripple counter with reset latch.

15. (a) Describe the problem which may occur using the reset method for modulo counters which is caused by unequal reset-to-output delay times. (b) Explain how the circuit in Fig. P9.2 prevents this problem from occurring.

Section 9.6

16. (a) Draw the schematic of the equivalent Q_A output circuit for the TTL 74LS193. (b) Describe this circuit.

Section 9.7

17. *Pulmonary Function Analyzer.* The schematic for part of the nitrogen analyzer section of a pulmonary function analyzer, pictured in Fig. 9.1, is shown in Fig. P9.3. As the percentage of nitrogen (N_2) in a breath sample increases, the light intensity from the photo (Geisler) tube increases and its resistance decreases. This results in more current to the differential amplifier U_1, which has a digitally controlled feedback resistance. As a technician helping to develop the pulmonary function analyzer, you are asked to check the operation of the auto gain circuit. Your first step is to study the circuit diagram and notations to understand how the circuit works. Explain how the auto gain circuit controls the gain of U_1.

18. *Programmable Frequency Divider* (Appendix F). Study the N2 Phase Detector schematic (Fig. F.4) and the description of the programmable divider circuit (Section F.2.2). Explain how the programmable divider works when thumbwheel digits 3, 4, and 5 are set to 2, 4, and 6, respectively. [*Hint*: Digit 3 in Fig. F.4 is denoted as digit 3–8; the 8421 coded values are digit 3–8 = 0, digit 3–4 = 0, digit 3–2 = 1, and digit 3–1 = 0.]

19. *Cascading Counters.* Asynchronous counters are cascaded by connecting the Q output from the MSB to the clock input of the next counter as shown in Fig. P9.4(a). Synchronous counters can be cascaded by ripple clocking (Fig. P9.4b), or synchronously with ripple carry between stages (Fig. P9.4c), or synchronously with parallel carry between stages (Fig. P9.4d). Some counters, such as the TTL 74161, have internal gating, which eliminates the need for external gates for the parallel carry method. [*Note*: E = enable input, C_0 = carry output.] Discuss cascading of synchronous counters in terms of (1) minimum clock pulse width, (2) speed, and (3) loading of inputs or outputs.

20. Show the logic block symbol for a divide-by-100 frequency divider with a square wave output using two TTL 7490 counters.

21. Explain in detail how the DVM in Fig. 9.24 would display 1.2 V.

Section 9.9

22. Calculate the maximum delay time that can be achieved using a 2048-bit MOS shift register at a clock rate of 10 kHz.

23. *Microcomputer* (Appendix G). Refer to sheet 1 of the schematic diagram (Fig. G.5) and Section G.3.1. Explain how the four-phase clock generator works.

24. *Numerical Control Drill Press.* The pulse waveform supplied to the stepper motor in Example 9.2 is shown in Fig. P9.5. Draw the logic diagram for the digital circuit that will generate this waveform.

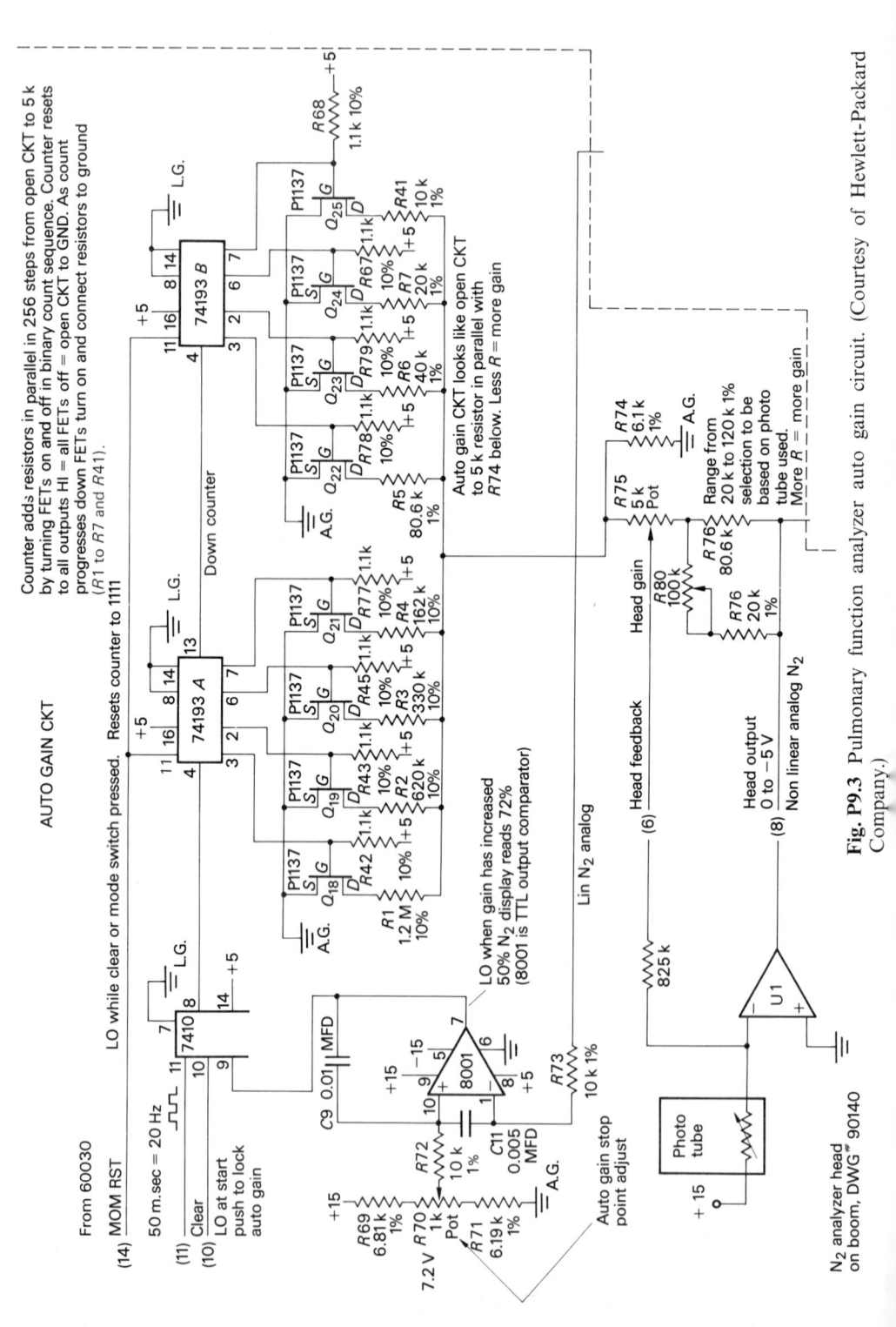

Fig. P9.3 Pulmonary function analyzer auto gain circuit. (Courtesy of Hewlett-Packard Company.)

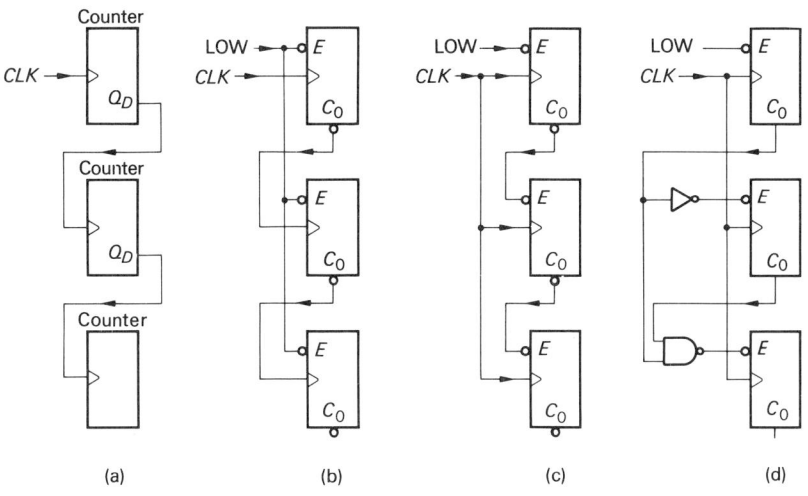

Fig. P9.4 Cascading counters: (a) asynchronous; (b) synchronous (ripple clocking); (c) synchronous (ripple carry); (d) synchronous (parallel carry).

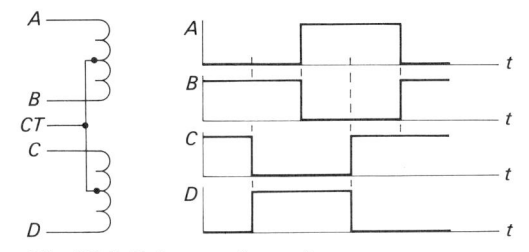

Fig. P9.5 Pulse waveforms for a stepper motor.

25. *Pulse Coded Modulation (PCM) Digital Communication System.* A PCM encoder converts analog signals into an eight-bit natural binary coded output. These pulses modulate a microwave transmitter for transmitting control signals to an earth satellite by means of the 94-foot dish antenna shown in Fig. 5.23. Figure P9.6 shows the block diagram and waveforms for the PCM encoder. It consists of a full-wave rectifier, sampler, eight-bit ADC, latch, and eight-bit shift register. (a) Draw the logic block diagram for an eight-bit shift register using four-bit 74S194 registers. (b) Explain the purpose of the shift register.

26. *Pseudo Random Sequence Generator (PRSG).* Figure P9.7(a) shows a PRSG made from a three-stage shift register and an EX-OR gate. The feedback provided by the EX-OR gate is responsible for the random outputs from flip-flops 1, 2, and 3 as shown in the count table in P9.7(a). Note that $2^3 - 1 = 7$ counts occur, but their order is random for seven counts. EX-OR inputs B and C are permanently LOW, for this three-stage example. Note that the 000 state never occurs; if it did, the PRSG would lock up and cease to count. A LOW on the START input produces a nonzero starting output of 001. Figure P9.7(b) shows the connections for PRSGs from $n = 3$ to $n = 20$ stages. The length of each sequence is $2^n - 1$, as shown. (a) Verify the count table for the three-stage PRSG

286   COUNTERS AND REGISTERS

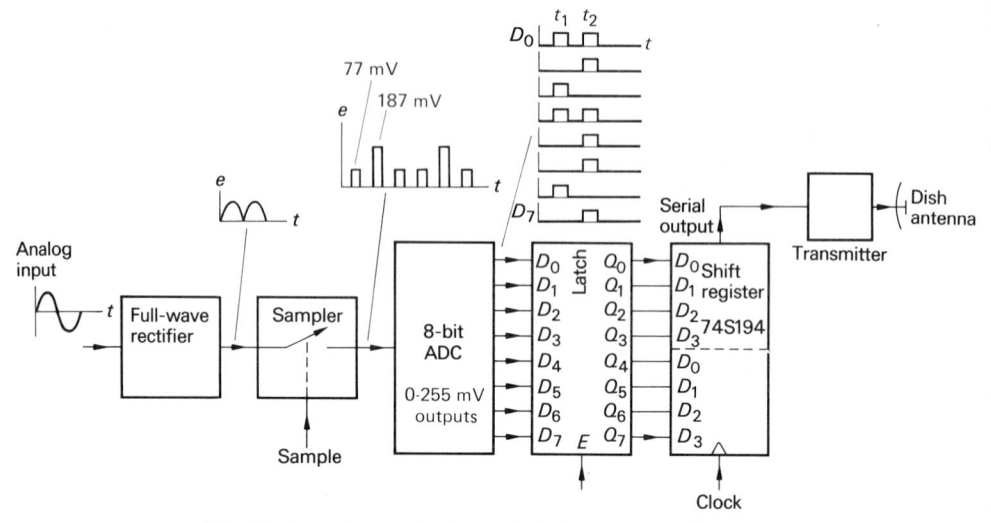

Fig. P9.6 PCM encoder for a digital communication system.

(a) n = 3 logic diagram:

Q_3	Q_2	Q_1	Count
0	0	1	1
0	1	0	2
1	0	1	5
0	1	1	3
1	1	1	7
1	1	0	6
1	0	0	4
0	0	1	1

(b) connections:

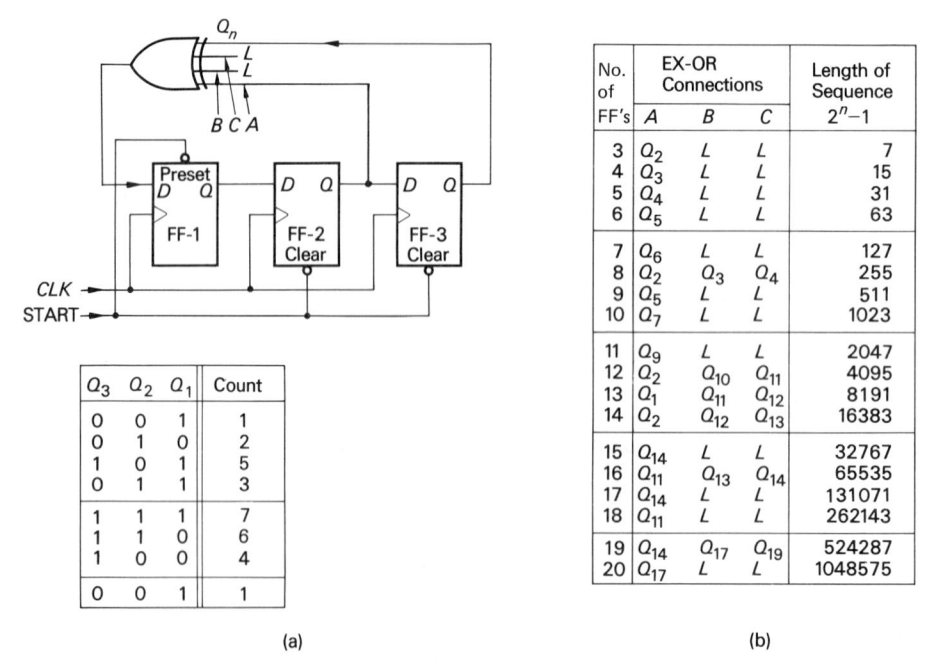

No. of FF's	EX-OR Connections A	B	C	Length of Sequence 2^n-1
3	Q_2	L	L	7
4	Q_3	L	L	15
5	Q_4	L	L	31
6	Q_5	L	L	63
7	Q_6	L	L	127
8	Q_2	Q_3	Q_4	255
9	Q_5	L	L	511
10	Q_7	L	L	1023
11	Q_9	L	L	2047
12	Q_2	Q_{10}	Q_{11}	4095
13	Q_1	Q_{11}	Q_{12}	8191
14	Q_2	Q_{12}	Q_{13}	16383
15	Q_{14}	L	L	32767
16	Q_{11}	Q_{13}	Q_{14}	65535
17	Q_{14}	L	L	131071
18	Q_{11}	L	L	262143
19	Q_{14}	Q_{17}	Q_{19}	524287
20	Q_{17}	L	L	1048575

Fig. P9.7 Pseudorandom sequence generator: (a) $n = 3$ logic diagram; (b) connections for $n = 3$ to $n = 20$ stages.

by drawing a timing diagram for each output. (b) Show how this PRSG can be used as a frequency divider. (c) Try generating a sequence of pseudorandom numbers using a digital computer.

27. *CRC Generator.* Cyclic redundancy check (CRC) generators are used in data communications systems and in data storage systems to detect errors in received data. The logic diagram of a serial CRC generator is shown in Fig. P9.8 for the generator polynomial $P(x) = 1 + x + x^8 + x^{15}$. Using a library reference on CRC generators (a) explain how a CRC generator detects errors, (b) give an example of polynomial division, (c) explain how the circuit in Fig. P9.8 works, (d) using a digital computer show an example of polynomial division.

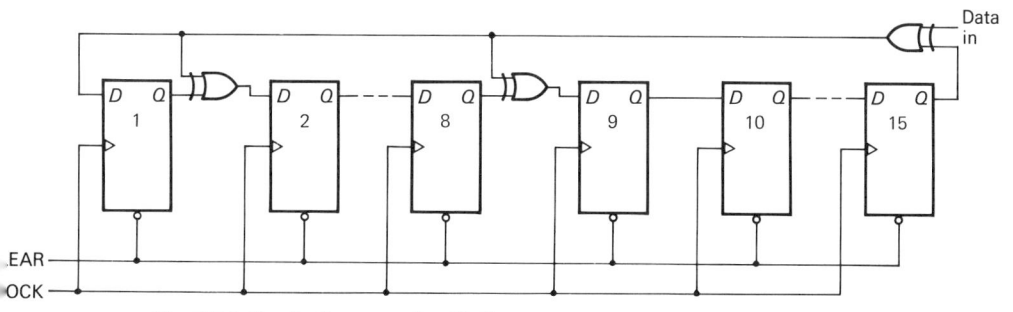

Fig. P9.8 Logic diagram of a CRC generator using a modified register.

Section 9.10

28. (a) Draw the logic diagram for a fully decoded three-bit Johnson counter. (b) Draw the Q outputs and decoder output waveforms.

29. (a) Draw the logic diagram (flip-flops) of a four-stage ring counter. (b) Draw a timing diagram showing the output waveforms if only a single 1-bit is in the counter. (c) Compare the advantages and disadvantages of a ring counter versus a Johnson counter.

REFERENCES

1. *Motors, Motor Generators, Synchros, Resolvers, Electronics, Servos.* 10th edition. Singer-General Precision, Inc., 1969.
2. S. J. Bailey. "Stepper Motors Respond to Direct Digital Command." *Control Engineering,* pp. 46–49, January 1974.
3. Robert L. Morris and John R. Miller (eds.). *Designing with TTL Integrated Circuits.* New York: McGraw-Hill, 1971.
4. Don Aldridge. "Building Your Own Digital Voltmeter." *Electronics,* pp. 132–133, December 6, 1973.

5. Bob Parsons. "SN 7497 Binary Rate Multiplier." Bulletin CA-160, Texas Instruments Inc., May 1971.
6. John B. Peatman. *The Design of Digital Systems.* New York: McGraw-Hill, 1972.
7. William E. Wickes. *Logic Design with Integrated Circuits.* John Wiley & Sons, 1968.
8. Jacob Millman and Christos C. Halkias. *Integrated Electronics: Analog and Digital Circuits and Systems.* New York: McGraw-Hill, 1972.
9. *Digital, Linear, MOS Applications.* Signetics Corporation, 1973.
10. J. Maynard. "Linear Sequence Generators." *Digital Design,* pp. 28–29, August 1973.
11. John Springer. "A Random-Pattern Generator for Testing Digital Delay Elements." *EDN,* pp. 48–50, November 5, 1973.
12. Arthur C. Erdman. "Operation of Programmable Frequency Dividers." *Computer Design,* pp. 110–112, May 1974.
13. William N. Carr and Jack P. Mize. *MOS/LSI Design and Applications.* New York: McGraw-Hill, 1972.
14. Peter Alfke and Ib Larsen (eds.). *The TTL Applications Handbook.* Fairchild Semiconductor, August 1973.

10: SEMICONDUCTOR MEMORIES

10.1 INTRODUCTION

A **memory** is a device or medium in which information can be stored and from which information can be retrieved. Memories are used to store information temporarily and over long periods of time in systems such as computers, in calculators, and in industrial controllers. Memories are also used in electronic instruments such as oscilloscopes and digital multimeters.

There are many types of memories, for example punched cards, punched paper tape, magnetic tape, magnetic disk, magnetic core, magnetic bubbles [1], optical devices [1], charge coupled devices (CCD) [2], and semiconductor devices. Only semiconductor memories will be discussed in this chapter. A **semiconductor memory** is an MSI or LSI device made from bipolar transistors or MOS transistors.

Memories vary in their capacity to store information; there may be only a few bits or there may be thousands of bits stored in one IC. Figure 10.1 shows

Fig. 10.1 Semiconductor memory system. (Courtesy of Prime Computer Company.)

a semiconductor memory that consists of eight rows and sixteen columns of LSI memory chips. Each IC can store 4096 bits, so the total capacity of this memory board is 8 × 16 × 4096 = 524,288 bits. A 1024-bit memory is commonly called a 1K memory, a 2048-bit memory is called 2K memory, etc.

In this chapter we shall treat read-only memories, programmable logic arrays, and read/write memories from an applications point of view.

10.2 THE READ-ONLY MEMORY

The following sections will discuss the basic principles of read-only memories (ROMs) and then give examples of their applications.

10.2.1 Basic Principles of ROMs

A **read-only memory** (ROM) is a memory that stores information in permanent or semipermanent form. Entering data into a ROM is called **programming** the ROM. Once a ROM is programmed, data can be read out of it any time it is needed. ROMs are available in the TTL, CMOSL, and ECL families. They vary in capacity from 256 bits to 65,536 bits, with larger capacities soon to become available.

Figure 10.2(a) shows an eight-bit ROM arranged in four words, each word having two bits—in other words, a 4W × 2b memory (also called simply a 4 × 2 memory). There are four rows labeled X1, X2, X3, and X4, and two columns, Y1 and Y2. The X and Y lines are not connected at their intersections. A diode is located at the intersection of each X and Y line. For example, diode D_{11} is located at the intersection of row X1 and column Y1. Diode D_{12} has been open-circuited, so no connection exists between lines X1 and Y2.

If a positive voltage (HIGH) is applied to input X1, sufficient to forward-bias D_{11}, then current flows from X1 through D_{11} and R_1 to ground, producing a HIGH output at Y1. Thus memory location X1–Y1 has a HIGH stored in it. Since no current can flow through the open-circuited D_{12}, Y2 = 0 V (LOW); i.e., memory location X1–Y2 has a LOW stored in it. Thus the word stored at the location corresponding to the address X1 is Y1 = H, Y2 = L, or HL. The words corresponding to addresses X2, X3, and X4 are HH, LL, and HL, respectively. Each memory location, represented by the presence or absence of a diode, is called a **memory (storage) cell**, since it stores a HIGH or a LOW.

Figure 10.2(b) shows the general form of a 4W × 4b ROM. It consists of an address decoder, 16 storage cells, and four output buffer amplifiers and resistors. This 4 × 4 ROM has four bit lines Y1, Y2, Y3, and Y4, and four word lines X1, X2, X3, and X4. The four word lines (rows) are selected by addressing the two-input decoder. The truth table in Fig. 10.2(c) shows the word corresponding to each address. For example, the word corresponding to address A_1 = H and A_0 = L is LHLL. Note that if the addresses are considered as binary numbers, then the ROM stores the square of these numbers. For example, the

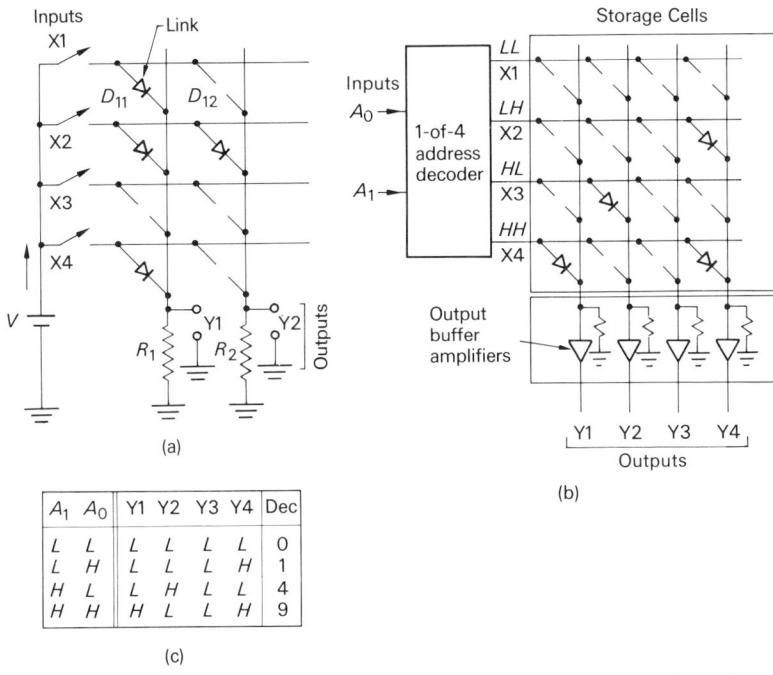

Fig. 10.2 Read-only memory: (a) basic eight-bit ROM; (b) ROM with address decoder and output buffers; (c) table of squares.

square of 0 is 0, the square of 1 is 1, the square of 2 is 4, and the square of 3 is 9. Thus a ROM may serve as a table look-up for squares. (See Section 10.2.6.)

The elements linking the X and Y lines in Fig. 10.2(a) are commonly either bipolar transistors or MOS transistors. The principles discussed above apply equally to any type of link. In general, bipolar links are faster and have higher drive capability, while MOS links occupy less space and consume less power. Note that the cells of the ROM can be read out in any order, i.e., randomly. Hence a ROM is an example of a **random-access memory**. A random access memory can also be read out in any predetermined order, such as sequentially for example. A discussion of the circuits used in ROM cells is found in the references [3; 4, p. 196].

The connecting links in a ROM can be programmed by the manufacturer to suit a particular application. Such ROMs are called **mask programmed** or **custom programmed ROMs** because of the special photographic mask that is prepared by the manufacturer for producing the ROM. The cost of such a mask is relatively high and is justified only when a large number (typically thousands) of identical ROMS are required. The masks are prepared from a table, similar to the one shown in Fig. 10.2(c), supplied by the customer. Frequently the table is prepared on computer cards or punched tape, for ease in preparing an error-free photographic mask. Mask programmed ROMs cannot be reprogrammed.

Manufacturers offer a variety of standard programmed ROMs that perform such tasks as sine look-up, code conversion, and character generation.

When only a few nonstandard ROMs are needed, the user may instead choose ROMs that he can program himself. Such ROMs are called **field programmable ROMs**, or simply **programmable ROMs** (pROMs). Bipolar pROMs are programmed by applying sufficient current to open-circuit the ROM links, or to short-circuit a diode junction in some types. A **reprogrammable ROM** is one that can be reprogrammed a number of times by the user. For example, a ROM that uses floating-gate MOS transistors can be programmed by applying a high voltage, and then cleared by applying ultraviolet (UV) radiation (for about 10 minutes). **Electrically alterable ROMs** (EAROMS) are pROMs which can be erased electrically (at a speed on the order of 1 sec for 1K bits) and then electrically reprogrammed. They can be made from a **metal nitride oxide silicon** (MNOS) process [20]. Programming a ROM requires special procedures such as specific voltage amplitudes and pulse widths, so it is unlikely that a memory cell would be accidentally programmed during normal use. There are several types of links that can be reprogrammed, as discussed in the references [3, p. 72]. Some of the pROMs as supplied by the manufacturer have all HIGHs stored in their memory cells, whereas others have all LOWs. pROMs can be programmed manually, or

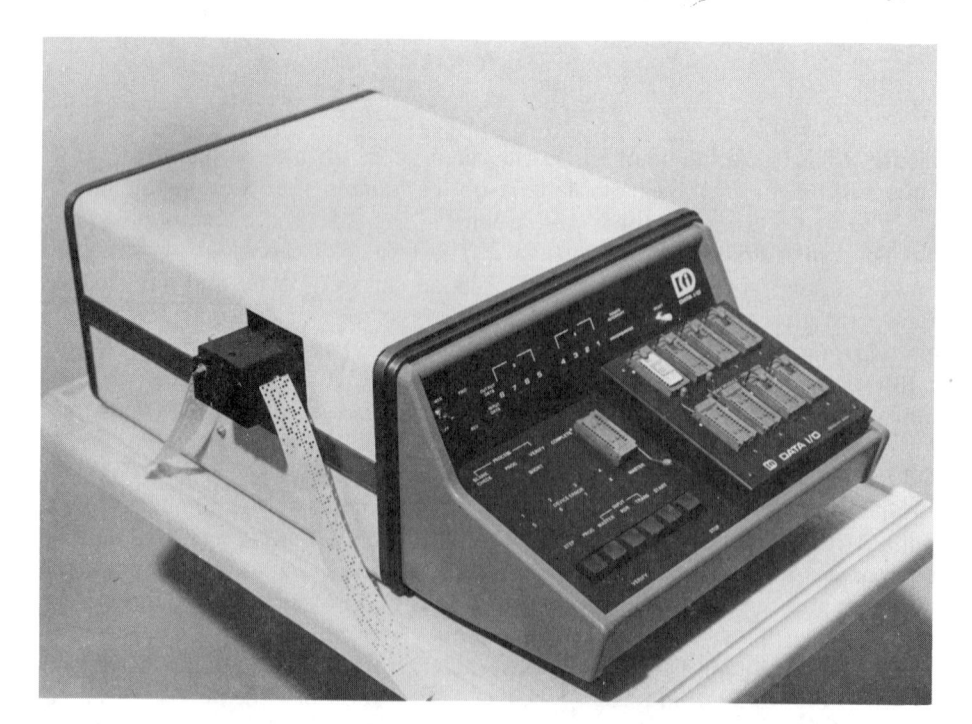

Fig. 10.3 ROM programmer using a minicomputer. (Courtesy of Intel Corporation.)

10.2　　　　　　　　　　　　　　　　　　　THE READ-ONLY MEMORY　　293

automatically using a teletype input to a computer. The disadvantages of pROM compared to ROM are lower bit density and greater cost; the advantage, its re-programmability.

Figure 10.3 shows a ROM programmer that has the bit pattern to be programmed in the ROM stored on a punched paper tape.

The main advantages of ROMs over other types of memories are their low cost, low power consumption, high speed, and nonvolatility. A **nonvolatile memory** is one that retains its stored data even when the power supply is turned off. The main disadvantage of ROMs is that it is costly and time-consuming to reprogram them if a programming error is made.

10.2.2 Electrical Characteristics of ROMs

Figure 10.4 shows the functional block diagram and word-select table for the TTL 74187 256 × 4 custom programmable ROM. Address inputs D, E, F, G, and H select one of 32 rows of memory cells. Address inputs A, B, and C select one of 8 columns for each of the four outputs Y_1, Y_2, Y_3, and Y_4. The word-select table shows which four-bit word is selected by each set of eight address inputs. Input A is the LSB and input H is the MSB. When one or both enable inputs are HIGH, all outputs become HIGH.

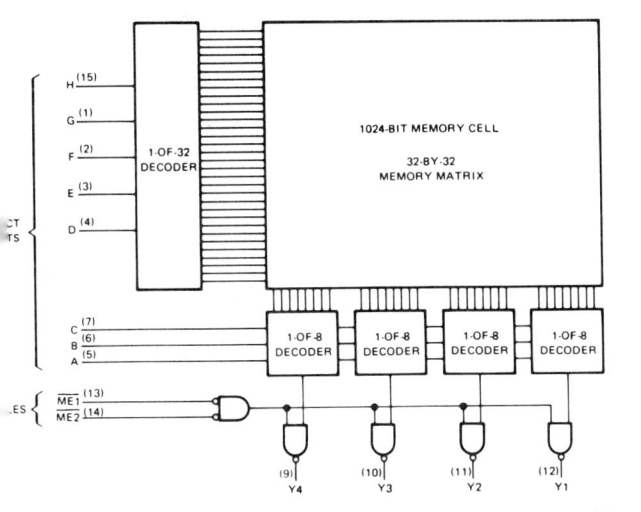

WORD-SELECT TABLE

WORD	H	G	F	E	D	C	B	A
0	L	L	L	L	L	L	L	L
1	L	L	L	L	L	L	L	H
2	L	L	L	L	L	L	H	L
3	L	L	L	L	L	L	H	H
4	L	L	L	L	L	H	L	L
5	L	L	L	L	L	H	L	H
6	L	L	L	L	L	H	H	L
7	L	L	L	L	L	H	H	H
8	L	L	L	L	H	L	L	L
Words 9 thru 250 omitted							
251	H	H	H	H	H	L	H	H
252	H	H	H	H	H	H	L	L
253	H	H	H	H	H	H	L	H
254	H	H	H	H	H	H	H	L
255	H	H	H	H	H	H	H	H

Fig. 10.4 TTL 74187 256W × 4b custom programmable ROM. (Courtesy of Texas Instruments Incorporated.)

Figure 10.5 shows the electrical specifications of the 74187 ROM. The **address** (or **select**) **access time** is the delay between the time addresses arrive at the address input and the time that word is at the output of the memory. The maximum address access time for the 74187 is 60 ns. For comparison, the maximum address access times for some other ROMs are: Signetics 2580 MOS 8K (2048 × 4)

recommended operating conditions

	SN54187 MIN	SN54187 NOM	SN54187 MAX	SN74187 MIN	SN74187 NOM	SN74187 MAX	UNIT
Supply voltage, V_{CC}	4.5	5	5.5	4.75	5	5.25	V
High-level output voltage, V_{OH}			5.5			5.5	V
Low-level output current, I_{OL}			16			16	mA
Operating free-air temperature, T_A (see Note 2)	−55		125	0		70	°C

electrical characteristics over recommended operating free-air temperature range (unless otherwise noted)

PARAMETER		TEST CONDITIONS†		MIN	TYP‡	MAX	UNIT
V_{IH}	High-level input voltage			2			V
V_{IL}	Low-level input voltage					0.8	V
V_I	Input clamp voltage	V_{CC} = MIN,	I_I = −12 mA			−1.5	V
I_{OH}	High-level output current	V_{CC} = MIN, V_{IL} = 0.8 V,	V_{IH} = 2 V, V_{OH} = 5.5 V			40	μA
V_{OL}	Low-level output voltage	V_{CC} = MIN, V_{IH} = 2 V, V_{IL} = 0.8 V	I_{OL} = 12 mA			0.4	V
			I_{OL} = 16 mA			0.45	
I_I	Input current at maximum input voltage	V_{CC} = MAX,	V_I = 5.5 V			1	mA
I_{IH}	High-level input current	V_{CC} = MAX,	V_I = 2.4 V			40	μA
I_{IL}	Low-level input current	V_{CC} = MAX,	V_I = 0.4 V			−1	mA
I_{CC}	Supply current	V_{CC} = MAX,	See Note 3		92	130	mA
C_o	Off-state output capacitance	V_{CC} = 5 V, f = 1 MHz	V_O = 5 V,		6.5		pF

†For conditions shown as MIN or MAX, use the appropriate value specified under recommended operating conditions.
‡All typical values are at V_{CC} = 5 V, T_A = 25°C.

switching characteristics, V_{CC} = 5 V, T_A = 25°C

PARAMETER		TEST CONDITIONS	MIN	TYP	MAX	UNIT
t_{PLH}	Propagation delay time, low-to-high-level output from enable	C_L = 30 pF to GND, R_{L1} = 300 Ω to V_{CC}, R_{L2} = 600 Ω to GND, See Note 4		20	30	ns
t_{PHL}	Propagation delay time, high-to-low-level output from enable			20	30	
t_{PLH}	Propagation delay time, low-to-high-level output from select			40	60	ns
t_{PHL}	Propagation delay time, high-to-low-level output from select			40	60	

Fig. 10.5 Electrical specifications for the TTL 74187 ROM. (Courtesy of Texas Instruments Incorporated.)

ROM—950 ns; Intel 3304 Schottky bipolar 4K (1024 × 4) ROM—65 ns; Signetics 10139 ECL 256-bit (32 × 8) pROM—20 ns; Motorola 14524 CMOSL 1K (256 × 4) ROM—1800 ns (typical). The Intel 2708 is an 8K pROM (1K × 8) with a worst-case access time of 500 ns, 100 sec programming time for 8K bits, and typical power dissipation of 97 μW/bit. The Intel 2316A 16K(2K × 8) ROM has an access time of 0.85 μs and maximum power dissipation of 500 mW. The Nitron NC7010 is a 1024 × 1 EAROM with a typical power dissipation of

10.2 THE READ-ONLY MEMORY 295

24 mW which can be electrically cleared in 2 sec typically. It has a 20-μs access time.

The outputs of the 74187 have open collectors for expanding the memory size in a manner that we shall now describe.

10.2.3 Memory Expansion

ROMs with open outputs can be wired together to increase the word size of the memory as shown in the following example.

Example 10.1 *Memory Expansion.* Draw the logic (block) diagram for a 512W × 4b memory using two 256W × 4b TTL 74187 ROMs with open collectors.

Solution. The word size of a memory can be easily increased by using the enable input as an address input. Figure 10.6(a) shows the expanded memory with the enable input acting as address input A_8. If A_8 is LOW, any one of 256 words in only ROM-1 can be selected with address inputs A_0–A_7. If A_8 is HIGH, only ROM-2 is selected (due to the inverter) and any of its 256 words can be addressed with inputs A_0–A_7.

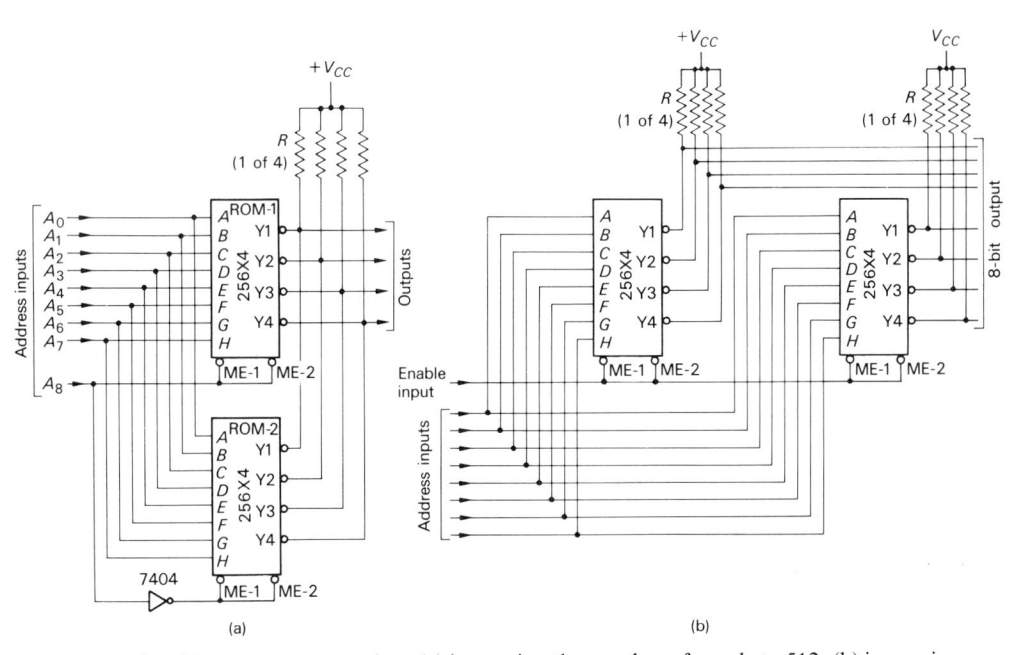

Fig. 10.6 ROM memory expansion: (a) increasing the number of words to 512; (b) increasing the number of bits to 8.

296   SEMICONDUCTOR MEMORIES   10.2

The outputs in Fig. 10.6(a) are wire-ANDed together through pull-up resistors R to V_{CC}. The value of R can be calculated from the formulas for open output gates given in Chapter 3. Although this is a wire-AND connection, only one ROM is connected to the outputs at one time. The open collector configuration avoids the use of output control gates and their attendant delays. ∎

In memory systems the overall access time and timing of inputs is important, and not just the timing of the individual ICs. The system access time must include delays introduced by (1) address gates, (2) external wiring and discrete components, and (3) capacitive loading.

Note that in Fig. 10.6(a), the delay time for address input A_8 in ROM-2 will be about 10 ns greater than that for the other address inputs, due to the NOT gate. However, in the specifications for the enable input (Fig. 10.5) the enable-to-output delay time is only 30 ns as compared with 60 ns for the address inputs. Therefore, unless many ROMs are connected together, external address decoding gates will not reduce the memory speed through propagation delays.

The external pull-up resistors for open output memories should be as small as possible to avoid reducing the memory speed due to large RC time constants. The layout and wiring of the ROMs should be done so as to minimize capacitive loading. The power supply should be adequately filtered and capable of delivering the required current during transient operation.

The number of bits per word can be increased by using a common enable line and not wiring the outputs together. Figure 10.6(b) shows a 256W × 8b ROM made from two 256W × 4b ROMs.

Some ROMs have three-state outputs for connecting the outputs to a memory bus. Three-state outputs provide the convenience of being able to wire the outputs together without sacrificing speed because of external pull-up resistors. Control signals must ensure that only one three-state ROM output is connected to the same bus at any one time. ROMS are generally designed so their output disable times are shorter than their output enable times in order to simplify this timing problem.

The five main types of applications for which ROMs are used are: (1) random logic generation, (2) microprogramming, (3) table look-up, (4) code conversion and generation, and (5) character generation.

10.2.4 Random Logic Generation

Any Boolean function can be programmed in a ROM. Consider programming the AND function $F = AB$, whose truth table and logic symbol are shown in Fig. 10.7(a). A ROM with four rows and one column is used, as shown in Fig. 10.7(b). For each input combination of A and B that results in a HIGH output in the truth table, an unbroken link, shown as a dot, is placed in the ROM array. For example, when the address input is $AB = HH$, the output F is HIGH, so there must be an unbroken link corresponding to the address $AB = HH$. This ROM now

10.2 THE READ-ONLY MEMORY   297

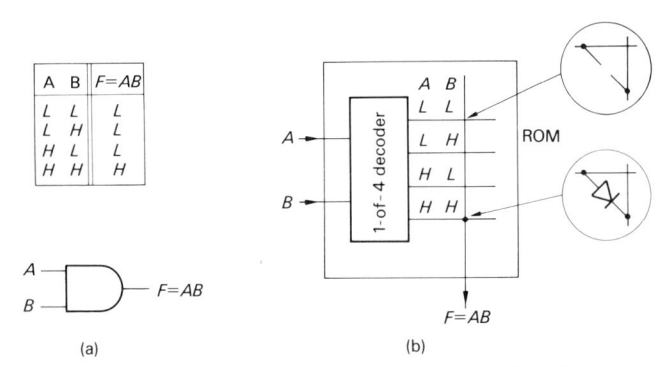

Fig. 10.7 ROM logic generation of the AND function.

performs the same logic as an AND gate. For example, if its inputs are $A = L$, $B = H$, then the ROM output is a LOW. If the inputs are $A = B = H$, then the ROM output is a HIGH.

The ROM used in Fig. 10.7 has two inputs and only one output line, and hence can generate only one function of two variables (A and B). A ROM with N outputs and M inputs can generate N functions of M variables simultaneously. The following example illustrates the ROM implementation of a half-adder which has two variables and two outputs.

Example 10.2 *ROM Half-Adder.* Show the ROM implementation of a two-bit half-adder.

Solution. A two-bit half-adder has two inputs A and B, a sum output $S = A \oplus B$, and a carry output $C = AB$ as shown in Fig. 10.8(a). The truth table for the half-adder is also shown in Fig. 10.8(a). The half-adder can be implemented with a 4×2 ROM as shown in Fig. 10.8(b). The carry output is the AND function.

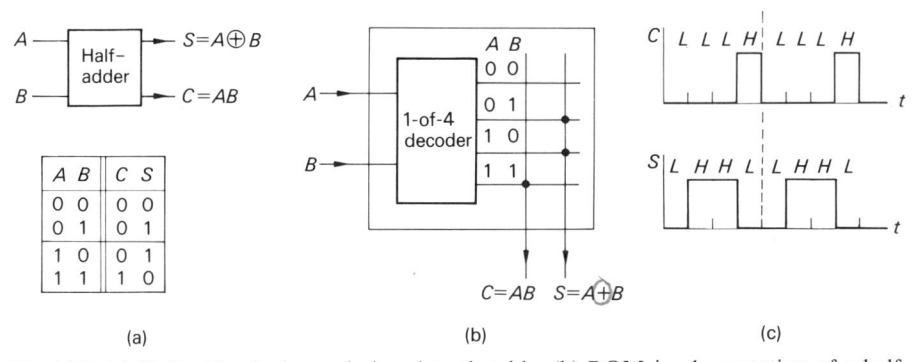

Fig. 10.8 (a) Half-adder logic symbol and truth table; (b) ROM implementation of a half-adder; (c) ROM pulse generator.

The sum output is the EX-OR function since S is a 1 when $AB = 01$ or when $AB = 10$. Hence a link (dot) is shown for these values of A and B in Fig. 10.8(b). ∎

A ROM can also be used to generate any desired pulse train, as shown by the next example.

Example 10.3 *ROM Pulse Generator.* Use a ROM to generate the pulse train shown in Fig. 10.8(c).

Solution. Each column in the ROM can generate one pulse train if the address inputs of the ROM are addressed in sequence repeatedly. The first pulse train is *LLLH*, so a ROM with four rows is programmed with an *LLLH*. Similarly, the second column in the ROM is programmed with an *LHHL*. The programmed ROM is shown in Fig. 10.8(b). ∎

The above examples illustrate the principles of programming a single ROM to generate a desired logic function. As long as IC devices are available to perform the AND function, binary addition, etc., it would not be economical to use a ROM.

In general one has four options for implementing a given logic system: (1) IC devices (gates, flip-flops, counters, etc.), (2) custom IC circuits, (3) a ROM, or (4) a microcomputer. Implementation using IC devices offers the greatest flexibility. Custom IC circuits, produced by semiconductor manufacturers, offer low cost in large volume. ROMs and pROMs can be programmed in a relatively short time and are somewhat flexible. Microcomputers can be programmed to perform any function and can be inexpensive for more complex logic systems.

The ease with which the ROM can be tested is one of its greatest assets. Verifying the correct operation of a complex logic array requires a special tester for each array or an expensive general-purpose testing machine, whereas the ROM implementation requires only a very simple standard tester that checks the memory cell contents and/or the output logic functions.

The next example illustrates the use of a ROM in the control logic for a modification of the vending machine described in Chapter 2.

Example 10.4 *Control Logic for a Vending Machine.* Develop the ROM implementation of a 10¢ coffee vending machine that will dispense coffee and the correct change for any 10¢-or-larger combination of one or two nickels, one dime, or one quarter.

Solution. The logic (block) symbol for the coffee machine is shown in Fig. 10.9(a). The truth table for the coffee machine shown in Fig. 10.9(b) has input columns for all possible inputs: two nickels N_1 and N_2, a dime D, or a quarter Q. There is one output column for coffee C, and there are four output columns for the change. A 1 in the truth table means a coin inserted, coffee dispensed, or change returned. A 0 has the opposite meaning. The ROM is easily programmed by putting a dot, representing a 1, in each row and column in the ROM, corresponding to each 1 in the truth table, as shown in Fig. 10.9(c).

10.2 THE READ-ONLY MEMORY

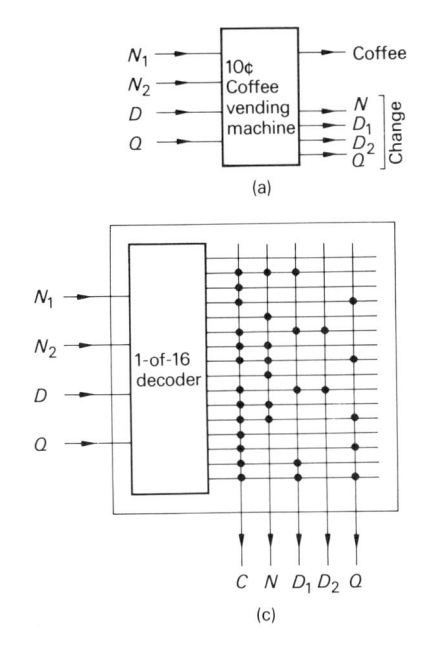

Fig. 10.9 Control logic for a vending machine: (a) logic block diagram; (b) truth table; (c) programmed ROM.

Inputs				Outputs				
N_1	N_2	D	Q	C	N	D_1	D_2	Q
0	0	0	0	0	0	0	0	0
0	0	0	1	1	1	1	0	0
0	0	1	0	1	0	0	0	0
0	0	1	1	1	0	0	0	1
0	1	0	0	0	1	0	0	0
0	1	0	1	1	0	1	1	0
0	1	1	0	1	1	0	0	0
0	1	1	1	1	1	0	0	1
1	0	0	0	0	1	0	0	0
1	0	0	1	1	0	1	1	0
1	0	1	0	1	1	0	0	0
1	0	1	1	1	1	0	0	1
1	1	0	0	1	0	0	0	0
1	1	0	1	1	0	0	0	1
1	1	1	0	1	0	1	0	0
1	1	1	1	1	0	1	0	1

This ROM implementation of the coffee machine should be compared with the logic gate implementation discussed in Section 2.3 and in the problems in Chapter 4. Additional features can easily be added to the ROM implementation, for example, returning two nickels instead of one dime change when there are no dimes left in the machine, or adding a delay time for the cup to fall in place before filling it with coffee. ∎

Sequential circuits can also be made with ROMs by feeding back some ROM outputs through strobed latches or flip-flops into some of the ROM inputs [16].

*10.2.5 Microprogramming

Microprogramming is a technique used for controlling computer-like systems by storing programs in ROMs rather than using special logic circuits. A **microprogram** is a sequence of instructions that controls the flow of data along data paths in the digital system.

Figure 10.10 shows a block diagram of a microprogrammed system which contains a logic circuit, two ROMs (a sequence ROM and a control ROM), an arithmetic logic unit (ALU), and registers. The sequence ROM is essentially a counter that can produce a variety of count sequences which become address inputs to the control ROM. An input command sets the sequence ROM to an initial state. The control ROM converts the address provided by the sequence

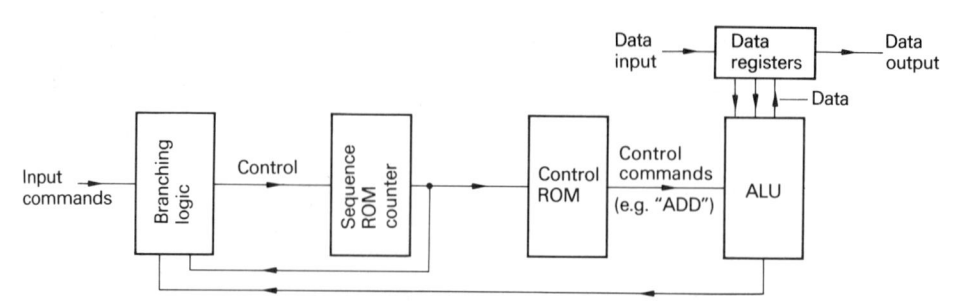

Fig. 10.10 Block diagram of a general microprogrammed system.

ROM into a series of control commands to be used by a device, such as an ALU. An example of an ALU command is "ADD the two numbers stored in the data registers." The control branching logic determines when the state of the sequence ROM is to be changed, and produces an instruction that gives a new command to the ALU.

The main advantages of a microprogrammed system are: (1) the instructions can be changed by simply plugging in a new ROM without having to change the design of the system, (2) the ROMs replace a lot of the system logic, and (3) there is increased speed and reduced cost for repetitive functions.

A microprogrammed computer is described in Appendix G and a further discussion of ROMs used for microprogramming will be found in the references [5; 4, p. 205; 21].

*10.2.6 Table Look-Up

In a look-up table, the address is related by some function to the word stored at that address. For example, in the square look-up table shown in Fig. 10.2(c), the output Y is related by the square function to the inputs X; i.e., $Y = X^2$.

Figure 10.11 shows part of an 8-bit, 0° to 90° sine look-up table. Only angles 0° through 44° are shown. The outputs $B_1, B_2, \ldots, B_8$ (LSB) are in fractional form and have the weights $\frac{1}{2}, \frac{1}{4}, \ldots, \frac{1}{256}$ respectively.

Example 10.5 *Sine Look-Up ROM.* Find the sine of 19° using the sine look-up table in Fig. 10.11.

Solution. In Fig. 10.11 find the angle nearest 19° which is 18.98°. The sine of 18.98° is .01010011. Referring to Appendix B we find that this is

$$.0000 + .2500 + .0000 + .0625 + .0000 + .0000 + .0078 + .0039 = 0.3242,$$

compared with 0.32557 for the more exact value of the sine of 19°, an error of only 0.437%. More accuracy is possible with a greater number of inputs and outputs. ∎

ADDRESS REFERENCE	FUNCTION INPUT DEGREES	FUNCTION INPUT RADIANS	CODE OUTPUT B_8	B_7	B_6	B_5	B_4	B_3	B_2	B_1
0	00	000	0	0	0	0	0	0	0	0
1	70	012	1	1	0	0	0	0	0	0
2	1 41	025	0	1	1	0	0	0	0	0
3	2 11	037	1	0	0	1	0	0	0	0
4	2 81	049	0	0	1	1	0	0	0	0
5	3 52	061	1	1	1	1	0	0	0	0
6	4 22	074	1	1	0	0	1	0	0	0
7	4 92	086	0	1	1	0	1	0	0	0
8	5 63	098	1	0	0	1	1	0	0	0
9	6 33	110	0	0	1	1	1	0	0	0
10	7 03	123	1	1	1	1	1	0	0	0
11	7 73	135	0	1	0	0	0	1	0	0
12	8 44	147	1	0	1	0	0	1	0	0
13	9 14	160	0	0	0	1	0	1	0	0
14	9 84	172	0	0	1	1	0	1	0	0
15	10 55	184	1	1	1	1	0	1	0	0
16	11 25	196	0	1	0	0	1	1	0	0
17	11 95	209	1	0	1	0	1	1	0	0
18	12 66	221	0	0	0	1	1	1	0	0
19	13 36	233	1	1	0	1	1	1	0	0
20	14 06	245	0	1	1	1	1	1	0	0
21	14 77	258	1	0	0	0	0	0	1	0
22	15 47	270	0	0	1	0	0	0	1	0
23	16 17	282	1	1	1	0	0	0	1	0
24	16 88	295	0	1	0	1	0	0	1	0
25	17 58	307	1	0	1	1	0	0	1	0
26	18 28	319	0	0	0	0	1	0	1	0
27	18 98	331	1	1	0	0	1	0	1	0
28	19 69	344	0	1	1	0	1	0	1	0
29	20 39	356	1	0	0	1	1	0	1	0
30	21 09	368	0	0	1	1	1	0	1	0
31	21 80	380	1	1	1	1	1	0	1	0
32	22 50	393	0	1	0	0	0	1	1	0
33	23 20	405	1	1	1	0	0	1	1	0
34	23 91	417	1	1	1	0	0	1	1	0
35	24 61	430	0	1	0	1	0	1	1	0
36	25 31	442	1	0	1	1	0	1	1	0
37	26 02	454	0	0	0	0	1	1	1	0
38	26 72	466	1	1	0	0	1	1	1	0
39	27 42	479	0	1	1	0	1	1	1	0
40	28 13	491	0	0	0	1	1	1	1	0
41	28 83	503	1	1	0	1	1	1	1	0
42	29 53	515	0	1	1	1	1	1	1	0
43	30 23	528	0	0	0	0	0	0	0	1
44	30 94	540	1	1	0	0	0	0	0	1
45	31 64	552	0	1	1	0	0	0	0	1
46	32 34	565	1	0	0	1	0	0	0	1
47	33 05	577	1	1	0	1	0	0	0	1
48	33 75	589	0	1	1	1	0	0	0	1
49	34 45	601	1	0	0	0	1	0	0	1
50	35 16	614	1	1	0	0	1	0	0	1
51	35 86	626	0	1	1	0	1	0	0	1
52	36 56	638	0	0	0	1	1	0	0	1
53	37 27	650	1	1	0	1	1	0	0	1
54	37 97	663	1	0	1	1	1	0	0	1
55	38 67	675	0	0	0	0	0	1	0	1
56	39 37	687	0	1	0	0	0	1	0	1
57	40 08	699	1	0	1	0	0	1	0	1
58	40 78	712	1	1	1	0	0	1	0	1
59	41 48	724	1	0	0	1	0	1	0	1
60	42 19	736	0	0	1	1	0	1	0	1
61	42 89	749	0	1	1	1	0	1	0	1
62	43 59	761	0	0	0	0	1	1	0	1
63	44 30	773	1	1	0	0	1	1	0	1

Fig. 10.11 Sine look-up table for 0 to 44°. (Courtesy of National Semiconductor.)

Cosine and arctangent look-up tables are also commerically available. Look-up tables for any function can be produced by the manufacturer using mask programming, or by the user using pROMs.

*10.2.7 Code Conversion and Generation

Let's develop a ROM which converts an 8421 binary code input into an excess-3 output. The excess-3 BCD code, discussed in Section 5.6, is obtained by adding 3_{10} to each decimal digit in a given decimal number and converting each digit to binary using the 8421 code. For example, the decimal number 2 in excess-3 is $2 + 3 = 5$ or 0101_2. The table in Fig. 10.12(a) shows the excess-3 code for the decimal numbers 0 through 9. The programmed ROM is shown in Fig. 10.12(a) for converting 0 through 9 in 8421 code, to the corresponding excess-3 code. Numbers larger than 9 can be converted by using one ROM for each decimal digit. Thus to convert 473_{10} from 8421 to excess-3 would require three 10×4-bit ROMs.

The excess-3 code is mathematically related to the 8421 code, by adding 3 to the 8421 number. Some codes are not mathematically related. For example, there is no way of mathematically determining the bits for any ASCII code character given the bits in the EBCDIC code. Thus arithmetic logic cannot be used to convert from ASCII to EBCDIC, so one must resort to some type of table. A ROM is ideally suited for conversions between nonmathematically related codes.

Some commonly available IC code converters are: EBCDIC to ASCII, ASCII to EBCDIC, ASCII to Hollerith (used in computer punched cards), and Hollerith to ASCII. Frequently code converters have a mode control which

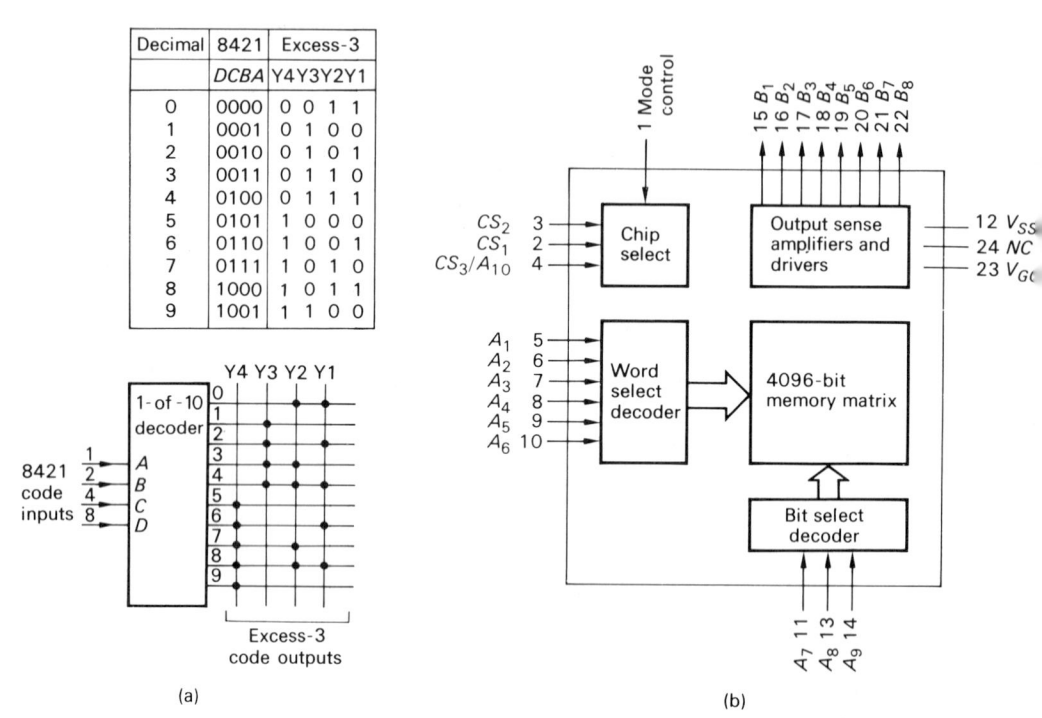

Fig. 10.12 Read-only memories: (a) excess-3 code converter; (b) EBCDIC/ASCII MOS code converter (Mostek 2503) (courtesy of Mostek Corp.).

allows the conversion to take place in either direction. Chip enable inputs are provided to control when a conversion is to take place. Figure 10.12(b) shows the block diagram for an EBCDIC/ASCII code converter with nine address inputs and eight outputs. Three chip select inputs control the conversion. The address access time for this Mostek 2503 is 400 ns.

*10.2.8 Character Generation

Alphameric characters can be displayed using a 5 × 7 or larger dot matrix display as shown in Section 6.7. The display can be made from devices such as LEDs, ink-spray printers, and cathode ray tube (CRT) displays. ROMs can be used to generate a set of characters to be displayed by any of the above devices.

Figure 10.13 shows the logic symbol for a ROM and dot matrix which will generate the logic signals for, and display, any one of 64 5 × 7 characters. The

10.2   THE READ-ONLY MEMORY   303

(a)

(b)

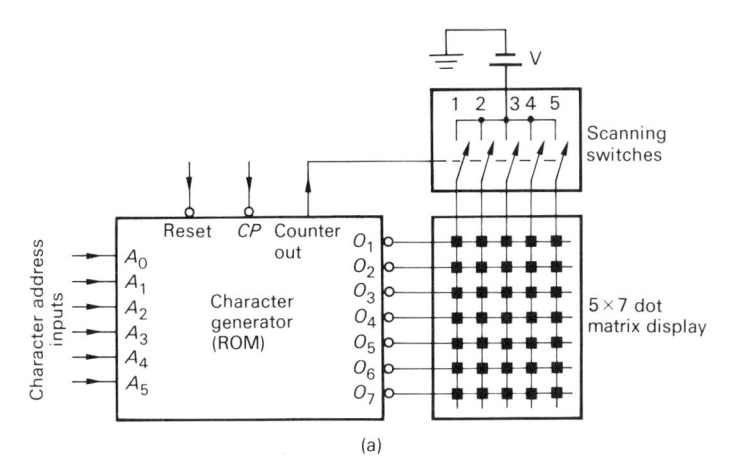

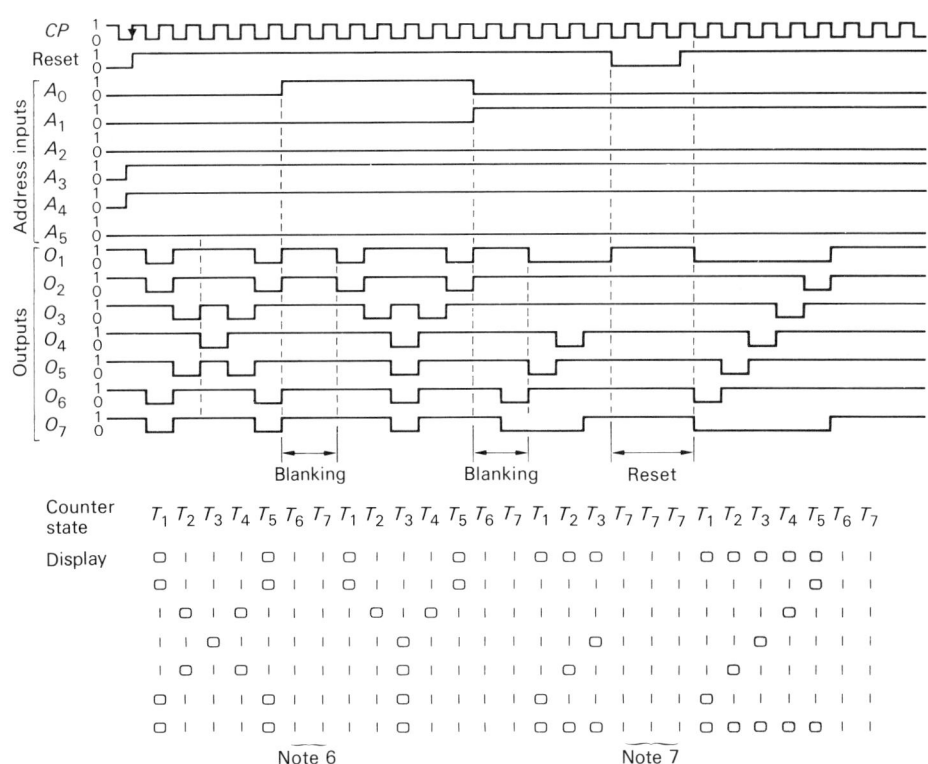

Notes
6. Last two counter states (count mode control = HIGH ⇒ MOD 7) provide blanking.
7. Counter is Reset to the last state.

Fig. 10.13 (a) Logic symbol and (b) timing diagram for a 3257 64-character 5 × 7 dot matrix character generator. (Courtesy of Fairchild Camera and Instrument Corporation.)

six address inputs (A_0–A_5) select a block in memory which contains the program for generating a specific character. Data representing the first column of a selected character is available at the data output (O_1–O_7) at the trailing edge of the next clock pulse (CP) after the reset input goes HIGH. The remaining four columns of the character are then automatically selected in sequence by an internal counter in the ROM. The scanning switches in Fig. 10.13(a) are closed one at a time in sequence by the internal ROM counter output. The last state of the counter T_5 is used to inactivate the outputs of the ROM to provide one or two blank spaces between characters for multicharacter displays. After one character is generated, a new address is applied to inputs A_0–A_5 to select the next character to be displayed and the above process is repeated, generating one column of the character at a time.

Figure 10.13 shows a timing diagram for generating of the characters X, Y, and Z. The timing diagram shows a clock input CP, six address inputs which select the desired characters, and seven active LOW outputs (O_1–O_7) which supply the proper signal for each of seven rows in the display unit. The count states of the internal counter (count-out) are also shown as T_1 through T_7. The count is reset to state T_7 as long as the reset input is LOW. The address for the character X, using a standard ASCII font (Fig. 6.18), is 000110 ($A_0 A_1 \cdots A_5$) as shown in Fig. 10.13. For counter state T_1, the outputs are 0011100, which forms the first column of the character X. The internal counter then automatically advances to state T_2 at the trailing edge of the next clock pulse and the ROM produces a 1101011 output for forming the second column in the character X. This process continues until the entire pattern shown in Fig. 10.13 is generated. Note that the character Z is not completely generated the first time because the reset input goes LOW during its generation.

The Fairchild 3257 character generator has the following characteristics: 1000 ns typical character address to output access time, clock frequency dc to 1 MHz, typical power dissipation 715 mW, power supplies required +5 V and −12 V, interfaces directly with TTL logic. It can be used for CRT displays, billboard displays, and LED matrix displays.

*10.3 PROGRAMMABLE LOGIC ARRAYS (PLAs)

A **programmable logic array** (PLA) is in an array of storage elements whose address inputs are programmable.

Let's illustrate the principles of a PLA for the half-adder. The logic symbol and truth table for the half-adder are shown in Fig. 10.14(a). The half-adder has two input variables A and B. Therefore, the PLA must have two inputs as shown in Fig. 10.14(b). Inverters U_1 and U_2 inside the PLA generate the complements $\bar{A}$ and $\bar{B}$.

The PLA consist of two distinct arrays of cells. The product array consists

10.3   PROGRAMMABLE LOGIC ARRAYS (PLAs)   305

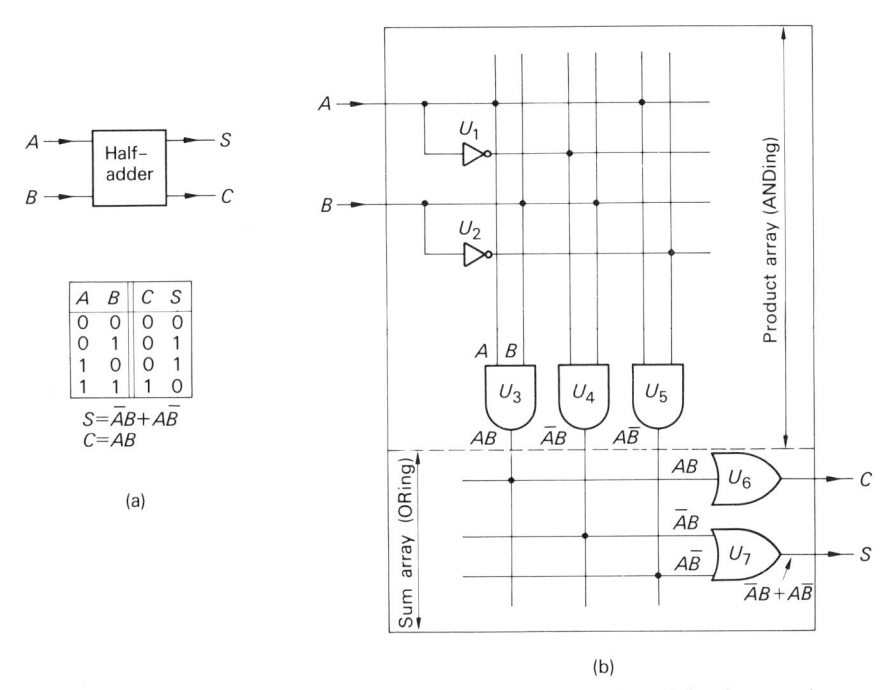

Fig. 10.14 Half-adder: (a) logic symbol and truth table; (b) PLA implementation.

of the input lines A and B, their complements, and AND gates U_3, U_4, and U_5. The sum array consists of the data output lines C and S and OR gates U_6 and U_7.

Let's see how the carry function $C = AB$ is generated. Gate U_3 performs the required ANDing of A and B. The inputs to U_3 are supplied by connecting the two links (shown as a dot) from inputs A and B as in Fig. 10.14(b). When the link in the sum array is connected, the carry function appears at the output of OR gate U_6.

The sum function $S = A\bar{B} + \bar{A}B$ contains two terms ($A\bar{B}$ and $\bar{A}B$), each of which is called a **product term**. The sum function is obtained as follows. The product term $A \cdot \bar{B}$ is generated by connecting links from the A and B address lines to the input of AND gate U_5. The product term $\bar{A} \cdot B$ is generated similarly by AND gate U_4. Then by connecting links in the sum array, these two terms are ORed by gate U_7. The result is $S = A\bar{B} + \bar{A}B$ as required. Note that the address inputs were decoded in the product array as required by the user and not as a standard two-input decoder would decode them.

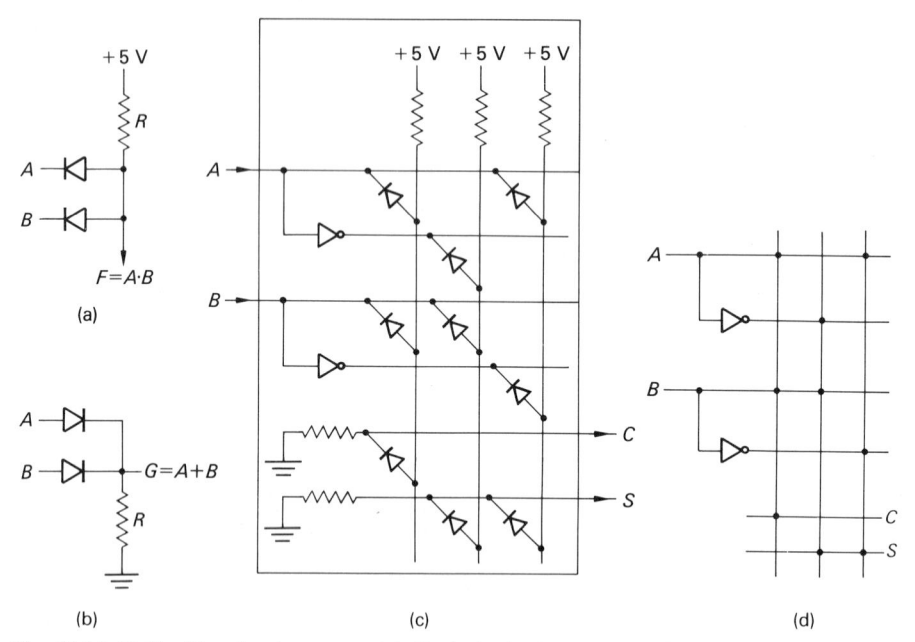

Fig. 10.15 Half-adder development: (a) diode logic AND gate; (b) diode logic OR gate; (c) PLA using diode logic AND and OR gates; (d) simplified representation of the PLA in (c).

Replacing the gates in Fig. 10.14(b) by diode links results in the equivalent PLA shown in Fig. 10.15(c). Its operation can be easily verified by referring to Fig. 10.15(a, b). To simplify the drawing of PLAs, the resistors and power supply voltages are omitted, and the diodes are indicated by dots as shown in Fig. 10.15(d). For clarity one should think of Fig. 10.15(d) in terms of ANDing taking place along vertical (product term) lines and ORing taking place along horizontal (sum term) lines. This is analogous to ANDing the columns across a row and ORing the rows in a truth table.

Let's compare the PLA and ROM implementation of the half-adder. The PLA has no input address decoder; its input decoding is performed by programming the product array. The ROM has a built-in address decoder, so the user has no choice of address decoding. The PLA implementation of the half-adder requires 12 address cells while the ROM requires 8 cells. The PLA in Fig. 10.15(c) has two inputs, two outputs, and can generate up to three product terms (each having two variables).

The fact that PLAs have programmed addresses leads to three special cases: (1) addresses that remain unprogrammed in the PLA, (2) a single address that corresponds to more than one word in the PLA, and (3) a single word in the PLA that has more than one address. These characteristics of a PLA are useful in applications such as microprogrammed systems [8, 9].

The next example shows the advantage of a PLA over a ROM when a large number of variables are involved.

Example 10.6 A microcomputer requires that 48 one-bit words be stored in a ROM having one output in order to implement its set of instructions. A total of 16 input variables are required. Determine the word size of a ROM used to implement the instructions. What configuration of PLA would be required?

Solution. Since 16 input variables are required the ROM would have to have $2^{16} = 65{,}536$ (see Appendix B) words, even though only 48 words are used. The large number of input variables would make the use of a ROM very inefficient. A PLA would require only 16 inputs, one output, and 48 product terms. ■

PLAs frequently have several internal flip-flops, along with the AND/OR logic, which makes the design of sequential PLA circuits easier than using external flip-flops. Let's design a synchronous decade counter with 8421 coded outputs, using a PLA that has internal JK flip-flops. The J and K inputs to the four flip-flops of a decade counter can be determined either from its timing diagram or mathematically [4, p. 244]; they are

$$J_A = H, \qquad K_A = H; \qquad \text{(LSB)}$$
$$J_B = Q_A\bar{Q}_D, \qquad K_B = Q_A\bar{Q}_D;$$
$$J_C = Q_AQ_B, \qquad K_C = Q_AQ_B;$$
$$J_D = Q_AQ_BQ_C, \qquad K_D = Q_A. \qquad \text{(MSB)}$$

Here J_A and K_A are the inputs to the LSB flip-flop A, etc., and Q_A is the output of flip-flop A.

Referring to Fig. 10.16, we first draw the four flip-flops A, B, C, and D. Then we draw the Q and $\bar{Q}$ outputs (generated by inverters) of each flip-flop, placing them in rows. Next we draw eight vertical lines. (Note that two vertical lines are required for each flip-flop output, one for the Q output and one for $\bar{Q}$.)

The J and K inputs to FF-A are connected to a permanent HIGH. Next the ANDing connections are made. The Q_A and $\bar{Q}_D$ outputs are ANDed with the J and K inputs to FF-B as shown in Fig. 10.16. After all the flip-flop outputs are connected to the correct flip-flop inputs we make the ORing connections. The Q_A output is OR connected to the A output line, Q_B to the B line, etc. The result is a synchronous decade counter.

Additional examples of PLAs are found in the references [10, 11, 12, 13].

PLAs are available in mask programmed or field programmed form. They may be made from either MOS or bipolar technology. The TI (Texas Instruments) TMS2000 JC MOS PLA has 17 inputs and 18 outputs, with eight internal JK flip-flops. The National DM 7575 bipolar PLA has 14 inputs and can generate 96 product terms, each containing as many as 14 variables. It has eight outputs and hence can produce eight distinct output functions. It has a typical access time of 90 ns and a typical power dissipation of 550 mW. The Intersil IM 5200

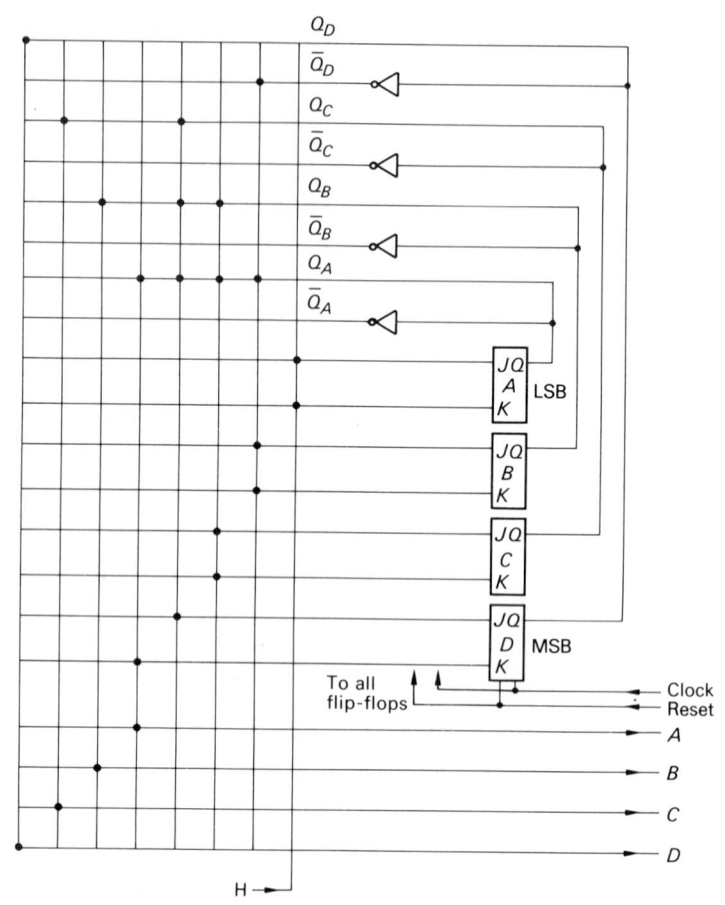

Fig. 10.16 PLA implementation of a synchronous decade counter with 8421 coded outputs.

field programmable PLA (FPLA) has 14 inputs, 8 outputs, and can generate 48 product terms. It can be designated by the notation 14 × 8 × 48. The Signetics 82S101 is a 16 × 8 × 48 FPLA with tristate outputs and a maximum access time of 50 ns.

10.4 THE READ/WRITE MEMORY

A **read/write (R/W) memory** is a memory in which externally selected information can be stored and from which stored information can be retrieved at will. New information can replace previously stored information rapidly and easily as compared with a ROM, for example. Read/write memories are usually called **random-access memories** (RAMs) since their access time is independent of the location in which the data is stored. However, this is an inaccurate term since other types of memories, such as ROMs, can also be accessed randomly.

Read/write memories have many applications such as in computers (where they store information before and after it is processed), calculators, digital processing equipment, digital filters, computer terminals, electronic cash registers, and traffic light controllers.

Read/write memories are available in the TTL, ECL, MOSL, and CMOSL families. They vary in capacity from 16 bits to 16,384 bits, with larger capacities to be offered in the near future. **Word-organized** memories have M words with N bits per word, N data inputs, and N data outputs. Word-organized memories are commonly small memories, such as a $16W \times 4b$ memory. They have only one address decoder, which has M outputs. **Bit-organized** memories have only one data input and one data output. They are more commonly used in larger memories such as a $1024W \times 1b$ memory. Bit-organized memories use two address decoders to address an array of memory cells. The array of cells is usually square, for example a 16×16 array. Memory systems are made from arrays of IC memory chips and can vary in size from a few thousand to more than a million bits.

A read/write memory consists of an array of memory elements called **memory cells**. Each memory cell is made from one or more transistors. **Bipolar memories** use bipolar transistors, whereas **MOS memories** use NMOS or PMOS transistors. Bipolar memories are available in the TTL (standard and Schottky series) and ECL families, and have relatively small capacities (less than 1K bits) due to their relatively large cell size. MOS memories are available in the MOSL and CMOSL families and have large capacities (greater than 1K bits) due to their relatively small cell size.

10.4.1 Bipolar Memory Cells

In this section we shall describe bipolar memory cells from a functional viewpoint. The actual cell circuits are discussed in the references [14] and in the problems.

There are two types of bipolar memory cells. The first type is essentially a strobed D latch, shown in Fig. 10.17(a). This type of cell is often used in **register**

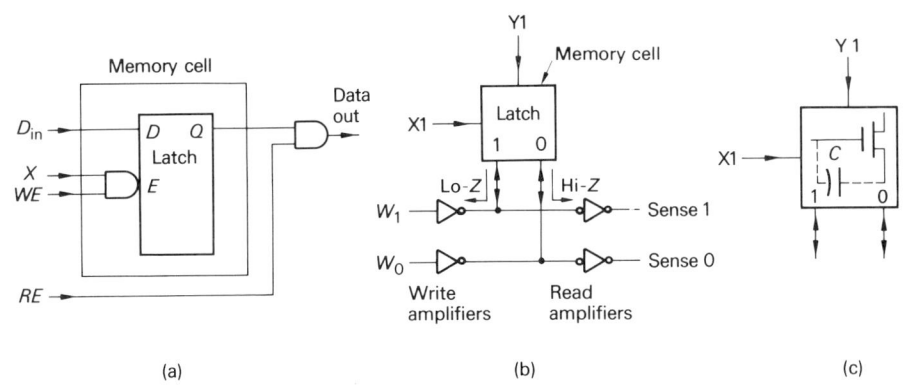

Fig. 10.17 Memory cells: (a) bipolar D latch; (b) bipolar latch with XY addressing; (c) MOS cell with capacitor storage.

files, which are small high-speed memories, such as the TTL 74170 4 × 4 register file. The memory cell in Fig. 10.17(a) has a data input, a gated enable input, and a gated Q output. To write a HIGH or a LOW bit into this memory cell, the cell is first addressed by making the write enable (WE) input HIGH and the address input X HIGH, thus enabling the latch. Then if $D_{in} = H$, a HIGH will be stored in the latch. Similarly, if X = HIGH, WE = HIGH, and $D_{in} = L$, a LOW will be stored in the latch. When either WE or X goes LOW, data stored in the latch will remain there and will not be affected if new data is present at the D input. Once a HIGH or a LOW bit is stored in the latch it can be read out by putting the read enable input HIGH, which enables the output AND gate. Note that the address decoding is done in the memory cell itself, rather than with an external decoder. This type of cell is used in word-organized memories.

The second type of bipolar memory cell is also a latch. It has several different forms; we shall use the form shown in Fig. 10.17(b) to explain its operation. This type of cell is used in memories such as the TTL 7481A 16 × 1 memory. It differs from the first type of memory cell in the following ways. It has two address inputs, X_1 and Y_1, both of which must be HIGH in order to address the cell. This is called X-Y addressing. It has two data lines, labeled 1 and 0, which are used for both input and output. Write and read (also called sense) amplifiers external to the memory cell are used for writing data into the cell, and for reading data out of the cell. Note that in this type of memory cell the decoding gates are external to the cell. The sink impedance of the write amplifiers (Lo-Z) is less than the input impedance of the sense amplifiers (Hi-Z).

To read a 0-bit from the bipolar cell shown in Fig. 10.17(b) the cell is first addressed by putting $X_1 = H$ and $Y_1 = H$ with both write inputs LOW (i.e. $W_0 = L$ and $W_1 = L$). This causes the 1-line of the cell to act like an open circuit so the sense-1 (S_1) amplifier output goes HIGH due to the LOW at W_1. The 0-line of the cell becomes grounded due to the 0 stored in the latch, so S_0 becomes LOW. Thus a 0-bit is read out of the cell.

To write a 0-bit into the cell in Fig. 10.17(b) it is addressed by putting $X_1 = H$ and $Y_1 = H$. The write-0 input is made a 1 while the write-1 input is made a 0 ($W_0 = 1$ and $W_1 = 0$). The outputs of the write amplifiers cause the latch to reset, i.e., to store a 0. Reading and writing of a 1-bit is similar to that for a 0-bit. The circuit schematic and truth table for this type of cell are given in the problems.

Data can be stored in a latch type memory cell indefinitely, as long as power is supplied to the latch; hence this type of memory is called a **static memory**. Data can be read out at any time without destroying the stored data. This type of readout is called a **nondestructive readout** (NDRO). When a magnetic core memory is read out, its stored data is destroyed in the process, i.e., it has a **destructive readout** (DRO). Core memories are operated in an NDRO mode by writing data back into the memory after each read operation. All semiconductor R/W memories are **volatile**, i.e., the stored information is lost if the power is removed from a cell. Magnetic core memories are nonvolatile, since stored data is retained even if the power is removed.

10.4.2 MOS Memory Cells

MOS memory cells can be either static or dynamic. A static MOS memory cell uses a latch similar to those used in bipolar cells. Static MOS cell circuits are discussed in the problems and in the references [14]. They require as many as eight transistors per cell, which limits the density of static cells on a memory chip. Also, their power consumption is high, which further limits their density on one chip.

A dynamic memory cell is made from one or more MOS transistors. Basically, data is written into a cell by charging or not charging the gate capacitor of a MOS transistor. Figure 10.17(c) illustrates the basic idea of a dynamic cell. The dotted lines show the gate capacitor of a MOS transistor, which forms part of the memory cell. One or more MOS transistors may be used in each cell. The MOS transistors may or may not be connected together as a latch. A dynamic cell is read by sensing the presence or absence of a charge.

Due to leakage currents in the MOS transistor, the charge gradually leaks off the capacitor and the stored data is lost. Hence the contents in each memory cell must be periodically **refreshed** by a refresh circuit every few milliseconds. The refresh period determines the minimum frequency at which a dynamic memory cell must be clocked. The circuitry of dynamic memory cells is covered in the references [3, 15].

Comparing the MOS dynamic memory with the MOS static or bipolar static memory we see that the dynamic memory offers (1) smaller volume per cell, i.e., greater density, (2) higher speed than a MOS static memory, (3) lower total power dissipation, because only leakage currents flow, and (4) lower cost. On the other hand, with a static memory, (1) no refresh circuitry is necessary, (2) no extra power supplies are needed for refreshing, (3) the bipolar static cell is faster, and (4) in general, fewer power supply voltages are needed.

10.4.3 Bipolar Memories

Let's turn our attention now to the specifications and timing diagram for a word-organized and a bit-organized bipolar memory.

Word-Organized Memory Figure 10.18 shows the logic symbol and block diagram for a word-organized 16W × 4b Schottky bipolar read/write memory. Relatively small fast memories such as this are useful in scratch pad applications. A **scratch pad** memory is a small memory used for temporary storage of data, such as in the central processing unit (CPU) of a digital computer. A **cache** memory is a small-capacity, high-speed memory used between main memory and the CPU of a computer. It holds selected information from main memory chosen such that the information the CPU is most likely to need is in the cache, thereby increasing computer speed. It has four address inputs, which are fed to four buffer amplifiers and then decoded into 16 lines. Thus, each address selects one of 16 words. There are four data inputs ($D_1 - D_4$) and four complemented data outputs ($\overline{O}_1 - \overline{O}_4$). Only a single 5-V power supply is required. The chip select

PIN CONFIGURATION

ADDRESS INPUT A₁	1	16	V_CC
CHIP SELECT C̄S̄	2	15	A₂ ADDRESS INPUT
WRITE ENABLE W̄Ē	3	14	A₃ ADDRESS INPUT
DATA INPUT D₁	4	13	A₀ ADDRESS INPUT
DATA OUTPUT Ō₁	5	12	D₂ DATA INPUT
DATA INPUT D₃	6	11	Ō₃ DATA OUTPUT
DATA OUTPUT Ō₂	7	10	D₄ DATA INPUT
GND	8	9	Ō₄ DATA OUTPUT

PIN NAMES

D₁–D₄	DATA INPUTS	C̄S̄	CHIP SELECT INPUT
A₀–A₃	ADDRESS INPUTS	Ō₁–Ō₄	DATA OUTPUTS
W̄Ē	WRITE ENABLE	V_CC	POWER (+5V)

LOGIC SYMBOL

BLOCK DIAGRAM

TRUTH TABLE

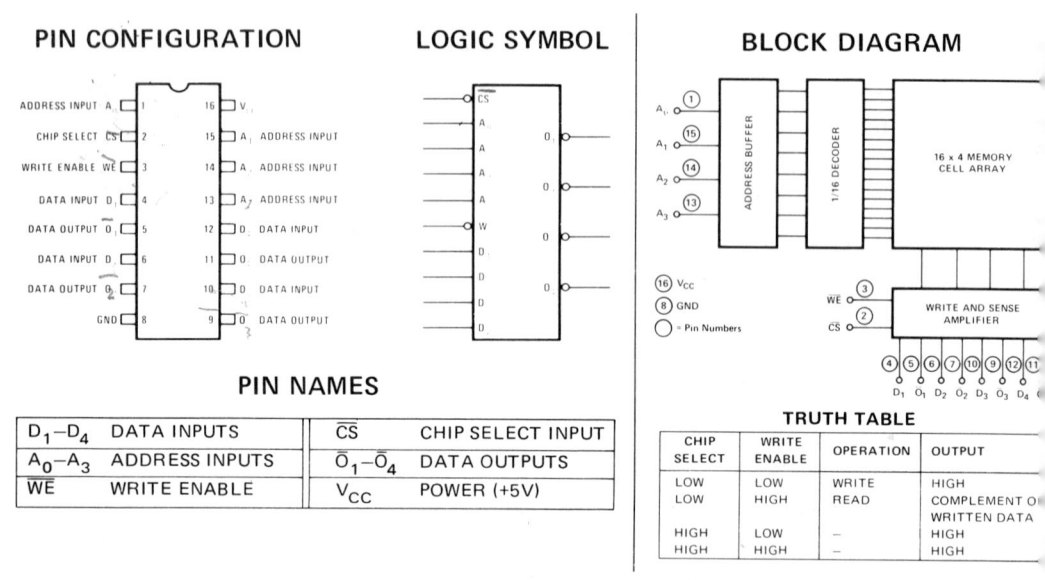

CHIP SELECT	WRITE ENABLE	OPERATION	OUTPUT
LOW	LOW	WRITE	HIGH
LOW	HIGH	READ	COMPLEMENT OF WRITTEN DATA
HIGH	LOW	–	HIGH
HIGH	HIGH	–	HIGH

Fig. 10.18 Word-organized 16W × 4b Schottky bipolar 3101A read/write memory. (Courtesy of Intel Corporation.)

input CS must be LOW to read or write. When $\overline{CS}$ is HIGH all outputs are forced HIGH and the read and write operations are disabled. When $\overline{CS} = L$ and the write enable input WE is LOW, data at the D inputs enters the cells selected by the address inputs and all outputs are forced HIGH (inactive state). When $\overline{CS}$ is LOW and WE is HIGH, data can be read from the outputs as the complement of the data that entered the addressed cells. The outputs have open collectors for ease in memory expansion.

Figure 10.19 shows the ac characteristics and timing diagrams for Fig. 10.18. Let's discuss the read operation first. There are two read cycle timing diagrams. The upper diagram applies if the chip select input goes LOW (active) within 18 ns after all four address inputs are selected. In this case the memory operates in the **address mode**, and the outputs will be valid from one address access time t_{A-} after the addresses become stable until a time t_{A+} after the addresses change. If the chip select input goes LOW at any time greater than 18 ns after the address inputs are selected, the memory operates in the **chip select mode**. In the chip select mode (lower read cycle timing diagram), the outputs become valid from one chip-select-to-output delay time t_{S-} after $\overline{CS}$ goes LOW until t_{S+} after $\overline{CS}$ goes HIGH. The address mode results in the least address-to-output delay time and should be used when maximum speed is required.

Next let's study the write cycle timing diagram in Fig. 10.19. As soon as the address inputs are stable, the $\overline{CS}$ input is made LOW. The $\overline{WE}$ input can then go LOW any time after the address inputs are stable. The $\overline{WE}$ pulse must have a minimum width of $t_{WP} = 25$ ns. The input data must be stable at the

10.4 THE READ/WRITE MEMORY

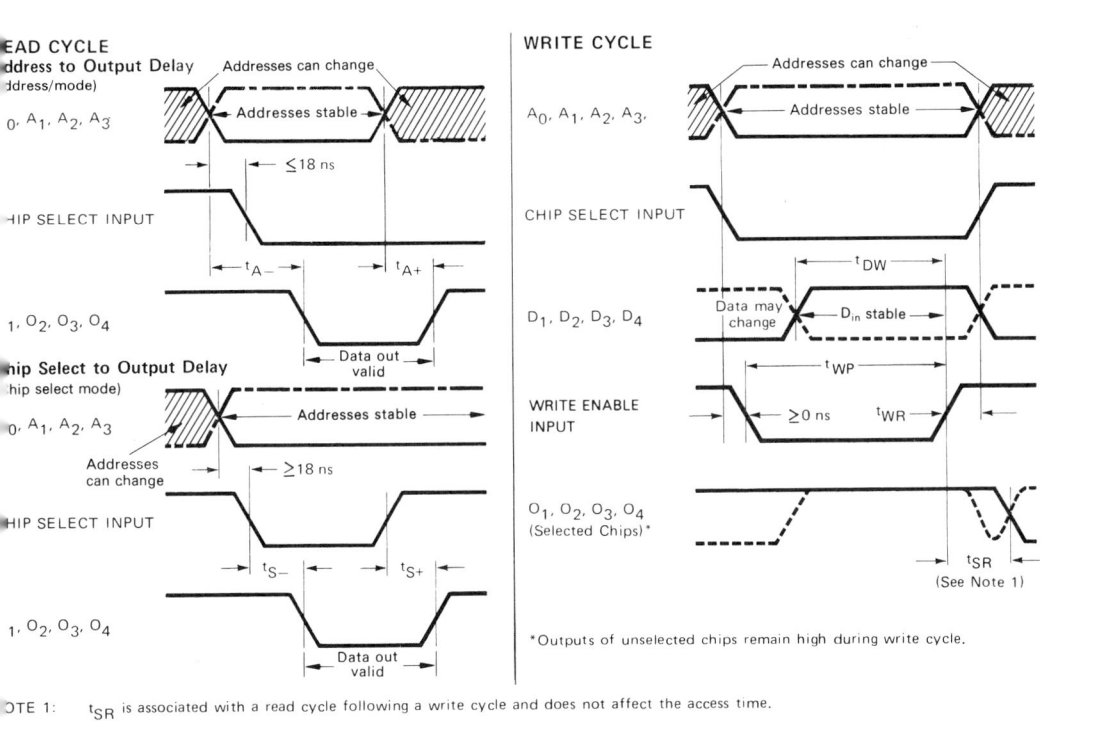

[Timing diagrams for READ CYCLE and WRITE CYCLE]

NOTE 1: t_{SR} is associated with a read cycle following a write cycle and does not affect the access time.

A.C. Characteristics $T_A = 0°C$ to $+75°C$, $V_{CC} = 5.0V \pm 5\%$

READ CYCLE

SYMBOL	PARAMETER	3101A LIMITS (ns) MIN.	3101A LIMITS (ns) MAX.	3101 LIMITS (ns) MIN.	3101 LIMITS (ns) MAX.
t_{S+}, t_{S-}	Chip Select to Output Delay	5	17	5	42
t_{A-}, t_{A+}	Address to Output Delay	10	35	10	60

CAPACITANCE[2] $T_A = 25°C$

C_{IN}	INPUT CAPACITANCE (All Pins)	10 pF maximum
C_{OUT}	OUTPUT CAPACITANCE	12 pF maximum

WRITE CYCLE

SYMBOL	TEST	3101A LIMITS (ns) MIN.	3101A LIMITS (ns) MAX.	3101 LIMITS (ns) MIN.	3101 LIMITS (ns) MAX.
t_{SR}	Sense Amplifier Recovery Time		35		50
t_{WP}	Write Pulse Width	25		40	
t_{DW}	Data-Write Overlap Time	25		40	
t_{WR}	Write Recovery Time	0		5	

NOTE 2: This parameter is periodically sampled and is not 100% tested. Condition of measurement is f = 1 MHz, V_{bias} = 2V, V_{CC} = 0V, and $T_A = 25°C$.

Fig. 10.19 Timing diagram for the 3101A 16W × 4b Schottky bipolar read/write memory. (Courtesy of Intel Corporation.)

D inputs at least $t_{DW} = 25$ ns before the $\overline{WE}$ pulse becomes HIGH. All four outputs will be HIGH during a write operation because $\overline{WE}$ is LOW (see truth table in Fig. 10.18). If a read cycle immediately follows a write cycle the outputs will have a spike in them due to the sense amplifier recovery time t_{SR}. This recovery

FUNCTION TABLE

FUNCTION	INPUTS		OUTPUT
	MEMORY ENABLE[†]	WRITE ENABLE	
Write (Store Complement of Data)	L	L	High Impedance
Read	L	H	Stored Data
Inhibit	H	X	High Impedance

H = high level, L = low level, X = irrelevant
[†]For memory enable: L = all ME inputs low;
H = one or more ME inputs high.

(a)

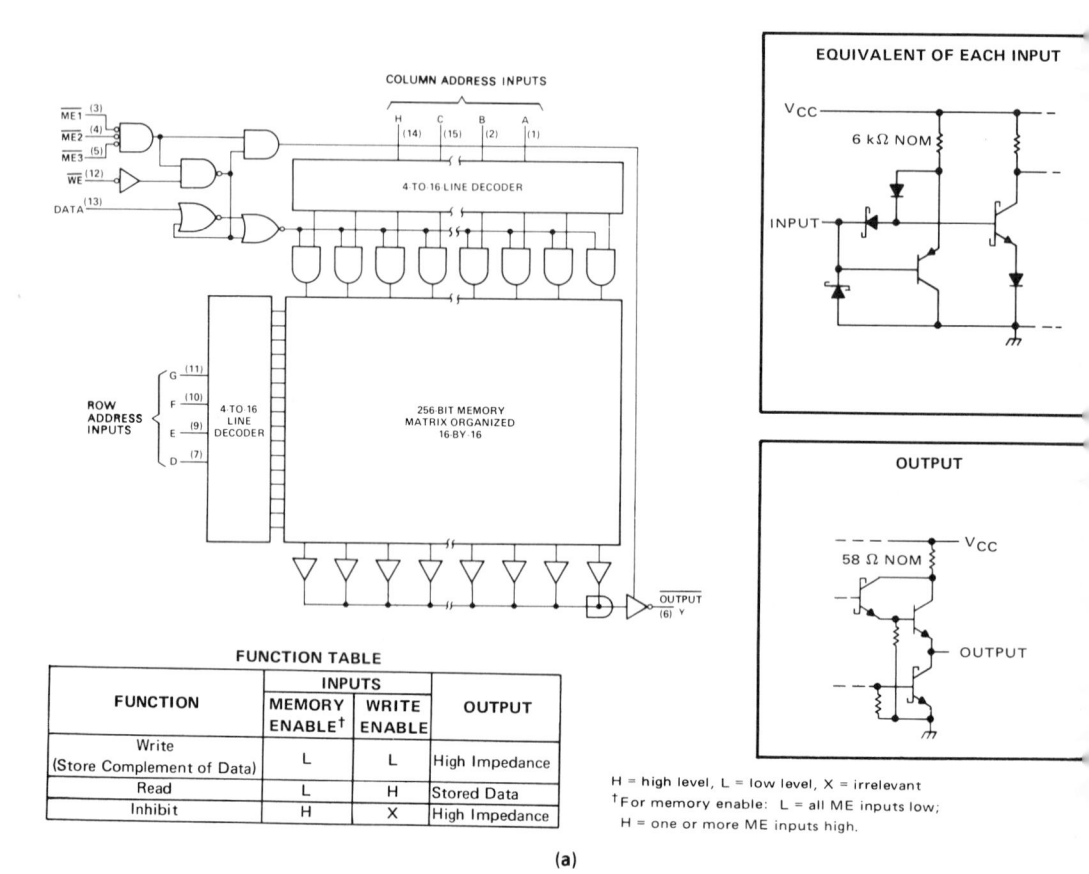

Fig. 10.20 Schottky bipolar 74S200 256W × 1b read/write memory: (a) functional block diagram and equivalent input/output circuits; (b) specifications and timing diagrams. (Courtesy of Texas Instruments Incorporated.)

time is 35 ns maximum, measured relative to the rising edge of the $\overline{WE}$ pulse, as shown in Fig. 10.19.

Bit-Organized Memory Figure 10.20(a) shows the functional block diagram of a bit-organized 256W × 1b Schottky bipolar memory with three-state outputs. The memory cells are arranged in a 16 × 16 array. A given memory cell is addressed by selecting the row X and column Y in which that cell resides with the two 1-of-16 address decoders. This method of selecting a given memory cell is called **coincident selection** or **X-Y selection**, since it selects a cell at the intersection, or coincidence, of a row (X) and a column (Y). Address inputs *A*, *B*, *C*, and *H* select one of 16 columns. Address inputs *D*, *E*, *F*, and *G* select one of 16 rows. Data can be written in when all memory enable inputs are LOW (i.e., $\overline{ME}_1 = \overline{ME}_2 = \overline{ME}_3 = L$) and $\overline{WE} = L$. The output is in its high-impedance state in this case. Data can be read out when $\overline{ME}_1 = \overline{ME}_2 = \overline{ME}_3 = L$ and

10.4 THE READ/WRITE MEMORY

recommended operating conditions

		SN54S200 MIN	SN54S200 NOM	SN54S200 MAX	SN74S200 MIN	SN74S200 NOM	SN74S200 MAX	UNIT
Supply voltage, V_{CC}		4.5	5	5.5	4.75	5	5.25	V
High-level output current, I_{OH}				−5.2			−10.3	mA
Low-level output current, I_{OL}				8			12	mA
Width of write-enable pulse, t_w		50			40			ns
Setup time, t_{setup}	Address-to-write enable	0			0			ns
	Data-to-write enable	0			0			
	Memory-enable-to-write-enable	0			0			
Hold time, t_{hold}	Address-from-write-enable	10			10			ns
	Data-from-write-enable	10			10			
	Memory-enable-from-write-enable	0			0			
Operating free-air temperature, T_A		−55		125	0		70	°C

PARAMETER		TEST CONDITIONS	SN54S200 TYP‡	SN54S200 MAX	SN74S200 TYP‡	SN74S200 MAX	UNIT
t_{PLH} Propagation delay time, low-to-high-level output	Access times from address	C_L = 15 pF, R_L = 560 Ω (SN54S'), 400 Ω (SN74S'), See Figure 1	33	70	33	50	ns
t_{PHL} Propagation delay time, high-to-low-level output			29	70	29	50	
t_{ZH} Output enable time to high level	Access times from memory enable		21	45	21	35	ns
t_{ZL} Output enable time to low level			20	45	20	35	
t_{ZH} Output enable time to high level	Sense recovery times from write enable		19	50	19	40	ns
t_{ZL} Output enable time to low level			17	50	17	40	
t_{HZ} Output disable time from high level	Disable times from memory enable	C_L = 5 pF, R_L = 560 Ω (SN54S'), 400 Ω (SN74S'), See Figure 1	7	30	7	20	ns
t_{LZ} Output disable time from low level			9	30	9	20	
t_{HZ} Output disable time from high level	Disable times from write enable		13	40	13	30	ns
t_{LZ} Output disable time from low level			16	40	16	30	

‡All typical values are at V_{CC} = 5 V, T_A = 25°C.

PARAMETER MEASUREMENT INFORMATION

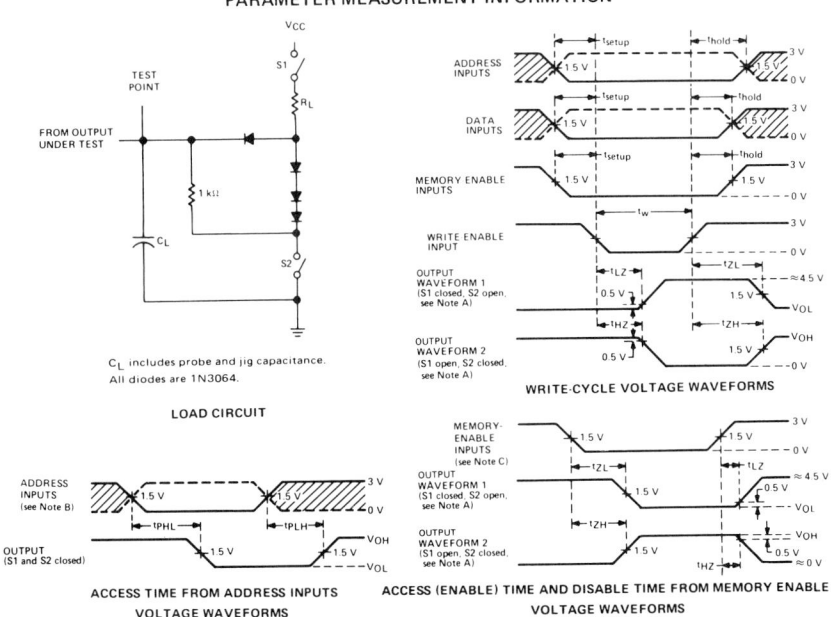

LOAD CIRCUIT

C_L includes probe and jig capacitance. All diodes are 1N3064.

WRITE-CYCLE VOLTAGE WAVEFORMS

ACCESS TIME FROM ADDRESS INPUTS VOLTAGE WAVEFORMS

ACCESS (ENABLE) TIME AND DISABLE TIME FROM MEMORY ENABLE VOLTAGE WAVEFORMS

READ - CYCLE VOLTAGE WAVEFORMS

NOTES: A. Waveform 1 is for the output with internal conditions such that the output is low except when disabled. Waveform 2 is for the output with internal conditions such that the output is high except when disabled.
B. When measuring delay times from address inputs, the memory-enable inputs are low and the write-enable input is high.
C. When measuring delay times from memory-enable inputs, the address inputs are steady-state and the write-enable input is high.
D. Input waveforms are supplied by pulse generators having the following characteristics: $t_r ⩽ 2.5$ ns, $t_f ⩽ 2.5$ ns, PRR ⩽ 1 MHz, and $Z_{out} ≈ 50$ Ω.

(b)

Fig. 10.20 Continued

$\overline{WE} = H$. The three-state output is in its high-impedance state when one or more memory enable inputs are HIGH, independent of the $\overline{WE}$ input.

Figure 10.20(b) shows the specifications and timing diagrams for the 74S200. For the write cycle, the address, data, and memory enable inputs must be set up $t_{\text{setup}} = 0$ ns before the write enable input goes LOW. The write enable pulse must stay LOW for a minimum of $t_W = 40$ ns. The output disable times, shown in the write-cycle voltage waveforms in Fig. 10.20(b), measured relative to the falling edge of the $\overline{WE}$ pulse, are $t_{LZ(\max)} = 30$ ns, $t_{HZ(\max)} = 30$ ns. These times are measured with the output of the memory connected to the load shown in Fig. 10.20(b). The output sense recovery times t_{ZL} and t_{ZH} ($t_{ZL(\max)} = t_{ZH(\max)} = 40$ ns) are measured relative to the rising edge of the $\overline{WE}$ pulse. Figure 10.20(b) also shows the read-cycle voltage waveforms. The access (enable) times and the disable times are measured relative to the memory enable inputs with the address inputs stable and $\overline{WE} = H$.

For memory expansion, the three-state outputs can be connected to a bus. However, two outputs at opposite logic levels must not be applied to a common bus simultaneously. To minimize this possibility the output disable times are made shorter than the output enable times. The high capacitive drive capability of this memory allows memory expansion to 65,536 words.

Following are some additional examples of bipolar read/write memories. The Fairchild ECL 10410 has 256 bits (256 × 1) with a typical access time of 20 ns and power dissipation of 1.8 mW/bit. The TTL 74LS170 is a 16 bit (4 × 4) memory with a typical address time of 27 ns and typical power dissipation of 7.8 mW/bit.

10.4.4 Static MOS Memories

In this section we shall discuss the specifications and timing diagrams for a typical static MOS memory. As we indicated earlier, MOS memories generally have a lower power dissipation, greater density, and lower speed than bipolar memories.

A static MOS memory does not require refreshing to retain its stored data and can operate with only one power supply. Figure 10.21 shows the block diagram, specifications, and timing diagram for a static 1024 × 1 MOS read/write memory with a three-state output. It uses NMOS transistors and operates from a single +5-V power supply. It has a typical power dissipation of 150 mW, and its inputs and outputs are directly compatible with TTL. The manufacturer also offers a low-standby-power version that dissipates only 42 mW maximum in standby. The lower power dissipation in this low-power version is achieved by reducing the supply voltage and the chip enable input voltages. When the chip enable input is HIGH, the three-state output goes to the high-impedance state and the data input is electrically disconnected from the internal input data bus. The addresses are, however, decoded independent of the $\overline{CS}$ input. The read/write input is held HIGH for read operations and LOW for write operations.

Let's look now at the read cycle timing diagram in Fig. 10.21. There are two methods for reading out data. The first method is a chip enable mode in which

10.4 THE READ/WRITE MEMORY

PIN CONFIGURATION — 2102A

LOGIC SYMBOL — 2102A

BLOCK DIAGRAM

PIN NAMES

D_{IN}	DATA INPUT	$\overline{CE}$	CHIP ENABLE
$A_0 - A_9$	ADDRESS INPUTS	D_{OUT}	DATA OUTPUT
R/W	READ/WRITE INPUT	V_{CC}	POWER (+5V)

Symbol	Parameter	Min.	Typ.[1]	Max.	Unit
READ CYCLE					
t_{RC}	Read Cycle	250			ns
t_A	Access Time			250	ns
t_{CO}	Chip Enable to Output Time			130	ns
t_{OH1}	Previous Read Data Valid with Respect to Address	40			ns
t_{OH2}	Previous Read Data Valid with Respect to Chip Enable	0			ns
WRITE CYCLE					
t_{WC}	Write Cycle	250			ns
t_{AW}	Address to Write Setup Time	20			ns
t_{WP}	Write Pulse Width	180			ns
t_{WR}	Write Recovery Time	0			ns
t_{DW}	Data Setup Time	180			ns
t_{DH}	Data Hold Time	0			ns
t_{CW}	Chip Enable to Write Setup Time	180			ns

READ CYCLE

① 1.5 VOLTS
② 2.0 VOLTS
③ 0.8 VOLTS

WRITE CYCLE

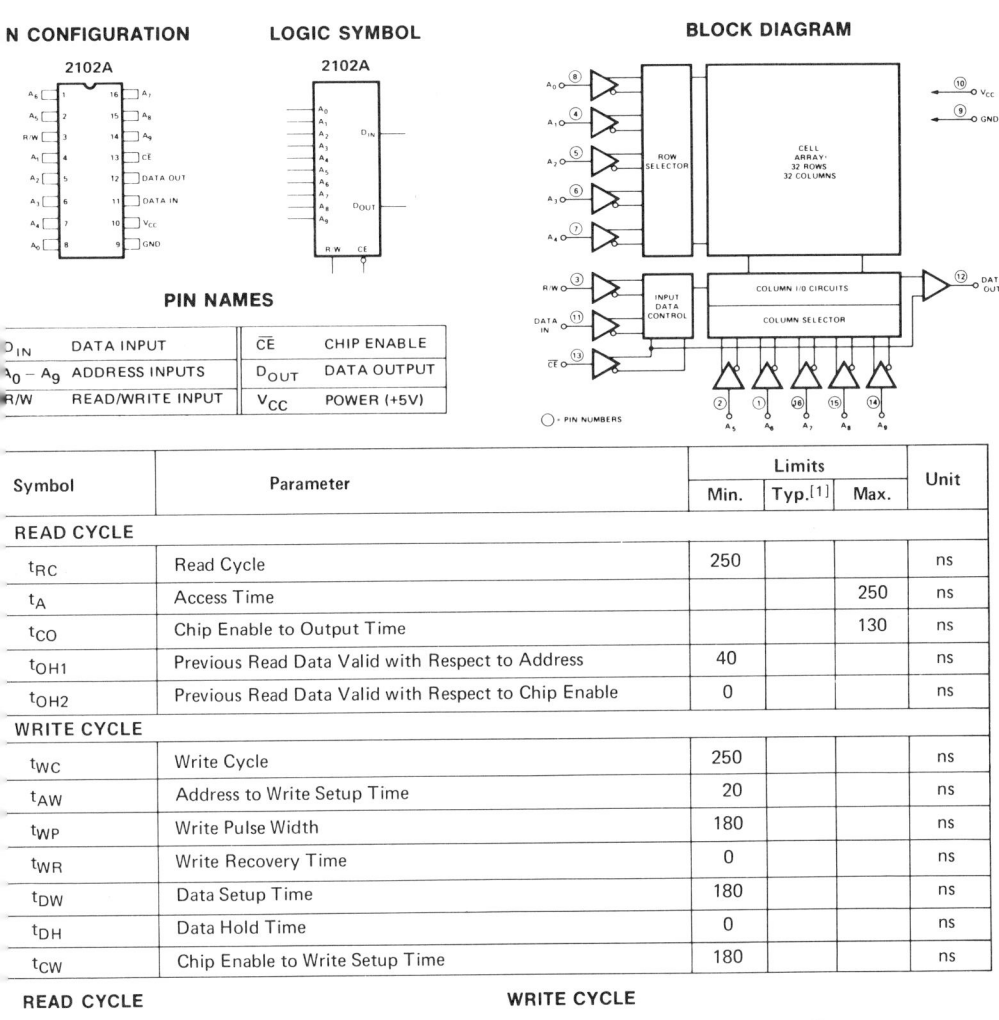

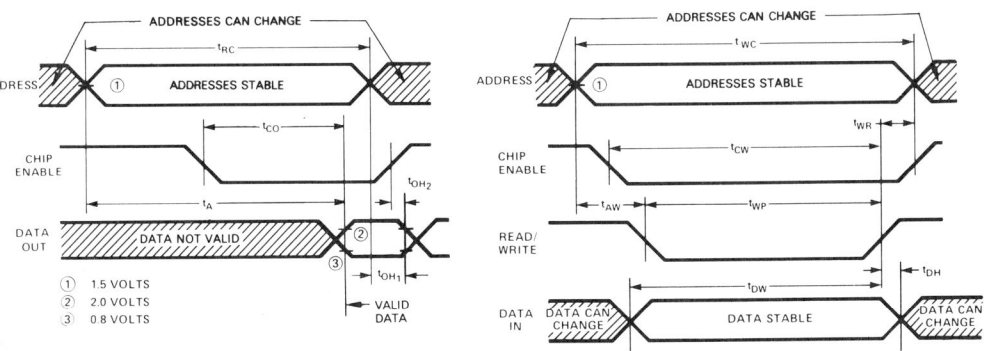

Fig. 10.21 Static MOS 1024W × 1b 2102A-2 read/write memory. (Courtesy of Intel Corporation.)

the read/write input is maintained HIGH during the entire read operation. The addresses are set up first, then when data is to be read out the chip enable input is put LOW. Valid data then appears at the data output D_{out} a time no later than $t_{CO} = 130$ ns (maximum) later. Data read out for a given address will remain valid at the output for a hold time $t_{OH_2} = 0$ ns after the chip enable input goes HIGH. This method is the one illustrated in the timing diagram in Fig. 10.21. The second method for reading out data keeps the chip enable input LOW during the entire read cycle and uses the read/write input for controlling the read operation. The data output in this case must not be OR-tied to other memory outputs because with more than one IC enabled excessive currents could damage other ICs. With this method valid data appears at the data output within time $t_A = 250$ ns maximum after the inputs are addressed. For a given address, data will remain valid at the output for one hold time $t_{OH_1} = 40$ ns (minimum) after that address changes to a new address. By maintaining the read/write input HIGH, a series of read cycles can be accomplished with this method.

The write timing diagram is also given in Fig. 10.21. A sequence of write cycles cannot be performed by keeping the chip enable and read/write input LOW and cycling through the desired addresses. However, the chip enable input can be kept LOW and a series of continuous write operations performed if the read/write input is pulsed, provided the address-to-write setup time $t_{AW} = 20$ ns (minimum) and write pulse width $t_{WP} = 180$ ns (minimum) times are adhered to.

Another example of a static MOS memory is the Motorola CMOSL 14505 64 × 1 read/write memory. It has a three-state output. A strobe input is kept inactive (LOW), except when reading or writing, in order to minimize the power dissipation. The quiescent power dissipation is only 150 nW ($V_{DD} = 5$ V), but increases to about 4 mW for a 1-MHz strobe frequency. The maximum read cycle time for this memory is 750 ns and the maximum write cycle time is 700 ns ($V_{DD} = +5$ V).

Next let's look at some of the properties of dynamic memories.

10.4.5 Dynamic MOS Memories

Recall that a dynamic memory cell must be periodically refreshed. This makes dynamic memories somewhat more difficult to use than static memories, in some applications. Dynamic memories have lower standby power dissipation, higher speed, and greater memory cell density than static memories; however, most dynamic memories require more than one power supply voltage.

Figure 10.22(a) shows the block diagram of a 4096 × 1 dynamic NMOS read/write memory with a three-state output. The memory cells are arranged in a 64 × 64 array. The rows and columns are addressed by two 6-line-to-64-line address decoders. This memory has a chip enable input CE, write enable input $\overline{WE}$, chip select input $\overline{CS}$, data input D_{in}, and complementary data output $\bar{D}_{out}$. When the CE input is LOW, the data output is in the high-impedance state.

Refreshing is accomplished by performing a read cycle on each of the 64 row address inputs at least once every 2 ms. During refreshing the $\overline{CS}$ input can be

10.4 THE READ/WRITE MEMORY 319

either HIGH or LOW and the *CE* input must be as shown in the read timing diagram, Fig. 10.22(b). Refreshing can also be accomplished during a write or a read-modify/write cycle. In this case $\overline{CS}$ must be held HIGH to avoid writing into the memory during the refresh operation. A **read-modify/write cycle** is simply a read cycle followed immediately by a write cycle, with the same addresses for both cycles.

The external refresh circuitry basically consists of a clock, or refresh oscillator, and a refresh address generator. The clock is used to generate a refresh request, and the refresh address generator produces the X-addresses necessary for refreshing. An entire row of cells in a chip is refreshed simultaneously, so the Y-columns in the memory array do not have to be addressed. Refresh circuitry is developed in the problems.

Let's discuss the read cycle timing diagram in Fig. 10.22(b). The addresses must be set up a time $t_{AC} = 0$ ns before the *CE* input goes HIGH. Addresses must be held stable for a time $t_{AH(min)} = 50$ ns after the rising edge of the *CE* pulse. The minimum *CE* pulse width during a read cycle is $t_{CE} = 230$ ns. The write enable $\overline{WE}$ input must be set up $t_{WC} = 0$ ns before the rising edge of the *CE* pulse and be held $t_{WL} = 0$ ns after the falling edge of *CE*. The output becomes valid one address access time $t_{ACC} = 200$ ns (maximum) after the address inputs become stable, or a time $t_{CO} = 180$ ns (maximum) after the *CE* input goes HIGH. The read cycle time $t_{CY} = 400$ ns and is measured from the rising edge of one *CE* pulse to the rising edge of the next *CE* pulse.

The write cycle timing diagram in Fig. 10.22(b) is similar to those already discussed and will be pursued further in the problems.

(a)

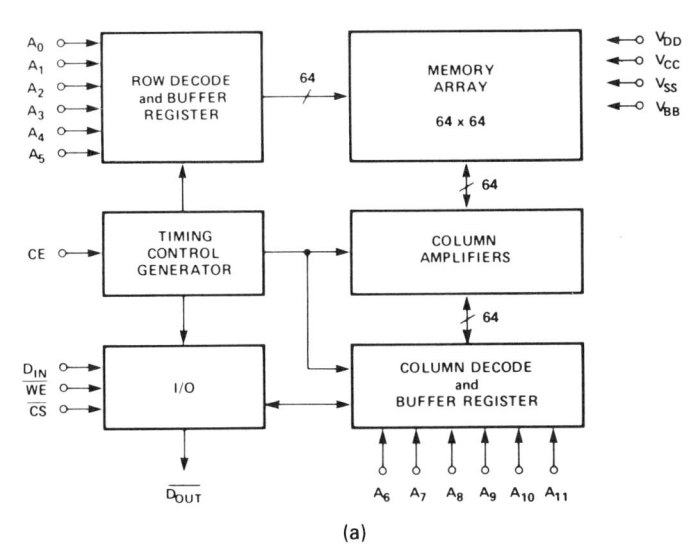

Fig. 10.22 Dynamic 4096W × 1b 2107B read/write memory: (a) block diagram. (Courtesy of Intel Corporation.) (Continued on pages 320–321)

Read and Refresh Cycle [1] (Numbers in parentheses are for minimum cycle timing in ns)

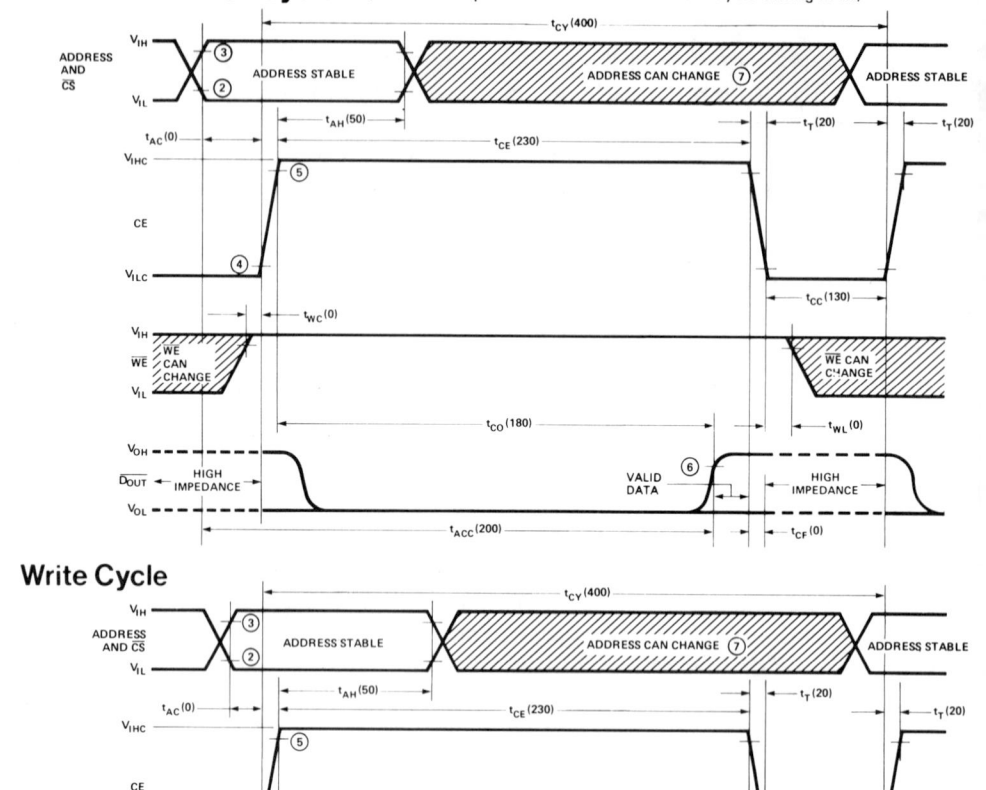

Write Cycle

NOTES: 1. For Refresh cycle row and column addresses must be stable before t_{AC} and remain stable for entire t_{AH} period.
2. V_{IL} MAX is the reference level for measuring timing of the addresses, $\overline{CS}$, $\overline{WE}$, and D_{IN}.
3. V_{IH} MIN is the reference level for measuring timing of the addresses, $\overline{CS}$, $\overline{WE}$, and D_{IN}.
4. V_{SS} +2.0V is the reference level for measuring timing of CE.
5. V_{DD} −2V is the reference level for measuring timing of CE.
6. V_{SS} +2.0V is the reference level for measuring the timing of $\overline{D_{OUT}}$.
7. During CE high typically 0.5mA will be drawn from any address pin which is switched from low to high.

(b)

Fig. 10.22 (continued) (b) Read and write cycle timing diagram. (Courtesy of Intel Corporation.)

10.4 THE READ/WRITE MEMORY

READ, WRITE, AND READ MODIFY/WRITE CYCLE

Symbol	Parameter	Min.	Max.	Unit
t_{REF}	Time Between Refresh		2	ms
t_{AC}	Address to CE Set Up Time	0		ns
t_{AH}	Address Hold Time	50		ns
t_{CC}	CE Off Time	130		ns
t_T	CE Transition Time	10	40	ns
t_{CF}	CE Off to Output High Impedance State	0		ns

READ CYCLE

Symbol	Parameter	Min.	Max.	Unit
t_{CY}	Cycle Time	400		ns
t_{CE}	CE On Time	230	3000	ns
t_{CO}	CE Output Delay		180	ns
t_{ACC}	Address to Output Access		200	ns
t_{WL}	CE to $\overline{WE}$	0		ns
t_{WC}	$\overline{WE}$ to CE on	0		ns

WRITE CYCLE

Symbol	Parameter	Min.	Max.	Unit
t_{CY}	Cycle Time	400		ns
t_{CE}	CE On Time	230	3000	ns
t_W	$\overline{WE}$ to CE Off	150		ns
t_{CW}	CE to $\overline{WE}$	100		ns
t_{DW} [2]	D_{IN} to $\overline{WE}$ Set Up	0		ns
t_{DH}	D_{IN} Hold Time	0		ns
t_{WP}	$\overline{WE}$ Pulse Width	50		ns

(c)

Fig. 10.22 (continued) (c) ac characteristics. (Courtesy of Intel Corporation.)

The 2107B requires three power supplies $V_{CC} = +5$ V, $V_{BB} = -5$ V, and $V_{DD} = +12$ V. The operating power dissipation for a 400-ns cycle (read or write) and CE pulse width of 320 ns is $P_D = 660$ mW typically. When the chip enable input is inactive (LOW), that is, when the device is in the **standby mode**, only leakage currents (I_{DD_1}, I_{CC_1}, I_{BB}, I_{LO}, and I_{L_1}) flow (see Fig. 10.25) and the power consumed is called the **standby power**. When the 2107B is in the standby mode ($CE = L$) the standby power dissipation is typically 1.4 mW. The next example shows how to calculate the total power dissipation in the standby mode.

Example 10.7 *Dynamic Memory Standby Power.* Calculate the total power dissipated by a 2107B 4K dynamic memory in the standby mode.

Solution. The total power dissipation during standby P_{TS} will be the sum of power dissipated in standby while being refreshed P_{RF} plus the standby power when no refreshing occurs P_{DS}. The refresh standby power is the product of the duty cycle for the refresh operation and the operating power dissipation $P_{DO} = 660$ mW (as noted above). The duty cycle (DC) is the product of the minimum time required to perform a refresh operation on one address and the total number of addresses N_A divided by the minimum refresh period, i.e.,

$$DC = \frac{t_{\text{cycle}} \times N_A}{T_{\text{refresh}}} = \frac{400 \text{ ns} \times 64}{2 \text{ ms}} = 1.28\%.$$

So the refresh power is

$$P_{RF} = \text{DC} \times P_{DO} = 0.0128 \times 660 \text{ mW} = 8.45 \text{ mW}.$$

Finally, the total power is

$$P_{TS} = P_{RF} + P_{DS} = (8.45 + 1.4) \text{ mW} = 9.85 \text{ mW}. \blacksquare$$

Switching current transients are appreciable, especially for large memories. Figure 10.23 shows the current spikes in the 2107B power supply lines. Because of these current spikes, memory power supplies should be adequately filtered. In an expanded 2107 memory system, every second chip should have a 0.1-μF decoupling capacitor between V_{DD} and ground. The alternate chips should have a 0.1-μF capacitor between V_{BB} and ground. Decoupling between V_{CC} and ground is required only on chips located on the periphery of the memory system. High-frequency, low-inductance capacitors should be used. In addition, for each 36 chips, a 100-μF tantalum (or equivalent) capacitor should be connected between V_{DD} and ground, close to the memory array.

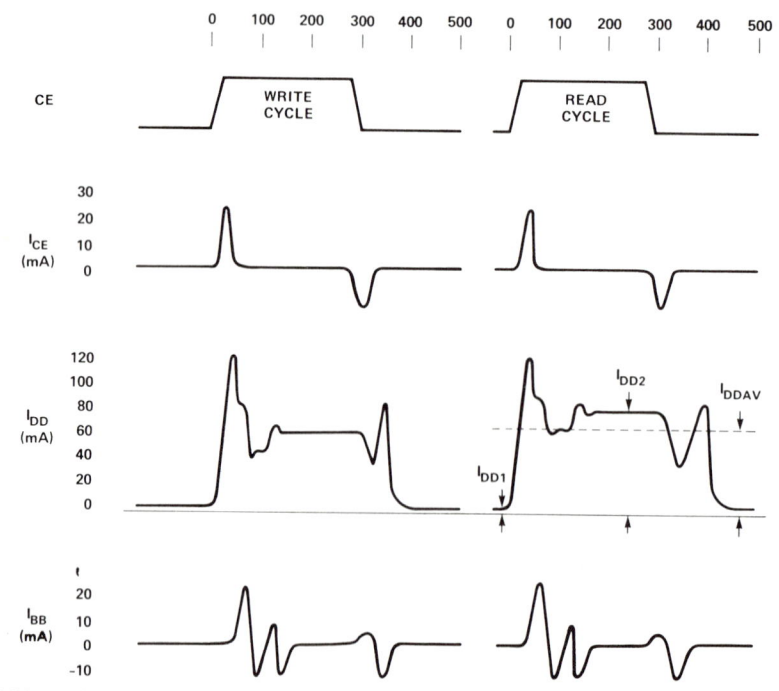

Fig. 10.23 Typical current transients in the 2107B 4K × 1 memory. (Courtesy of Intel Corporation.)

*10.4.6 Memory Expansion [22, 23]

It is often necessary to expand the size of a memory by combining several IC memories. Memory expansion may be accomplished by increasing the bit size (number of bits per word), the word size (total number of words), or both. The basic technique for read/write memory expansion is essentially the same as for ROM expansion.

Increasing the Number of Bits Per Word Consider expanding a 64W × 4b memory to a 64 × 8 memory. This can be done by arranging two 64 × 4 memories in parallel as shown in Fig. 10.24. Notice that address inputs are in parallel and that there is a common $\overline{CS}$ and a common WE line. External resistors R are used to pull up the open collector outputs to V_{CC}.

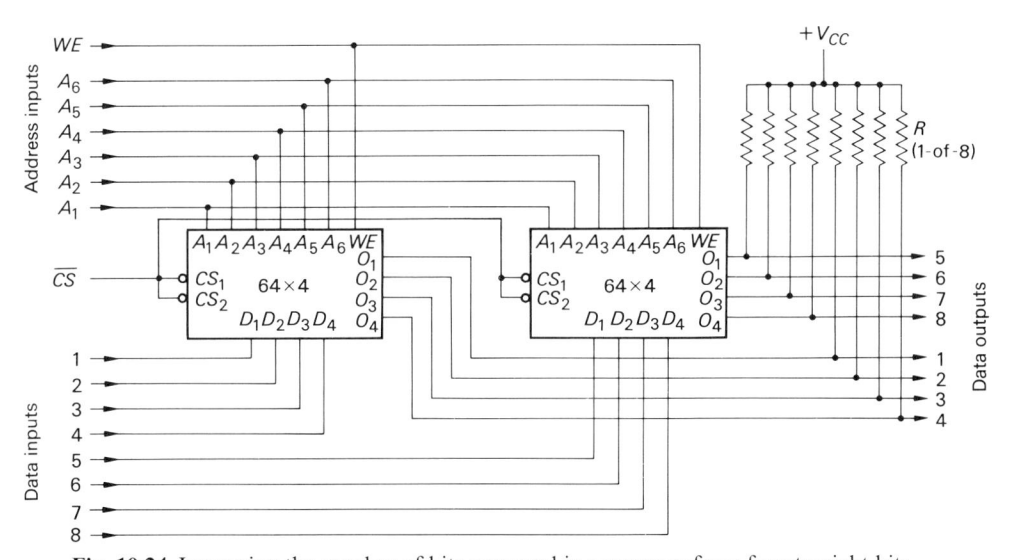

Fig. 10.24 Increasing the number of bits per word in a memory from four to eight bits.

Memories can be expanded to almost any desired capacity by connecting more ICs in parallel, the capacity is limited only by the overall system speed and current requirements. As the number of chips in a memory system increases, more current will be required to drive parallel inputs, and drive amplifiers may then be required. To aid in the design of larger memories most memory specification sheets give the current requirements of each input and output. Figure 10.25 shows the current requirements for the 2107B 4096 × 1 memory.

324 SEMICONDUCTOR MEMORIES   10.4

$T_A = 0°C$ to $70°C$, $V_{DD} = +12V \pm 5\%$, $V_{CC} = +5V \pm 5\%$, V_{BB} [1] $= -5V \pm 5\%$, $V_{SS} = 0V$, unless otherwise noted.

Symbol	Parameter	Limits Min.	Limits Typ.[2]	Limits Max.	Unit	Conditions
I_{LI}	Input Load Current (all inputs except CE)		.01	10	µA	$V_{IN} = V_{IL\ MIN}$ to $V_{IH\ MAX}$
I_{LC}	Input Load Current		.01	10	µA	$V_{IN} = V_{IL\ MIN}$ to $V_{IH\ MAX}$
$|I_{LO}|$	Output Leakage Current for high impedance state		.01	10	µA	$CE = V_{ILC}$ or $\overline{CS} = V_{IH}$ $V_O = 0V$ to $5.25V$
I_{DD1}	V_{DD} Supply Current during CE off[3]		110	200	µA	$CE = -1V$ to $+.6V$
I_{DD2}	V_{DD} Supply Current during CE on		80	100	mA	$CE = V_{IHC}$, $T_A = 25°C$
$I_{DD\ AV1}$	Average V_{DD} Current		55	80	mA	Cycle time = 470ns, $t_{CE} = 300$ns
$I_{DD\ AV2}$	Average V_{DD} Current		27	40	mA	Cycle time = 1000ns, $t_{CE} = 300$ns, $T_A = 25°C$
I_{CC1} [4]	V_{CC} Supply Current during CE off		.01	10	µA	$CE = V_{ILC}$ or $\overline{CS} = V_{IH}$
I_{BB}	V_{BB} Supply Current		5	100	µA	
V_{IL}	Input Low Voltage	-1.0		0.6	V	$t_T = 20$ns
V_{IH}	Input High Voltage	2.4		$V_{CC}+1$	V	
V_{ILC}	CE Input Low Voltage	-1.0		+1.0	V	
V_{IHC}	CE Input High Voltage	$V_{DD}-1$		$V_{DD}+1$	V	
V_{OL}	Output Low Voltage	0.0		0.45	V	$I_{OL} = 2.0$mA
V_{OH}	Output High Voltage	2.4		V_{CC}	V	$I_{OH} = -2.0$mA

NOTES:
1. The only requirement for the sequence of applying voltage to the device is that V_{DD}, V_{CC}, and V_{SS} should never be .3V more negative than V_{BB}.
2. Typical values are for $T_A = 25°C$ and nominal power supply voltages.
3. The I_{DD} and I_{CC} currents flow to V_{SS}. The I_{BB} current is the sum of all leakage currents.
4. During CE on V_{CC} supply current is dependent on output loading, V_{CC} is connected to output buffer only.

Fig. 10.25 The dc characteristics of the 2107B memory. (Courtesy of Intel Corporation.)

The next example illustrates the calculation of currents in expanded memories.

Example 10.8 *Expanded Memory Current Requirements.* Calculate the total maximum input current required to drive each address input in a 16K × 9 memory using 2107B 4K × 1 memory chips.

Solution. A total of 36 chips will be required. Each address input will be connected in parallel, so the total current required per address input will be 36 × I_{in}. From Fig. 10.25 this gives

$$I_T = 36 \times I_{LI} = 36 \times 10\ \mu A = 360\ \mu A. \blacksquare$$

Increasing the Total Number of Words Consider the use of two 64 × 4 memories to increase the total number of words to 128, while keeping a constant four bits per word. Figure 10.26(a) shows how this can be done by wire-ANDing the data outputs. The chip select input *CS* effectively is made part of the address inputs

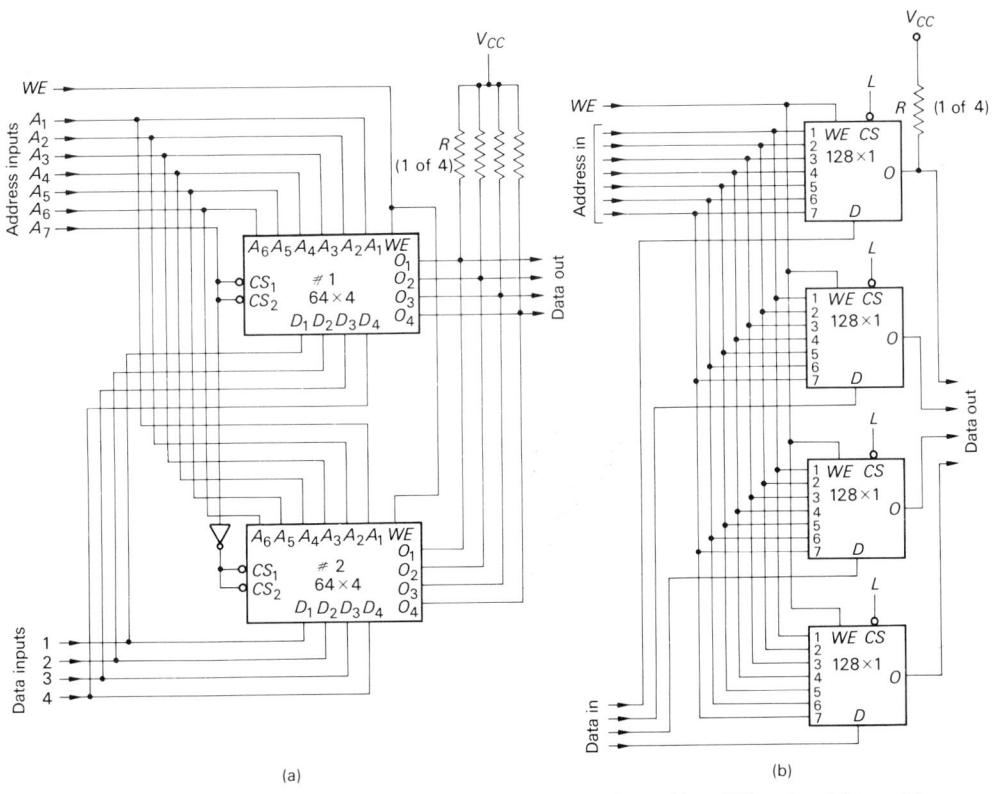

Fig. 10.26 Expanding the number of words in a memory from 64 to 128, using (a) two 64 × 4 chips, (b) four 128 × 1 chips.

as shown. When $A_7 = \overline{CS} = H$, memory number 2 is selected, and when $A_7 = \overline{CS} = L$, memory number 1 is selected.

The address, data, and WE inputs are connected in parallel. The data outputs have open collectors and are wire-ANDed together using four external pull-up resistors R.

By using the two chip select inputs $\overline{CS}_1$ and $\overline{CS}_2$ independently, $2^2 = 4$ chips can be selected, giving a 256-word memory. Word expansion using 2^N ICs would require N separate chip select inputs per IC and 2^N inverters, or a 1-of-2^N decoder. For the case of two ICs ($N = 1$), only one inverter is required (as shown in Fig. 10.26a). Chip select times are less than address access times, so unless many CS inputs are decoded, no additional address access time for the entire system will occur due to decoding.

Figure 10.26(b) shows another way of obtaining a 128 × 4 memory, using

326   SEMICONDUCTOR MEMORIES

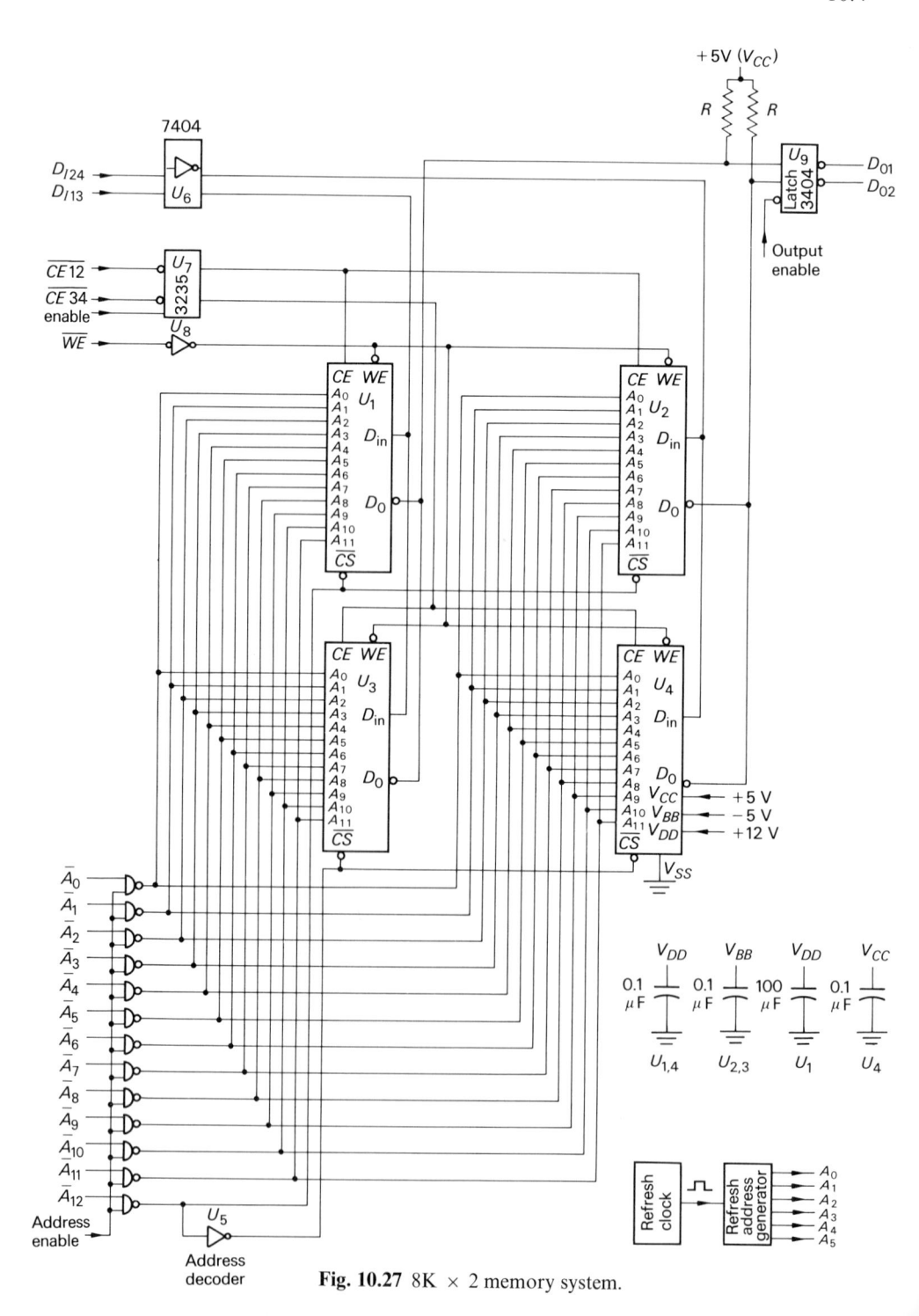

Fig. 10.27 8K × 2 memory system.

four 128 × 1 memories. Note in this case that all four CS inputs are kept permanently active (LOW), hence all four chips draw operating power all the time the memory is in use.

The method used for expanding a memory depends primarily upon factors such as cost, capacities available, and speed. Generally, the 64 × 4 is the largest four-bit/word memory available, due to the number of pins required for larger word sizes. The following configurations are commonly available in bit organized memories: 256 × 1, 512 × 1, 1024 × 1, 2048 × 1, 4096 × 1, and 16,384 × 1.

The next example illustrates memory expansion where both the word size and the number of bits are increased.

Example 10.9 *Word and Bit Memory Expansion.* Draw the logic diagram for an 8K × 2 memory using dynamic 2107B 4K × 1 memories. Include a block diagram of the refresh circuitry, data input 7404 buffers, CE 3235 driver, WE driver data output latch, and address strobed NAND gate drivers. Also indicate the location and value filter capacitors required.

Solution. Figure 10.27 shows how four 2107B 4K × 1 memory chips are wired to produce an 8K × 2 memory system. All address inputs are in parallel. Address $\bar{A}_{12}$ is obtained by using an address decoder U_5 and the chip select inputs. Addresses are enabled simultaneously through an address enable input on the address NAND gates. Data inputs are buffered with inverter gates. Data outputs for each column of ICs are wire-ANDed through a pull-up resistor R to $V_{CC} = +5$ V. Outputs are latched through a 3404 strobed latch.

All write enable inputs $\overline{WE}$ are tied together and buffered with an inverter gate. The chip enable inputs require +12 V ($V_{DD} = \pm 1$ V), so a 3235-chip enable driver is used to enable either U_1 and U_2 or U_3 and U_4. Refreshing is accomplished by performing a read operation on the row addresses $\bar{A}_0 - \bar{A}_5$. Refresh circuitry will be discussed in the problems. ■

*10.5 MISCELLANEOUS MEMORIES

To conclude this chapter we shall mention briefly several types of memories that are not as widely used as the ROM and read/write memories.

Content-Addressable Memories (CAMs) A **CAM**, also called an **associative memory**, is a read/write memory that can store data, read out stored data, and search for stored data with specified characteristics. As an example, consider the 5W × 8b CAM illustrated in Fig. 10.28(a). Suppose it is desired to find all names stored in words 1 through 5 that begin with the letter *D*, where *D* is coded in binary as 1110. In an ordinary read/write memory, word 1 would be retrieved from memory and compared with the search word to see if it matches. Word 2 would then be retrieved and compared, etc., until all words have been compared. This is a very slow process for a memory which contains, say, 10,000 words.

The CAM searches all words in memory in parallel (simultaneously) to locate those words that match the word in the search register, for all bit positions

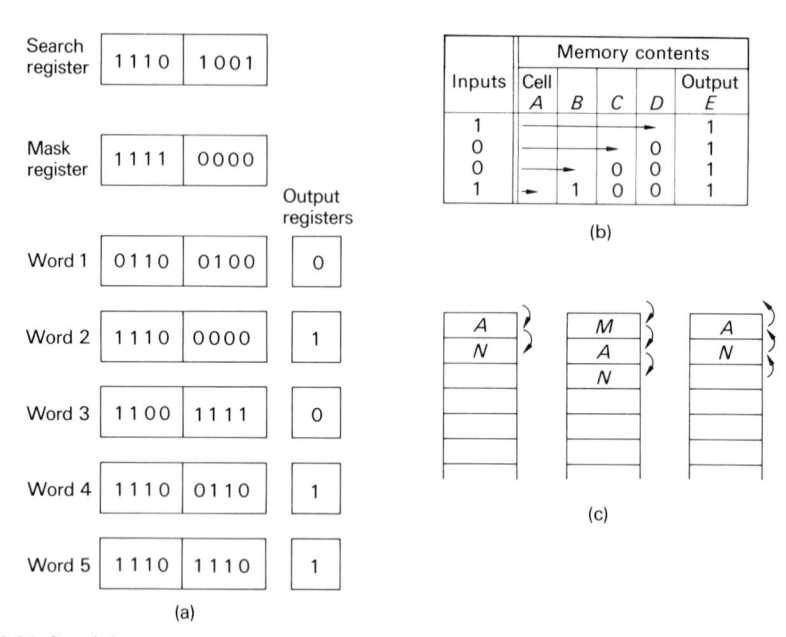

Fig. 10.28 Special-purpose memories: (a) content addressable memory block diagram; (b) FIFO memory; (c) LIFO memory.

in the mask register that have a 1 in them. Thus the CAM in Fig. 10.28(a) finds a match between the first four bits of the search register (viz. 1110) and words 2, 4, and 5. Each word has an output register that stores a 1 if that word matches the search word.

After a search operation, the matched words can be read out. A second search could then be initiated that would, say, search for names whose second letter was an E.

CAMs are very useful for applications where information is stored according to its content, such as in control memories. CAM cells are more complicated than ordinary memory cells and are treated in the references [4, p. 222; 16, p. 4–1]. The main disadvantages of CAMs are their more complex cell structure, the large number of pins required for a given memory size, and their high cost. Arrays of CAMs and other applications are also discussed in the references [17, p. 5–7].

The Intel 4 × 4 Schottky bipolar CAM is packaged in a 24-pin DIP and has a maximum match time of 30 ns.

First-In First-Out (FIFO) Memory A **FIFO** memory is a sequential memory arranged so that the first word to be stored is the first word available to be read out. In a serial-in to serial-out shift-right shift register, data that first enters the shift register from the left is the first data available at the output; however, in the conventional shift register, data that enters gets shifted only one bit position to the right with each clock pulse. In a FIFO memory, data that enters from the left is immediately shifted all the way to the right and is available to be read out.

Figure 10.28(b) illustrates the shifting of four data bits in a 5-bit FIFO memory. The main advantage of the FIFO memory over an ordinary memory is that it allows buffering between two digital systems that operate at different bit rates.

Last In/First Out (LIFO) A **LIFO** memory is a read/write memory, operated as a sequential memory, in which the last item to be stored is the first one available to be read out. LIFO memories are also called **push-down stack** memories since the data appears to be in a stack with the last data pushed down on top of the previous data, as shown in Fig. 10.28(c). LIFO memories are discussed further in the references [19, p. 7–13; 18].

Write-While-Read Memory A **write-while-read** memory is a read/write memory from which the contents of any previously addressed word can be read while writing a new word into the same or another memory location. The Signetics 82S21 is an example of a 32 × 2 bipolar Schottky write-while-read memory which uses strobed latches on the data outputs to hold a data word.

SUMMARY

In this chapter we have discussed semiconductor memories and some of their applications. A read-only memory (ROM) is a memory whose contents are not rapidly varied or perhaps cannot be varied at all once programmed. ROMs are used for the following main applications: (1) table look-up, (2) code conversion and generation, (3) character generation, (4) random logic generation, and (5) microprogramming.

Programmable logic arrays (PLAs) are similar to ROMs except the user programs the address decoder as needed for a given application. They are especially efficient for programming functions that involve a large number of variables and few product terms.

Read/write (R/W) memories are memories into which information can be stored on command and from which data can be read. Read/write memories are arrays of static or dynamic memory cells. Timing of memory inputs is necessary for proper memory operation. Memories can be expanded to handle more words and/or more bits per word.

In addition to the above there exist several special-purpose memories such as the CAM, FIFO, LIFO, and write-while-read memories.

PROBLEMS

Section 10.2.1

1. Briefly explain the difference between a mask programmed, a programmable, a reprogrammable and an electrical erasable ROM.

Section 10.2.3

2. Draw the logic diagram for a 128W × 8b ROM made from 64W × 4b ROMs. Each 64 × 4 ROM has active HIGH open-collector outputs, and one active LOW enable input.

Section 10.2.4

3. *Majority Voter Circuit.* Implement the majority voter circuit of Example 4.5 using a ROM. Indicate ROM cell connections with dots.
4. *Digital Door Lock.* Implement the digital door lock of Example 4.1 with a ROM. Indicate ROM cell connections with dots.
5. *Microcomputer.* (a) Draw the logic diagram for a TTL 7488A ROM that will generate the CLK*, CLK, CQ0, CQ1, C81, C23, C45, and C67 pulse waveforms shown in Fig. G.8 of Appendix G. (b) Show the ROM implementation of these functions. Use dots to indicate ROM cell connections. (c) Using a TTL data book, prepare a computer program for programming your ROM. (d) Punch out a deck of computer cards for your ROM program and make a computer listing of these cards.

Section 10.2.5

6. An ADD instruction to an MM 5750 (Fig. G.6) ALU of a microprogrammed computer consists of the following sequence of microinstructions: 1001[A ← (R6)], 1011[A ← (R4)], 0011[ADD], and 1100[(R3) ← (ALU)], where ( ) means "the contents of." Show the ROM implementation of these microinstructions. Use a counter to supply a sequence of addresses starting at ROM address 8. Include block symbols for the four-bit ALU and input data registers R4 and R6 and data output register R3.
7. *Microcomputer.* (a) Using a data book and referring to the microcomputer described in Appendix G, explain how the control ROM (CROM) works. (b) Use a simplified block diagram in your explanation.

Section 10.3

8. What is the main difference between a ROM and a PLA?
9. (a) Draw the PLA implementation of a 16-line-to-1-line multiplexer. (b) How many ROM cells would this implementation require?
10. Draw the PLA implementation of a seven-segment decoder.

Section 10.4.1

11. *Bipolar Memory Cell Circuit.* Figure P10.1 shows a typical three-emitter bipolar XY-addressable memory cell and its truth table. (a) Explain how this circuit works as a memory cell latch (store function) when both address inputs are kept LOW. (b) Explain how a HIGH stored in the latch (i.e., Q_0 is on, Q_1 is off) can be read out, when both address lines are HIGH and $W_0 = W_1 = 0$, by detecting current flow in the read-write lines with the sense amplifiers. (c) Explain how a LOW can be written into this cell, which previously stored a 1, when both address inputs are HIGH, by putting $W_0 = 1$ and $W_1 = 0$. [*Hint:* Refer to the general description of this cell in Section 10.4.1.]

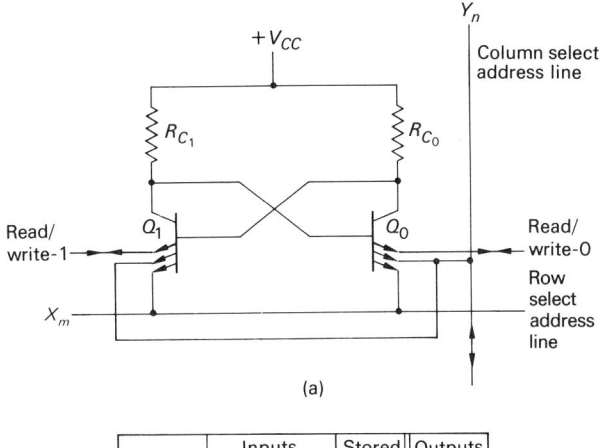

Fig. P10.1 Bipolar memory cell: (a) logic diagram; (b) function table.

Section 10.4.2

12. *Static MOS Memory Cell.* Figure P10.2 shows a static MOS storage cell. Data is stored as charge on the gate of transistor Q_3 or Q_4. By definition, a LOW is stored in the memory cell if Q_3 is on, and a HIGH is stored if Q_4 is on. Resistors R_p represent the path for parasitic leakage currents ($\sim$ pA). Explain how this cell works. [*Hint*: Transistors Q_3 and Q_4 form a latch. A cell is addressed by making the row select address line HIGH. During a write operation the data line being written on is held HIGH and the other data line is held LOW.]

Section 10.4.3

13. Explain the difference between the address mode and the chip select mode for reading from a read/write memory.
14. Explain the meaning of the read cycle output waveforms 1 and 2 in Fig. 10.20(b).

Section 10.4.4

15. With the aid of a timing diagram, explain how a series of continuous write operations can be performed on a 2102A-2 1024 × 1 static MOS memory.

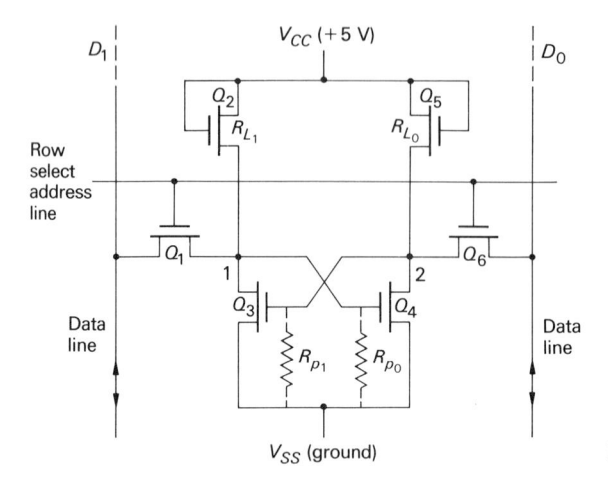

Fig. P10.2 Static MOS memory cell.

Section 10.4.5

16. With the aid of a timing diagram explain the write cycle of a 2107B 4K × 1 read/write memory.
17. Draw the logic diagram for refreshing a 2107B dynamic 4K × 1 read/write memory.
18. What causes current transients in the supply lines of a dynamic read/write memory?
19. Calculate the total standby power, including refreshing power, for a 16K × 8 memory made from 4K × 1 2107B memories. Use the minimum cycle time and the maximum refresh time. Assume all rows in the memory system are refreshed simultaneously.

Section 10.4.6

20. Draw the logic block diagram for a 16K × 8 memory system using 4K × 1 2107B ICs. Detailed address lines should be omitted. Show latched address inputs, address decoder, chip enable driver (block symbol), and latched outputs.
21. *Microcomputer.* Refer to the microcomputer memory schematics in Appendix G (Fig. G.7). Draw a logic diagram of read/write ICs G2A, G2B, and ROM IC 3H. Show all inputs and outputs in detail, i.e., do not combine address and data lines into one line as is done in Appendix G. Label each input line that comes from nonmemory ICs, showing clearly what circuit its signal comes from. Label all output lines similarly.

Section 10.5

22. Briefly explain the operation of a CAM.
23. Briefly explain a FIFO memory.
24. Briefly explain a LIFO memory.

REFERENCES

1. G. D. Feth. "Memories are Bigger, Faster—and Cheaper." *IEEE Spectrum*, pp. 28–35, November 1973.
2. Laurence Altman. "Charge-Coupled Devices Move in on Memories and Analog Signal Processing." *Electronics*, pp. 91–101, August 8, 1974.
3. Sidney Davis. "Selection and Application of Semiconductor Memories." *Computer Design*, pp. 65–77, January 1974.
4. William N. Carr and Jack P. Mize. *MOS/LSI Design and Application*. New York: McGraw-Hill, 1972.
5. Richard Percival. *The Application of ROMs*. AN-61, National Semiconductor, June 1972.
6. *Trig Function Generators*. MB-10, National Semiconductor, January 1970.
7. Gene Carter and Dale Mrazek. *The Systems Approach to Character Generators*. AN-40, National Semiconductor, June 1970.
8. George Reyling. "PLAs Enhance Digital Processor Speed and Cut Component Count." *Electronics*, pp. 109–114, August 8, 1974.
9. Ken Gorman. "The Programmable Logic Array: A New Approach to Microprogramming." *EDN*, pp. 68–75, November 20, 1973.
10. Ken Gorman and Paul Kaufman. "Low-Cost Minicomputer Opens Up Many New System Opportunities." *Electronics*, p. 111, June 7, 1973.
11. Dale Mrazek and Mel Morris. *How to Design with Programmable Logic Arrays*. AN-89, National Semiconductor, August 1973.
12. Kent Andres. "MOS Programmable Logic Arrays." Bulletin CA-158, Texas Instruments Corp., October 1970.
13. Ury Priel and Phil Holland. "Applications of High Speed Programmable Logic Array." *Computer Design*, pp. 94–96, December 1973.
14. Jacob Millman and Christos C. Halkias. *Integrated Electronics: Analog and Digital Circuits and Systems*. New York: McGraw-Hill, 1972.
15. Wallace B. Riley. "Semiconductor Memories are Taking Over Data-Storage Applications." *Electronics*, pp. 75–90, August 2, 1973.
16. *The Intel Memory Design Handbook*. Intel Corp., August 1973.
17. *Signetics Digital, Linear, MOS Applications*. Signetics Corp., 1973.
18. Rod Burns and Don Savitt. "Microprogramming, Stack Architecture Ease Minicomputer Programmer's Burden." *Electronics*, pp. 95–97, February 15, 1973.
19. Peter Alfke and Ib Larsen (eds.). *The TTL Applications Handbook*. Fairchild Semiconductor, August 1973.
20. *Non-Volatile Memories*. AN-1, Nitron Co., Cupertino, Ca. 95014, June 1974.
21. *Microprogramming Guide*. Hewlett-Packard Co., Cupertino, Ca., 1972.
22. Dick Brunner. "Designing Minicomputer Memory Systems with 4-Kilobit n-MOS Memories." *Computer Design*, pp. 61–68, July 1975.
23. Eugene R. Hnatek. "4-Kilobit Memories Present a Challenge to Testing." *Computer Design*, pp. 117–125, May 1975.

APPENDIX A: Summary of Boolean Algebra Definitions, Laws, and Rules

Let $F, G, H, \ldots$, be any Boolean variables or functions.

I. Definitions

1. If $F \neq 0$ then $F = 1$.
2. If $F \neq 1$ then $F = 0$.

II. Laws (Postulates)

1. $0 \cdot 0 = 0$
2. $0 \cdot 1 = 0$
3. $1 \cdot 0 = 0$
4. $1 \cdot 1 = 1$
5. $0 + 0 = 0$
6. $0 + 1 = 1$
7. $1 + 0 = 1$
8. $1 + 1 = 1$
9. $\overline{0} = 1$
10. $\overline{1} = 0$

III. Rules (Theorems)

1. $F \cdot 1 = F$ (Uniqueness)
2. $F \cdot 0 = 0$
3. $F \cdot G = G \cdot F$ (Commutative)
4. $F + G = G + F$ (Commutative)
5. $F \cdot (G \cdot H) = (F \cdot G) \cdot H$ (Associative)
6. $F + (G + H) = (F + G) + H$ (Associative)
7. $F \cdot (G + H) = F \cdot G + F \cdot H$ (Distributive)
8. $F + 1 = 1$
9. $F + 0 = F$ (Uniqueness)
10. $F \cdot F = F$ (Idempotent)
11. $F + F = F$ (Idempotent)
12. $F \cdot \overline{F} = 0$
13. $F + \overline{F} = 1$

14. $(\bar{\bar{F}}) = F$
15. $\overline{F + G} = \bar{F} \cdot \bar{G}$ (DeMorgan)
    a) $F + G = \overline{\bar{F} \cdot \bar{G}}$
    b) $\overline{F + G + H + \cdots} = \bar{F} \cdot \bar{G} \cdot \bar{H} \cdots$
16. $\overline{F \cdot G} = \bar{F} + \bar{G}$ (DeMorgan)
    a) $F \cdot G = \overline{\bar{F} + \bar{G}}$
    b) $\overline{F \cdot G \cdot H \cdots} + \bar{F} + \bar{G} + \bar{H} + \cdots$
17. $F + F \cdot G = F$ (Absorption)
18. $F \cdot (F + G) = F$ (Absorption)
19. $F + \bar{F} \cdot G = F + G$ (Redundancy)
20. $F \cdot (\bar{F} + G) = F \cdot G$
21. $(F + G) \cdot (F + H) = F + G \cdot H$
22. $F \cdot G + \bar{F} \cdot G = G$
23. $(F + G) \cdot (F + \bar{G}) = F$
24. $F \cdot G + G \cdot H + \bar{F} \cdot H = F \cdot G + \bar{F} \cdot H$
25. $(F + G)(\bar{F} + H) = F \cdot H + \bar{F} \cdot G$
26. a) $F(A, B, \ldots, C) = A \cdot F(1, B, \ldots, C) + \bar{A} \cdot F(0, B, \ldots, C)$ (Expansion)
    b) $F(A, B, \ldots, C) = [A + F(0, B, \ldots, C)] \cdot [\bar{A} + F(1, B, \ldots, C)]$

APPENDIX B: The Powers of 2

2^n	n	2^{-n}
1	0	1.0
2	1	0.5
4	2	0.25
8	3	0.125
16	4	0.062 5
32	5	0.031 25
64	6	0.015 625
128	7	0.007 812 5
256	8	0.003 906 25
512	9	0.001 953 125
1 024	10	0.000 976 562 5
2 048	11	0.000 488 281 25
4 096	12	0.000 244 140 625
8 192	13	0.000 122 070 312 5
16 384	14	0.000 061 035 156 25
32 768	15	0.000 030 517 578 125
65 536	16	0.000 015 258 789 062 5
131 072	17	0.000 007 629 394 531 25
262 144	18	0.000 003 814 697 265 625
524 288	19	0.000 001 907 348 632 812 5
1 048 576	20	0.000 000 953 674 316 406 25
2 097 152	21	0.000 000 476 837 158 203 125
4 194 304	22	0.000 000 238 418 579 101 562 5
8 388 608	23	0.000 000 119 209 289 550 781 25
16 777 216	24	0.000 000 059 604 644 775 390 625
33 554 432	25	0.000 000 029 802 322 387 695 312 5
67 108 864	26	0.000 000 014 901 161 193 847 656 25
134 217 728	27	0.000 000 007 450 580 596 923 828 125
268 435 456	28	0.000 000 003 725 290 298 461 914 062 5
536 870 912	29	0.000 000 001 862 645 149 230 957 031 25
1 073 741 824	30	0.000 000 000 931 322 574 615 478 515 625
2 147 483 648	31	0.000 000 000 465 661 287 307 739 257 812 5
4 294 967 296	32	0.000 000 000 232 830 643 653 869 628 906 25
8 589 934 592	33	0.000 000 000 116 415 321 826 934 814 453 125
17 179 869 184	34	0.000 000 000 058 207 660 913 467 407 226 562 5
34 359 738 368	35	0.000 000 000 029 103 830 456 733 703 613 281 25
68 719 476 736	36	0.000 000 000 014 551 915 228 366 851 806 640 625
137 438 953 472	37	0.000 000 000 007 275 957 614 183 425 903 320 312 5
274 877 906 944	38	0.000 000 000 003 637 978 807 091 712 951 660 156 25
549 755 813 888	39	0.000 000 000 001 818 989 403 545 856 475 830 078 125
1 099 511 627 776	40	0.000 000 000 000 909 494 701 772 928 237 915 039 062 5
2 199 023 255 552	41	0.000 000 000 000 454 747 350 886 464 118 957 519 531 25
4 398 046 511 104	42	0.000 000 000 000 227 373 675 443 232 059 478 759 765 625
8 796 093 022 208	43	0.000 000 000 000 113 686 837 721 616 029 739 379 882 812 5
17 592 186 044 416	44	0.000 000 000 000 056 843 418 860 808 014 869 689 941 406 25
35 184 372 088 832	45	0.000 000 000 000 028 421 709 430 404 007 434 844 970 703 125
70 368 744 177 664	46	0.000 000 000 000 014 210 854 715 202 003 717 422 485 351 562 5
140 737 488 355 328	47	0.000 000 000 000 007 105 427 357 601 001 858 711 242 675 781 25
281 474 976 710 656	48	0.000 000 000 000 003 552 713 678 800 500 929 355 621 337 890 625
562 949 953 421 312	49	0.000 000 000 000 001 776 356 839 400 250 464 677 810 668 945 312 5
1 125 899 906 842 624	50	0.000 000 000 000 000 888 178 419 700 125 232 338 905 334 472 656 25
2 251 799 813 685 248	51	0.000 000 000 000 000 444 089 209 850 062 616 169 452 667 236 328 125
4 503 599 627 370 496	52	0.000 000 000 000 000 222 044 604 925 031 308 084 726 333 618 164 062 5
9 007 199 254 740 992	53	0.000 000 000 000 000 111 022 302 462 515 654 042 363 166 809 082 031 25
18 014 398 509 481 984	54	0.000 000 000 000 000 055 511 151 231 257 827 021 181 583 404 541 015 625
36 028 797 018 963 968	55	0.000 000 000 000 000 027 755 575 615 628 913 510 590 791 702 270 507 812 5
72 057 594 037 927 936	56	0.000 000 000 000 000 013 877 787 807 814 456 755 295 395 851 135 253 906 25
144 115 188 075 855 872	57	0.000 000 000 000 000 006 938 893 903 907 228 377 647 697 925 567 626 953 125
288 230 376 151 711 744	58	0.000 000 000 000 000 003 469 446 951 953 614 188 823 848 962 783 813 476 562 5
576 460 752 303 423 488	59	0.000 000 000 000 000 001 734 723 475 976 807 094 411 924 481 391 906 738 281 25
1 152 921 504 606 846 976	60	0.000 000 000 000 000 000 867 361 737 988 403 547 205 962 240 695 953 369 140 625
2 305 843 009 213 693 952	61	0.000 000 000 000 000 000 433 680 868 994 201 773 602 981 120 347 976 684 570 312 5
4 611 686 018 427 387 904	62	0.000 000 000 000 000 000 216 840 434 497 100 886 801 490 560 173 988 342 285 156 25
9 223 372 036 854 775 808	63	0.000 000 000 000 000 000 108 420 217 248 550 433 400 745 280 086 994 171 142 578 125
18 446 744 073 709 551 616	64	0.000 000 000 000 000 000 054 210 108 624 275 221 700 372 640 043 497 085 571 289 062 5
36 893 488 147 419 103 232	65	0.000 000 000 000 000 000 027 105 054 312 137 610 850 186 320 021 748 542 785 644 531 25
73 786 976 294 838 206 464	66	0.000 000 000 000 000 000 013 552 527 156 068 805 425 093 160 010 874 271 392 822 265 625
147 573 952 589 676 412 928	67	0.000 000 000 000 000 000 006 776 263 578 034 402 712 546 580 005 437 135 696 411 132 812 5
295 147 905 179 352 825 856	68	0.000 000 000 000 000 000 003 388 131 789 017 201 356 273 290 002 718 567 848 205 566 406 25
590 295 810 358 705 651 712	69	0.000 000 000 000 000 000 001 694 065 894 508 600 678 136 645 001 359 283 924 102 783 203 125
1 180 591 620 717 411 303 424	70	0.000 000 000 000 000 000 000 847 032 947 254 300 339 068 322 500 679 641 962 051 391 601 562 5
2 361 183 241 434 822 606 848	71	0.000 000 000 000 000 000 000 423 516 473 627 150 169 534 161 250 339 820 981 025 695 800 781 25
4 722 366 482 869 645 213 696	72	0.000 000 000 000 000 000 000 211 758 236 813 575 084 767 080 625 169 910 490 512 847 900 390 625

APPENDIX C: Schematic Diagrams and Troubleshooting Aids

Table C.1   Logic Symbols*

Symbol		Symbol	
AND	⟹—	Schmitt trigger	
OR	⟹—	Monostable	1 ⎍
NOT	▷∘—		
NAND	⟹∘—	Flip-flop	FF
		Delay element	
NOR	⟹∘—	DOT-AND	
EX-OR	⟹—	DOT-OR	
EX-NOR	⟹∘—	Symbol extension	
		Symbol extension	

Negation indicator	∘	The presence (or absence) of the negation indicator means that signal input or output is active when it is LOW (HIGH). Generally, signal inputs and outputs which have the negation indicator should have a bar over their name.
Dynamic indicator	▷	The presence of the dynamic indicator means that input becomes active on the LOW to HIGH transition of that signal input. Its absence means that input is static, i.e., it is active any time it is HIGH.
Dynamic indicator and negation indicator	∘▷	The combined dynamic indicator and negation indicator means that input becomes active on the HIGH to LOW transition of that signal input.

* *IEEE Standard Graphic Symbols for Logic Diagrams (Two-State Devices).* IEEE Std. 91–1973, ANSI Y32. 14–1973, April 6, 1973.

Table C.1 is a summary of the logic symbols used in this text. These symbols can be drawn with a template like the one shown in Fig. C.1.

Fig. C.1 Standard logic symbols drafting template, half size.

Figure C.2 illustrates typical notation used on schematic diagrams found in service manuals that accompany electronic equipment.

Figure C.3 shows a portion of a **troubleshooting flow diagram**, an aid that is extremely useful in troubleshooting. Note that an oscilloscope is needed for observing waveforms and that a logic comparator, such as the one shown in Fig. 6.6., is used to check digital ICs. The physical layout of the printed circuit board (Fig. C.4) aids in locating ICs, discrete components, and test points.

340   SCHEMATIC DIAGRAMS AND TROUBLESHOOTING AIDS

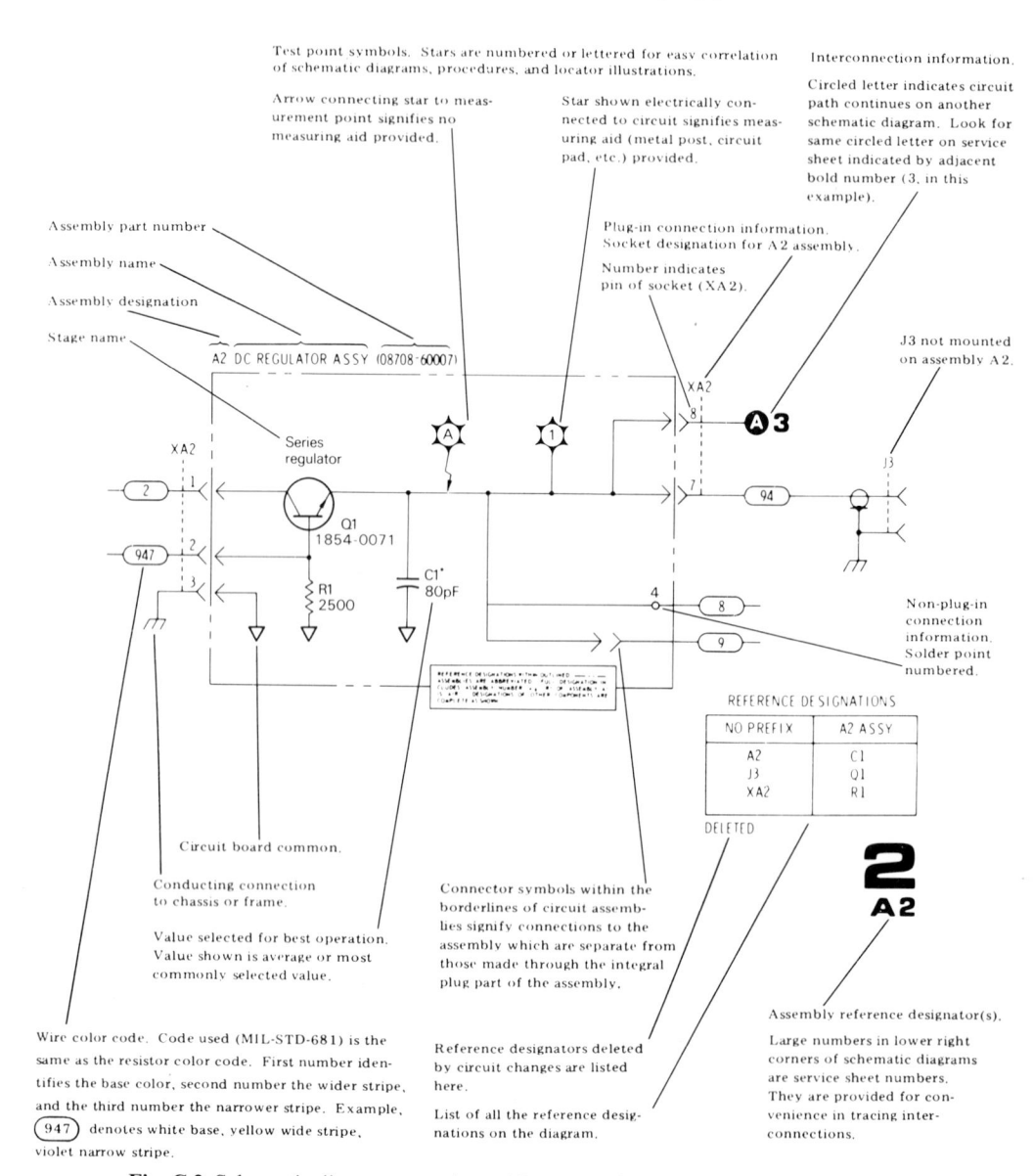

Fig. C.2 Schematic diagram notations. (Courtesy of Hewlett-Packard Company.)

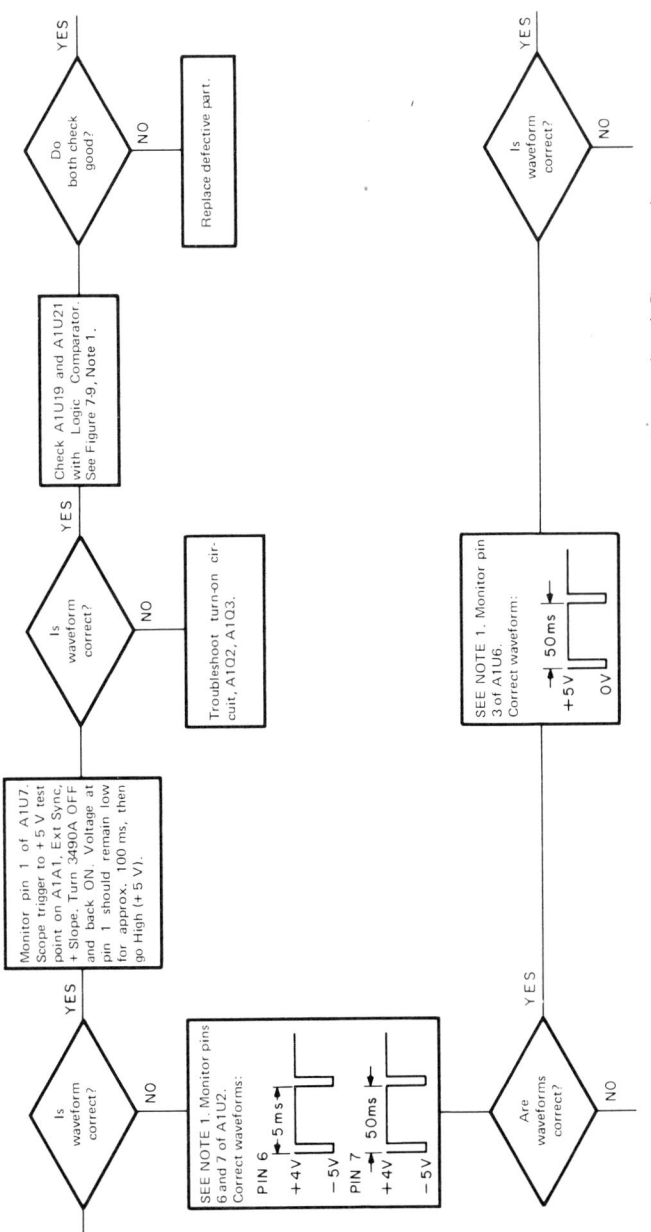

SCHEMATIC DIAGRAMS AND TROUBLESHOOTING AIDS 341

Fig. C.3 Troubleshooting flow diagram. (Courtesy of Hewlett-Packard Company.)

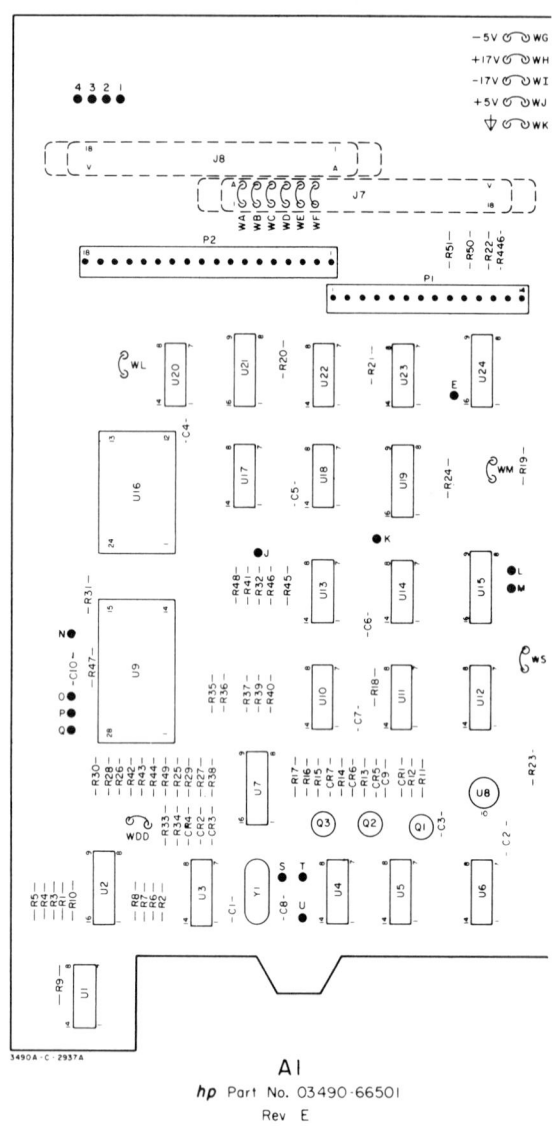

Fig. C.4 Physical layout diagram. (Courtesy of Hewlett-Packard Company.)

APPENDIX D: Digital IC Device Characteristics

D.1 TTL DEVICE CHARACTERISTICS*

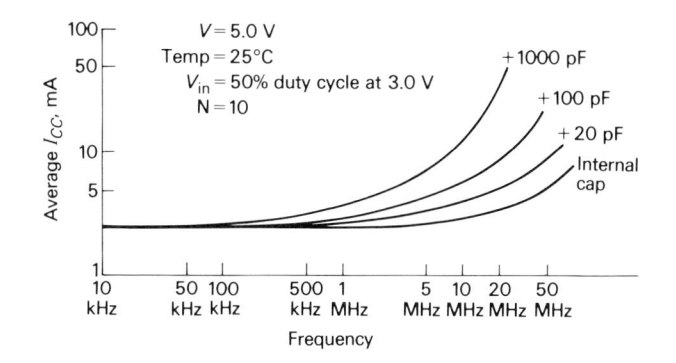

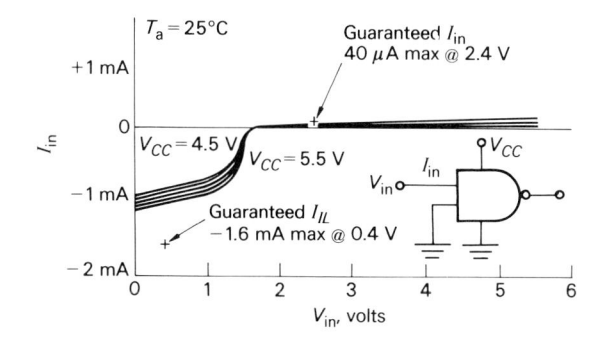

*All curves reproduced by Courtesy of Texas Instruments Incorporated.

SERIES 54/74 TRANSISTOR-TRANSISTOR LOGIC

TYPICAL CHARACTERISTICS†§

FIGURE A1 — OUTPUT VOLTAGE vs INPUT VOLTAGE

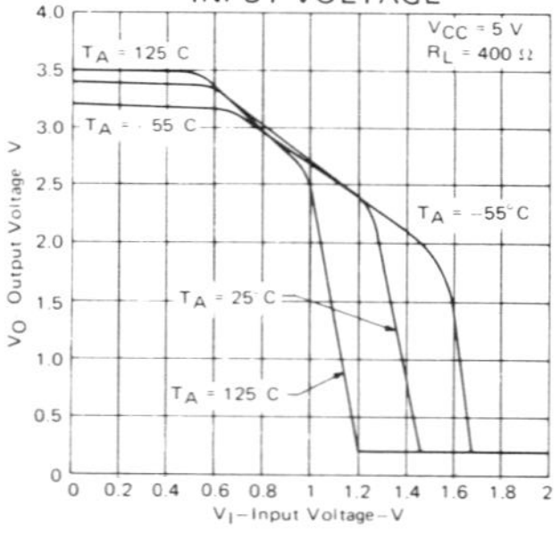

FIGURE A2 — HIGH-LEVEL OUTPUT VOLTAGE vs HIGH-LEVEL OUTPUT CURRENT

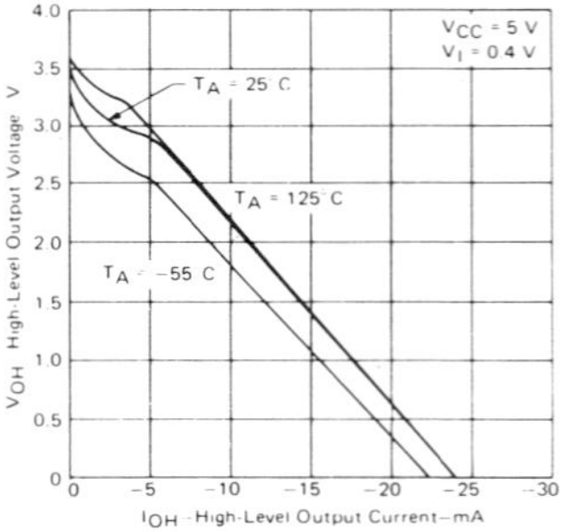

FIGURE A3 — LOW-LEVEL OUTPUT VOLTAGE vs LOW-LEVEL OUTPUT CURRENT

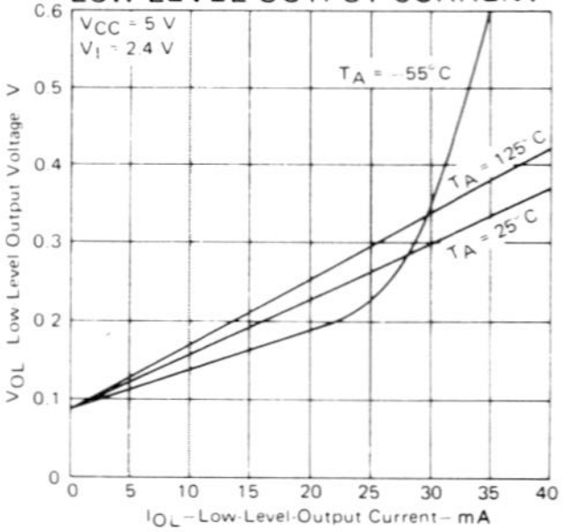

FIGURE A4 — AVERAGE PROPAGATION DELAY TIME vs FREE-AIR TEMPERATURE

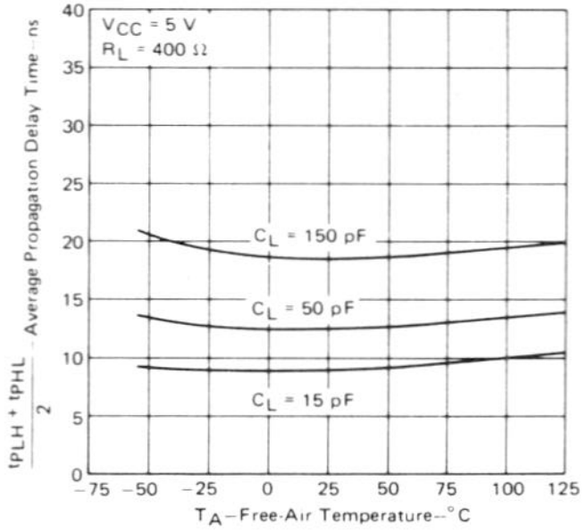

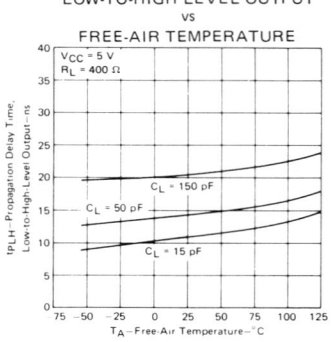

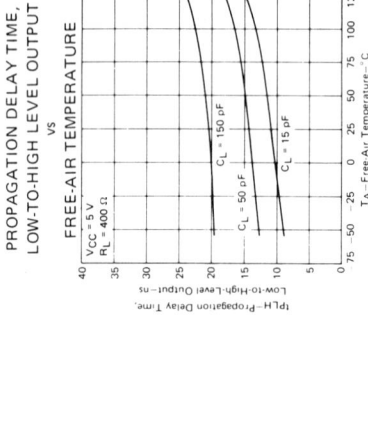

FIGURE A5 — PROPAGATION DELAY TIME, LOW-TO-HIGH LEVEL OUTPUT vs FREE-AIR TEMPERATURE

FIGURE A6 — PROPAGATION DELAY TIME, HIGH-TO-LOW-LEVEL OUTPUT vs FREE-AIR TEMPERATURE

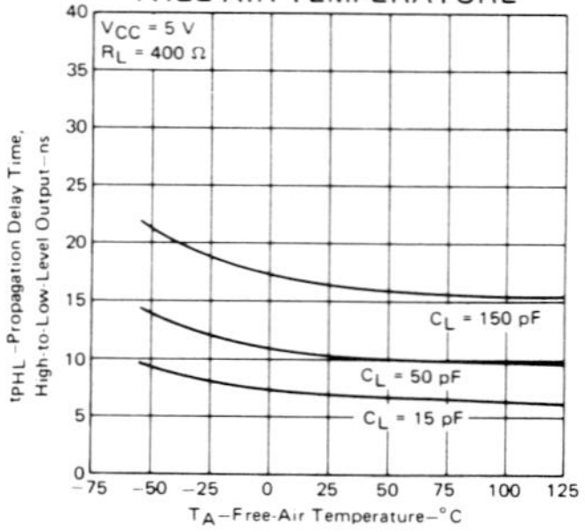

† Data for temperatures below 0°C and above 70°C are applicable for Series 54 circuits only.
§ Data as shown are applicable specifically for the NAND gates with totem-pole outputs.

SCHOTTKY-CLAMPED LOW-POWER TRANSISTOR-TRANSISTOR LOGIC

TYPICAL CHARACTERISTICS†§

FIGURE D1 — OUTPUT VOLTAGE vs INPUT VOLTAGE

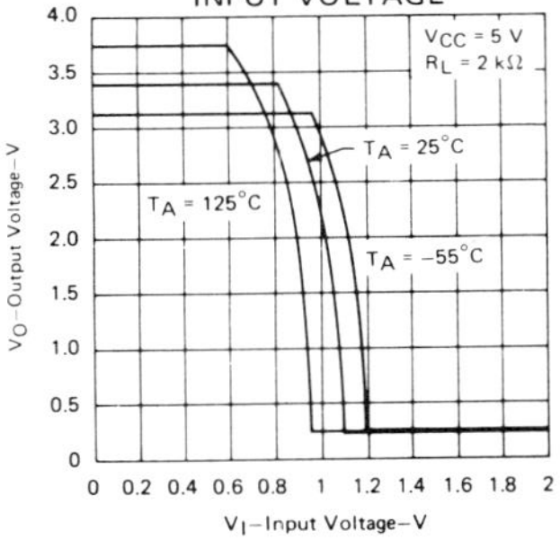

FIGURE D2 — HIGH-LEVEL OUTPUT VOLTAGE vs HIGH-LEVEL OUTPUT CURRENT

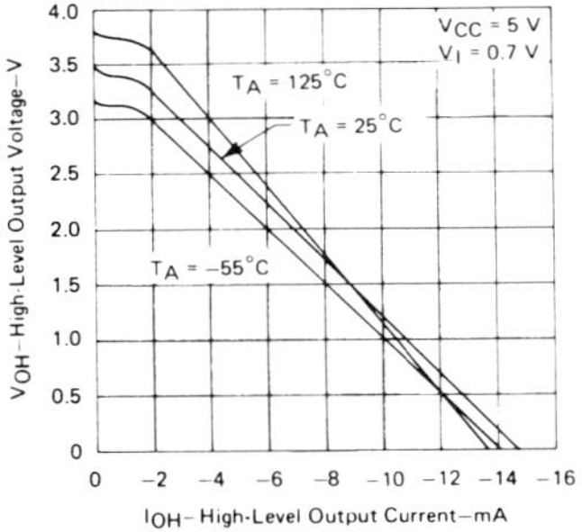

FIGURE D3 — LOW-LEVEL OUTPUT VOLTAGE vs LOW-LEVEL OUTPUT CURRENT

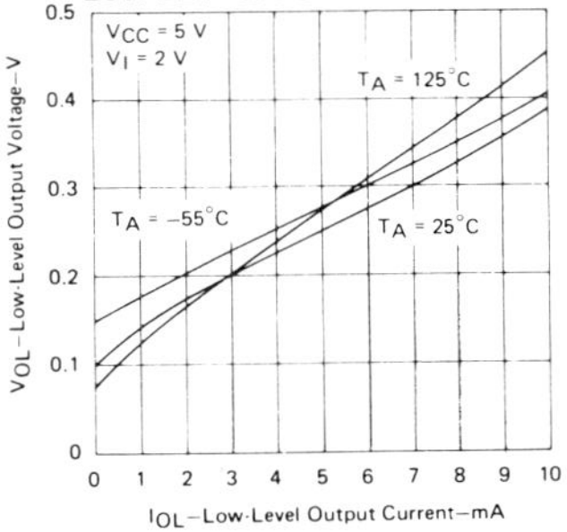

FIGURE D4 — POWER DISSIPATION PER GATE vs FREQUENCY

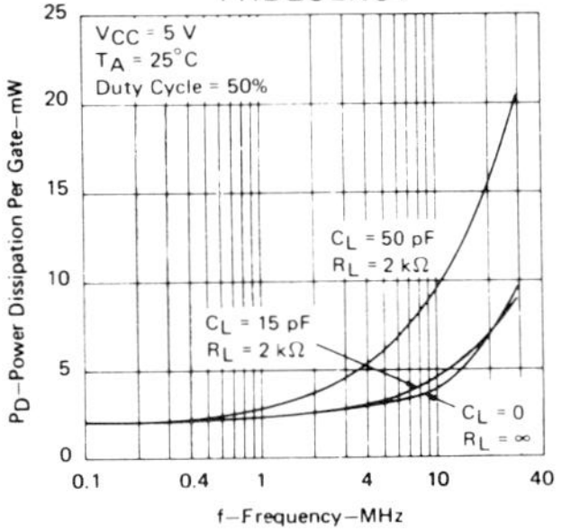

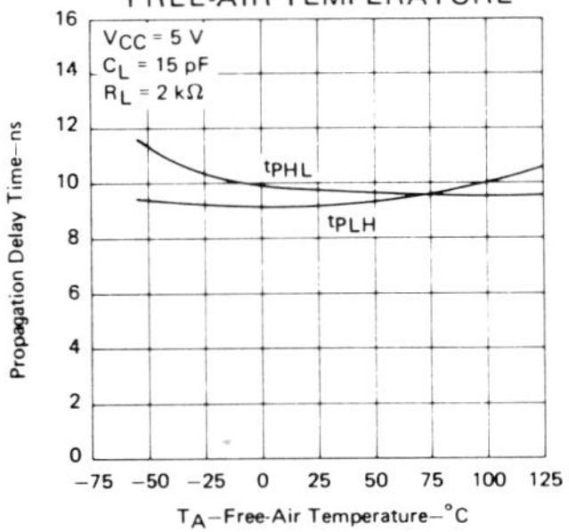

FIGURE D5 — PROPAGATION DELAY TIMES vs FREE-AIR TEMPERATURE

FIGURE D6 — PROPAGATION DELAY TIMES vs LOAD CAPACITANCE

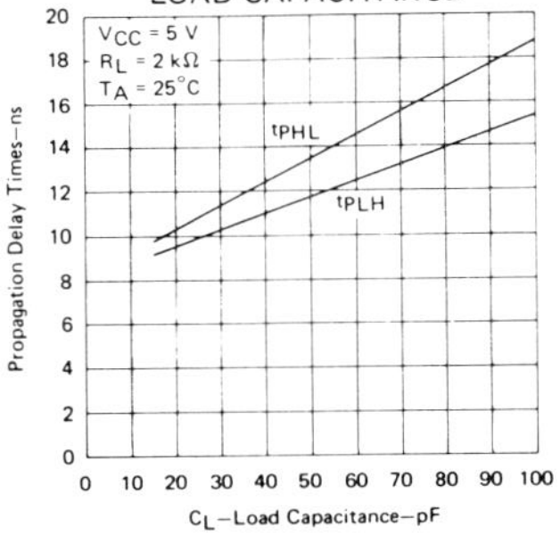

† Data for temperatures below 0°C and above 70°C are applicable for Series 54LS circuits only.
§ Data as shown are applicable specifically for the NAND gates with totem-pole outputs.

346 DIGITAL IC DEVICE CHARACTERISTICS D.1

SERIES 54S/74S
SCHOTTKY-CLAMPED TRANSISTOR-TRANSISTOR LOGIC

TYPICAL CHARACTERISTICS†§

OUTPUT VOLTAGE
vs
INPUT VOLTAGE 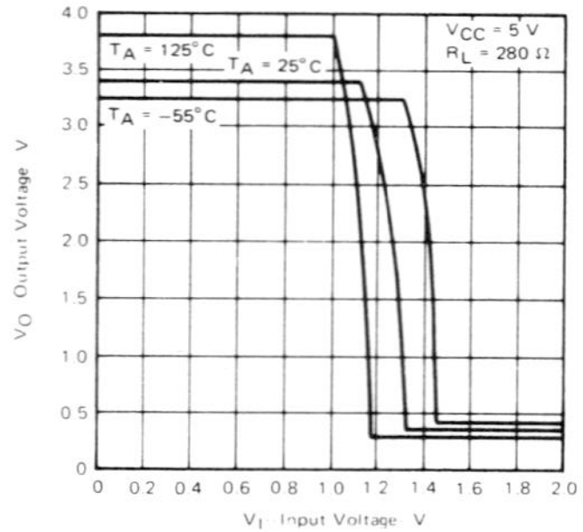
FIGURE E1

INPUT-CLAMPING-DIODE
FORWARD VOLTAGE
vs
FREE-AIR TEMPERATURE 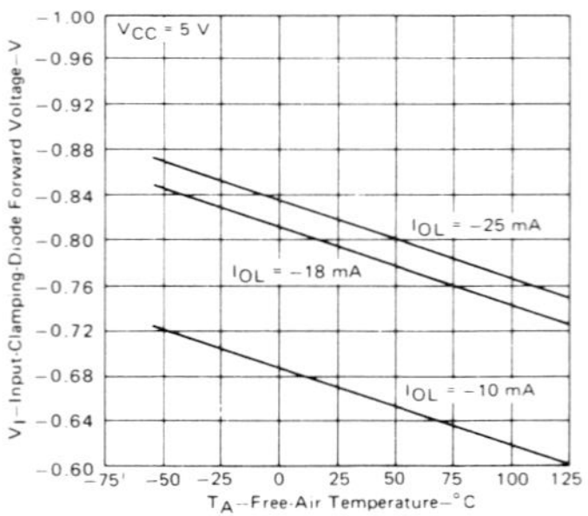
FIGURE E2

HIGH-LEVEL OUTPUT VOLTAGE
vs
HIGH-LEVEL OUTPUT CURRENT 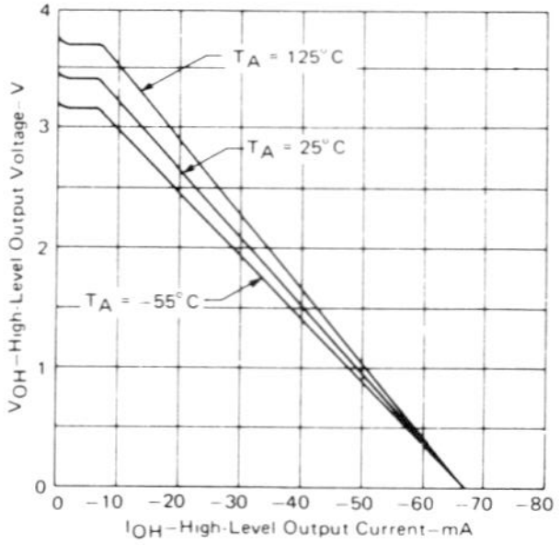
FIGURE E3

LOW-LEVEL OUTPUT VOLTAGE
vs
LOW-LEVEL OUTPUT CURRENT 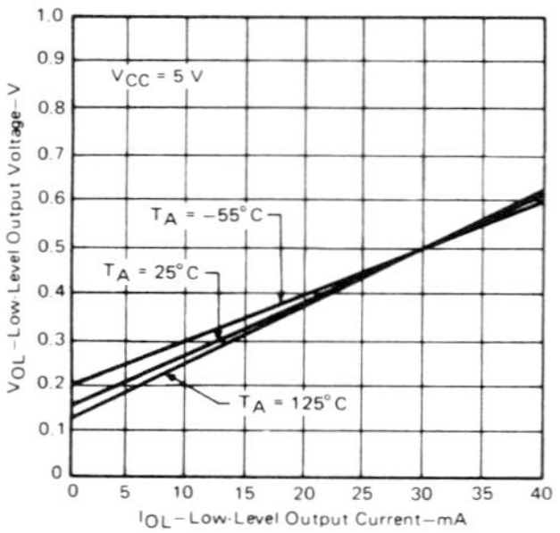
FIGURE E4

INPUT CURRENT
vs
INPUT VOLTAGE 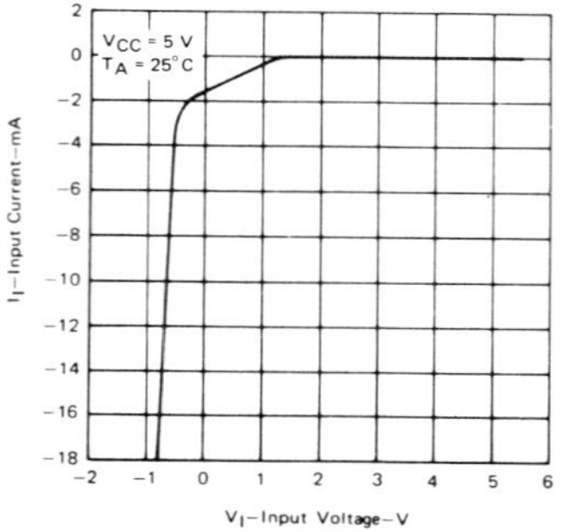
FIGURE E5

HIGH-LEVEL INPUT CURRENT
vs
FREE-AIR TEMPERATURE 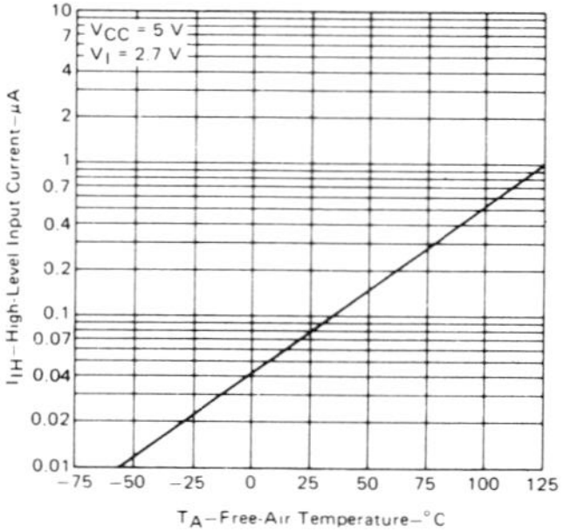
FIGURE E6

D.1 TTL DEVICE CHARACTERISTICS 347

SCHOTTKY-CLAMPED TRANSISTOR-TRANSISTOR LOGIC

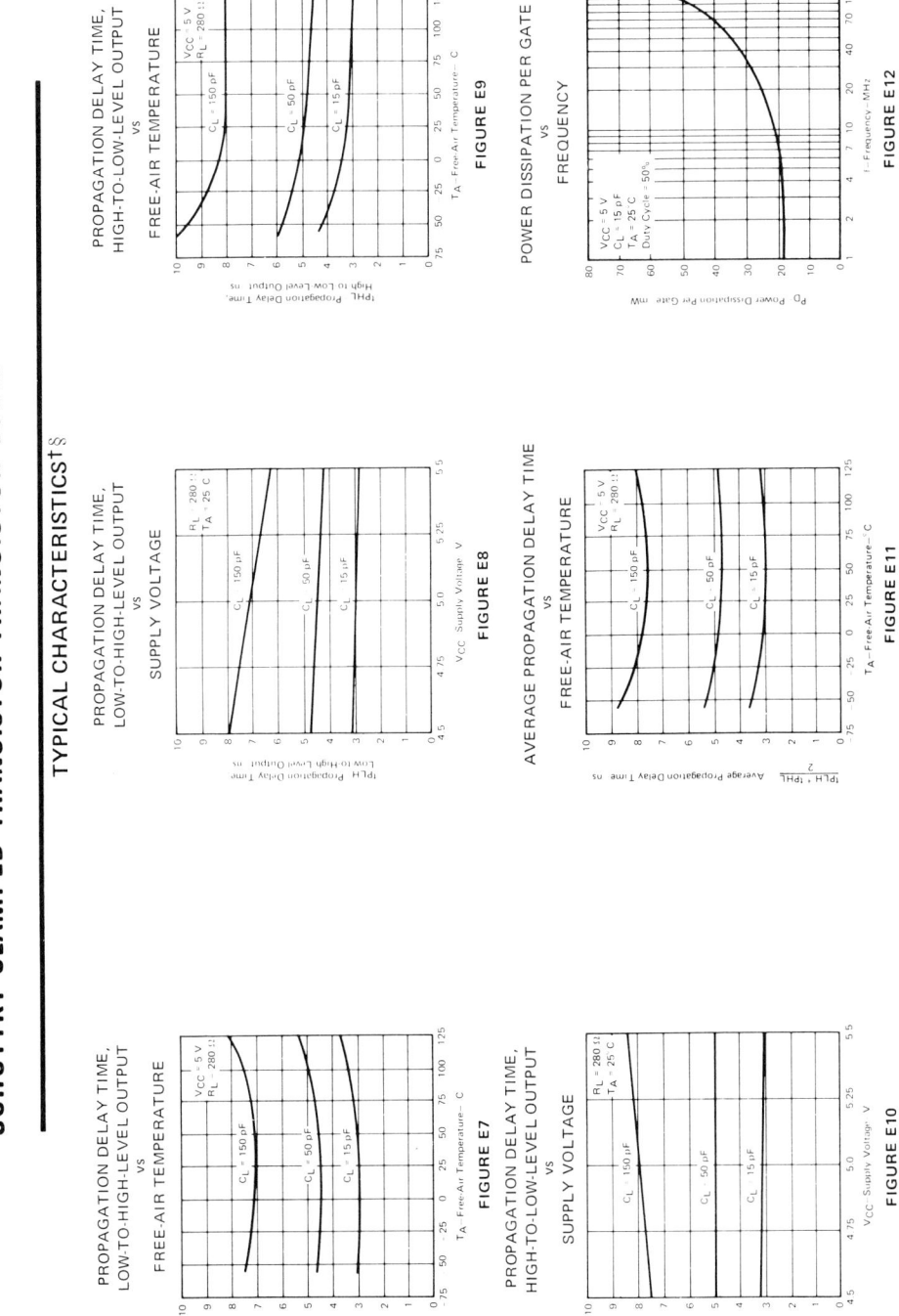

TYPICAL CHARACTERISTICS†§

FIGURE E7

FIGURE E8

FIGURE E9

FIGURE E10

FIGURE E11

FIGURE E12

†Data for temperatures below 0°C and above 70°C are applicable for Series 54S circuits only.
§Data as shown are applicable specifically for the NAND gates with totem-pole outputs.

D.2 CMOSL DEVICE CHARACTERISTICS*

2-INPUT "NOR-INVERTER"

FIGURE 3 – CURRENT AND VOLTAGE TRANSFER CHARACTERISTICS TEST CIRCUIT

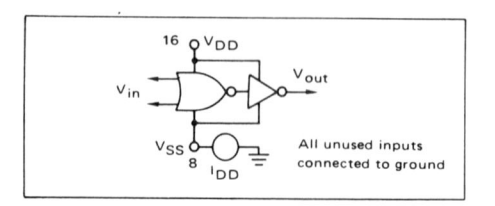

FIGURE 4 – TYPICAL VOLTAGE AND CURRENT TRANSFER CHARACTERISTICS

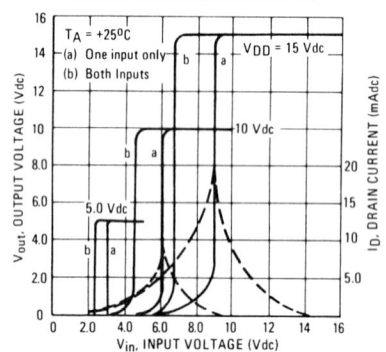

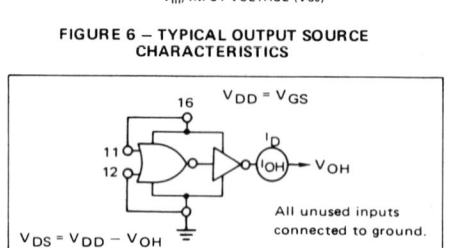

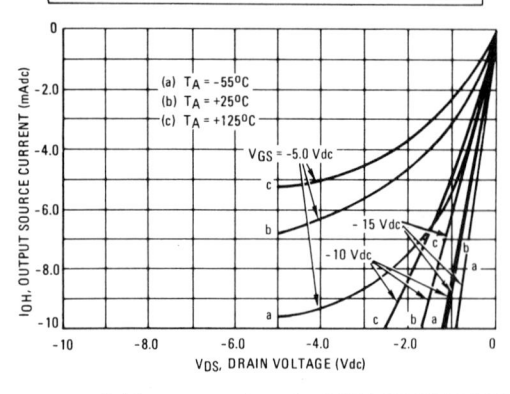

FIGURE 5 – TYPICAL VOLTAGE TRANSFER CHARACTERISTICS versus TEMPERATURE

FIGURE 6 – TYPICAL OUTPUT SOURCE CHARACTERISTICS

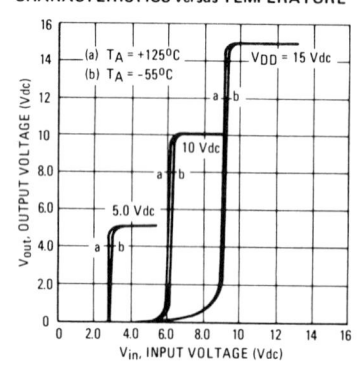

FIGURE 7 – TYPICAL OUTPUT SINK CHARACTERISTICS

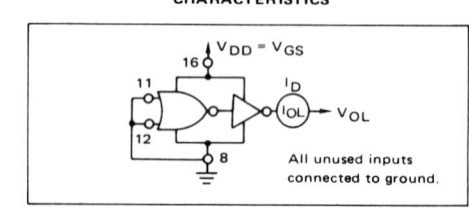

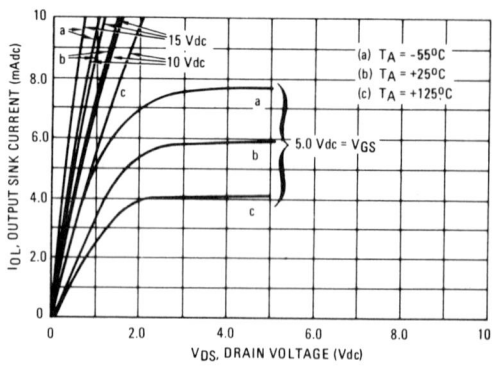

* All curves apply to the 14501 OR/NOR/NAND Gate and are reproduced by courtesy of Motorola Corporation.

D.2 CMOSL DEVICE CHARACTERISTICS 349

FIGURE 14 — TYPICAL RISE TIME versus LOAD CAPACITANCE

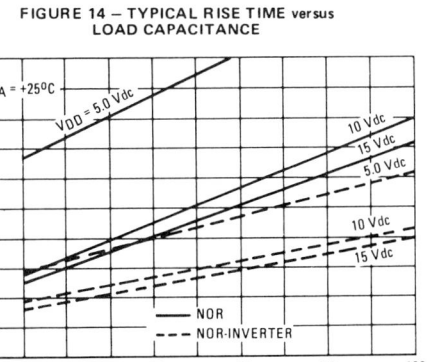

FIGURE 15 — TYPICAL FALL TIME versus LOAD CAPACITANCE

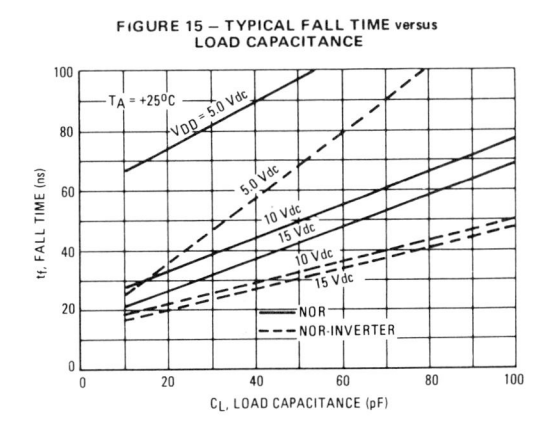

FIGURE 16 — TYPICAL TURN-ON DELAY TIME versus LOAD CAPACITANCE

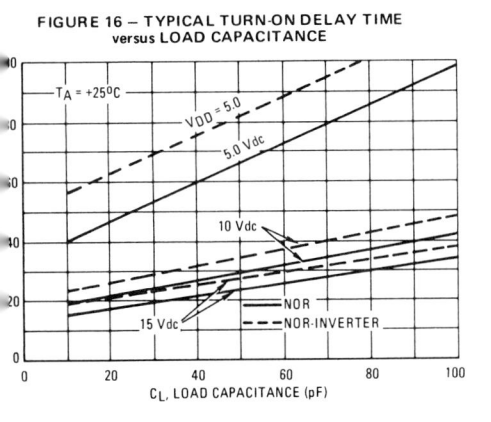

FIGURE 17 — TYPICAL TURN-OFF DELAY TIME versus LOAD CAPACITANCE

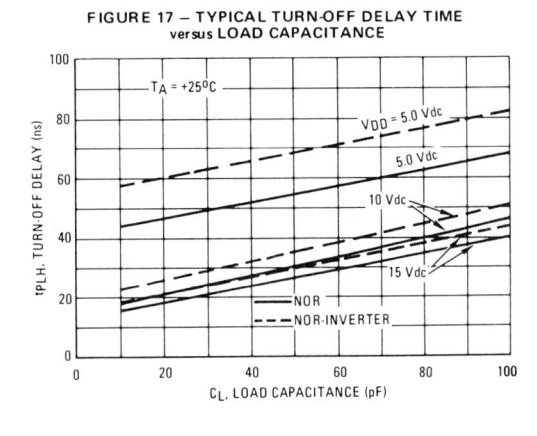

D.3 ECL DEVICE CHARACTERISTICS*

TYPICAL F10K SERIES CHARACTERISTICS

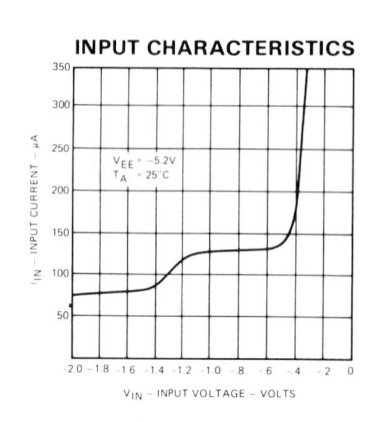

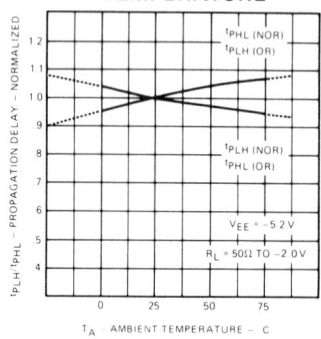

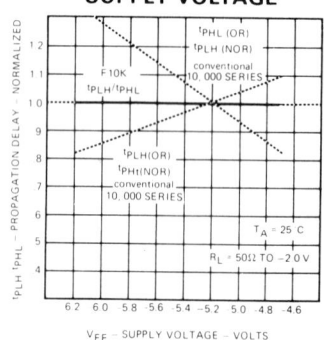

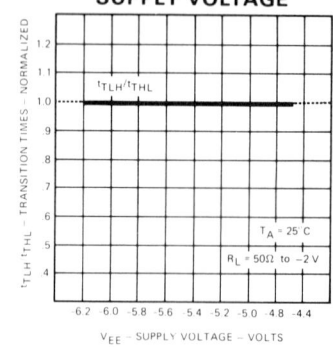

* All curves reproduced by courtesy of Fairchild Semiconductor.

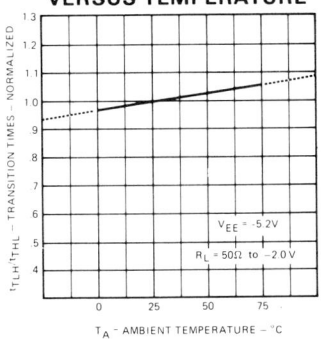

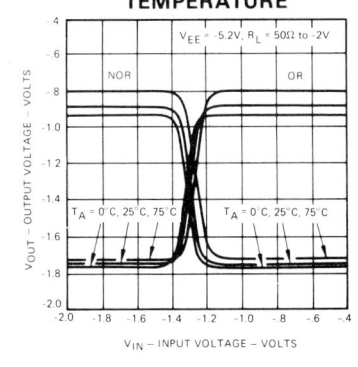

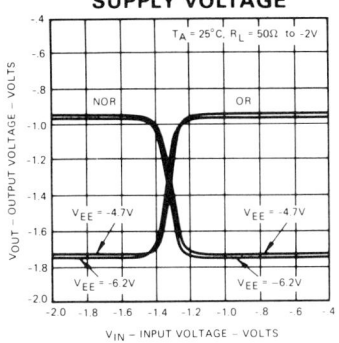

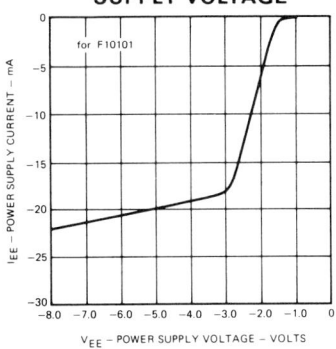

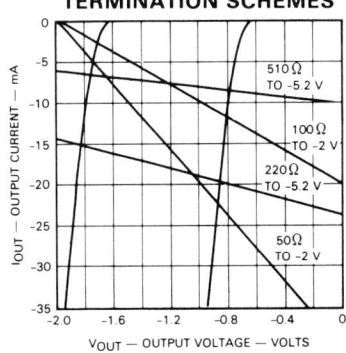

D.3   ECL DEVICE CHARACTERISTICS   351

APPENDIX E: Optoelectronics

E.1 THE ELECTROMAGNETIC SPECTRUM

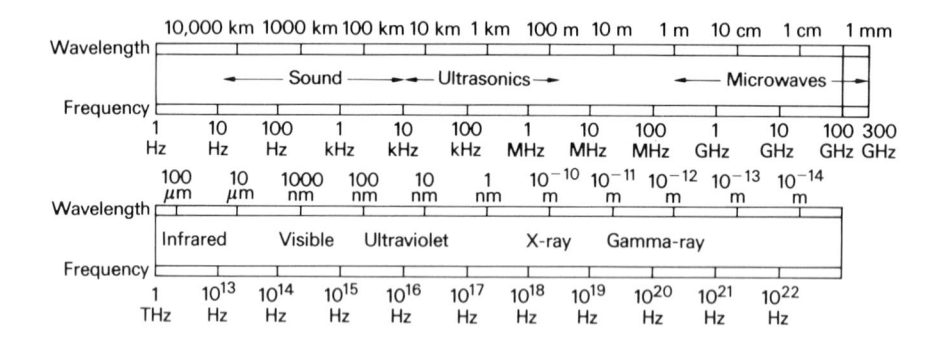

E.2 VISIBLE SPECTRUM

Color	Wavelength λ, nm
Violet	<450
Blue	450–500
Green	500–570
Yellow	570–590
Orange	590–610
Red	>610

E.3 RESPONSE CURVES*

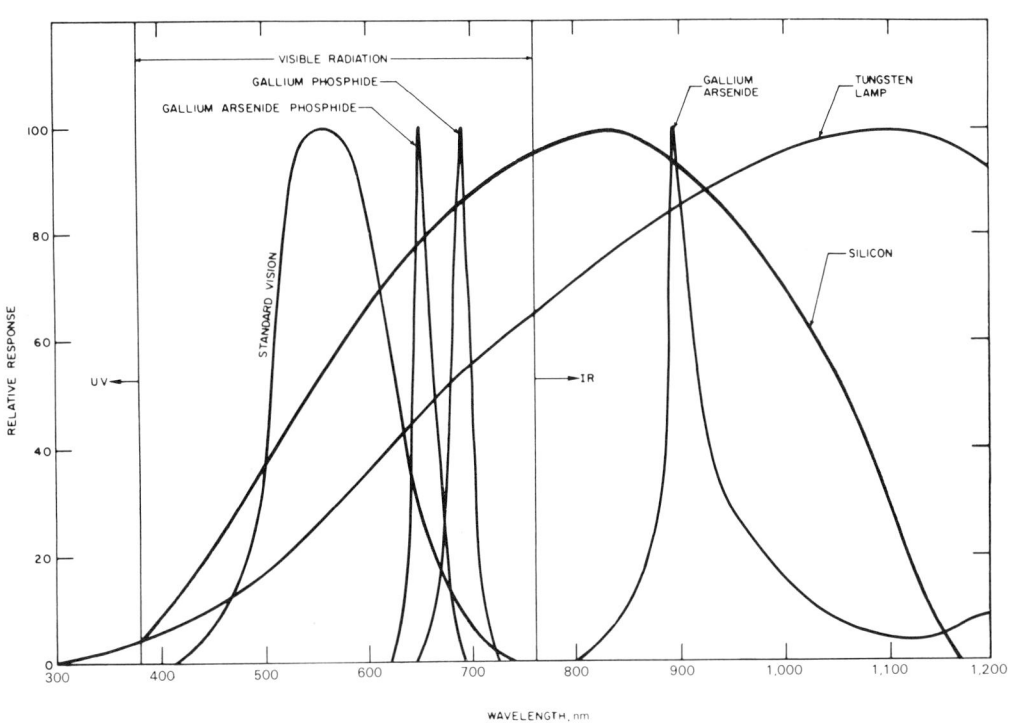

Performance

LED type	Color of light	Wavelength of light, nm	Relative optical sensitivity, lumens per optical Watt	Efficiency, percent	Visual merit, lumens/elect. Watt
GaP	Red	690	15	1	0.15
GaP	Green	565	590	0.01	0.05
GaAsP	Red	670	20	0.1	0.02
GaAsP	Red	660	40	0.03	0.012
GaAsP	Orange	610	330	0.001	0.003
GaAsP	Yellow	585	500	—	—

(Commercial device performance)

* Courtesy of Fairchild Semiconductor.

APPENDIX F: Synthesized Signal Generator—Description and Logic Diagrams*

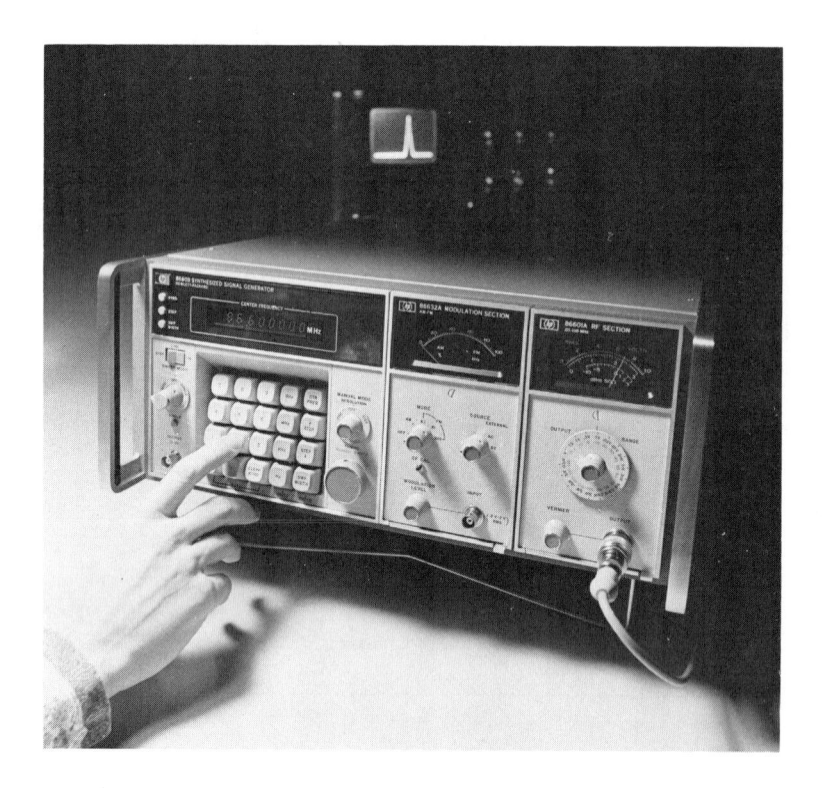

F.1 SUMMING LOOP 2 PHASE DETECTOR A12

F.1.1 Summing Loop 2 General

The purpose of Summing Loop 2 (SL2) is to generate digitally controlled rf signals in the range of 20.0001 to 30.0000 MHz in selectable 100 Hz increments. The difference frequency between the SL2 voltage controlled oscillator and the input from the N2 loop is phase locked to the divided-by-ten output of the N3 assembly. The output of SL2 is applied to SL1.

* Photograph, diagrams, and accompanying text reproduced by courtesy of Hewlett-Packard Company.

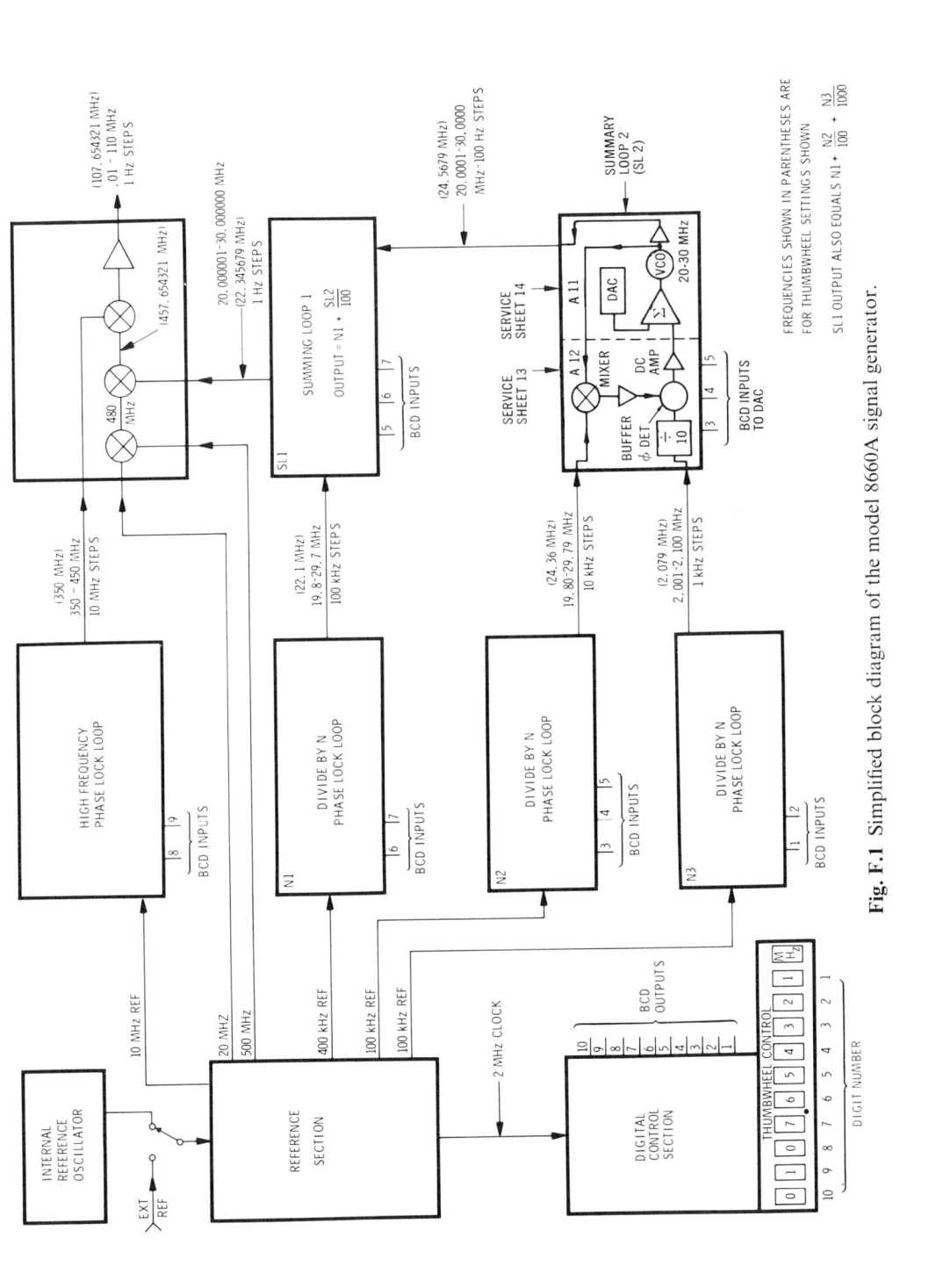

Fig. F.1 Simplified block diagram of the model 8660A signal generator.

F.1.2 Phase Detector

There are three signal inputs to the phase detector assembly. They are the output of the N2 voltage controlled oscillator, the divided by ten output of the N3 voltage controlled oscillator and the output of the SL2 voltage controlled oscillator.

The N2 and SL2 signals are mixed and the difference frequency is used as one input to the digital phase detector. The second input to the digital phase detector is the divided by ten input from the N3 assembly.

The output of the N3 voltage controlled oscillator is divided by ten in the N3 assembly and again divided by ten by U9. Q12 and NAND gate U7A shape the resulting pulses which vary in frequency (depending on programming to the N3 loop) from 0.2001 to 0.2100 MHz. The pulses at TP2 are negative-going.

The inputs from the N2 loop and the SL2 voltage controlled oscillator are applied to double balanced mixer E1 R and L ports. The difference signal from the X port is amplified by Q5 and Q4 and shaped by Q3, Q7 and NAND gates U4B and U4C. When the loop is phase locked the negative-going pulses at TP3 are at the same frequency as those at TP2. The pulses do not appear in time coincidence; they are received alternately.

U7B, U7D, U4A and U4D comprise a coincidence gate which inhibits signals that appear simultaneously at TP2 and TP3. Normally, when signals are not present, TP2 and TP3 are both high. When a signal appears at TP2, U7B pin 6 and U4D pin 13 go high. If there is no signal at TP3 U4D pin 12 is also high; U4D pin 11 goes low, and U1B pin 6 goes high. The positive pulse at TP5 drives the clock generator and the sense circuit or phase detector. When a signal appears at TP3, U4A pin 3 and U7D pin 12 go high. If there is no signal at TP2, U7D pin 13 is also high; U7D pin 11 goes low, and U7C pin 8 goes high. The positive pulse at TP9 drives the clock generator and the sense circuit or the phase detector. When signals appear at TP2 and TP3 at the same time U7D pin 13 and U4D pin 12 go low, U7D pin 11 and U4D pin 11 remain high, and the signals cannot reach TP5 or TP9.

U1A, U1C, U1D and U5C comprise a clock generator which clocks U2A and U2B each time a signal appears at TP5 or TP9. With no signals present TP5 and TP9 are low. When a positive pulse appears at TP9 U1A pin 3 goes low, U1D pin 11 goes high and a negative-going pulse appears at TP6. When a positive pulse appears at TP5 operation of the circuit is the same except that U1C pin 8 goes low (rather than U1A pin 3). Since a clock pulse is generated for each input, the pulse frequency at TP6 is the sum of the frequencies at TP5 and TP9.

Since the sense circuit does not function when the loop is locked, operation of the phase detector will be discussed first.

When the loop is phase locked U2A $\bar{Q}$ is held high to enable U3A and U3D. Assume that initially U2B $\bar{Q}$ is high, U3B pin 6 is low and U3C pin 8 is high. When a positive-going signal from TP9 appears at U3A pin 1, U3A pin 3 goes low and causes a change in state of flip-flop U3B/U3C; U3B pin 6 goes high and U3C pin 8 goes low. The high at U2B pin 12 sets the flip-flop and the positive-going trailing edge of the clock pulse causes U2B Q to go high. The following positive pulse from TP5 is applied to U3D pin 12, U3D pin 11 goes low and changes the state of flip/flop U3B/U3C. U3B pin 6 goes low and the clock pulse causes U2B $\bar{Q}$ to again go high. This sequence continues as long as the signals at TP5 and TP9 are received alternately.

The signals at TP5 and TP9 are applied to the sense circuit even when the loop is phase locked. They have no effect on the circuit because of the relationship of the Q and $\bar{Q}$ outputs of U2B to the incoming signals.

When U2B Q is high NAND gates U6A and U6C are enabled. When the signal from TP5 appears at U6C pin 9, U6C pin 8 goes low; flip/flop U5A/U5B does not change state because U5B pin 3 is low. The signal at U6B has no effect because U2B $\bar{Q}$ and U6B pin 4 are low.

When U2B $\bar{Q}$ is high NAND gates U6B and U6D are enabled. When the signal at TP9 appears at U6D pin 13, U6D pin 11 goes low; flip/flop U5A/U5B does not change state because U5B pin 3 is low. The signal at pin 1 of U6A has no effect on the circuit because U2B Q and pin 2 of U6A are low.

When two or more consecutive pulses from either input (TP5 or TP9) occur between pulses from the other input the sense circuit functions to disable the phase detector until the frequency error is corrected.

As an example of circuit operation assume that two pulses from TP9 (SL2 signal) are received between two pulses from TP5 (N3 signal) indicating that the SL2 frequency is high. When the first pulse from TP9 is received U3A pin 3 goes low, U3B pin 6 goes high to set U2B and the clock pulse causes U2B Q to go high. When the second consecutive pulse is received from TP9 U6A has been enabled by the high Q output of U2B. U6A pin 3 goes low and causes flip/flop U5A/U5B to change state. When the D input of U2A goes high the clock pulse causes U2A $\bar{Q}$ to go low and inhibit U3A and U3D. If a third SL2 signal is received prior to receipt of an N3 signal U6A pin 3 will again go low but will have no effect on flip/flop U5A/U5B because U5A pin 13 is low.

When an N3 pulse is received U2B Q is still high and U6C pin 8 will go low to change the state of flip/flop U5A/U5B. When the D input of U2A goes low the clock pulse causes U2A $\bar{Q}$ to go high and enable U3A and U3D. The propagation time of the signal through the sense circuit is long enough for the pulse from N3 (TP5) to have ended before U3D is enabled so the state of flip/flop U3B/U3C does not change.

The next pulse from SL2 will again cause U6A pin 3 to go low and change the state of flip/flop U5A/U5B. With the D input to U2A high again, the clock pulse again causes U2A $\bar{Q}$ to go low and inhibit U3A and U3D. The signal applied to U3A has no effect on flip/flop U3B/U3C because U3B pin 5 is low.

The sense circuit continues operation in the manner described above until two consecutive N3 pulses are received between two SL2 signals. When this occurs the first pulse causes U6C pin 8 to go low and change the state of flip/flop U5A/U5B. With the D input to U2A low the clock pulse will cause U2A $\bar{Q}$ to go

358   SYNTHESIZED SIGNAL GENERATOR

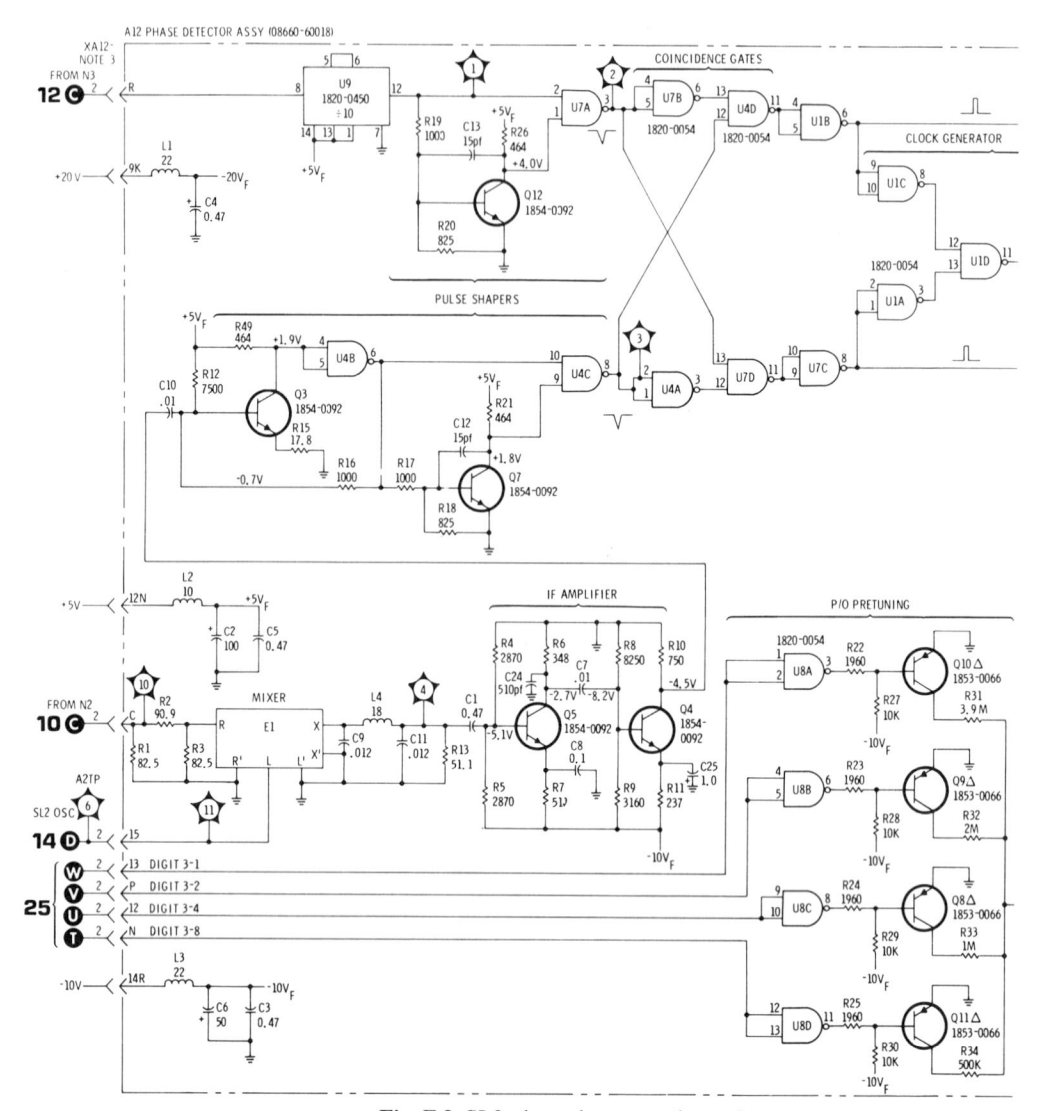

Fig. F.2 SL2 phase detector schematic.

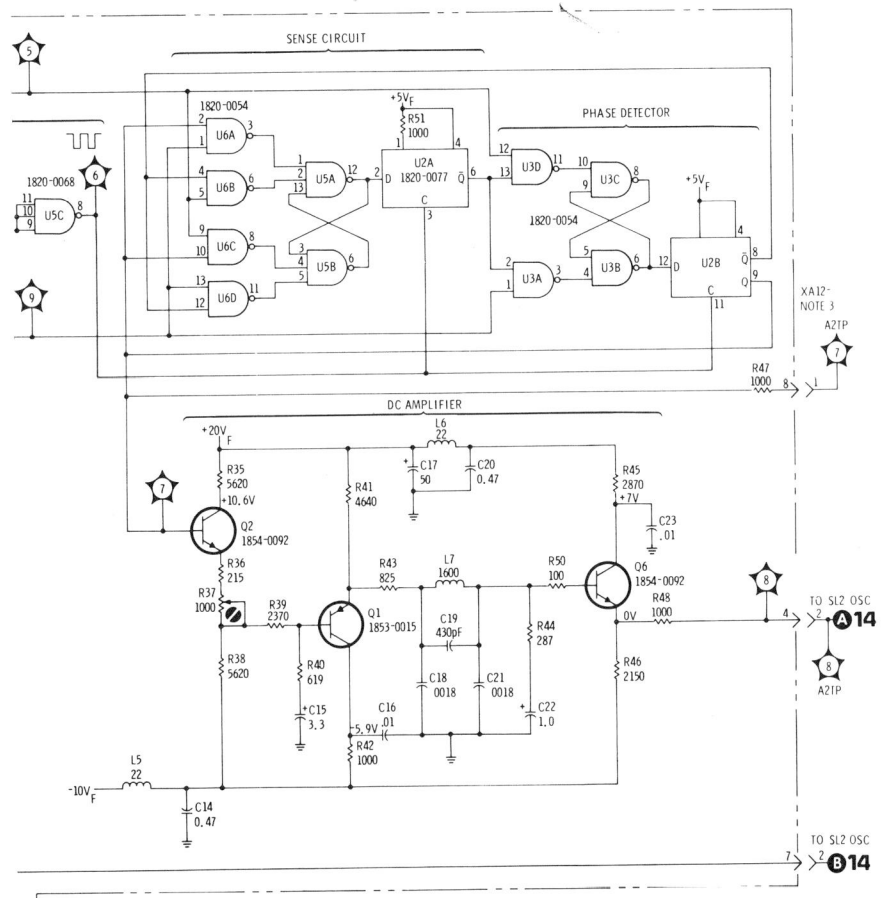

Reference Designation	HP Part Number	Qty	Description	Mfr Code	Mfr Part Number
A12R23	0698-0083		R:FXD MET FLM 1.96K OHM 1% 1/8W	28480	0698-0083
A12R24	0698-0083		R:FXD MET FLM 1.96K OHM 1% 1/8W	28480	0698-0083
A12R25	0698-0083		R:FXD MET FLM 1.96K OHM 1% 1/8W	28480	0698-0083
A12R27	0757-0442		R:FXD MET FLM 10.0K OHM 1% 1/8W	28480	0757-0442
A12R28	0757-0442		R:FXD MET FLM 10.0K OHM 1% 1/8W	28480	0757-0442
A12R29	0757-0442		R:FXD MET FLM 10.0K OHM 1% 1/8W	28480	0757-0442
A12R29	0698-0082		R:FXD MET FLM 464 OHM 1% 1/8W	28480	0698-0082
A12R30	0757-0442		R:FXD MET FLM 10.0K OHM 1% 1/8W	28480	0757-0442
A12R31	0683-3955	2	R:FXD COMP 3-9 MEGOHM 5% 1/4W	01121	CB 3955
A12R32	0683-2055	2	R:FXD COMP 2 MEGOHM 5% 1/4W	01121	CB 2055
A12R33	0683-1055	2	R:FXD COMP 1 MEGOHM 5% 1/4W	01121	CB 1055
A12R34	0698-3263		R:FXD MET FLM 500K OHM 1% 1/8W	28480	0698-3263
A12R35	0757-0200		R:FXD MET FLM 5.62K OHM 1% 1/8W	28480	0757-0200
A12R36	0698-3441		R:FXD MET FLM 215 OHM 1% 1/8W	28480	0698-3441
A12R37	2100-2633	2	R:VAR CERMET 1K OHM 10% LIN 1/2W	28480	2100-2633
A12R38	0757-0200		R:FXD MET FLM 5.62K OHM 1% 1/8W	28480	0757-0200
A12R39	0698-3150		R:FXD MET FLM 2.37K OHM 1% 1/8W	28480	0698-3150
A12R40	0757-0418		R:FXD MET FLM 619 OHM 1% 1/8W	28480	0757-0418
A12R41	0698-3155		R:FXD MET FLM 4.64K OHM 1% 1/8W	28480	0698-3155
A12R42	0757-0280		R:FXD MET FLM 1K OHM 1% 1/8W	28480	0757-0280
A12R43	0757-0421		R:FXD MET FLM 825 OHM 1% 1/8W	28480	0757-0421
A12R44	0698-3443		R:FXD MET FLM 287 OHM 1% 1/8W	28480	0698-3443
A12R45	0698-3151		R:FXD MET FLM 2.87K OHM 1% 1/8W	28480	0698-3151
A12R46	0698-0084		R:FXD MET FLM 2.15K OHM 1% 1/8W	28480	0698-0084
A12R47	0757-0280		R:FXD MET FLM 1K OHM 1% 1/8W	28480	0757-0280
A12R48	0757-0280		R:FXD MET FLM 1K OHM 1% 1/8W	28480	0757-0280
A12R49	0698-0082		R:FXD MET FLM 464 OHM 1% 1/8W	28480	0698-0082
A12R50	0757-0401		R:FXD MET FLM 100 OHM 1% 1/8W	28480	0757-0401
A12R51	0757-0280		R:FXD MET FLM 1K OHM 1% 1/8W	28480	0757-0280
A12U1	1820-0054		IC:TTL QUAD 2-INPT NAND GATE	01295	SN7400N
A12U2	1820-0077		IC:TTL DUAL D F/F	01295	SN7474N
A12U3	1820-0054		IC:TTL QUAD 2-INPT NAND GATE	01295	SN7400N
A12U4	1820-0054		IC:TTL QUAD 2-INPT NAND GATE	01295	SN7400N
A12U5	1820-0068		IC:TTL TRIPLE 3-INPUT POS NAND GATE	12040	SN7410N
A12U6	1820-0054		IC:TTL QUAD 2-INPT NAND GATE	01295	SN7400N
A12U7	1820-0054		IC:TTL QUAD 2-INPT NAND GATE	01295	SN7400N
A12U8	1820-0054		IC:TTL QUAD 2-INPT NAND GATE	01295	SN7400N
A12U9	1820-0450		IC:DIGITAL TTL	18324	N8290A
A13					
A13					

A13	08660-60012			28480	08660-60012
A13			BOARD ASSY:N2 OSCILLATOR		
A13C1	0180-0058		C:FXD AL ELECT 50 UF +75-10% 25VDCW	56289	30D506G025CC2-DSM
A13C2	0180-0228		C:FXD ELECT 22 UF 10% 15VDCW	56289	150D226X9015B2-DYS
A13C3	0180-0049		C:FXD ELECT 20 UF +75-10% 50VDCW	56289	30D206G050CC2-DSM
A13C4	0180-2207		C:FXD ELECT 100 UF 10% 10VDCW	56289	150D101X9010R2-DYS
A13C5	0150-0121		C:FXD CER 0.1 UF +80-20% 50VDCW	56289	5C50BIS-CML
A13C6	0150-0121		C:FXD CER 0.1 UF +80-20% 50VDCW	56289	5C50BIS-CML
A13C7	0150-0121		C:FXD CER 0.1 UF +80-20% 50VDCW	56289	5C50BIS-CML
A13C8	0160-3459		C:FXD CER 0.02 UF 20% 100VDCW	56289	C023F101H203MS22CDH
A13C9			NOT ASSIGNED		
A13C10	0180-0228		C:FXD ELECT 22 UF 10% 15VDCW	56289	150D226X9015B2-DYS
A13C11	0180-0116		C:FXD ELECT 6.8 UF 10% 35VDCW	56289	150D685X9035B2-DYS
A13C12	0180-0228		C:FXD ELECT 22 UF 10% 15VDCW	56289	150D226X9015B2-DYS
A13C13	0180-2210		C:FXD ELECT 2 UF +50-10% 150VDCW	28480	0180-2210
A13C14	0180-0374		C:FXD TANT. 10 UF 10% 20VDCW	56289	150D106X9020B2-DYS
A13C15	0160-2055		C:FXD CER 0.01 UF +80-20% 100VDCW	56289	C023F101F103Z522-CDH
A13C16	0160-0386		C:FXD CER 3.3+/-0.25 PF 500VDCW	72982	301-000-S2H0-339C
A13C17	0160-2204		C:FXD MICA 100PF 5%	72136	RDM15F101J3C
A13C18	0170-0082		C:FXD MY 0.01UF 20% 50VDCW	84411	601PE STYLE 1
A13C19	0121-0059		C:VAR CER 2-8 PF 300VDCW	28480	0121-0059
A13C20			NOT ASSIGNED		
A13C21	0160-2055		C:FXD CER 0.01 UF +80-20% 100VDCW	56289	C023F101F103Z522-CDH
A13C22	0160-0386		C:FXD CER 3.3+/-0.25 PF 500VDCW	72982	301-000-S2H0-339C
A13C23	0160-0386		C:FXD CER 3.3+/-0.25 PF 500VDCW	72982	301-000-S2H0-339C
A13C24	0160-2055		C:FXD CER 0.01 UF +80-20% 100VDCW	56289	C023F101F103Z522-CDH
A13C25	0160-2055		C:FXD CER 0.01 UF +80-20% 100VDCW	56289	C023F101F103Z522-CDH
A13C26	0160-2055		C:FXD CER 0.01 UF +80-20% 100VDCW	56289	C023F101F103Z522-CDH
A13C27	0160-2055		C:FXD CER 0.01 UF +80-20% 100VDCW	56289	C023F101F103Z522-CDH
A13C28	0160-3459		C:FXD CER 0.02 UF 20% 100VDCW	56289	C023F101H203MS22CDH
A13C29	0160-0163		C:FXD MY 0.033 UF 10% 200VDCW	56289	192P33392-PTS
A13CR1	1901-0040		DIODE:SILICON 50 MA 30 WV	07263	FDG1088
A13CR2			NOT ASSIGNED		
A13CR3	1901-0040	1	DIODE:SILICON 50 MA 30 WV	07263	FDG1088

Fig. F.3 Model 8660 signal generator: partial table of replaceable parts, assembly A12.

high and enable U3A and U3D. Again, because of propagation time through the sense circuit the pulse will have ended before U3D in enabled. The second consecutive N3 pulse again causes U6C pin 8 to go low but, because U5B pin 3 is low, no change in state occurs in flip/flop U5A/U5B. Since U3D is now enabled, U3D pin 11 goes low and causes flip/flop U3B/U3C to change state. With the D input to U2B low, the clock pulse causes U2B $\bar{Q}$ output to go high. Phase lock has been achieved and the loop will remain locked as long as pulses at the same frequency appear alternately at TP5 and TP9.

When the SL2 frequency is low U2B Q is low. When the SL2 frequency is high U2B Q is high.

DC amplifier Q2, Q1, Q6 and associated components filter the Q output of U2B and applies it to a summing circuit in the A11 assembly to precisely control the voltage controlled oscillator.

F.2 N2 PHASE DETECTOR ASSEMBLY A14

F.2.1 N2 Loop General Information

The purpose of the N2 loop is to generate digitally controlled rf signals in the range of 19.80 to 29.79 MHz in selectable 10 kHz increments. The voltage controlled oscillator is phase locked to a 100 kHz reference which is derived from the master oscillator in the reference section. The rf output from the N2 loop is applied to Summing Loop 2.

F.2.2 Programmable Divider Circuit

All of the integrated circuits in the A14 assembly are used to count down the input from the N2 voltage controlled oscillator.

When there is no BCD input to U5, U6 and U7 (all inputs low) the input from the oscillator will be 29.79 MHz; the programmable divider will divide by 2979 to provide a 10 kHz output. U5, U6 and U7 may be preset by thumbwheel digits 3, 4 and 5 and programmed to vary between counts of 1980 and 2979. Operation of the circuit is as follows:

Assume that initially there are no BCD inputs to U5, U6 and U7 (divide-by-ten decades) and they have all been preset to zero.

At the start of every count cycle, regardless of the BCD input, U1A pin 6 ($\bar{Q}$) and U1B pin 8 ($\bar{Q}$) are both low; U3 pin 6 ($\bar{Q}$), U4A pin 6 ($\bar{Q}$) and U4B pin 8 ($\bar{Q}$) are all high.

NAND gate U8C functions as a Schmitt trigger and provides pulses derived from the N2 voltage controlled oscillator output to clock U7 when AND gate U2B is enabled. U7 provides a divide-by-ten output to clock U6 and also provides A and C (binary 1 and 4) outputs to J inputs of JK flip-flop U3. The A and C outputs have no effect on U3 until the count down reaches 2975.

U6 provides a divide-by-ten output to clock U5 and also provides A, B and C (binary 1, 2 and 4) outputs to AND gates U2A and U2C. The A, B and C outputs have no effect on the circuit until the count down of 2970 is reached.

U5 provides a divide-by-ten output to clock U1A and also provides A and D outputs to NAND gate U8A. The A and D (binary 1 and 8) outputs have no effect on the circuit until the count down has reached 2900.

The D output of U5 (pin 12) goes low on the 1000th pulse input to U7 pin 8 and clocks U1A. One thousand input cycles later U1A is again clocked and the negative-going $\bar{Q}$ output of U1A (pin 6) clocks U1B. When U1B $\bar{Q}$ goes high it provides a high to AND gate U2A. The count down has reached 2000.

When the count down reaches 2900, U5 A and D outputs are high. NAND gate U8A pin 3 goes low and NAND gate U8B pin 6 goes high.

When the count down reaches 2970, U6 A, B and C outputs are high. The B and C outputs are applied to AND gate U2C pins 10 and 11, and since U2C pin 9 has been high since the count of 2900, U2C pin 8 goes high. The U6A output is applied to AND gate U2A, and since the other two inputs to U2A are high, U2A pin 12 goes high and is applied to U3 J input pin 3.

When the count down reaches 2975, U7 A and C high outputs are applied to U3 J input pins 4 and 5. Since U3 J pin 3 is now held high, the next input pulse from U8C will clock U3. Count coincidence at 2975 cycles has been achieved.

When the count down reaches 2976, U3 is clocked and the U3 $\bar{Q}$ output goes low. When U3 $\bar{Q}$ goes low, AND gate U2B is no longer enabled; the count, as far as U7, U6, U5 and U1 are concerned is ended. When U3 $\bar{Q}$ goes low it also sets U4A and U4B; the $\bar{Q}$ outputs go low and the $\bar{Q}$ outputs go high. When the $\bar{Q}$ output of U4B goes low it presets U7, U6, U5 and U1. When U7, U6, U5 and U1 are preset the J inputs to U3 are inhibited since the count is no longer at the coincident count of 2975.

When the U4B Q output goes high the leading edge of the pulse is used to generate the sampler pulse. The first pulse to the sampling phase detector is initiated by the 2976th input cycle. Since three more cycles are required to restart the count cycle, following sampler pulses will be 2979 cycles apart.

When the count down reaches 2977, U3 is again clocked and since the K input is high and the J input is low, $\bar{Q}$ will go high. This $\bar{Q}$ high is applied to the K input of U4A and to pin 4 of AND gate U2B. U2B will not be enabled because U4B $\bar{Q}$ is holding AND gate U2B pin 5 low.

When the count down reaches 2978 U4A is clocked because the K input is high. U4A $\bar{Q}$ goes high and is applied to the K input of U4B.

On the 2979th input cycle, U4B is clocked and the $\bar{Q}$ output goes high. When U4B $\bar{Q}$ goes high the preset pulse is ended and AND gate U2B is enabled. The next input cycle will initiate the count cycle.

When there is a preset input programmed into U7, U6 and U5, the terminal count is still 2979. However, the count down starts at the number programmed into the BCD inputs. As an example, if the binary input to U7, U6 and U5 is 999,

364 SYNTHESIZED SIGNAL GENERATOR

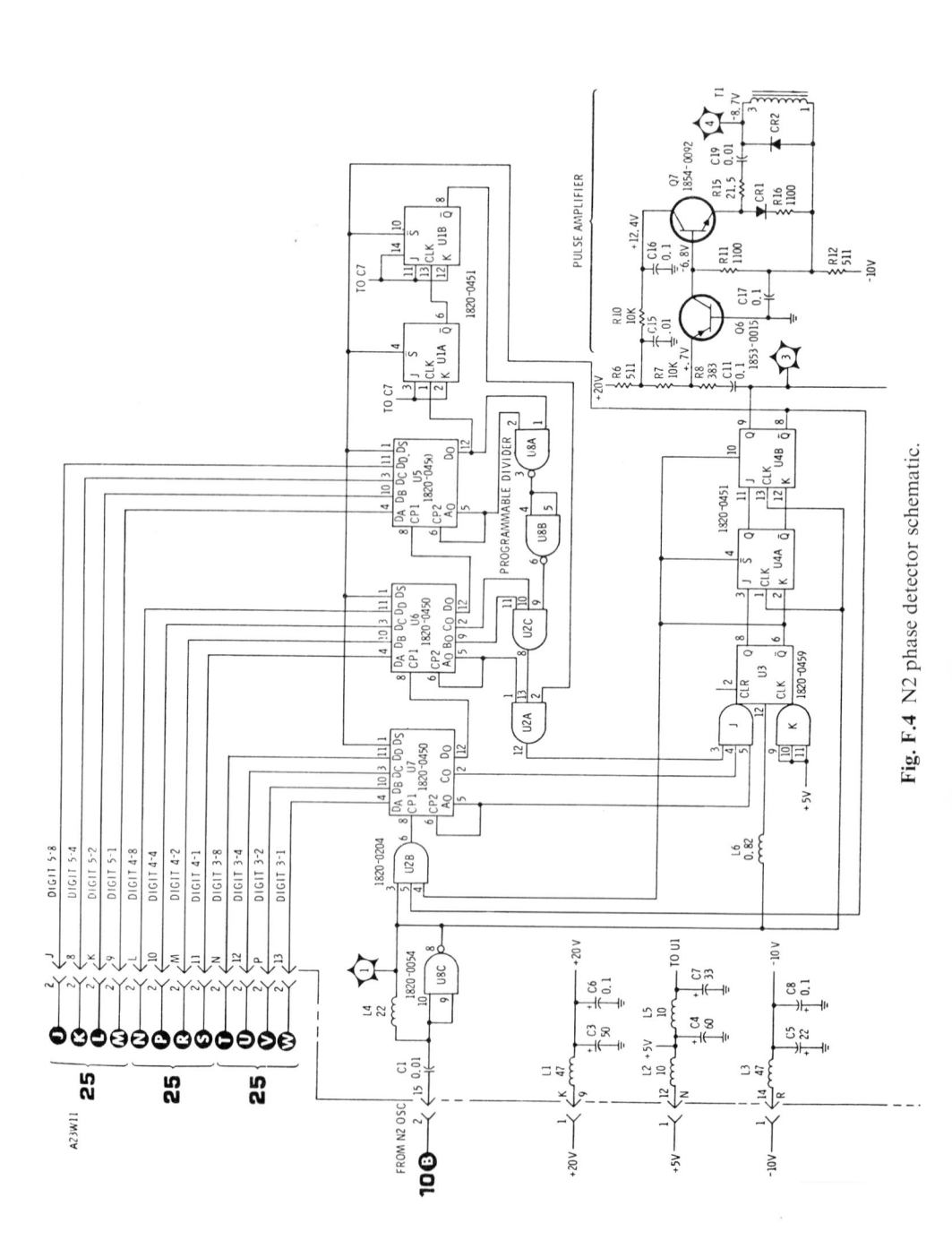

Fig. F.4 N2 phase detector schematic.

Reference Designation	HP Part Number	Qty	Description	Mfr Code	Mfr Part Number
A14R34	0757-0424		R:FXD MET FLM 1.10K OHM 1% 1/8W	28480	0757-0424
A14R35	0757-1094		R:FXD MET FLM 1.47K OHM 1% 1/8W	28480	0757-1094
A14R36	0757-0416		R:FXD MET FLM 511 OHM 1% 1/8W	28480	0757-0416
A14T1	08660-80001		TRANSFORMER:SAMPLER	28480	08660-80001
A14U1	1820-0451		IC:TTL DUAL J-K F/F	04713	MC3062P
A14U2	1820-0204		IC:TTL TRIPLE 3-INPT AND GATE	04713	MC3006P
A14U3	1820-0469		IC:DIGITAL TTL HI-SPEED F/F	01295	SN74H102N
A14U4	1820-0451		IC:TTL DUAL J-K F/F	04713	MC3062P
A14U5	1820-0450		IC:DIGITAL TTL	18324	N8290A
A14U6	1820-0450		IC:DIGITAL TTL	18324	N8290A
A14U7	1820-0450		IC:DIGITAL TTL	18324	N8290A
A14U8	1820-0054		IC:TTL QUAD 2-INPT NAND GATE	01295	SN7400N
A14U8					
A14	08660-60039	1	BOARD ASSY:N2 PHASE DETECTOR	28480	08660-60039

Fig. F.5 Model 8660 signal generator: partial table of replacement parts, assembly A14.

the first input cycle would cause the same digital circuit changes that the 1000th input cycle caused in the discussion above (U1A would be clocked for the first time). The frequency division would be 2979 minus 999, equal to division by 1980. The phase lock loop operation would result in an input frequency to the programmable divider of 19.80 MHz. When the 19.80 MHz is divided by 1980 the divider output would again be 10 kHz.

The output from U4B is always 10 kHz when the oscillator is phase locked.

APPENDIX G: Microcomputer—Description and Logic Diagrams

G.1 DIGITAL COMPUTER BASICS

The purpose of this section is to describe briefly the functions a digital computer performs and to explain its operation from a block diagram viewpoint.

The **digital computer** performs five main functions: (1) numerical computations, (2) information storage and retrieval, (3) logic operations, (4) manipulation of characters, and (5) modification of its own programs. Mathematical problems too time-consuming to be done by hand can be done very rapidly using a computer. Digital computers can store and retrieve information such as accounting records, airline ticket reservations, and the results of numerical calculations. Finally, a computer can perform logic operations, such as $A \cdot B$ and $A \oplus B$.

A microprogrammable computer consists of five main units as shown in Fig. G.1. Information flows between the **input-output (I/O) unit**, the memory unit, and the arithmetic logic unit (ALU). The control unit controls the flow of data between these three units. The **ALU** performs arithmetic and logical operations on the data. The memory unit stores data until it is operated on by the ALU. It also stores the results of such operations and instructions to the computer. The **control memory** stores sequences of control signals for use by the control unit. Data is entered into the I/O unit from peripheral devices such as teleprinters, magnetic tape units, magnetic disk units, or CRT terminals. The ALU and the control unit are sometimes combined in one device called the **central processing**

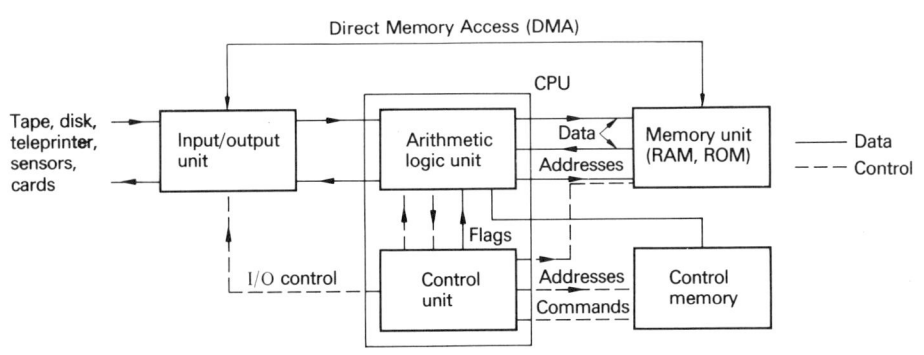

Fig. G.1 Microprogrammed digital computer block diagram.

unit (CPU). The **direct memory access** (DMA) option allows input information to be loaded directly into memory without passing through the ALU.

The next section gives a general description of one microprogrammable microcomputer, the National Semiconductor IMP-16C.

G.2 IMP-16C CONFIGURATION*

The IMP-16C is a 16-bit parallel processor. It is packaged on an $8\frac{1}{2}$-by-11-inch printed wiring card, which is shown in Fig. G.2. A 144-pin connector is located on the edge of the card for connecting the IMP-16C circuits to interfacing units.

The major functional units of the IMP-16C are shown in Fig. G.3 and are composed of the following:

- Central Processing Unit (CPU)
- Clock Generators
- Input Multiplexer
- Data Buffer
- Control Flags
- Conditional Jump Multiplexer
- On-Card Memory
- Address Latches

The CPU is configured around the National Semiconductor GPC/P (General-Purpose Controller/Processor) MOS/LSI devices. The MOS/LSI devices consist of one CROM (Control Read Only Memory) and four RALUs (Register and Arithmetic Logic Units). Each RALU handles 4 bits, and a 16-bit unit is formed by connecting four RALUs in parallel. A 4-bit-wide control bus is used by the CROM to communicate most of the control information to the RALUs.

The Clock Generator provides the MOS clock drivers and CPU timing signals. The system clock is distributed outside of the IMP-16C for synchronization of peripheral units with the IMP-16C.

External to the MOS/LSI circuits but still within the IMP-16C are control flags for both the IMP-16C and external interfacing circuits. These control flags are in addition to the status flags that are internal to the RALUs. Conditional branches are selected by the Conditional Jump Multiplexer.

Data from the user's peripheral devices and add-on memory are received by the Input Multiplexer. Data from the On-Card Memory are also processed through the Input Multiplexer en route to the Central Processing Unit.

Output data are made available from the 16-bit Data Buffer via the card-edge connector to the user's peripheral devices and add-on memory. A 16-bit address bus is also brought out to the card-edge connector for addressing both add-on memory and peripheral devices.

The memory on the IMP-16C card consists of 256 words of read/write memory and 512 words of read-only memory. The memory may be expanded to a maximum of 65,536 words.

---

* Diagrams and accompanying text for this and the remaining sections of Appendix G are reproduced, with modifications, by courtesy of National Semiconductor.

Table G.1 IMP-16C Parts List

Item	Description	Reference Designation	Part Number	Quantity
	Integrated Circuits			
1	MOS/LSI Register, Arithmetic, and Logic Unit (RALU)	1A, 2A, 1C, 2C	MM5750	4
2	MOS/LSI Control Read-Only Memory	3C	MM5751	1
3	Electrically Programmable 2048-bit Read-Only Memory (PROM)	3E, 3F, 3G, 3H	MM5203	4
4	Dual MOS Clock Driver	3A1, 3A2	MM0026	2
5	Quad 2-input AND Gate	3A3, 5A, 7B, 8C	DM74H08	4
6	4-bit Shift Register	4A	DM74195	1
7	Dual D Flip-Flop	6A, 5B, 7C, 8D	DM74H74	4
8	Quad 2-input AND Gate	7A	DM7408	1
9	Triple Differential Line Receiver	8A	DM10116	1
10	Quad 2-input OR Gate	3B	DM7432	1
11	Quad 2-input Multiplexer	4B	DD9322	1
12	Dual J-K Edge Triggered Flip-Flop	6B	74H103	1
13	TRI-STATE Hex Noninverting Buffer	4C, 4F, 4G, 4H	DM8095	4
14	Triple 3-input AND Gate	5C	DM74H11	1
15	Quad 2-input NOR Gate	6C	DM7402	1
16	256-bit Static Random Access Memory	1EA, 1EB, 2EA, 2EB, 1FA, 1FB, 2FA, 2FB, 1GA, 1GB, 2GA, 2GB, 1HA, 1HB, 2HA, 2HB	DM1101A2	16
17	Quad 2-input NAND Gate	7E, 7F, 8F	IM74H00	3
18	Hex Inverter	8E	DM74H04	1
19	TRI-STATE 8-channel Digital Multiplexer	8G, 8G	DM8121	2
20	Quad Latch	5E, 6E, 5F, 6F	DM7475	4
21	Quad 2-input Multiplexer	5G, 6G, 5H, 5H	DM8123	4
22	8-bit Addressable Latch	7H, 8H	DM9334	2
	Transistors			
23	Transistor	Q1, Q2	2N4258A	2
24	Transistor	Q3, Q4	2N4275	2
	Capacitors			
25	Capacitor, 33 μf, ±10%, 20V	C1, C2, C64		3
26	Capacitor, 0.01 μf, 50V	C5, C19, C58, C60		4
27	Capacitor, 27 pf, 500V	C8		1
28	Capacitor, 0.1 μf, 50V	C7, C10–C18, C20–C57, C63, C65–C67		52
29	Capacitor, 150 μf, ±10%, 15V	C9		1
30	Capacitor, 220 pf	C3, C4		2
31	Capacitor, 300 pf	C6		1
	Diodes			
32	Diode	CR1	IN4001	1
33	Diode	CR2–CR5	IN4454	4
	Resistors			
34	Resistor, 15 ohms, ±5%, 1/4W	R1–R4		4
35	Resistor, 10K, ±5%, 1/4W	R5–R8, R23		5
36	Resistor, 5.6K, ±5%, 1/4W	R9–R11		3
37	Resistor, 1K, ±5%, 1/4W	R12, R20		2
38	Resistor, 330 ohms, ±5%, 1/4W	R13, R18		2
39	Resistor, 220 ohms, ±5%, 1/4W	R14, R19		2
40	Resistor, 100 ohms, ±5%, 1/4W	R15, R17		2
41	Resistor, 39 ohms, ±5%, 1/4W	R16		1
42	Resistor, 300 ohms, ±5%, 1/4W	R21		1
43	Resistor, 120 ohms, ±5%, 1/4W	R22		1
44	Resistor, 2K, ±5%, 1/4W	R24		1
45	Resistor, 2.2K, ±5%, 1/4W	R25		1
	Crystal			
46	Crystal, 5.7143 Mhz	Y1		1

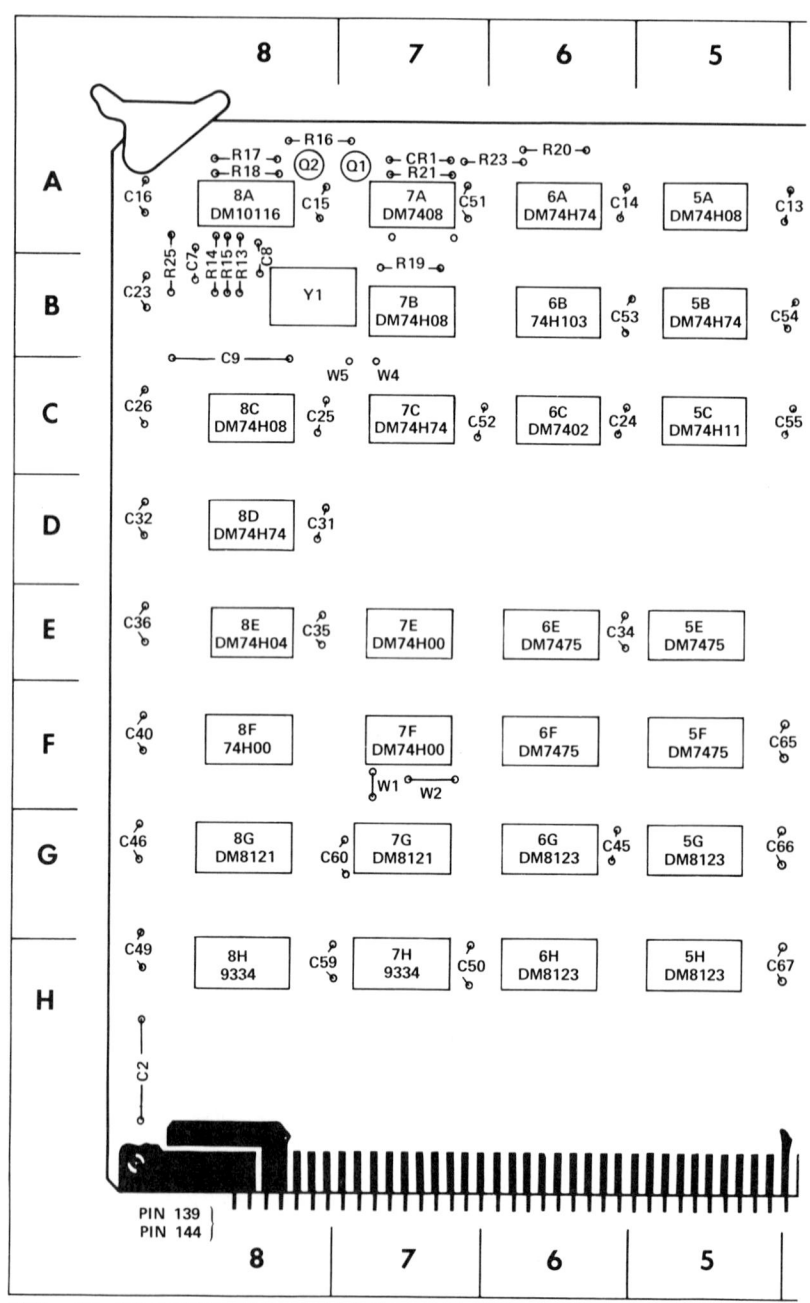

MICROCOMPUTER—DESCRIPTION AND LOGIC DIAGRAMS G.2

NOTE: (1) FOR DEVICES THAT ARE REMOVABLE, A LARGE DOT INDICATES THE LOCATION OF PIN 1.
(2) ODD-NUMBERED PINS ARE LOCATED ON COMPONENT SIDE.
(3) EVEN-NUMBERED PINS ARE LOCATED ON SOLDER SIDE.

G.2   IMP-16C CONFIGURATION   371

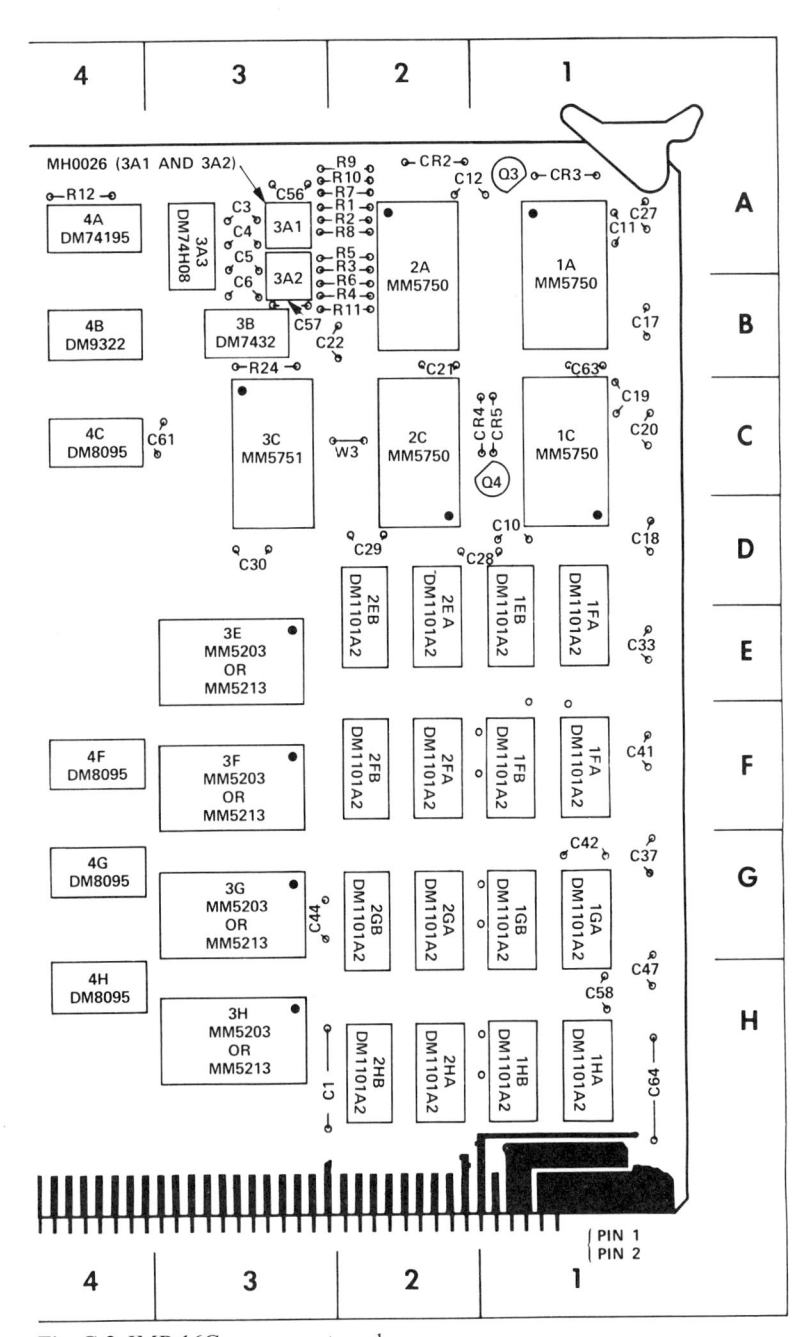

Fig. G.2 IMP-16C component card.

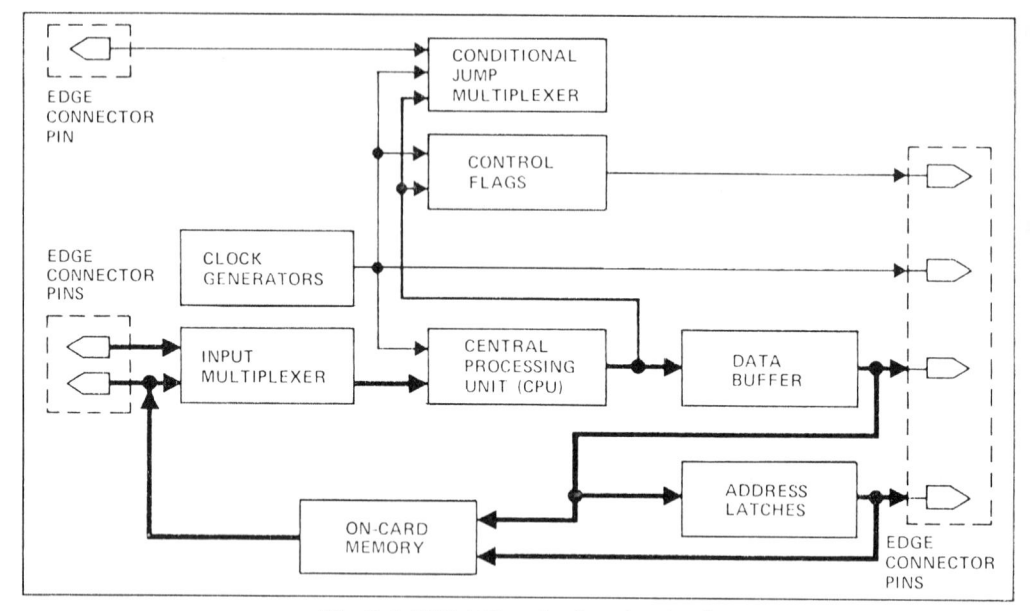

Fig. G.3 IMP-16C major functional units.

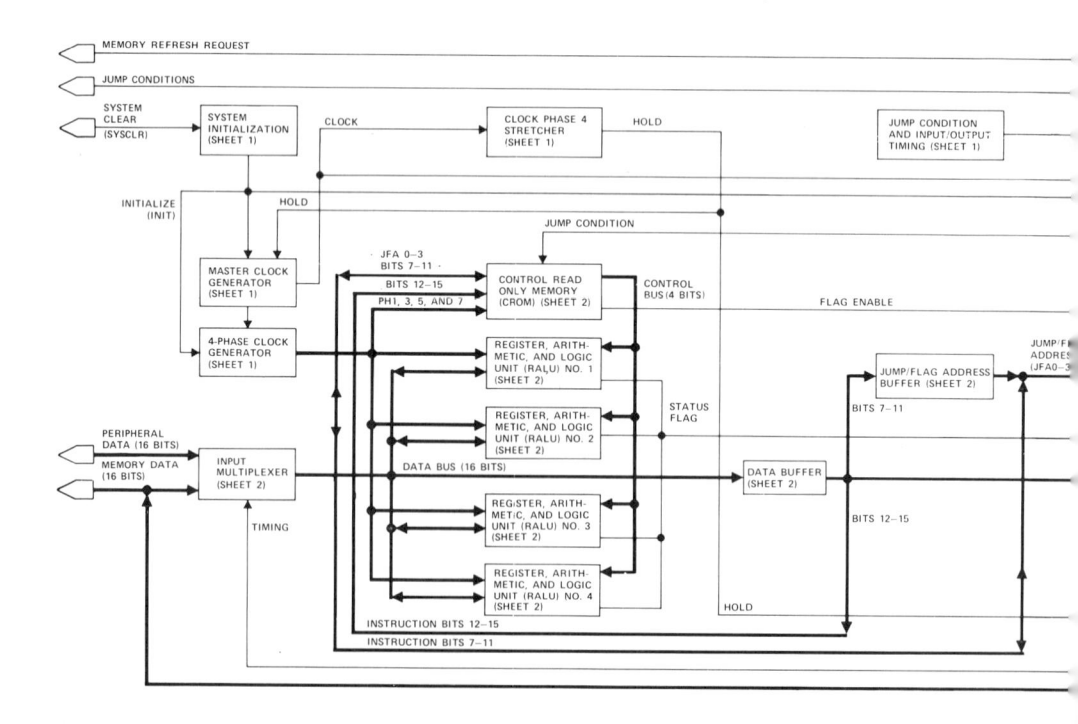

G.3 IMP-16C CIRCUIT DESCRIPTIONS

Figure G.4 is a functional block diagram of the IMP-16C. Details are given in the three-sheet schematic diagram shown in Figs. G.5, G.6, and G.7. You should refer frequently to the schematic while reading the individual circuit descriptions that follow. The heading for each circuit description bears a reference to the appropriate sheet to be consulted, and on the schematic each circuit is labeled and marked off with broken-line enclosures.

We suggest you familiarize yourself briefly with Figs. G.4 through G.7 before proceeding to the individual circuit descriptions.

G.3.1 Master Clock and 4-Phase Clock Generators—Sheet 1 (Fig. G.5)

The 4-phase clocks required for the CPU devices (one CROM and four RALUs) are generated with a shift register and two MH0026 clock drivers. The master clock signal is generated by a crystal oscillator circuit made from a DM10116 triple line receiver connected as an amplifier and a Schmitt trigger circuit. Two transistors, Q1 and Q2, provide level shifting to convert from ECL levels to TTL compatible logic signals.

The shift register DM74195 generates four clock signals, each of which lasts for two time periods. These signals are then logically gated to yield the odd phases

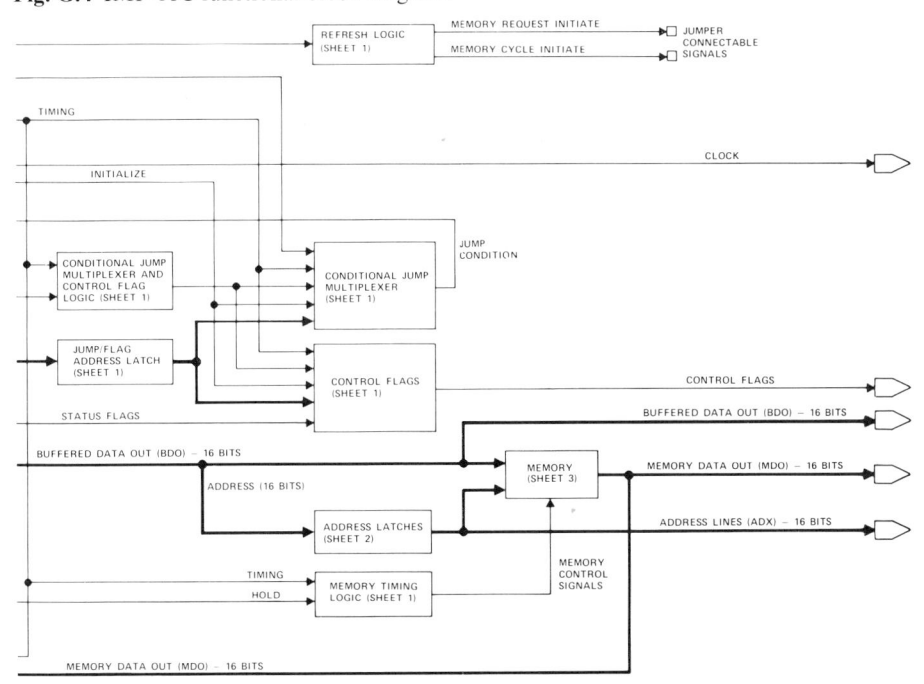

Fig. G.4 IMP-16C functional block diagram.

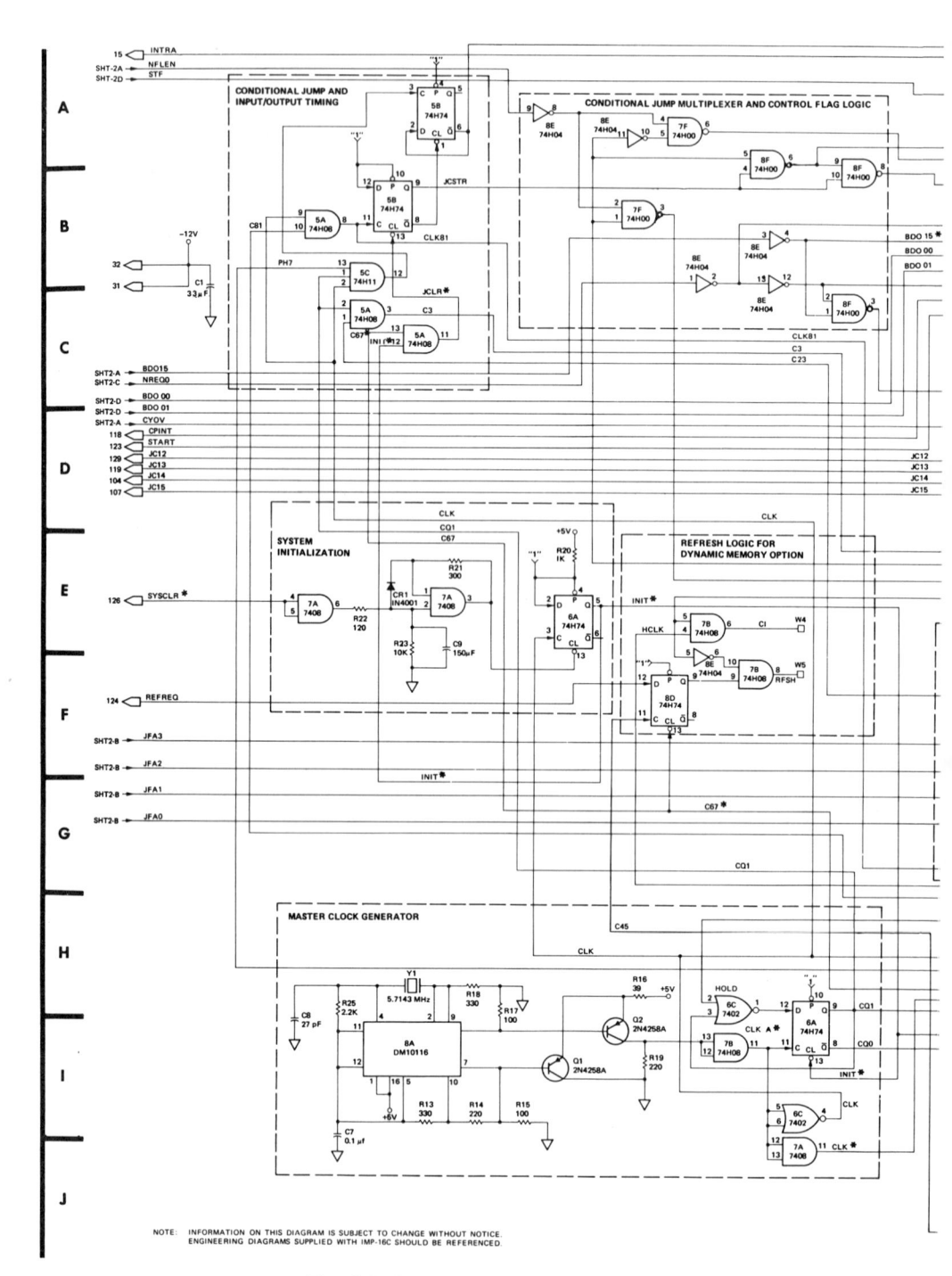

374   MICROCOMPUTER—DESCRIPTION AND LOGIC DIAGRAMS   G.3

Fig. G.5 IMP-16C schematic diagram—sheet 1 of 3.

G.3 IMP-16C CIRCUIT DESCRIPTIONS 375

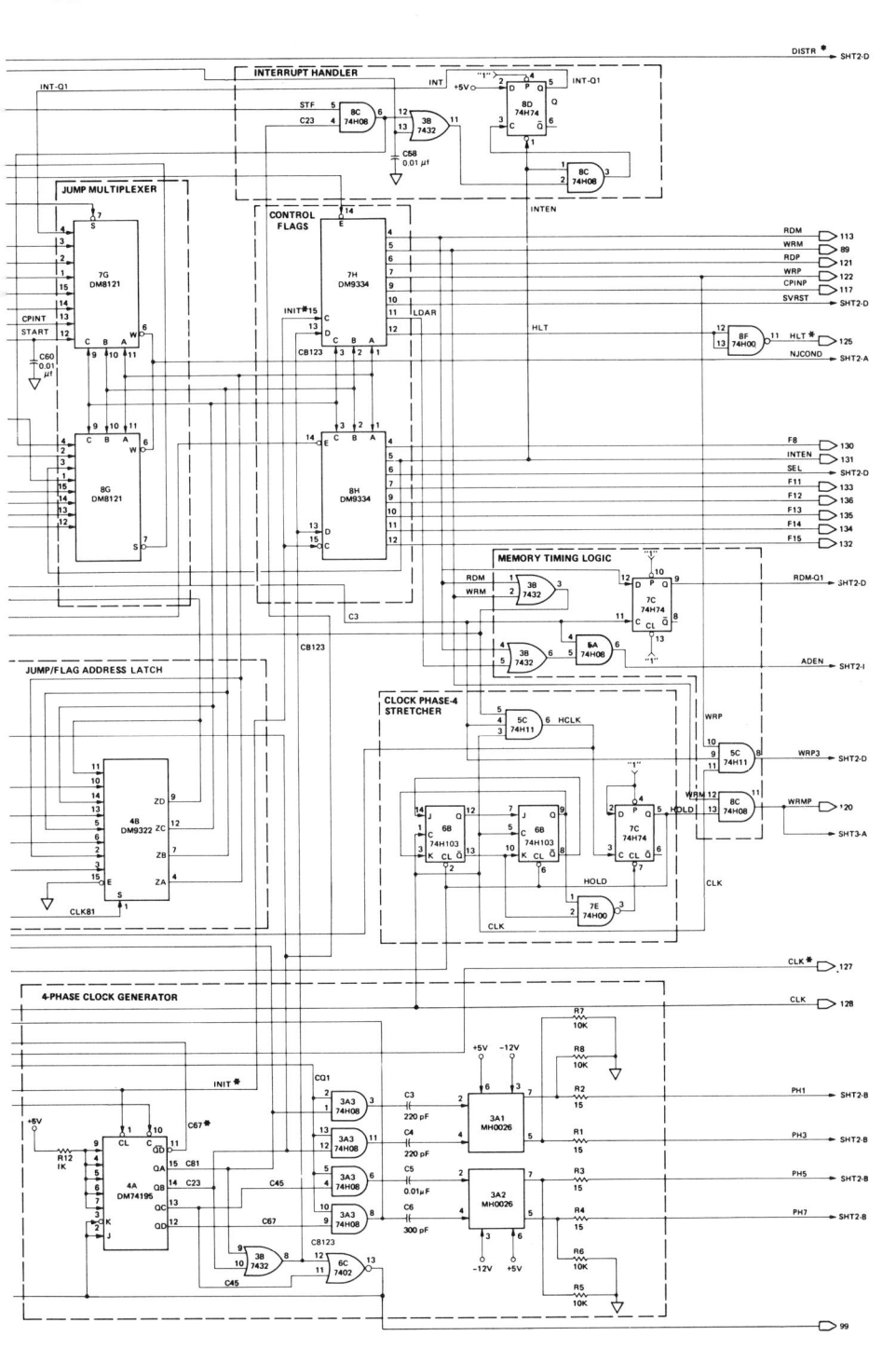

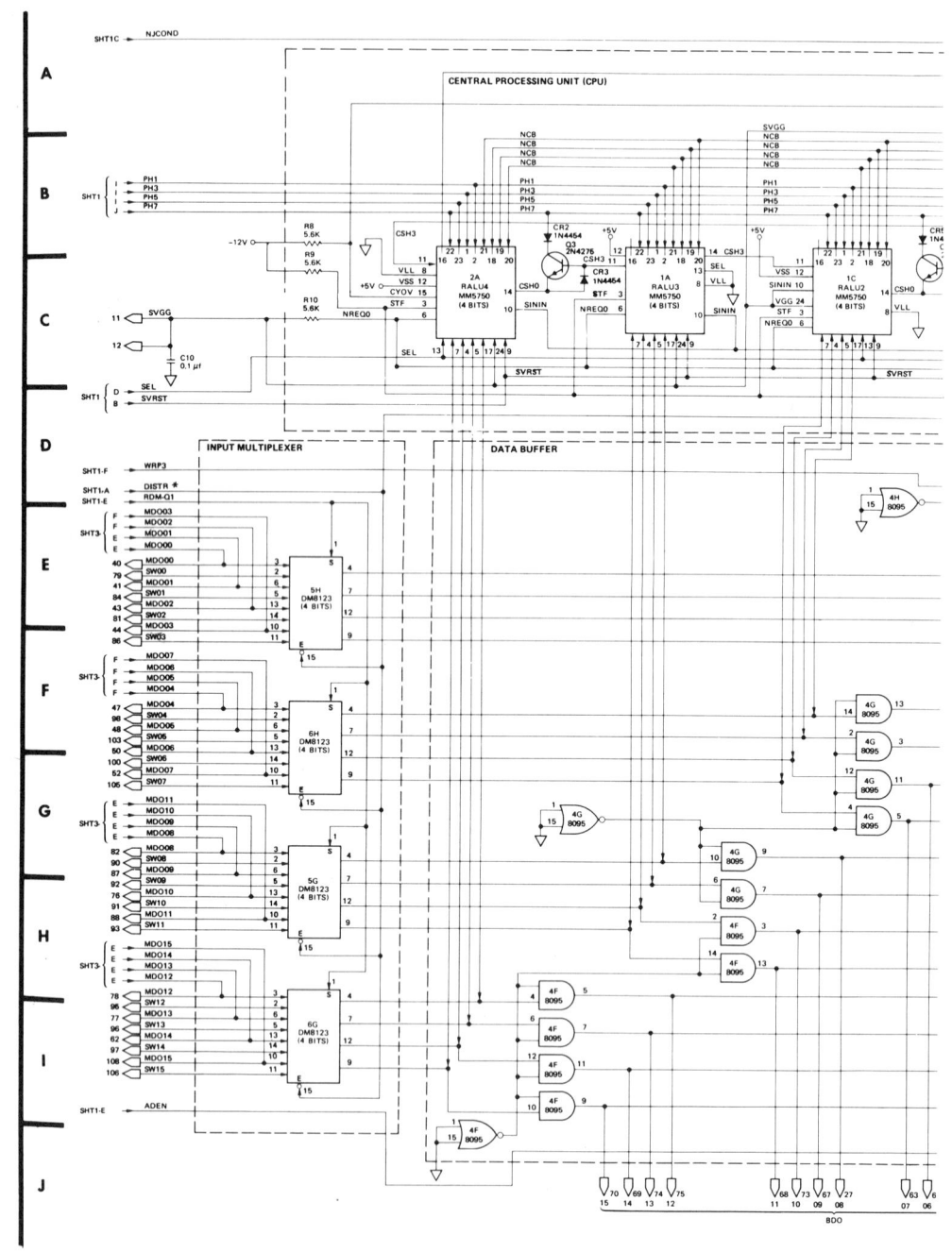

376 MICROCOMPUTER—DESCRIPTION AND LOGIC DIAGRAMS G.3

G.3 IMP-16C CIRCUIT DESCRIPTIONS 377

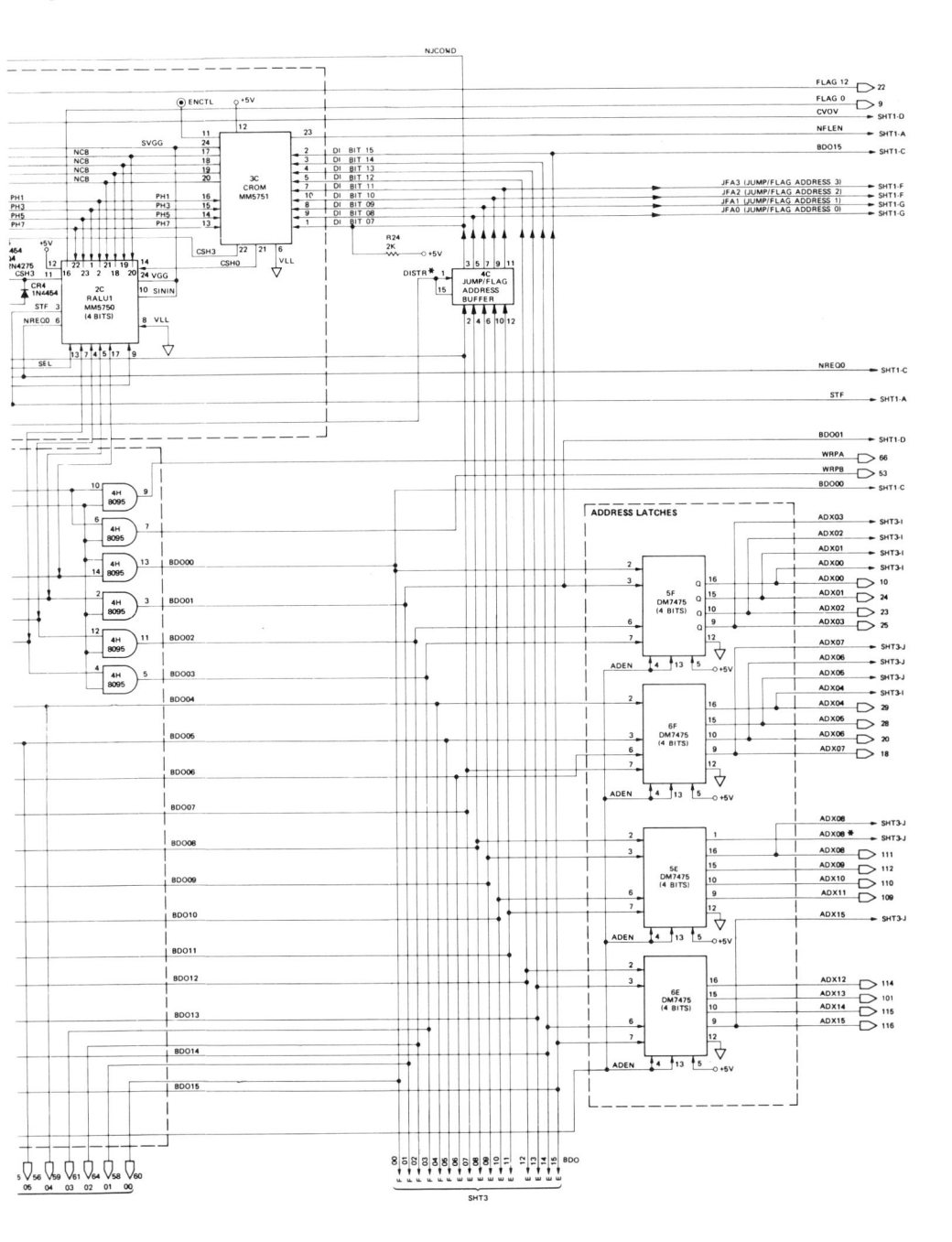

Fig. G.6 IMP-16C schematic diagram—sheet 2 of 3.

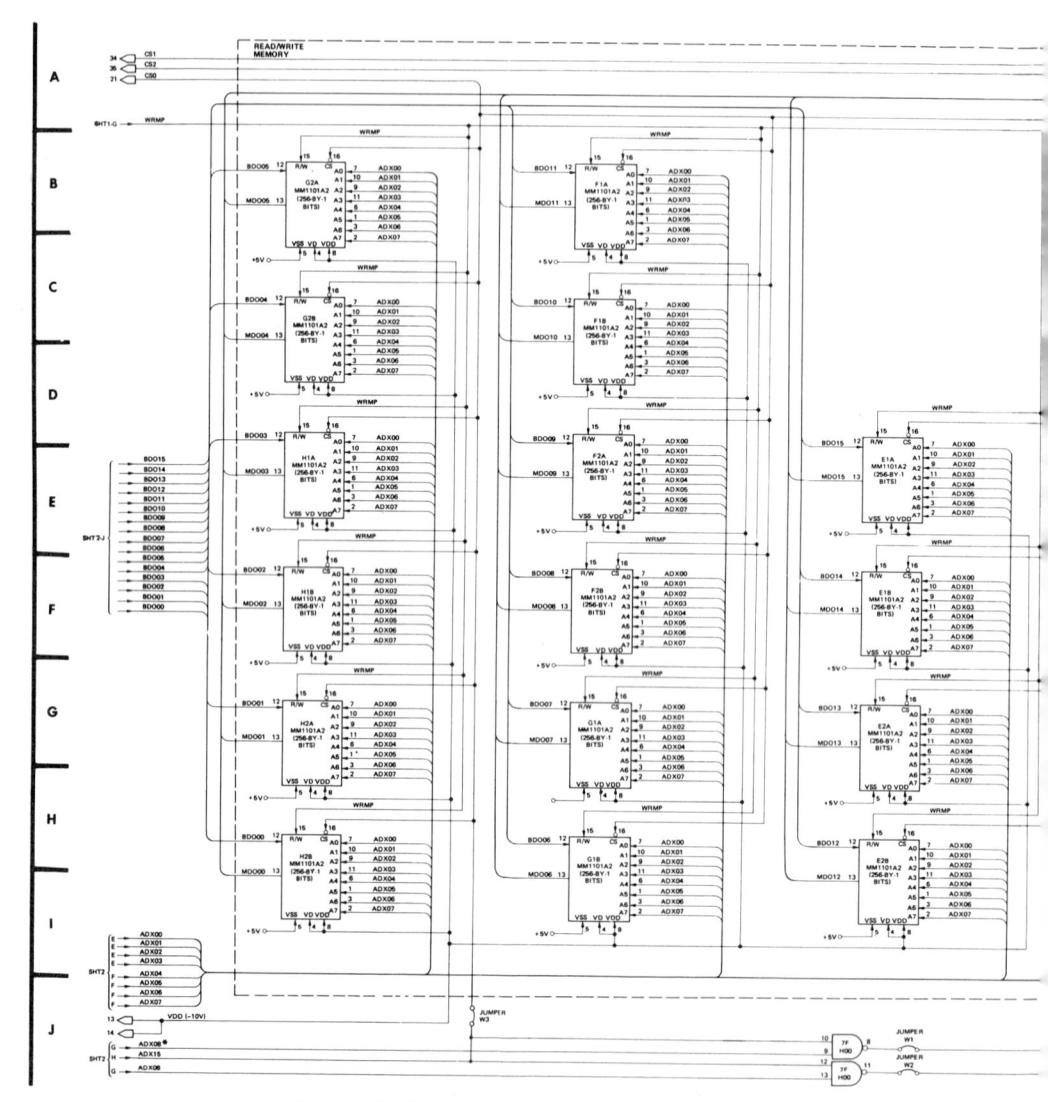

Fig. G.7 IMP-16C schematic diagram—sheet 3 of 3.

G.3 IMP-16C CIRCUIT DESCRIPTIONS 379

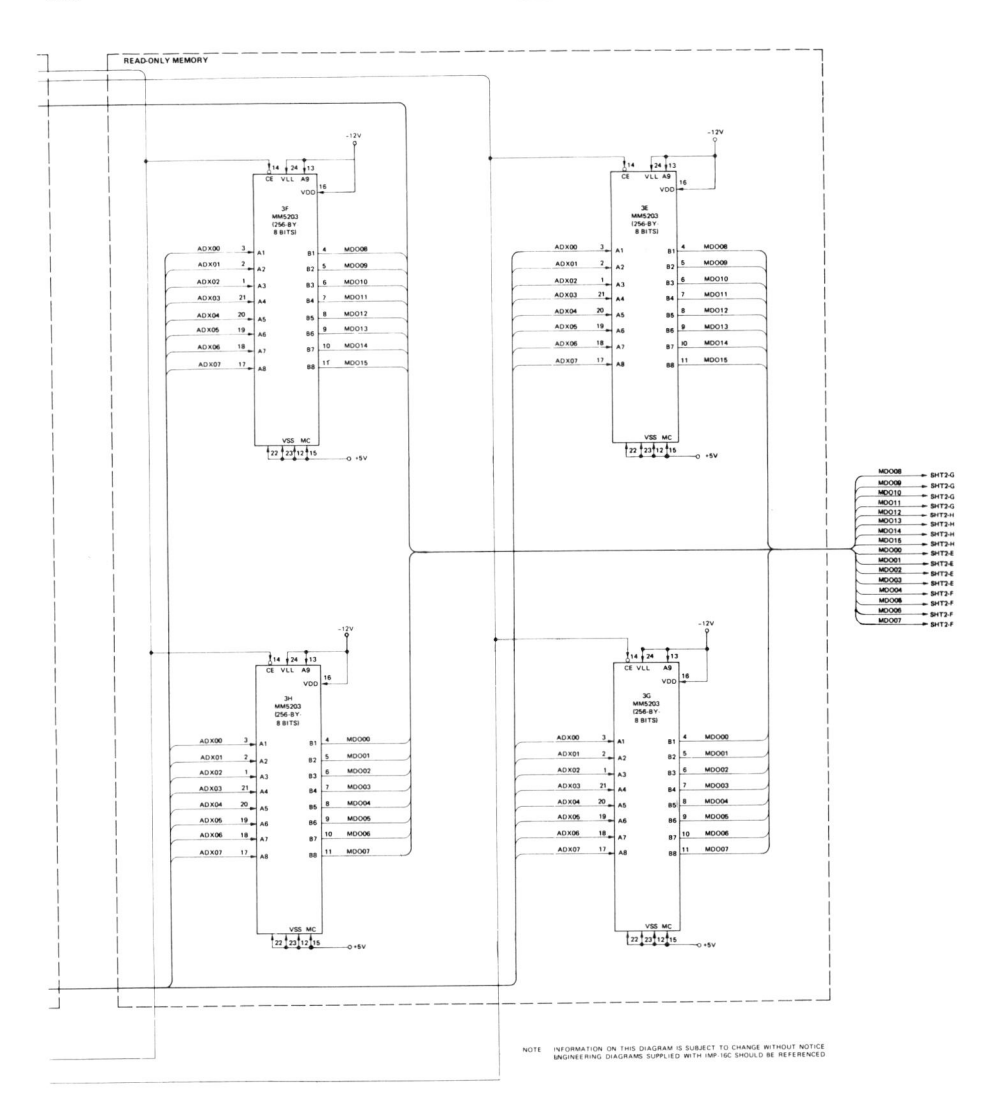

that drive the MH0026 clock driver devices. The MH0026 clock drivers are capable of driving 1000 pf loads with rise and fall times of 20 ns. The typical loading by the CROM and the RALUs is 215 pf (45 pf for each device). The resistors in the output lines of the MOS clock drivers damp out any possible clock overshoots by compensating for the inductance of the clock lines.

The DM74195 shift register outputs are also used to generate some of the other timing signals required in accordance with the timing diagrams shown in Fig. G.8. These signals are derived by gating the appropriate shift register outputs with various combinations of the master clock and the shift register input clock.

In Fig. G.8, the symbols used to designate clock periods have the following significance: the numbers following the letter C denote the specific time period for which the signal is valid. For example, C23 refers to a clock that is high during T2 and T3. Similarly, CLK81 refers to a signal derived from the logical AND of C81 and CLK. The MOS clocks and phase signals that drive the CPU circuits are shown in Fig. G.9.

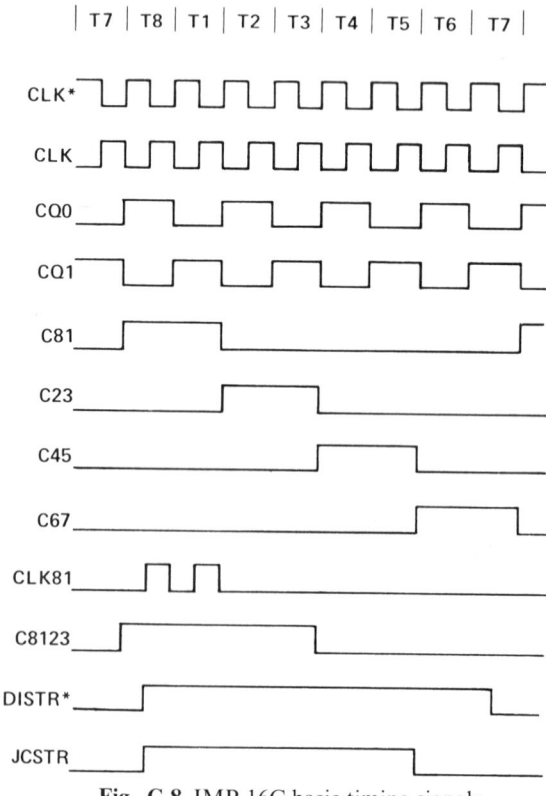

Fig. G.8 IMP-16C basic timing signals.

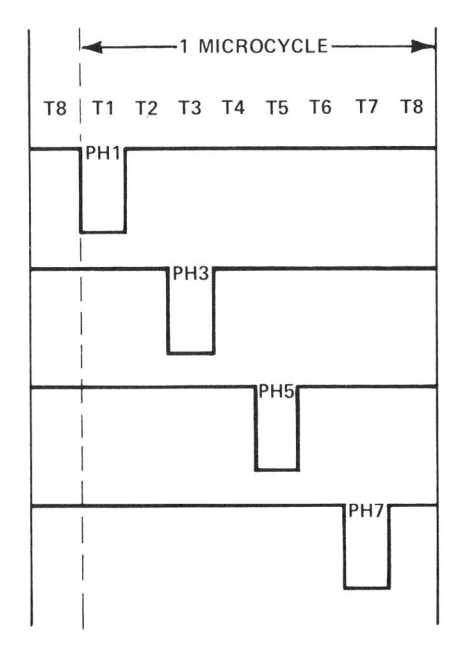

Fig. G.9 MOS clocks and phase signals.

G.3.2 MOS/LSI CPU Logic—Sheet 2 (Fig. G.6)

The CPU consists of one CROM and four RALU circuits driven by the 4-phase clocks. Control between the CROM and the RALUs is effected over the NCB (complemented control bus) lines. The DI (Data In) lines to the CROM serve the purpose of entering the instruction word bits 7 through 15 into the CROM. For sending out a 4-bit address to the Conditional Jump Multiplexer and the Control Flag latches, the lines JFA0 through JFA3 (bidirectional from the CROM) are used. The jump condition signal (NJCOND) enters the CROM at the same pin as bit 7 of the instruction word. The CROM has a flag enable signal* (NFLEN) that may be pulsed during T2 to set a particular control flag and/or may be pulsed at T6 to reset the flag.

The other signals that go to and come from the CPU indicate various status conditions and also perform certain auxiliary operations. The following paragraphs explain these functions.

During an instruction fetch, bits 0 through 7 are loaded into the RALU Memory Data Register, and bit 7 of the instruction word is extended through bit positions 8 through 15 of the Memory Data Register. In this way, the signed

---

* The prefix "N" to a signal name denotes logical complementation in the MOS/LSI CROM and RALUs. For signals generated external to these units, an asterisk (*) suffixed to the signal name denotes complementation.

displacement value "disp" [not discussed here] is extended for use in arithmetic operations for forming memory addresses and for immediate instructions. The SININ signal to the RALUs accomplishes the sign extension. For the two low-order RALUs (1 and 2), the SININ pins are permanently connected to a logic 0 (-12 V). For the two high-order RALUs (3 and 4), where bits 8 through 15 are located, the SININ pins are connected to the bit 7 output of the Buffered Data Out lines; this bit 7 is used in the two high-order RALUs (3 and 4) to effect the sign extension of "disp" of the instruction word.

The STF signal indicates a "stack-full" condition. When the bottom entry of the stack is filled with nonzero data, the STF line is a "1." The STF lines of all RALUs are tied together and connected to the Conditional Jump Multiplexer to allow testing for the stack-full condition. A similar scheme is used to detect a zero-result condition with the NREQ0 signal. The NREQ0 lines are tied together for all the RALUs; the NREQ0 signal is a "0" if the R-bus is zero as a result of the preceding machine cycle.

During T7 and T8, CSH3 and CSH0 are used to transfer shift data: for a left shift, the most significant bit is shifted out over CSH3, and the least significant bit is shifted in by CSH0; for a right shift, the converse is true.

Each RALU has four status flags, which are interfaced to the A- and R-buses. This provides a convenient means of saving status after an interrupt and for setting the status flags. For all except the most significant RALU (4), the status flags are general purpose and may be used for a variety of functions, depending on the application requirements. For the most significant RALU (4), the status flags have the following functions:

LINK Flip-Flop. When the SEL input to the RALU is "1," the Link Flag (L) is included in shift operations.

OVERFLOW Flag. When enabled (under control of the CROM), the Overflow Flag (OV) is set if an arithmetic overflow occurs during an add operation.

CARRY Flag. When enabled (under control of the CROM), the Carry Flag (CY) is set to the value of the carry bit out of the most significant ALU bit after an add operation.

FLAG Flip-Flop. This flag is available for general-purpose use.

These status flags may be loaded from the R-bus or stored onto the A-bus under control of the Save/Restore Flag (SVRST) input; this is used by the CROM to implement the PUSHF and PULLF instructions. The output of the general-purpose Flag is available at the Flag output pin; Carry and Overflow Flags are available at CYOV. The Select Flag (SEL) input is used to select the Carry or Overflow for output on CYOV and to determine whether the Link (L) is included in shift operations. General-purpose flags 0 and 12 are brought out to terminals on the IMP-16C edge connector as signals FLAG0 and FLAG12.

G.3.3 Control Flags and Conditional Jump Multiplexer Logic—Sheet 1 (Fig. G.5)

External to the CPU portion of the IMP-16C is the logic required to set and reset the control flags and to select one of 16 jump conditions.

The flag addresses sent out by the CROM are latched to keep them stable; this is done with a DM9322 multiplexer connected as a latch by feeding the outputs back to the second set of inputs. (This particular technique has been chosen here because the DM9322 has a smaller propagation delay than conventional DM7475 or DM74175 latches.) The CROM sends out the flag addresses at T1; these are latched in the DM9322 device during the latter half of T1 by the signal CLK81.

The latched addresses are then used to select one of 16 jump conditions in two DM8121 8-to-1 multiplexers. The complemented flag enable signal (NFLEN), which is low at T2 and then again at T6, enables the selection of a flag in one of the two DM9334 8-bit addressable latch devices. The data to the addressable latches comes from the signal C8123, which provides a logic "1" when NFLEN is low at T2 and a logic "0" when NFLEN is low again at T6. This allows setting and resetting of the flags under control of NFLEN at T2 and T6, respectively.

The outputs of the TRI-STATE DM8121 devices are tied directly to the jump condition (NJCOND) input of the CROM. This is the line that is tested during conditional jump operations; the testing is done during T2. The START and JC12 through JC15 inputs are user-supplied signals and could be asynchronously generated. Thus, the user must ensure that the logic levels for these signals are stable during T2.

The various conditions that can be tested are hardwired to the conditional jump multiplexer. Four user-assigned jump conditions and six user-assigned control flag lines are brought out to pins on the edge connector.

The more significant 8 of the 16 control flags may be set using the SFLG instruction and cleared or pulsed using the PFLG instruction. The less significant 8 flags may be modified by the CROM-resident microprograms. The assignment of these flags is listed in Table G.2.

Table G.2 Control Flags Modified by Microprogram

Flag Number	Signal Name	Function
0	RDM	Read Memory
1	WRM	Write Memory
2	RDP	Read Peripheral
3	WRP	Write Peripheral
4	CPINP	Control Panel Input Flag
5	SVRST	Save/Restore Status Flags
6	LDAR	Load Address Register
7	HLT	Set by HALT Instruction

G.3.4 Input Multiplexer, Data Buffer, and Address Latches—Sheet 2 (Fig. G.6)

The 16-bit bidirectional data bus from the RALU devices is used to transfer all information between the CPU and memory and peripheral units. This bus is buffered by passing all output signals through a set of TRI-STATE DM8095 hex-buffer circuits.

Input data destined for the CPU are passed through a set of input multiplexers such that data from a memory or peripheral unit may be switched in. The TRI-STATE DM8123 multiplexers are controlled by RDM, the read memory flag (delayed until T7 because data may be accepted into the CPU only at T7).

During T7, the data lines are used for input to the RALU from system memory or peripheral devices. The data receivers are "zeros catching," so the data lines must not be allowed to go negative during T7 unless the data input is to be a zero. Because of this zeros catching feature, the strobe signal for input data (DISTR*) is generated such that it occurs during the latter half of T7; this ensures that the data bits will be strobed in only when it is assured that they are stable.

When the data bus is sending out an address during read and write operations, the address is stored in a register consisting of four DM7475 latches. These address lines are brought out to terminals on the edge connector to be used for addressing peripheral devices and add-on memory. The signal that strobes in the address values (ADEN) is generated according to the logical equation ADEN = C3 (RDM + LDAR). Data from the RALU are sent out during part of T3 and all of T4 and are clocked into the latch by ADEN if the RDM and the LDAR flag is set. The RDM and LDAR flags are pulsed (set at T2 and reset at T6) during memory read and write operations.

G.3.5 Read/Write and Read-Only Memories—Sheet 3 (Fig. G.7)

When executing a memory read operation, the processor sends out an address on the RALU data input/output bus; this address into a latch starts coming up during T3, and it is assured of being valid during T4. In the IMP-16C, the ADEN signal strobes the address into a latch during T3 in order to catch it as soon as it becomes stable. Since the processor expects data back from memory at T7 of the same microcycle, this allows extra time (~ 0.7 μs) for memory access. If the processor is used with slow memories whose access times are longer than the interval between T4 and T7, it is necessary to stretch a clock period to allow for the slow access. For this purpose, the circuit clock phase-4 stretcher (sheet 1, Fig. G.5) is used to extend T4 for an additional two periods.

During read and write operations, a clock hold signal (HCLK) is developed during C3. The timing relations are given in Fig. G.10. This signal sets a flip-flop output (HOLD) such that a count-by-four circuit is enabled. After four counts of the master clock, the HOLD flip-flop is shut off. The delay provided by the counter circuit is used to inhibit phase 4 of the clock generator circuit.

The circuit labeled Refresh Logic for Dynamic Memory Option (sheet 1,

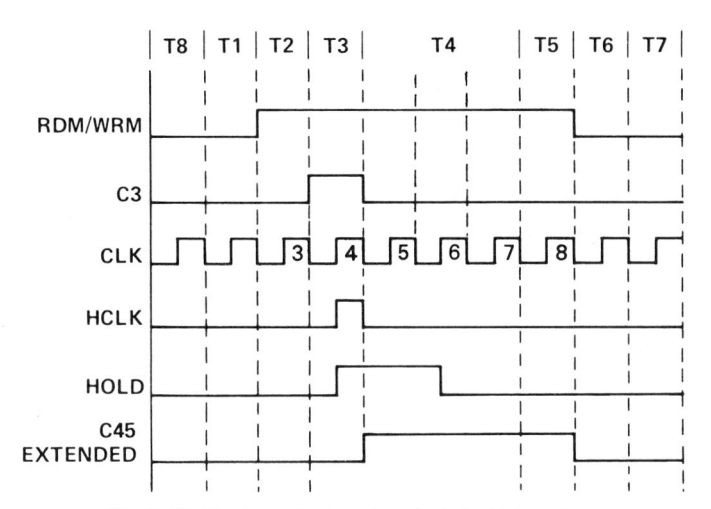

Fig. G.10 Timing relations for clock hold function.

Fig. G.5) consists of logic that can receive memory refresh requests and send out appropriate refresh orders and read/write cycle initiate signals.

G.3.6 Interrupt Handler—Sheet 1 (Fig. G.5)

The IMP-16C interrupt facility is handled through the conditional jump multiplexer inputs. Two interrupt inputs are provided. One is directly wired to the Conditional Jump Multiplexer and responds to a specific interrupt by jumping to a microprogram subroutine designed for control panel interrupts. The other interrupt is a general interrupt input (INTRA), which can be wired to the user's interrupting device.

The flip-flop output (INT-Q1) is set high whenever an external interrupt (INTRA) or a stack-full signal (STFL) is true simultaneously with the interrupt enable (INTEN) flag. INT-Q1 is wired to the interrupt input of the conditional jump multiplexer. The interrupt processing microprogram resets the interrupt enable (INTEN) flag to zero to disable any further interrupts, and control is transferred to the instruction stored in location 1 of main memory. The stack-full line is also wired to the jump condition multiplexer to permit testing for stack-full interrupts. If STFL causes such an interrupt, the bottom entry of the hardware stack is lost. In anticipation of this, the user can put a dummy word in the stack during his program initialization sequences.

G.3.7 System Initialization—Sheet 1 (Fig. G.5)

During startup, the IMP-16C is initialized so all the sequential logic is conditioned to known logic states. There are two aspects to initialization: startup of the TTL logic and initialization of the CPU MOS/LSI devices.

The System Initialization circuit shows the startup logic required for the TTL clocks. When power comes on, the output of flip-flop (INIT*) is forced to a logic "0" (independent of the state of the SYSCLR* input) by the RC timing circuit. Because this signal goes to the Clear inputs of all the other flip-flops in the clock-generator circuit, the system starts up in a cleared condition. The system may also be cleared at any other time by grounding the SYSCLR* input.

When the System Clear signal (SYSCLR*) goes high, INIT* comes up after a delay of a few hundred milliseconds (time constant R10C5). At this time, all the system clocks are enabled. For systems that do not have external initialization, the SYSCLR* input should be left continuously at a logic "1."

Startup for the MOS parts is achieved by controlling the application of the −12-volt supply. The CPU MOS/LSI devices receive −12 volts from the SVGG (switched VGG) line. This voltage must be turned on a few milliseconds before the clocks are started. During turn-off, the clocks are kept on for a few milliseconds after SVGG goes off. The timing relationship between SVGG and the Power-On Condition (POC) signal in the System Initialization circuit is shown in Fig. G.11.

Figure G.12 is a recommended circuit that may be used to effect system reinitialization to clear the CPU MOS/LSI devices without shutting down other circuits on the IMP-16 card. This circuit is not on earlier versions of the IMP-16C card and would have to be user-supplied and connected to the SVGG terminal pin (sheet 2, Fig. G.6).

If a special reinitialization circuit, such as discussed above, is not supplied, the SVGG pins on the IMP-16C card-edge connector should be connected to the −12-volt supply. With this latter setup, reinitialization is effected by turning off power to the IMP-16C for at least 5 seconds and then turning it on again.

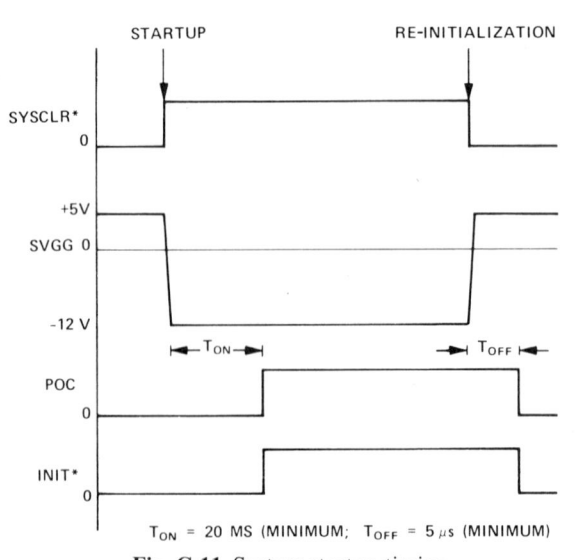

Fig. G.11 System startup timing.

G.3 IMP-16C CIRCUIT DESCRIPTIONS

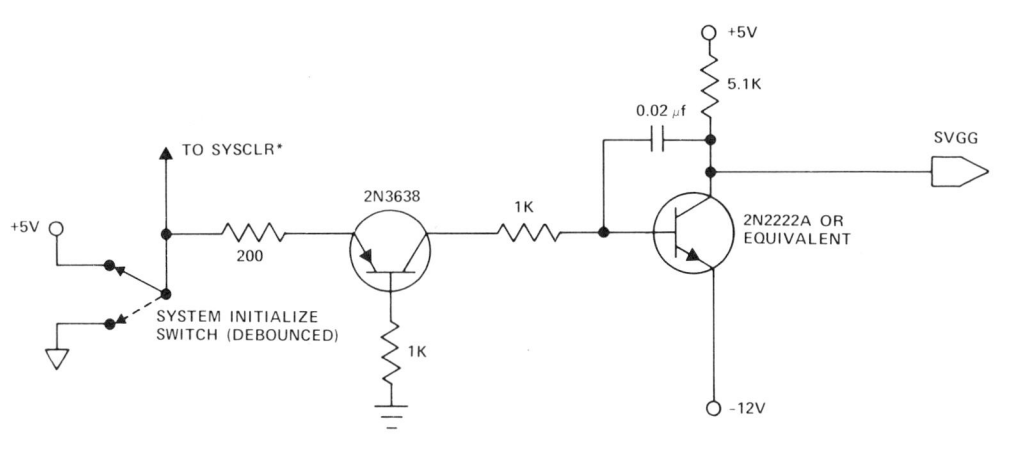

NOTE: BY TYING THE SYSCLR* PIN ON THE IMP-16C CARD TO A "SYSTEM INITIALIZE" BUTTON, THE SAME SIGNAL CAN BE USED TO CONTROL THE SWITCHED −12V (SVGG) FOR THE MOS/LSI DEVICES ON THE IMP-16C CARD. IN THIS CASE, IT IS REQUIRED THAT SYSCLR* BE LOW WHEN POWER IS APPLIED.

Fig. G.12 Recommended circuit for powering CPU MOS/LSI devices.

Table G.3 Nomenclature Used in IMPC-16 Circuit Schematics and Text

An asterisk (*) after a signal name (except a CROM or a RALU signal) denotes a complemented signal. Complementation is denoted for CROM and RALU signals by the prefix N as part of the signal names.

Signal Name	Description
ADEN	Address Enable Signal
ADX (0), (1), . . . , (15)	Address Lines to Memory
ALU	Arithmetic/Logic Unit
BDO (0), (1), . . . , (15)	Buffered Data-Out Lines
CI	Memory Cycle Initiate
CJMUX	Conditional Jump Multiplexer
CLK, CLK*	Master Clock Signals
CPINP	Control Panel Read Flag
CPINT	Control Panel Interrupt
CROM	Control Read-Only Memory
CSH0, CSH3	Carry/Shift Signal Lines (RALU)
CS0, CS1, CS2	Memory Chip-Select Lines
CYOV	Carry or Overflow Signal
C81, C23, C45, C67	Clock Signals, each lasting for two periods
DATA (0), (1), (2), and (3)	Data Bus Lines
DI (0), (1), (2), and (3)	Data Input Lines to the CROM
DISTR	Data Input Strobe
ENCTL	Enable Control Signal
START	Start or continue signal to restart operation

Table G.3 (*Continued*)

Signal Name	Description
FLAG0, FLAG12	User status flags
F8, F11, ..., F15	User status flags
HCLK	Clock control flags
HLT	Halt Flag
HOCSH	High-Order Carry/Shift Signal Line (CROM)
HOLD	Phase 4 Hold Signal
INIT*	Initialize Line (Complemented)
INTEN	Interrupt Enable Flag
INT-Q1	Delayed Interrupt Signal
INTRA	Interrupt Request Signal
JCLR	Jump-Strobe Clear
JCSTRA	Jump Condition Strobe Signal
JFA	Jump Flag Addresses
LDAR	Load Address Register Flag
LOCSH	Low-Order Carry/Shift Signal Line (CROM)
MDO (0), (1), ..., (15)	Memory Data Out
MUX	Multiplexer
NCB (0), (1), (2), and (3)	Complemented Control Bus Lines
NFLEN	Flag Enable Signal Line (Complemented)
NJCOND	Jump Condition Input Line (Complemented)
NREQ0	Register Equal Zero Signal (Complemented)
POC	Power-On Condition
PROM	Electrically Programmed Read-Only Memory
RALU	Register/Arithmetic/Logic Unit
RDM	Read Memory Flag
RDM-Q1	Delayed Read Memory Flag
RDP	Read Peripheral Flag
RFREQ	Memory Refresh Request Signal
RFSH	Memory Refresh Initiate
SEL	Select Flag
SININ	Sign-In Signal Line
STF	Stack Full Signal Line
STFL	Stack Full Interrupt Signal
SVRST	Save/Restore Flag
SW	Switch Data (Input Port for Peripheral Data)
SYSCLR*	System Clear (Complemented)
V_{GG}, V_{SS}	Supply Voltages
WRM	Write Memory Flag
WRMP	Write Memory Pulse
WRP	Write Peripheral Flag
WRPA	Write Peripheral Strobe A
WRPB	Write Peripheral Strobe B
WRP3	Write Peripheral Strobe
PH1, PH3, PH5, and PH7	Clock Phase Times 1, 3, 5, and 7. (Each of these phase times corresponds to a clock pulse: T1, T3, T5, and T7.)

ANSWERS TO SELECTED PROBLEMS

Chapter 1

5. HIGH and LOW

7.

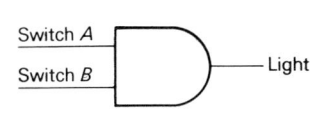

Switch A	Switch B	Light
Up	Up	Off
Up	Down	Off
Down	Up	Off
Down	Down	On

9.

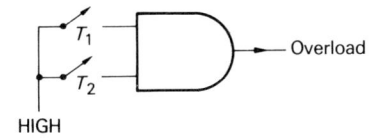

13. An IC chip is a semiconductor on which IC components are formed.
15. An SSI device is an IC containing 1 to 12 logic gates.
17.

2^0	1	2^6	64
2^1	2	2^7	128
2^2	4	2^8	256
2^3	8	2^9	512
2^4	16	2^{10}	1024
2^5	32		

19. (a) 2, (c) 6, (e) 9, (g) 23, (i) 63, (k) 42
20. (a) 110, (c) 1100, (e) 101000, (g) 10001100
21. (a) 100, (c) 1101, (e) 100, (g) 100111

Chapter 2

1. Traffic light control logic, digital computers, digital voltmeters
3. See Fig. 2.2(a).

ANSWERS TO SELECTED PROBLEMS

5.

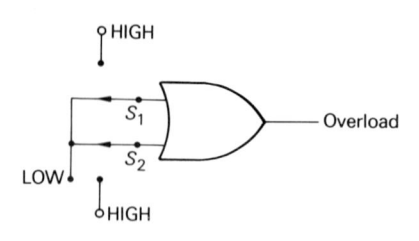

7. See Fig. 2.5(b).

9.

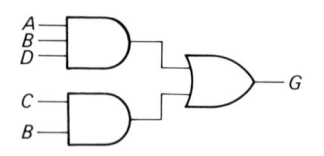

11.

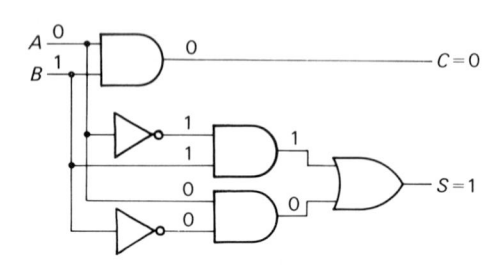

13. See Fig. 2.11.

15.

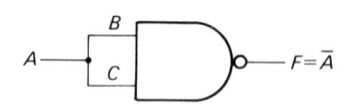

If $A = L$, then $B = C = L$, so $F = H$. If $A = H$, then $B = C = H$, so $F = L$. Thus $F = \bar{A}$.

17.

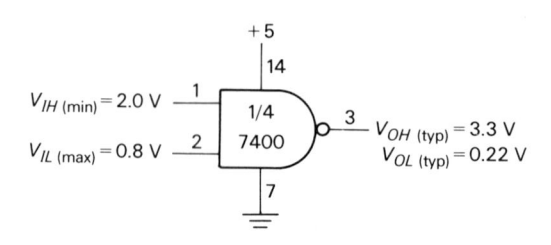

CHAPTER 3   391

19.

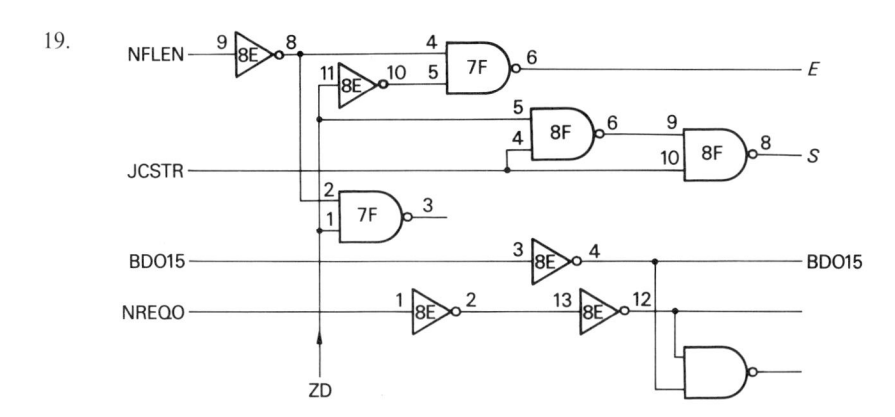

Note: Pin 14 of all ICs is +5 V, Pin 7 is ground. $E = \overline{\overline{NFLEN} \cdot \overline{ZD}}$

23. See Fig. 2.18(a, b) for the cases $R = L, S = H$ and $R = H, S = L$.

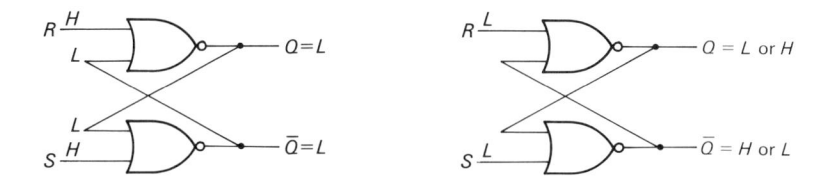

If $R = H$ and $S = H$, then Q and $\bar{Q}$ are both forced LOW since the output of a NOR gate is LOW if any input is HIGH. If $R = L$ and $S = L$ the inputs above cannot determine the output of a NOR gate, so the outputs cannot change from what they were previous to applying these inputs.

Chapter 3

1. (b) The diode is reverse-biased because its anode voltage is less positive than its cathode voltage.
3. See Fig. 3.4(a).
5. Binary voltage levels have a range of voltage values to allow for device variations and noise.
7. The following table shows the state of each transistor in the accompanying 7408 AND gate schematic for the cases any input, say A, is LOW and both inputs are HIGH.

A	B	Q_1	Q_2	Q_3	Q_4	Q_5	Q_6	Output
L	X	On	Off	Off	On	Off	On	L
H	H	Off	On	On	Off	On	Off	H

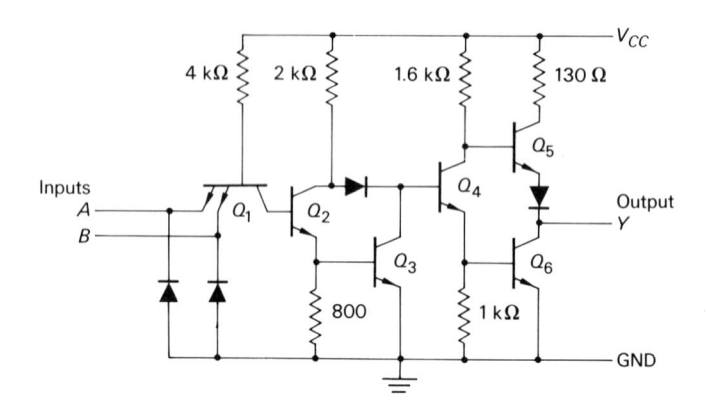

9. (a) Current spiking in totem pole outputs is caused by (1) unequal conduction times for the totem pole transistors, and (2) the charging of load capacitances.
   (b) Power supply decoupling capacitors are used to prevent current spikes from propagating through the system.
   (c) $I_{CCH} = 14.5$ mA, $I_{CCL} = 4.8$ mA.
11. 111 ns.
13. 70 mA, 385 mW.
15. For I_{IH} there are N separate current paths for an N-emitter transistor, so $I_{IN} = N \times I_{IL}$. For I_{IL} all N emitters are sourced by only one resistor so $I_{IN} = 1 \times I_{IL}$.
17. Wire-ANDing saves one or more levels of gates, which results in a reduction in the overall delay time and total power dissipation.
19. When $G = L$ the control emitter of Q_1 is HIGH, so only A will determine the output of Q_1. If $A = L$, Q_2 is off, D_2 is reverse-biased, so $F = H$. If $A = H$, Q_2 is on, D_2 is reverse-biased, and Q_3 is on, so $F = L$.
21. Keep interconnecting wires short, use decoupling capacitors, use a ground plane.
23. (a) Low power dissipation, greater PCB density, greater longevity, lower amplitude current spikes, easier to interface. (b) F.O. $(H) = 25$, F.O. $(L) = 12.5$. (c) At 1 MHz: $P_D(C_L = 15$ pF$) = 2$ mW, $P_D(C_L = 50$ pF$) = 3$ mW; at 30 MHz: $P_D(C_L = 15$ pF$) = 10$ mW, $P_D(C_L = 50$ pF$) = 20$ mW.

25.

	Bipolar	MOS
Advantages	Fast Not susceptible to electric fields	Small Easily fabricated Cheaply fabricated Very high input impedance
Disadvantages	Medium size Hard to fabricate Costly to fabricate Lower input impedance	Slow Susceptible to electric fields

CHAPTER 4   393

27. High noise environment, portable.
31. A 14001 can drive 42.5 and 0.5 standard gates in the HIGH and LOW cases, and 85 and 2.2 low-power Schottky gates in the HIGH and LOW cases respectively.
35. (a) Refer to Fig. 3.26(b). If A or B is HIGH then Q_1 or Q_3 is on, Q_2 is off, and $V_{OR} = V_{BQ5} - V_{BEQ5} = -0.05 \text{ V} + (-0.8)\text{V} = -0.85 \text{ V}$.

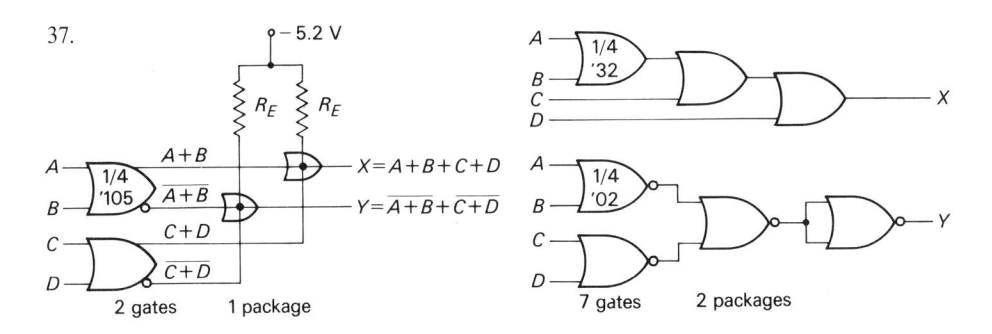

37.

Chapter 4

1. For 128 input combinations 7 inputs will be required ($2^7 = 128$). Let the correct combination be $A = B = C = L, D = F = G = H$.

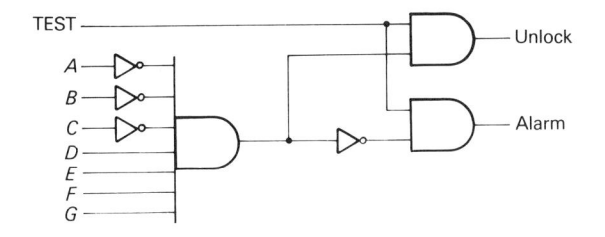

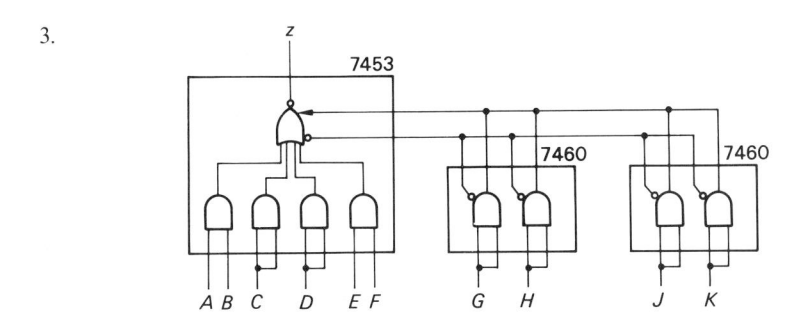

3.

5. (a) 75 ns, (b) 18.5 ns, (c) 22.5 ns.

394 ANSWERS TO SELECTED PROBLEMS

7.

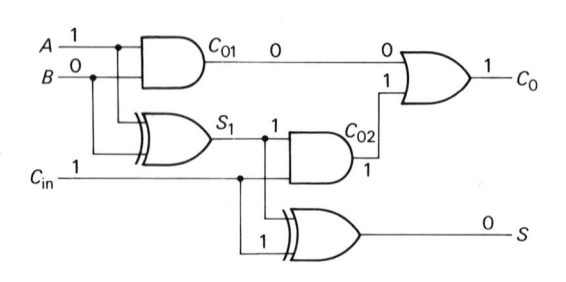

9.

Inputs				Outputs	
N_1	N_2	D	Q	Coffee	Change
0	0	0	0	0	0
0	0	0	1	1	1
0	0	1	0	1	0
0	0	1	1	1	1
0	1	0	0	0	0
0	1	0	1	1	1
0	1	1	0	1	0
0	1	1	1	1	1
1	0	0	0	0	0
1	0	0	1	1	1
1	0	1	0	1	0
1	0	1	1	1	1
1	1	0	0	1	0
1	1	0	1	1	1
1	1	1	0	1	0
1	1	1	1	1	1

Since there are fewer 0's than 1's for the coffee (C) function let's use $\bar{C}$, i.e.:

$$\bar{C} = \bar{N}_1\bar{N}_2\bar{D}Q + \bar{N}_1 N_2 \bar{D}\bar{Q} + N_1 \bar{N}_2 \bar{D}\bar{Q}$$
$$= \bar{D}\bar{Q}[\bar{N}_1\bar{N}_2 + \bar{N}_1 N_2 + N_1 \bar{N}_2]$$
$$= \bar{D}\bar{Q}[\bar{N}_1 + N_1 \bar{N}_2]$$
$$= (\bar{D}\bar{Q})(\bar{N}_1 + \bar{N}_2)$$

so $C = \overline{(\bar{C})} = \overline{(\bar{D}\bar{Q})(\bar{N}_1 + \bar{N}_2)}$
$$= \overline{(\bar{D} + \bar{Q})} + \overline{(\bar{N}_1 + \bar{N}_2)}$$
$$= D + Q + N_1 N_2.$$

11.

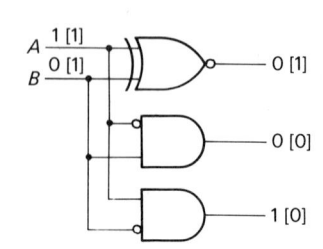

CHAPTER 4   395

13. (d)

A	B	$\bar{A}$	$\bar{B}$	$\bar{A} \cdot \bar{B}$	$A + B$	$\overline{A + B}$
0	0	1	1	1	0	1
0	1	1	0	0	1	0
1	0	0	1	0	1	0
1	1	0	0	0	1	0

(Note: $\bar{A} \cdot \bar{B} = \overline{A + B}$)

15.

16.

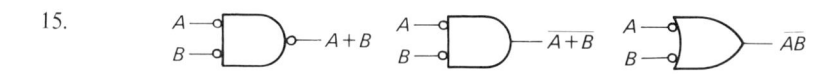

17. (a)

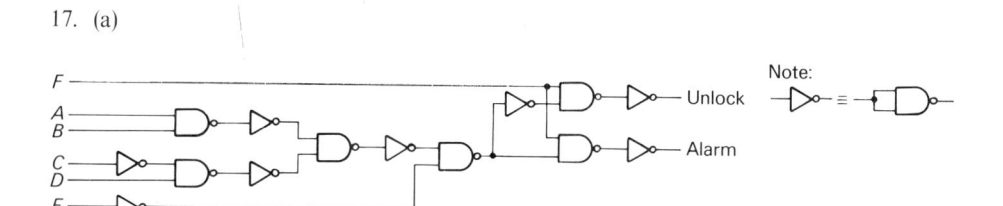

19.

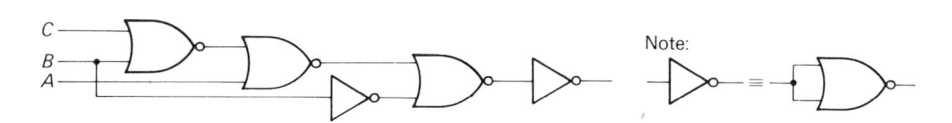

21.

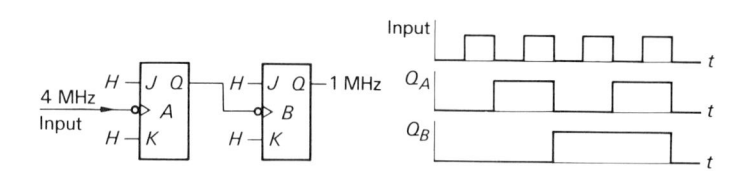

23.

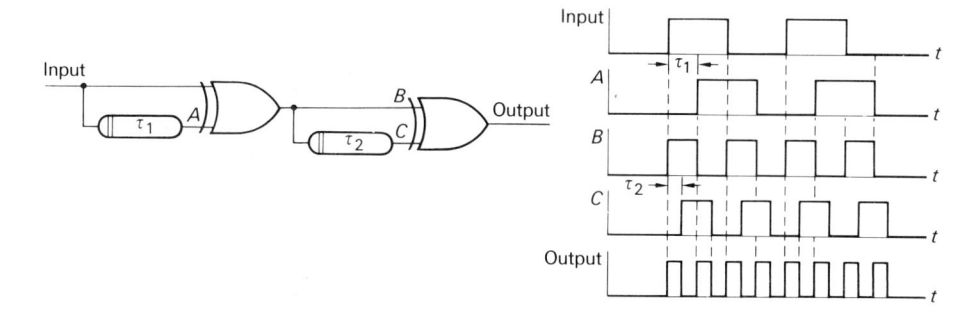

396 ANSWERS TO SELECTED PROBLEMS

25.

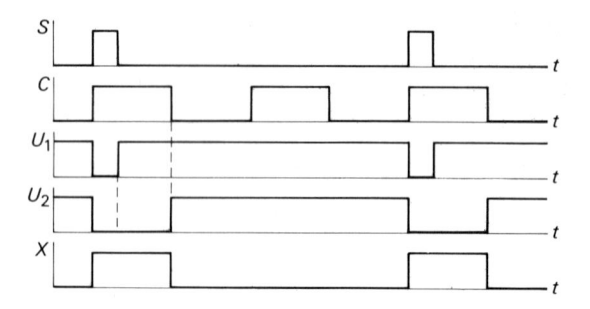

Waveforms S and C are synchronized. The output goes HIGH, and has a width equal to the C pulse, each time an S pulse occurs.

27. (a)

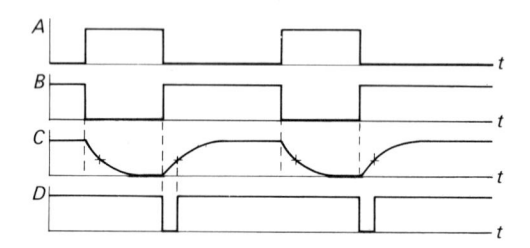

(b) This circuit produces an output pulse having width $R_1 C_1$ at the trailing edge of each input pulse.

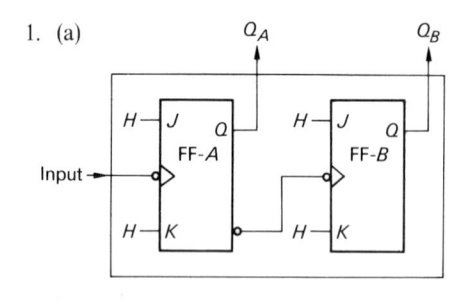

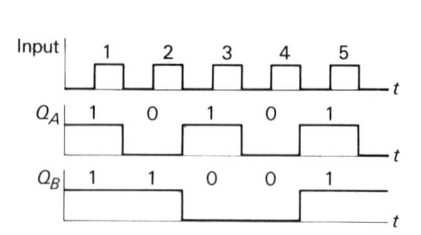

Chapter 5

1. (a)

3. $A < B$ is HIGH, $A = B$ is LOW, $A > B$ is LOW.
5. (b) 11101110.

CHAPTER 5   397

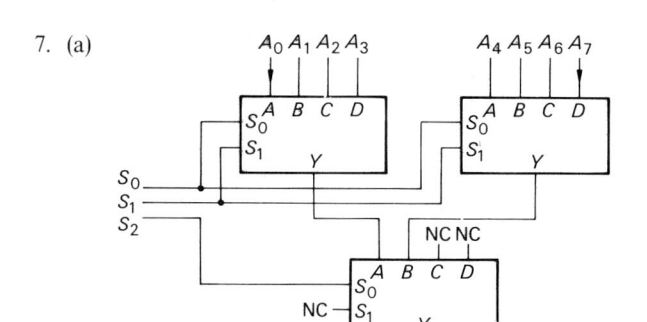

7. (a)

Address inputs		Output
S_0	S_1	Y
L	L	A
L	H	B
H	L	C
H	H	D

9. 65.

11.

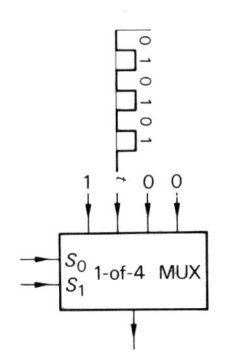

13. (a) Fewer transmission lines, therefore cheaper. (b) Slower.

15. (a)   (b) Edge pin #44 and row F of sheet 3.

19. (b) 11101 (nat), 0010 1001 (8421); (d) 110001000 (nat), 0011 1001 0010 (8421).
21. (b) 1010.

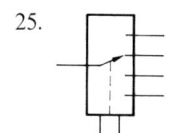

25.

398 ANSWERS TO SELECTED PROBLEMS

27.

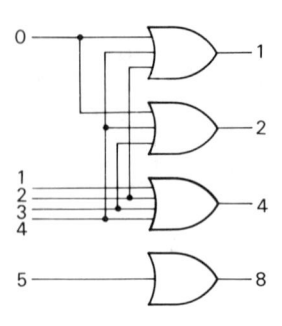

[Figure: OR gates combining inputs 0–5 to outputs 1, 2, 4, 8]

29. (a)

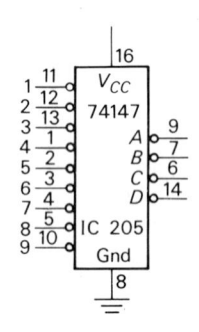

[Figure: 74147 priority encoder IC 205 with pin connections]

(b) If potentiometer R276 in the preset tuning circuit is set to receive channel 5, a jumper wire would be connected from the anode of diode D201 to the 5 input (pin #2) of IC 205. Similarly, if R278 was set for channel 9, a jumper could be connected from D203 to the 9 input (pin #10) of IC 205. Then if the UP channel selector was depressed (assuming the counter starts from 0) the priority encoder produces the 8421 coded 0101 (channel 5) and 1001 (channel 9), in sequence. These BCD outputs can then be used to drive a digital display.

30. (b)

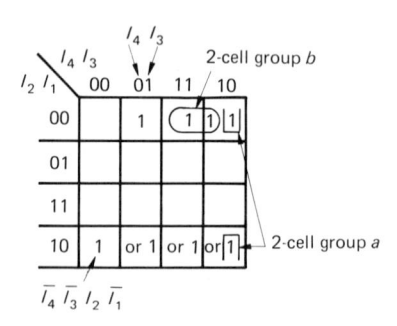

[Karnaugh map with 2-cell group a and 2-cell group b indicated; axes $I_2 I_1$ and $I_4 I_3$; entries showing 1s in the 00 row at columns 01, 11, 10 and in the 10 row at columns 00, 01, 11, 10 with "or 1" notations; bottom axis labeled $\bar{I}_4 \bar{I}_3 I_2 \bar{I}_1$]

Chapter 6

1. Digital voltmeter, electronic calculator, digital wrist watch.
3. About 20% of the 0° intensity.

CHAPTER 7   399

4. (a)

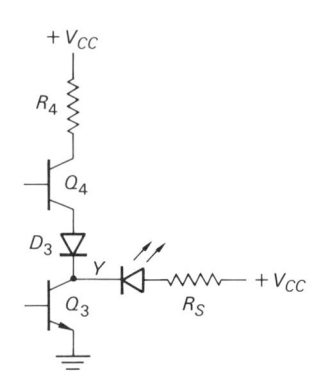

(b) When the output Y is HIGH, Q_4 is on and Q_3 is off. For the LED to light, current would have to flow from the LED through D_3, Q_4, and R_4. This would cause D_3 to be reverse-biased, hence only leakage current can flow when Y is HIGH, so the LED is off.

9. (a) $R = \dfrac{V_{OH} - V_{LED}}{I_{LED}} = \dfrac{5\text{ V} - 1.5\text{ V}}{9\text{ mA}} = 388\ \Omega$.

11. Invalid inputs are inputs for which there are no corresponding useful outputs.

13. $E_x = TE_{ref} = (\tfrac{1}{2} \times 4)10\text{ V/s} = 20\text{ V}$.

15. The interlock circuit generates a reset pulse for the 14518 frequency counter once for each heartbeat.

17. Referring to Fig. 6.15(a), the number 26 can be displayed by addressing the decoder so that segments a, b, d, e, and g are on (decimal 2) and closing scanner switch 1. Then address segments c, d, e, f, and g (decimal 6) and close scanner switch 2.

19. Referring to Fig. 6.17(c), the number 4 can be generated as follows: Apply a HIGH to X_1 and connect Y_1 and Y_3 to a LOW. Then $X_2 = H$ and $Y_1 = Y_3 = L$; $X_3 = H$ and $Y_1 = Y_2 = Y_3 = Y_4 = L$; $X_4 = H$ and $Y_3 = L$; $X_5 = H$ and $Y_3 = L$; $X_6 = H$ and $Y_3 = L$; and $X_7 = H$ and $Y_3 = L$.

Chapter 7

1.

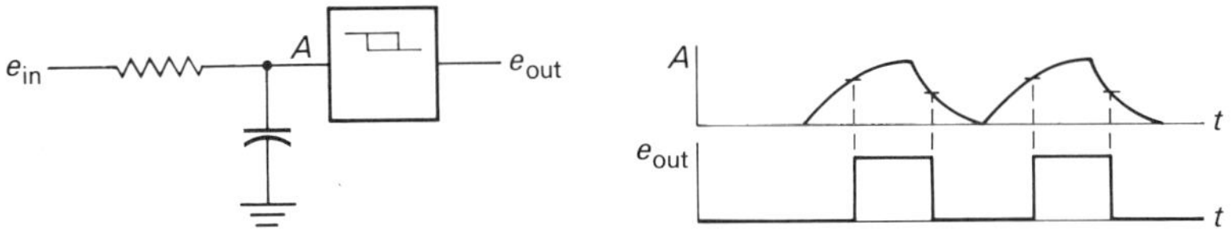

3.

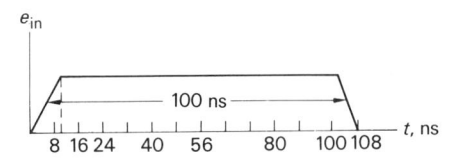

400   ANSWERS TO SELECTED PROBLEMS

4. prf = 100 KHz, DC = 20%.
7. (1) Keep R and C lead lengths short, (2) decouple power supply, (3) keep Q output away from C, (4) keep the monostable away from noise sources.

9.

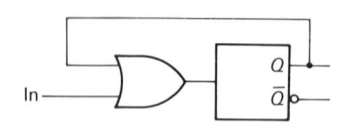

11.

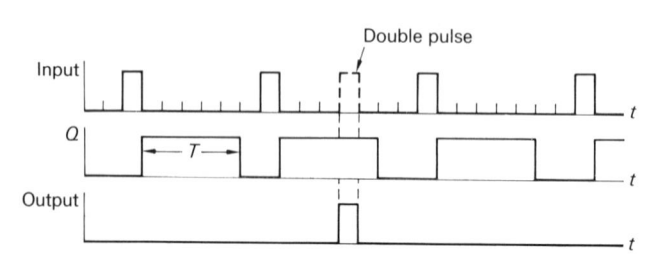

13.

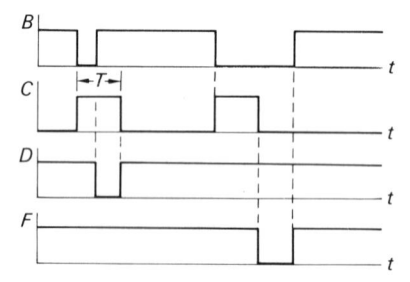

17. A portion of an amplified input signal is fed in phase back into the input thereby sustaining the input signal.

19.

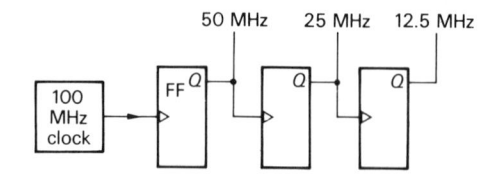

Chapter 8

2.

R	S	Q_{n+1}	$\bar{Q}_{n+1}$
L	L	H	H
L	H	L	H
H	L	H	L
H	H	Q_n	$\bar{Q}_n$

CHAPTER 8   401

3.

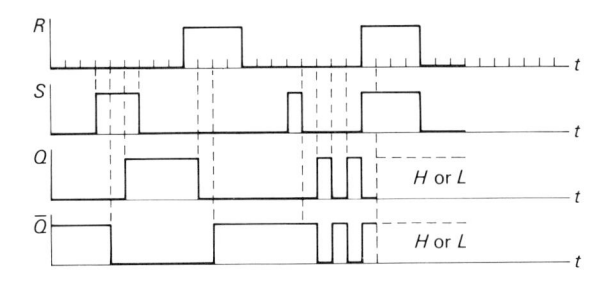

[Timing diagram showing signals R, S, Q, Q̄ with "H or L" labels on Q and Q̄ pulses]

5. No oscillations can occur.

7.

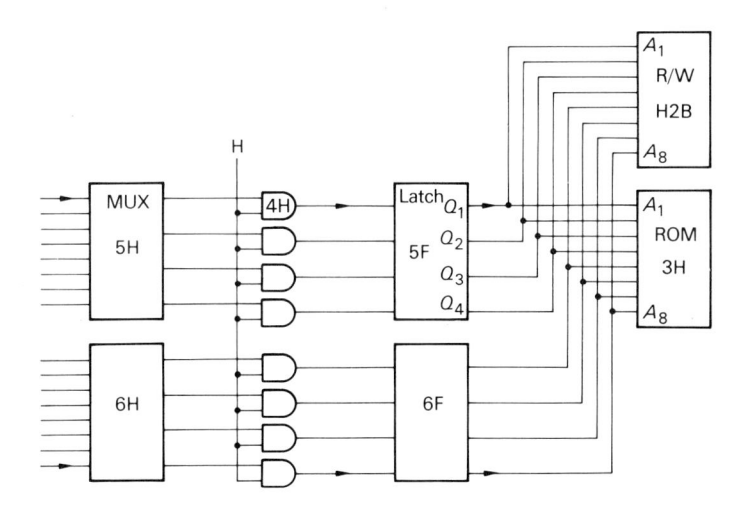

[Circuit diagram with MUX 5H, 6H feeding AND gates through 4H inverter with H input, into Latch 5F (Q1,Q2,Q3,Q4) and 6F, connecting to ROM 3H with A1–A8, R/W, H2B signals]

9. When enable pulse 3 goes HIGH R is already HIGH and S is LOW, so $\bar{Q}$ goes HIGH $2t_p$ after the rising edge of CP-3 due to the delays through gates B and D. Q goes LOW one delay time after $\bar{Q}$ goes HIGH. The description is similar for enable pulse 4.

13. (a)

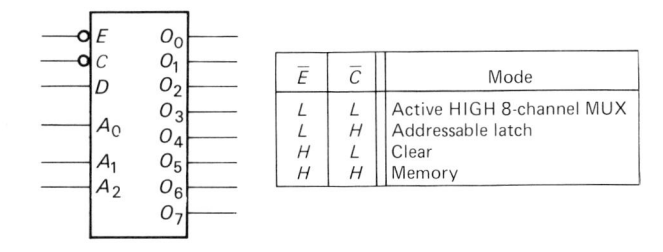

[Block diagram with inputs E (bubble), C (bubble), D, A0, A1, A2 and outputs O0–O7]

$\bar{E}$	$\bar{C}$	Mode
L	L	Active HIGH 8-channel MUX
L	H	Addressable latch
H	L	Clear
H	H	Memory

(b) As a latch, data at the D input is written into the latch selected by the address inputs (A_0, A_1, A_2) when $E = L$.
(c) As a 1-of-8 decoder, the address inputs select one out of eight outputs which will have the same state as the D input.

402 ANSWERS TO SELECTED PROBLEMS

16. 5.

26. The edge triggered flip-flop is insensitive to clock pulse width, as long as it is greater than the minimum, and to duty cycle, but is more sensitive to clock transition times than master/slave flip-flops.

30. 66 ns.

Chapter 9

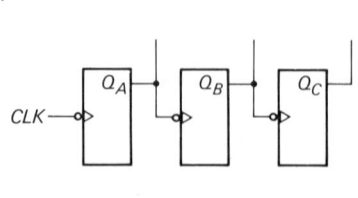

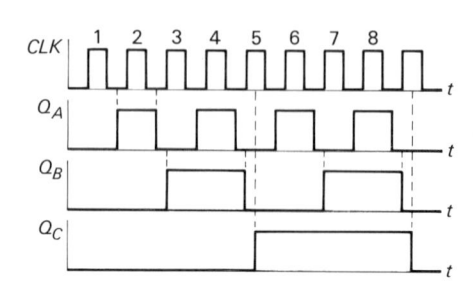

1.

3. An asynchronous counter is often cheaper and consumes less power than a synchronous counter. Synchronous counters are faster, have less tendency for decoding spikes, and their rising-edge clocks are compatible with other MSI devices.

5. The z-axis logic diagram is the same as the x-axis diagram.

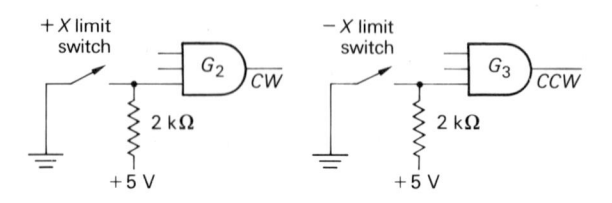

6. (b) 6–2–3–4–0–1–2, etc.

8.

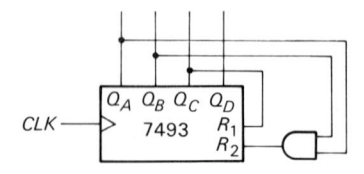

15. (a) The fastest flip-flop may cause the reset pulse to disappear before the slower flip-flops have had time to reset. (b) The latch maintains the reset pulse for the duration of one clock pulse.

22. 205 ms.

CHAPTER 10   403

25. (a)

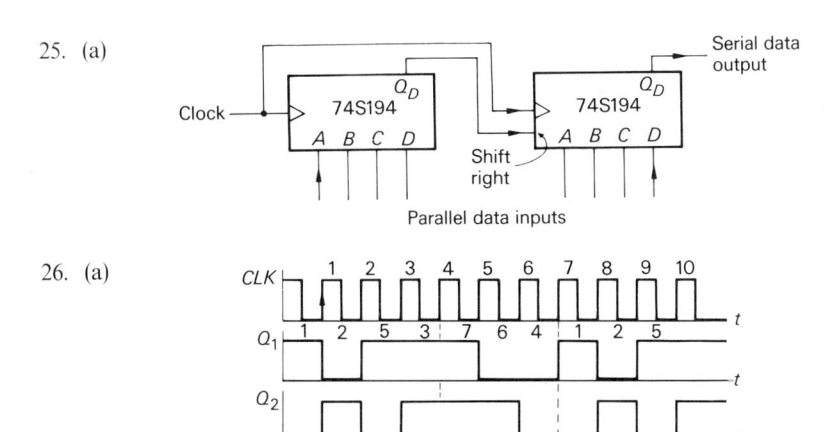

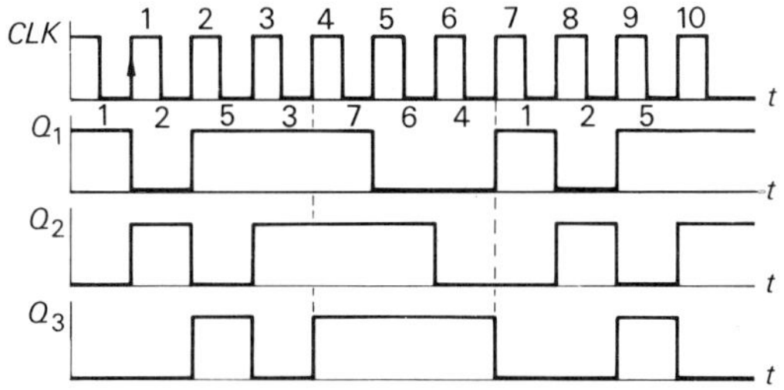

[Figure: Two 74S194 shift registers cascaded, Clock input to first, Q_D of first connects to shift right of second, Serial data output from Q_D of second, both with Parallel data inputs A B C D]

26. (a)

[Timing diagram: CLK with 10 pulses, Q_1, Q_2, Q_3 waveforms with labels 1 2 5 3 7 6 4 1 2 5]

Chapter 10

1. A mask programmed ROM is programmed by the manufacturer. A programmable ROM (PROM) can be programmed by the user once. A reprogrammable PROM can be reprogrammed many times. An electrically erasable ROM can be erased electrically and electrically reprogrammed.

3.

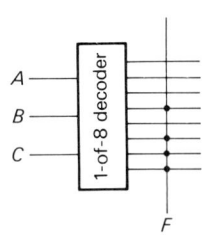

[Figure: 1-of-8 decoder with inputs A, B, C and output F showing diode matrix connections]

7. (a) The control ROM (CROM) contains a microprogram that is used to implement any of the standard instructions. The programmer writes his program using these standard instructions. The program is stored in the R/W memory. The CROM decodes and executes each instruction. The operations performed by the RALU (addition, AND, etc.) are determined by the four-bit control bus.

(b)

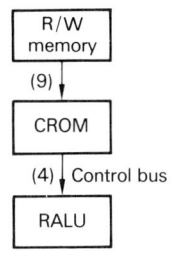

[Block diagram: R/W memory → (9) → CROM → (4) Control bus → RALU]

8. A ROM has a built-in address decoder; the PLA decoding is programmed by the user.
14. Waveform 1 is the write-cycle output waveform of the memory with a load capacitance of C_L and pull-up resistor R_L. The output is LOW except when the memory is disabled. The output-to-disable times are denoted t_{LZ} and t_{ZL}. Waveform 2 is similar to waveform 1.
18. Surges in the supply currents due to charging and discharging capacitances in the memory as chips are enabled, disabled, and refreshed.
22. A CAM is a read/write memory which can read in data, write out data, and locate words in its memory which match a word in its search register.

INDEX

Numbers in boldface indicate pages that contain the main discussion of that entry. Topics that are discussed or explored in examples or problems are indicated by (Ex) or (Pb), respectively, following their page number.

Active LOW output, 33
Active pull-up, 56; *see also*
 Transistor-transistor logic
Adder, half-, 103 (Ex)
 ROM, 297 (Ex)
 full, 120 (Pb)
Addition, binary, 19–20
 rules, 20
Address access time, 293
Alarm, 181 (Pb)
Analog circuits, compared with digital, 10–12
Analog electronic system, 4
Analog voltages, 4
AND decision, 6
AND-OR-INVERT (AOI) gate, 100–102
Anode, 44; *see also* Diode(s)
Antilock circuits, 280
Applications, of gates, **95–123**
 of *RC* elements, 117–118
Arithmetic logic unit, 367; *see also*
 Computer(s)
Assembly line monitor, 125 (Ex)
Associative memory, 327
Astable multivibrator, 198
Asynchronous input, 220; *see also*
 Flip-flop(s)
Asynchronous system, 199
Asynchronous counter, *see* Counter(s)

Base, 17; *see also* Number(s)
Beat frequency indicator, 181 (Pb)

Binary addition, 19–20
 rules, 20
Binary circuit, 5
Binary codes, *see* Code(s)
Binary decoder, 33, 141
Binary half-adder, 30 (Ex)
Binary logic, 6
Binary number system, 17
Binary system, 5
Binary variables, 7
Boole, George, 108
Boolean algebra, 108
 definitions, laws, and rules, 335–336
Boolean expression, 108
Buffer/driver gate (TTL), 66
Buffering, 34
Bus, data, 66
 bidirectional, 68

Cache memory, 311
Candela, 158
Capacitive loading (TTL), 61–62
Card reader, 121 (Pb), 97 (Ex)
Carry flag, 382; *see also* Computer(s)
Cascading, of gates, 98–100
 of counters, 283
 of shift registers, 276–277
Cathode, 44; *see also* Diode(s)
Character, alphameric, 130, **139**
Character generation (ROM), 302–304
Characteristic curve, of diode, 46
 of transistor, 48

406 INDEX

Characteristics of IC devices,
  CMOSL, 76-81; see also
    Complementary MOS logic
  ECL, 84-89; see also Emitter-coupled
    logic
  TTL, 56-61; see also
    Transistor-transistor logic
Charge storage (diode), 46
Chip, 12
Circuits, digital versus analog, 10-12
  discrete, 12
  integrated, see Integrated circuits
  antilock (counter), 280
Clear input (flip-flop), 220
Clock, crystal controlled, 204-205
  definition, 113
  IC, 201-204
  logic gate, 199-201
  monostable, 208 (Pb)
  multiphase, 203
Clock generator, 123 (Pb)
Clock skew, 238
Clock stretcher, 225 (Ex)
Code(s), alphameric, 130
  alpha-numeric, 130
  BCD, 140
  binary, 139-141
  biquinary (2-5), 263
  conversion and generation (ROM),
    301-302
  definition, 130
  8421, 139
  excess-3 BCD, 140
  one's complement, 140
  table, 139
  two's complement, 140
  unweighted, 139
Coincidence detector, 107, 121 (Pb)
Coincident selection memory, 314
Collector-dotting (TTL), 66
Common anode LED display, 167
Common cathode LED display, 167
Common emitter configuration, 47; see
  also Transistor(s)
Comparator, digital, 106-107, **125-129**
  troubleshooting, 162 (Ex), 181
  voltage, 187

Comparison of digital and analog
  circuits, 10-12
Complement, 29
Complementary MOS logic (CMOSL).
  74-81
  current spiking, 77
  delay time, 78
  device characteristics, 348-349
  impedance, 78
  latch-up, 80
  loading, 78-79
  noise immunity, 78
  short circuit current, 77
  silicon-on-sapphire (SOS), 78
  SCR condition, 80
  three-state devices, 79-80
  transfer curves, 77
  power dissipation, 77-78
  unused inputs, 78
  wire-ORing, 79-80
Computer(s)
  address latch, 240 (Pb)
  arithmetic logic unit (ALU), 367
  control circuit, 241 (Pb)
  control ROM, 367
  control unit, 367
  digital, 367, 150 (Pb)
  direct memory access (DMA), 368
  flag, carry, 382
    flip-flop, 382
    overflow, 382
  initialization, 385-386
  interrupt handler, 385
  microcomputer, logic diagram and
    description, 367-388
    problems, 150, 151, 283, 330, 332
  microinstruction, 330 (Pb)
  peripherals, 367
  programming problems, 183, 208, 244
Connection diagram, 42 (Pb)
Content addressable memory (CAM),
  327
Control circuit, automobile, 7 (Ex), 9
  (Ex)
Control memory (ROM), 367; see also
  Computer, ROM
Control system, 151 (Pb)

Control unit, 367; *see also* Computer(s)
Conversion, of numbers, *see* Number(s)
Count states, 247
Counter(s), antilock circuit in, 280
  asynchronous, 247-251
    definition, 247
    down, 253-254
    modulo, 258-267
    ripple, 248
  binary, 124, **247**
  cascading, 270-272, 283 (Pb)
  count states, 124, **247**
  decoding, 250
  decoding spikes, 250
  divide-by-N, 250
  down, 149 (Pb), 253-255
  electrical characteristics, 267-270
  frequency divider, 250
  hexadecimal, 247
  illegal states, 280
  modulo
    feedback method, 259-261
    preset method, 264-267
    reset method, 261-264
  Moebius, 279
  mod-9, 267 (Ex)
  synchronous, 247, 251-253
    modulo, 258-267
  up, 149 (Pb), 253
  programmable, 265
  ring, 280
  semisynchronous, 260
  shift, 279
  shift register, 279-280
  speed, 250
  state diagram, 280, 281 (Pb)
  twisted ring, 279
CRC generator, 287 (Pb)
Current, conventional, **23**, 52
  conventions, 52
  dark, 160
  gain, 47
  short circuit, CMOSL, 77
    ECL, 87
    TTL, 56-57
  sink, 52
  source, 52

  spiking
    CMOSL, 77
    ECL, 87
    TTL, 57
Cut-in voltage, 45; *see also* Diode(s)
Cutoff region, 48; *see also* Transistor(s)

Dark current, 160
Darlington configuration (TTL), 69
Data link, 130
Data lockout (counter), 233
Data input (latch), 216
Data selector, 133
Decimal and binary equivalents, 18
    (table)
Decimal number system, *see* Number(s)
Decoder, 141-144
  binary, 33, **141-144**
Decoding spikes, 250; *see also*
    Counter(s)
Delay element, 122 (Pb)
Delay line (registers), 276 (Ex)
Delay time, CMOSL, 78
  ECL, 87-88
  propagation, 37
  total (TTL), 62 (Ex)
  transistor, 50
  TTL, 58-59
  turn-on, 37
  turn-off, 37
DeMorgan's rules, **109-111**, 335-336
Demultiplexer, 144-147
Depletion mode, 72; *see also*
    Transistor(s), MOS
Destructive readout (memory), 310
Detector, coincidence, 107
  error, 191 (Ex)
  odd-bit, 103
Device, 5
Differential amplifier, 82-83
Digit drivers, 175
Digital automobile ignition lock, 43 (Pb)
Digital circuits, 44-94
  compared with analog, 10-12
Digital comparator, 106 (Ex)
Digital computer, *see* Computer(s)

408 INDEX

Digital data communication system
  basic, 144 (Ex)
  eight-channel, 147 (Ex)
  problem, 151
Digital electronic system, 4
Digital electronics, principles of, 4-6
Digital ICs, 12-16; *see also* Integrated circuits
  device characteristics, 343-351
Digital transmission system, 132 (Ex)
Digital voltages, 4
Digital voltmeter (DVM), examples, 168, 270
  ramp type, 169
  readout, 182 (Pb)
Diode(s), anode, 44
  cathode, 44
  characteristic curve, 46
  charge storage, 46
  cut-in voltage, 45
  forward-biased, 45
  forward current, 45
  junction, 44
  leakage current, 46
  logic (DL), 70
  photo, 160
  reverse-biased, 45
  reverse current, 45
  reverse recovery time, 46
  Schottky (barrier), 46
  transistor logic (DTL), 70
Direct inputs, to flip-flop, 220
  to latch, 214
Direct memory access (DMA), 368; *see also* Computer(s)
Disabled device (TTL), 67
Displays, LED, 155-183
  dot matrix, 176-180
    font, 177
    horizontal scanning, 180
    multicharacter, 179
    single-character, 176
    vertical scanning, 180
    X-Y addressing, 177
  seven-segment
    common anode, 167
    common cathode, 167
    continuous method, 171
    decoder/driver, 164, **167-168**
    digit drivers, 175
    examples, 165, 166, 168, 170, 175
    failure, 176
    invalid inputs, 167
    multidigit, 168-176
    multiplexed multidigit, 172-176
    readout (DVM), 182 (Pb)
    ripple blanking input, 168
    ripple blanking output, 168
    single-digit, 163-167
Distributive rule, 108
Divide-by-N counter, 250
Door lock, digital, 95 (Ex), 120 (Pb), 122 (Pb), 330 (Pb)
Dot matrix, *see* Display(s)
Dot-OR, 66
Double negation, 107
Drain, 71; *see also* Transistor(s), MOS
Duty cycle, 189
Dynamic indicator, 114

Electrical characteristics of logic families, 89 (table)
Electromagnetic spectrum, 352
Electronic system, 4
Emitter, 46; *see also* Transistor(s)
Emitter-coupled logic (ECL),
  characteristics of, 84-89
  current spiking, 87
  delay time, 87-88
  device characteristics, 350-351
  differential amplifier, 82-83
  family, 81-89
  impedance, 87
  loading, 88
  noise immunity, 88
  nonsaturated switching, 83
  OR/NOR gate schematic, 83-84
  power dissipation, 87
  pull-down resistor (input), 87
  series gating, 89
  short circuit current, 87
  threshold voltage, 85
  transfer curve, 84-85
  unused inputs, 88

wire-ORing, 88-89
wired collectors, 89
1OK series, 83-84
Emitter follower configuration, 47; see also Transistor(s)
Encoders, 148-149
  priority, 149
Enhancement mode, 72; see also Transistor(s), MOS
Even parity, 130
Expand inputs, 100; see also Gates, AOI

Fall time, 50; see also Transistor(s)
Fan-in, 64
Fan-out, 36, **62**, 63 (Ex)
First-in/first-out (FIFO) memory, 328
Flag flip-flop, 382; see also Computer(s)
Flip-flop(s), 8-9, **210-245**
  asynchronous input, 220
  capacitively coupled, 236-237
  clear input, 220
  clock frequency, 223
  clocking methods, 228-237
  D, 220-224
  data lockout, 233
  direct input, 220
  edge triggered, 233-235
  flag, 382; see also Computer(s)
  JK, 114, 122 (Pb), **224-227**
  master/slave, 229-233
  one's catching, 232
  preset input, 220
  pulse triggered, 229
  race condition, 237
  synchronous input, 221
  truth table summary (table), 239
Floating input (TTL), 60
Flow diagram, 339
Font, 177
Forward-biased, 45; see also Diode(s)
Forward current, 45; see also Diode(s)
Free-running multivibrator, 198
Frequency, programmable, 283 (Pb)
Frequency divider chain, 203
Frequency dividers, **113-115**, 115 (Ex), 122 (Pb), 250; see also Counter(s)
Frequency doubler, 117 (Ex)

Frequency quadrupler, 122 (Pb)
Functionally complete, 111

Gate(s), 71, **23-43**; see also Transistor(s), MOS
  AND, 6-8, **26-29**
  AND-OR-INVERT (AOI), 100-102
  applications, **95-123**
  buffer/driver (TTL), 66
  cascaded, 98-100
  EXCLUSIVE-NOR (EX-NOR), 103
  EXCLUSIVE-OR (EX-OR), 102
  INCLUSIVE-OR, 103
  Inverter, 29
  NAND, 32, 34
  NOR, 37-40
  NOT, 29-31
  OR, 23-26
  schematic, TTL AND, 392
  summary, 41 (table)
  universal, 111
Glitch, 189

Half-adder, 30 (Ex), 103 (Ex)
Heart-rate monitor, 170 (Ex), 182 (Pb)
HIGH (H), 5
Hold time, 220
Horizontal scanning, 180; see also Displays, LED dot matrix
Hysteresis curve (Schmitt trigger), 187
Hysteresis voltage (Schmitt trigger), 185

Idempotent logic rule, 108
Illegal states (counter) 280
Impedance, CMOSL, 78
  ECL, 87
  TTL, 58
Implementation, NAND/NOR rules, 112
  table, 110
Indeterminate state, 211
Information, 129
  theory, 129
  waveform representation of, 129-130
Input load factor (TTL), 64
Integrated circuits (ICs), 12
  chip, 12
  digital, 12-16

dual in-line package (DIP), 12
monolithic, 12
package, 12
Interrupt circuit, 223 (Ex)
Interrupt handler, 385; *see also*
Computer(s)
Invalid inputs
decoder, 143
seven-segment display, 167
Inverter, 29; *see also* NOT gate
I/O unit, 367; *see also* Computer(s)
Isolate, 34

Karnaugh map, 154 (Pb)

Large-scale integration (LSI), 14
Last-in/first-out (LIFO) memory, 329
Latch-up (CMOSL), 80
Latch, addressable, 241 (Pb)
  data inputs, 216
  direct inputs, 214
  hold time, 220
  indeterminate state, 211
  NAND, 43 (Pb)
  NOR, 43 (Pb)
  RS, 8, **210–215**
  setup time, 219
  strobed *D*, 216–220
  strobed *RS*, 118–119, **215–216**
Leakage current, 46; *see also* Diode(s)
Least significant bit (LSB), 17; *see also*
Number(s)
Level shifting, 83
Light-emitting diode (LED), applications, 158–163
  discrete, 156–158
  display, *see* Displays
  dot matrix display, *see* Displays
  driver, 158 (Ex)
  frequency response curves, 353
  luminous intensity, 158
  peak wavelength, 158
  seven-segment display, *see* Displays
Line driver, 147
Line receiver, 147
Load line, 48; *see also* Transistor(s)

Loading, CMOSL, 78–79
  ECL, 88
  TTL, 61–65
Lock system, 21 (Pb)
Logic, 6
  complementary MOS, *see*
    Complementary MOS logic
  emitter-coupled, *see* Emitter-coupled
    logic
  transistor-transistor, *see*
    Transistor-transistor logic
Logic diagram, 29
Logic expression, 108
Logic family, 44
Logic gate, *see* Gate(s)
Logic rules, 107–111, 335–336
  distributive, 108
  double negation, 107
  DeMorgan's, 109–111, 336
  idempotent, 108
Logic series, 44
Logic symbols, 338 (table)
  drafting template, 339
Logic variables, 7
LOW (L), 5
Low-power Schottky series, 70–71; *see
also* Transistor-transistor logic
Luminous intensity, 158

Majority voter circuit, 105 (Ex), 109
  (Ex), 330 (Pb)
Medium-scale integration (MSI), 14
Memory, 287–333
  address access time, 293
  associative, 327
  bipolar read/write, 309–316
    bit-organized, 314
    cell, 309–310, 330 (Pb)
    cell circuit, 331 (Pb)
    word-organized, 311
  cache, 311
  circulating, 277
  content addressable (CAM), 327
  coincident selection, 314
  example, 38
  first-in/first-out, 328
  last-in/first-out, 328

INDEX 411

MOS read/write, 316-322
  cell, 311
  cell circuit, 332 (Pb)
  dynamic, 318-322
  standby mode, 321
  standby power, 321 (Ex)
  static, 316-318
  refresh, 311
nonvolatile, 293
push-down stack, 329
random access, 291, **308**
read-only memory (ROM), 290-304
  cell, 290
  character generator, 302-304
  code conversion and generation, 301-302
  custom programmed, 291
  electrically alterable (EAROM), 292
  expansion, 295-296
  field programmed, 292
  half-adder, 297 (Ex)
  mask programmed, 291
  metal nitride oxide silicon (MNOS), 292
  microprogram, 299
  programmable, 292
  pulse generator, 298 (Ex)
  random logic generation, 296-299
  reprogrammable, 292
  sine look-up, 300 (Ex)
  table look-up, 300-301
  vending machine control logic, 298
read/write (R/W), 308-329
  address mode, 312
  bipolar, 309-316
  bit-organized, 309
  cache, 311
  cell, 309
  chip select model, 312
  destructive readout, 310
  expansion, 323-327
  MOS, 316-322
  nondestructive read-out (NDRO), 310
  register file, 309
  scratch pad, 311
  word-organized, 309

  read-modify/write cycle, 319
  semiconductor, 289
  shift register, 277-279
  volatile, 310
  write-while-read, 329
  X-Y selection, 314
Metal nitride oxide silicon (MNOS), 292
Metal-oxide semiconductor (MOS) transistor, 71-74; *see also* Transistor(s)
Microcomputer, logic diagram and description, 367-388; *see also* Computer(s)
  problems, 42, 91, 150, 151
Microinstruction, 330; *see also* Computer(s)
Microprocessor, 14
Microprogram, 299; *see also* ROM
Microprogramming, 299-300
Minimization, 154 (Pb)
Modulus, 247; *see also* Counter(s)
Monolithic IC, 12
Monostable, 190-197
  leading-edge triggered, 192
  multivibrator, 190
  nonretriggerable, 193
  one-shot, 191
  retriggerable, 193-197
  trailing-edge triggered, 192
Most significant bit (MSB), 17; *see also* Number(s)
MSI devices and applications, 124-154
Multiple-emitter transistors, 51
Multiplexers, cascading, 137
  function generation, 138
  pulse generators, 137-138
Multiplexing, 133
Multipliers, binary-rate, 272
  decade-rate, 272
  rate, 272
Multivibrator, astable, 198
  free-running, 198
  monostable, 190

NAND/NOR implementation rules, 112
N-channel transistor (NMOS), 72; *see also* Transistor(s), MOS

412 INDEX

Negative-going threshold voltage, 184
Negative logic, 51
Number(s)
  base, 17
  binary, 17
  conversion from decimal to binary, 16
  decimal, 17
  decimal-binary equivalents, 18
  least significant bit (LSB), 17
  most significant bit (MSB), 17
  positional notation, 17
  scientific notation, 17
  system, 17
  weight, 17
Numerical controlled drill press, 256 (Ex), 281 (Pb), 283 (Pb)
Noise, 10
Noise discriminator, 118 (Ex), 122 (Pb)
Noise immunity, CMOSL, 78
  ECL, 88
  TTL, 59–60
Nondestructive readout, 310
Nonsaturated switching (ECL), 83

Odd parity, 130
Odd-bit detector, 103
One-shot, 191; *see also* Monostable
One's catching flip-flop, 232
One's complement, 140
Open collector devices, TTL, 65–66
  gates, TTL, 65
Optical tachometer, 165 (Ex)
Optoelectronics, 352–353
Overflow flag, 382; *see also* Computer(s)

Parallel transmission, 133
Parity, even, 130
  odd, 130
Parity bit, 130
Parity checker/generator, 130–133
Parity generator, 132
Parity tree, 131
Passive pull-up, 56; *see also* Transistor-transistor logic
P-channel transistor (PMOS), 72; *see also* Transistor(s), MOS
Period, 188; *see also* Pulse
Peripherals (computer), 367

Phase detector, 123 (Pb)
Photodiode, 160
  dark current, 160
Phototransistor, 160
Positional notation, 17; *see also* Number(s)
Positive-going threshold voltage, 184
Positive logic, 51
Power dissipation, CMOSL, 77–78
  ECL, 87
  TTL, 57–58
Powers of 2, 337
Preset input, 220; *see also* Flip-flop(s)
Priority control system, 148 (Ex)
Priority encoder, 149
Product term (PLA), 305
Propagation delay time, *see* Delay time
Programmable logic array (PLA), 304–308
  compared with ROM, 306, 307 (Ex)
  field programmed, 307
  half-adder, 304
  mask programmed, 307
  product array, 304
  product term, 305
  sum array, 305
  synchronous decade counter, 307
Pseudorandom sequence generator (PRSG), 276, 285 (Pb)
Pull-down resistor (ECL), input, 87
  output, 83
Pull-up resistor (TTL), 66
Pulmonary function analyzer, 247 (Ex), 283 (Pb)
Pulse(s), 36; *see also* Pulse train(s)
  characteristics, 188–189
    duty cycle, 189
    period, 188
    pulse repetition frequency (prf), 188
    pulse repetition time (prt), 188
    reference level, 188
  falling edge, 37
  glitch, 189
  ideal, 37
  leading edge, 37
  rising edge, 37
  spike, 189

INDEX 413

Pulse coded modulation, 285 (Pb)
Pulse detector, double, 206 (Pb)
Pulse edge detector, 117 (Ex)
Pulse generator, 194 (Ex), 298 (Ex)
Pulse height discriminator, 187 (Ex)
Pulse shaper, 190 (Ex)
Pulse-shaping circuits, 115–116
Pulse stretcher, 116
Pulse train(s), 129
   rectangular wave, 189
   square wave, 189
Pulse width, 37
   discriminator, 188, 206 (Pb)
   modulation, 207 (Pb)
Push-down stack memory, 329

Quine-McClusky minimization, 154 (Pb)

Race condition, 237
RC differentiator, 116
RC element, 117–118
RC integrator, 116
Read-modify/write memory cycle, 319
Recommended operating characteristics (TTL), 36
Rectangular wave, 189
Reference level, 188; *see also* Pulse
Refresh memory, 311
Register files (memory), 309
Registers, 272–279
   buffer, 273
   cascaded, 276–277
   circulating memory, 277
   conversion time, 275 (Ex)
   counter, 279–280
   delay line, 276
   dynamic MOS, 277
   memories, 277–279
   Moebius, 279
   MOS, 277
   pseudorandom sequence generator, 276
   recirculating, 277
   sequence generator, 276
   static MOS, 277
   shift, 272, 279
   shift-right, 274
   twisted ring, 279

Reset, 8, **211**
Resistive loading (TTL), 62–65
Reverse-biased (diode), 45
Reverse current (diode), 45
Reverse recovery time (diode), 46
Ripple blanking, 168
Rise time, transistors, 50

Safety system, 28 (Ex)
Saturated MOS transistor, 73
Saturation region, transistor, 48
Schematic diagrams, 338–342
Schmitt trigger, 184–188
   hysteresis curve, 187
   hysteresis voltage, 185
   threshold voltage
      positive-going, 184
      negative-going, 184
   transfer characteristics, 186
Schottky, Walter, 46
Schottky (barrier) diode, 46
Schottky TTL series, 68–70
Scientific notation, 17
SCR CMOSL condition, 80
Scratch pad memory, 311
Sense circuit, 123 (Pb)
Sequence generators (register), 276
Sequential device, 210
Serial transmission, 133
Series gating, ECL, 89
Set, 8, 211
Setup time, 219
Seven-segment displays, *see* Displays
Short circuit current, CMOSL, 77
   ECL, 87
   TTL, 56–57
Signal generator, logic diagram and description, 354–366
   problems, 43, 91
Silicon-on-sapphire (SOS), CMOSL, 78
Sine look-up (ROM), 300 (Ex)
Small-scale integration (SSI), 14
Source, MOS transistor, 71
Specification sheets, TTL, 34–37
Spectral distribution curve, 158
Speed power product, 93 (Pb)
Spike, 189

Square wave, 189
Stable oscillator, 204
State diagram, 280
Stepper motor, 256
Stepping angle, 256
Storage time, transistor, 50
Strobe input, 141
Substrate, MOS transistor, 71
Subsystem, 124
Switch contact bounce, 212
Switching, nonsaturated (ECL), 83
Switching characteristics, TTL, 36
Switching time, transistor, 49–50
Synchronous counters, *see* Counter(s)
Synchronous input (flip-flop), 221
Synchronous system, 198
System, analog, 4
  binary, 5
  digital, 4
  two-state, 5

Table look-up (ROM), 300–301
Table of combinations, 25
Three-state devices, CMOSL, 79–80
  TTL, 66–68
Threshold voltage, average, 56
  ECL, 85
  MOS, transistor, 72
Thumbwheel switch, 125
Time division multiplexing, 135
Timing diagram, 37
Totem pole (TTL), 56
Transfer characteristic, Schmitt trigger, 186
Transfer curves, CMOSL, 77
  ECL, 84–85
  TTL, 55–56
Transistor(s), base, 46
  bipolar, 46–48
  collector, 46
  common emitter configuration, 47
  current gain, 47
  cutoff region, 48
  delay time, 50
  emitter, 46
  emitter follower configuration, 47, **49**
  fall time, 50

load line, 48
MOS, 71–74
  depletion mode, 72
  drain, 71
  enhancement mode, 72
  gate, 71
  n-channel (NMOS), 72
  p-channel (PMOS), 72
  saturated, 73
  source, 71
  substrate, 71
  threshold voltage, 72
multiple emitter, 68
photo, 160
rise time, 50
saturation region, 48
Schottky, 68
storage time, 50
switching times, 49–50
Transistor-transistor logic (TTL), 50–71
  buffer/driver gate, 66
  characteristics of, 56–61
  collector dotting, 66
  current
    short circuit, 56–57
    spiking, 57
  Darlington configuration, 69
  device characteristics, 343–347
  disabled, 67
  delay times
    turn-on, 59
    turn-off, 59
  dot-OR, 66
  electrical characteristics, 36
  family, 50–71
  fan-in, 64
  fan-out, 36, **62**, 63 (Ex)
  floating inputs, 60
  impedance, 58
  input load factor, 64
  loading, 61–65
    capacitive, 61–62
    resistive, 62–65
  low-power Schottky series, 70–71
  noise immunity, 59–60
  noise margin, 60
  open collector devices, 65–66

open collector gates, 65
power dissipation, 57-58
pull-up
  active, 56
  passive, 56
  resistor, external, 66
recommended operating
  characteristics, 36
Schottky
  series, 68-70
  transistor, 68
specification sheets, 34-37
standard 54/74 series, 52-55
switching characteristics, 36
three-state devices, 66-68
totem pole, 56
tristate, 67
unit load, 63
unused gates, 60-61
unused inputs, 60-61
wire-AND, 66
wire-OR, 66
Transition region, transfer curve (TTL), 56
Transmission, parallel, 133
  serial, 130, **133**
Transmission line, 130
Transmission system, digital, 132
Triple modular redundancy (TMR), 104
Tristate device (TTL), 67
Troubleshooting, 12, 242 (Pb)
  comparator, 181 (Pb)
  flow diagrams, 339
Truth table, 6, **25**
  to logic diagram, 104-107
Turn-off delay, 37
  TTL, 59
Turn-on delay, 37
  TTL, 59
TV channel selector, example, 142
  problem, 151, 154
  schematic, 152
Two's complement, 140
Two-state system, 5

Unit load, 63
Universal gate, 111
Unused gates, TTL, 60-61
Unused inputs, CMOSL, 78
  ECL, 88
  TTL, 60-61

Vending machine, basic, 29 (Ex)
  control logic, ROM, 298 (Ex)
  foolproof, 121 (Pb)
Vertical scanning, dot matrix display, 180
Visible spectrum, 352
Volatile memory, 310
Voltage, analog, 4
  digital, 4
  hysteresis, Schmitt trigger, 185
  levels, 4
  threshold, MOS, 72
  transfer curve, TTL, 55
Voltage controlled oscillator (VCO), 202

Warning systems, automobile seat belt, 97-98 (Ex), 113 (Ex)
  bank teller, 25 (Ex)
  elevator, 42 (Pb)
  motor, 42 (Pb)
  oscillator, 200 (Ex)
  pressure, 214 (Ex), 240 (Pb)
  thermal motor overload, 21 (Pb)
Waveform generator, 185 (Ex)
Waveform representation of information, 129-130
Wavelength, peak, 158
Weight, **17**, 140; *see also* Number(s)
Wire-AND, TTL, 66
Wired collectors, ECL, 89
Wire-OR, CMOSL, 79-80
  ECL, 88-89
  TTL, 66, 79-80

X-Y addressing, dot matrix display, 177
X-Y selection memory, 314

Zero crossing detector, 185